LEVEL SETS AND EXTREMA OF RANDOM PROCESSES AND FIELDS

LEVEL SETS AND EXTREMA OF RANDOM PROCESSES AND FIELDS

JEAN-MARC AZAÏS
Université de Toulouse
Institut de Mathématiques
Laboratoire de Statistiques et Probabilités
Toulouse, France

MARIO WSCHEBOR
Universidad de la República
Centro de Matemática
Facultad de Ciencias
Montevideo, Uruguay

WILEY

A JOHN WILEY & SONS, INC., PUBLICATION

Published by John Wiley & Sons, Inc., Hoboken, New Jersey.
Published simultaneously in Canada.

Library of Congress Cataloging-in-Publication Data:

Azaïs, Jean-Marc, 1957–
Level sets and extrema of random processes and fields / Jean-Marc Azaïs, Mario Wschebor.
p. cm.
Includes bibliographical references and index.
ISBN 978-0-470-40933-6 (cloth)
1. Gaussian processes. 2. Level set methods. 3. Random fields. 4. Stochastic processes. I. Wschebor, Mario. II. Title.
QA274.4.A93 2009
519.2′3–dc22

2008033281

10 9 8 7 6 5 4 3 2 1

MIX
Paper from responsible sources
FSC® C013604

CONTENTS

PREFACE

This book is a result of long-term cooperation between the authors. Our aim has been to present, in book form, knowledge on two related subjects: extreme values of continuous parameter random fields and their level sets. These are broad subjects in probability theory and its applications. The book deals specifically with the situation in which the paths of the random functions are regular in the sense of having some degree of differentiability and the parameter set, some geometric regularity. One finds few books that treat these topics, even though they lead to interesting and difficult mathematical problems, as well as to deep connections with other branches of mathematics and a wide variety of applications in many fields of science and technology.

Our main references are two classical books: those of Cramér and Leadbetter (1967) and Leadbetter, Rootzen, and Lindgren (1983). The appearance of the recent book by Adler and Taylor (2007), which has several topics in common with ours, even though approached differently, may indicate that these subjects have recovered the wide interest that we believe they actually deserve.

We would like the book to be useful for people doing research in various fields as well as for postgraduate training. The mathematical reader looking for active research problems will find here a variety of open and interesting questions. These problems have a wide spectrum of difficulty, from those that are tractable by the methods contained here to others that appear to be more difficult and seem to require new ideas. We would be happy if the book were of interest to researchers in various areas of mathematics (probability and mathematical statistics, of course, but also numerical analysis and algorithmic complexity). At the same time, we believe the results should be useful to people using statistical modeling in engineering, econometrics, and biology, and hope to have made some contribution to building bridges in these directions.

Applications deserve a good deal of our attention. Concerning applications to problems outside mathematics, we recognize that our choice has been strongly dependent on the taste and experience of the authors: readers will find a section on genetics, one on inference on mixtures of populations, and another on statistical modeling of ocean waves. However, we have not included applications to mathematical physics or econometrics, in which the fine properties of the distribution of maximum of stochastic processes and fields play a central role.

We have included applications of random field methods to other parts of mathematics, especially to systems of equations and condition numbers of random matrices. This is a new field, even though some of the problems it considers are quite old, and has become a very important theme that mixes various branches of mathematics. One of the aims is to help understanding of algorithm complexity via the randomization of the problems that algorithms are designed to solve. One can also apply random field methods to a study of the condition number of systems of inequalities (as has been done by Cucker and Wschebor, 2003) or of polynomial systems of equations. In our view, these are very interesting subjects that are only beginning and we preferred not to include them in the book.

Numerical methods appear in various chapters. They are by no means simple but are crucial to be able to use the mathematical results, so we stress their importance. Some are solved, such as we do using the Matlab toolbox MAGP described in Chapter 9. Some of them appear to be difficult: Even in the simplest cases of Gaussian stationary one-dimensional parameter processes defined on an interval, numerical computation of the moments of the number of crossings of the paths with a given level presents difficult obstacles and is a source of interesting open problems.

Since our purpose is to reach readers of quite different backgrounds, we have attempted to make the book as self-contained as possible. The reader is expected to have attained at least the level of a postgraduate student with basic training in probability theory and analysis, including some elementary Fourier analysis. However, given our intention to limit the size of the book, at certain points we have not respected the rule of being self-contained. The following are the most relevant cases in which we use ideas and results and do not give proofs: Kolmogorov's extension theorem (Chapter 1); the definition and main properties of the Itô integrals and their basic properties, which are used in the first two chapters, including Itô's formula, quadratic variation and the exponential martingales; the convergence of empirical measures and asymptotic methods in statistics, of which we give a quick account without proofs of the results we need in Chapter 4; the co-area formula (Chapter 6) from integral geometry; ergodicity, which underlies a certain number of results in Chapter 10; and finally, computation of the density of the eigenvalues of random matrices, used in Chapter 8. In the applied examples, the nonmathematical background has not been discussed in detail, and other references should be consulted to go more deeply into the subjects. In all these cases, we refer the reader to other books or scientific papers, expecting that this will suffice for complete understanding.

ACKNOWLEDGMENTS

We thank Christine Cierco-Ayrolles, Céline Delmas, Cécile Mercadier, and Joaquín Ortega for allowing us to reproduce some tables and figures from their published work (see Chapters 4, 9, and 11), as well as for valuable discussions.

We also owe a debt of gratitude to numerous colleagues for talks, comments, and the many interactions that mathematical work generates every day. In particular, we want to thank Alan Genz, Marie Kratz, Michel Ledoux, Paul Doukhan and José León for remarks and comments that have contributed to improving the text. We are also grateful for ECOS Cooperation Action U03E01, which enabled us to cross the Atlantic Ocean in both directions in order to produce this book.

Jean-Marc Azaïs
Mario Wschebor

INTRODUCTION

The theory of stochastic processes is a powerful tool for the study of a vast diversity of problems in the natural sciences, in which randomness is required as a component to describe phenomena. In this book we consider two general classes of problems that arise when studying random models.

Let $\mathcal{X} = \{X(t) : t \in T\}$ be a stochastic process with parameter set T defined on a probability space $(\Omega, \mathcal{A}, \mathrm{P})$. We are interested primarily in the following subjects:

1. For each value u in the range space of $X(\cdot)$, understanding the properties of the level sets of the paths of $\mathcal{X}$, that is, the random sets $\{t \in T : X(t) = u\}$
2. Whenever the process has real values, studying and computing the distribution function of the random variable $M_T = \sup_{t\in T} X(t)$, that is, the function $F_{M_T}(u) = \mathrm{P}(M_T \le u), u \in \mathbb{R}$

These are classical subjects in probability theory and have been considered for a long time in a variety of contexts. Generally speaking, our framework will be continuous parameter processes, which means here that T is a subset of the real line (such as a finite or infinite interval) or of some higher-dimensional Euclidean space. When the parameter set T is multidimensional, we will call $\mathcal{X}$ a *random field*.

In most of the theory and the applications that we consider, the parameter set T will have some geometric regularity, such as being a manifold embedded in a

Level Sets and Extrema of Random Processes and Fields, By Jean-Marc Azaïs and Mario Wschebor

finite-dimensional Euclidean space or having a more general structure. As for the paths $t \rightsquigarrow X(t)$ of the stochastic function, we require that they satisfy regularity properties, such as differentiability of a certain order. We also need results on the supremum of random sequences, in which the geometry of the domain or the regularity of the paths does not play any role. This provides us with basic and useful ingredients (such as in Chapter 2), but the emphasis is on random functions possessing certain regularities.

For random level sets, our main tools are Rice formulas. Assume that T is a Borel subset of the Euclidean space $\mathbb{R}^d$ and $\mathcal{X}$ a stochastic process or field having regular paths, defined on some open set containing T and taking values in $\mathbb{R}^d$. For given $u \in \mathbb{R}^d$, denote by $N_u(X; T)$ the number of roots of $X(t) = u$ lying in the set T. Rice formulas allow one to express the kth factorial moment of the random variable $N_u(X; T)$ as an integral over T^k of a function that depends on the joint distribution of the process and its derivative and is evaluated at the k-tuples $(t_1, \dots, t_k) \in T^k$.

In fact, the main interest lies in the probability distribution of the random variable $N_u(X, T)$, which remains unknown. The authors are not aware of any nontrivial situation in which one can compute this distribution by means of a reasonable formula. Rice formulas appear to be a contribution to understanding the distribution of $N_u(X, T)$, giving a general expression for its moments. One can measure to what extent this remains an open subject by the fact that for the time being, no useful formula exists to compute the expectation of $f[N_u(X, T)]$, where $f : \mathbb{R} \to \mathbb{R}$ is a simple function: as, for example, $f(x) = x^\alpha$ when α is not an integer.

When the dimension d of the domain is strictly larger than that of the range space, say d', generically a random level set has dimension $d - d'$, so that the interesting questions do not refer to the number of points lying in a set, but should aim to understand its geometry, which is of course richer than in the $d = d'$ case. The natural questions concern the probabilistic properties of the geometric measure of the level set, its Euler–Poincaré characteristic, and so on. We give expressions and the corresponding proofs for the moments of the geometric measure of the level set under quite general conditions. Even though this is generally considered to be known, to our knowledge rigorous proofs are available only in special cases, such as real-valued random fields or particular probability laws. In Chapter 11 we present some applications to sea waves. We should say that the results we present are only a minor part of an important subject, which, however, remains largely unexplored.

The term *Rice's formula* honors the pioneering work of S. O. Rice, who computed the expectation of $N_u(X; T)$ for one-parameter Gaussian stationary processes in the 1940s and used it in telecommunications in an early stage of this discipline [this is formula (3.2)]. In the 1950s and 1960s one could find applications of Rice's formula to physical oceanography (e.g., in a series of papers by M. S. Longuett-Higgins and collaborators on the statistical properties of ocean waves, including some useful formulas in the multiparameter case).

Applications to other areas of random mechanics were developed somewhat later (see the book by Krée and Soize, 1983).

H. Cramér and M. R. Leadbetter's excellent book contains a systematic presentation of what was known at the time of publication (1967), together with connected subjects in probability theory and various applications. The book is on one-parameter processes, and most of the material concerns Gaussian stationary processes. We still use it as a reference on various subjects.

Around 1970, for one-parameter Gaussian processes, rigorous proofs of Rice formulas had already been given for certain classes of processes, with contribution, among others, by K. Ito (1964), D. Ylvisaker (1965), and Y. Belayev (1966, 1972b). Some of this work included non-Gaussian processes and the multiparameter case but to our knowledge, the first treatment of the multiparameter case in book form is Adler's *Geometry of Random Fields* (1981).

Our aim is to make a contribution to update the subject of Rice formulas, including the improvements that have taken place during the last decades in both the basic theory and in applications.

There is a part of probability theory that refers to level sets of random functions which are not differentiable. This may have started with Paul Lévy's definition of local time [see, e.g., Ikeda and Watanabe (1981) or Karatzas and Shreeve (1998) for a modern presentation of the subject] and has led to the study of the geometry of level sets of semimartingales or some other classes of one-parameter processes and similar problems for random fields. A short list of references on this subject could include books by Kahane (1985) and Revuz and Yor (1998) and a paper by Ehm (1981). We are not considering this type of problem in the book; our processes and fields have regular paths and for fixed height, almost surely, the level sets are nice sets. One can build up a bridge between these two worlds: Take a process with nondifferentiable paths, smooth it by means of some device, and try to reconstruct relevant properties of the geometry of the level sets of the original (irregular) paths from the regularized paths. This leads to asymptotic expansions which are interesting by themselves and have important applications [see, e.g., Jacod (1998, 2000) for polygonal and related approximations and Wschebor (2006) for a review without proofs].

With respect to the second main subject of this book, the general situation is that the computation of the distribution function of the supremum by means of a closed formula is known only for a very restricted number of stochastic processes (and trivial functions of them). The following is a list of one-dimensional parameter processes for which, as far as we know, an actual formula exists for the distribution of $M = M_{[0,T]}$:

- Brownian motion or Wiener process $\{W(t) : t \geq 0\}$, for which the distribution of M has, in fact, been known since the nineteenth century (Kelvin, D. André)
- Brownian bridge, $B(t) := W(t) - tW(1)$ $(0 \leq t \leq 1)$
- $B(t) - \int_0^1 B(s)\,ds$ (Darling, 1983)
- Brownian motion with a linear drift (Malmquist, 1954; Shepp, 1979)

- $\int_0^t W(s)\,ds + yt$ (McKean, 1963; Goldman, 1971; Lachal, 1991)
- Restriction to the boundary of the unit square of the Wiener sheet (Paranjape and Park, 1973)
- Each of the stationary Gaussian processes with covariance equal to:
  - $\Gamma(t) = e^{-|t|}$ (Ornstein–Uhlenbeck process; DeLong, 1981)
  - $\Gamma(t) = (1 - |t|)^+$, T a positive integer (Slepian process; Slepian, 1961; Shepp, 1971)
  - $\Gamma(t)$ even, periodic with period 2, $\Gamma(t) = 1 - \alpha|t|$ for $0 \le |t| \le 1$, $0 < \alpha \le 2$ (Shepp and Slepian, 1976)
  - $\Gamma(t) = \big(1 - |t|/(1-\beta)\big) \vee \big(-\beta/(1-\beta)\big)$, $|t| < (1-\beta)/\beta, 0 < \beta \le 1/2$, $T = (1-\beta)/\beta$ (Cressie, 1980)
  - $\Gamma(t) = \cos t$ (Berman, 1971b; Delmas, 2003)
  - $\Gamma(t) = [2|t| - 1]^2$, $T = 1$ (Cabaña, 1991)

Methods for finding formulas for the distribution of the supremum over an interval of this list of processes are ad hoc, hence nontransposable to more general random functions, even in the Gaussian context. Given the interest in the distribution of the random variable M_T, arising in a diversity of theoretical and technical questions, a large body of mathematics has been developed beyond these particular formulas.

A first method has been to obtain general inequalities for the distribution function $F_{M_T}(u)$. Of course, if one cannot compute, one tries to get upper or lower bounds. This is the subject of Chapter 2, which concerns Gaussian processes. The inequalities therein are essential starting points for the remainder of the book; a good part of the theory in the Gaussian case depends on these results.

However, generally speaking the situation is that these inequalities, when applied to the computation of the distribution of the random variable M_T, have two drawbacks. First, the bounds depend on certain parameters for which it is difficult or impossible to obtain sharp estimates, implying that the actual computation of probabilities can become inaccurate or plainly useless, for statistical or other purposes. Second, these bounds hold for general classes of random functions, but may become rough when applied to a particular stochastic process or field. Hence, a crucial question is to improve the estimations derived from the general theory contained in Chapter 2. This is one of the purposes of this book, which is attained to a certain extent but at the same time, leaves many open problems. This question is considered in Chapters 4, 5, 8, and 9.

A second method has been to describe the behavior of $F_{M_T}(u)$ under various asymptotics. This subject is considered in several chapters. In particular, a part of Chapter 8 is devoted to the asymptotic behavior of the tail $1 - F_{M_T}(u)$ for a random field defined on a fixed domain as $u \to +\infty$. For extensions and a diversity of results concerning asymptotic behavior as $u \to +\infty$ that are not mentioned here, we refer the reader to books (and the references therein) by Berman (1992a), Lifshits (1995), and Piterbarg (1996a). We are not considering in this book the so-called "small ball probabilities," consisting for example in the behavior of $\mathrm{P}(\sup_{t \in T}(X(t) - X(t_0)) \leq u)$ as $u \downarrow 0$ (see Li and Shao, 2001, 2004).

A third method consists of studying the regularity of the function $u \rightsquigarrow F_{M_T}(u)$. This is considered in Chapter 7.

Let us now give a quick overview of the contents of each chapter. This may help the reader to choose. Our advice on the order in which the various chapters might be read, especially based on the chain of prerequisites they require, is given in Figure I.1.

Chapter 1 contains the basic definitions of stochastic processes and Kolmogorov-type conditions, implying that, almost surely, the paths have a certain regularity property (continuity, Hölder condition, differentiability). These classical and well-known results are not optimal, but they are sufficient for most uses. We have included a reminder on the Gaussian distribution and its elementary properties, since a large part of the book is devoted to Gaussian processes and the reader may appreciate to have them in situ. We have not included Fernique's (1974) and Talagrand's (1987) beautiful results giving necessary and sufficient conditions for the continuity of the paths of Gaussian processes in stationary and nonstationary cases, respectively. We do not use them in the book.

Chapter 2 is about inequalities for Gaussian processes, related primarily to the distribution of the supremum. For comparison inequalities (which we call of Slepian type), the main result is the Li and Shao inequality (2002), which includes and improves on a large set of similar results that have been in use for 50 years or so, apparently starting with a paper by Plackett in 1954, motivated by statistical linear models with Gaussian noise.

The remainder of Chapter 2 is devoted to the classical upper bounds for the tails of the distribution of the supremum. They roughly say that if $\mathcal{X}$ is a centered Gaussian process with bounded paths, then $1 - F_{M_T}(u)$ is bounded above by some constant times a Gaussian density. This was known around 1975, due to a series of key contributions: Dudley (1967), Landau and Shepp (1970), Marcus and Shepp (1972), Fernique (1974), Sudakov and Tsirelson (1974), and Borell (1975). To present isoperimetric inequalities, the basis is a quite recent proof by Borell (2003) of the extension to a general case of Ehrhard's inequality (1983), a Minkowski-type inequality for the Gaussian measure in $\mathbb{R}^n$.

Rice formulas are proved in Chapters 3 and 6. Chapter 3 starts with a proof for Gaussian one-parameter processes which is very simple; then we consider the general non-Gaussian case. For various uses one only wants to know whether the moments of crossings are finite, or to give upper bounds for them, but without direct use of Rice formulas, since these can lead to nontractable calculations. This has been the motivation for a series of papers deducing bounds from hypothesis on the process, mainly in the Gaussian case, which are cited in Chapter 3. In the same chapter we state a simple general criterion to ensure finiteness and obtain some rough bounds for the moments. To illustrate this type of result, a corollary is that if $\mathcal{X}$ is a Gaussian process defined on a compact interval $[0, T]$ of the real

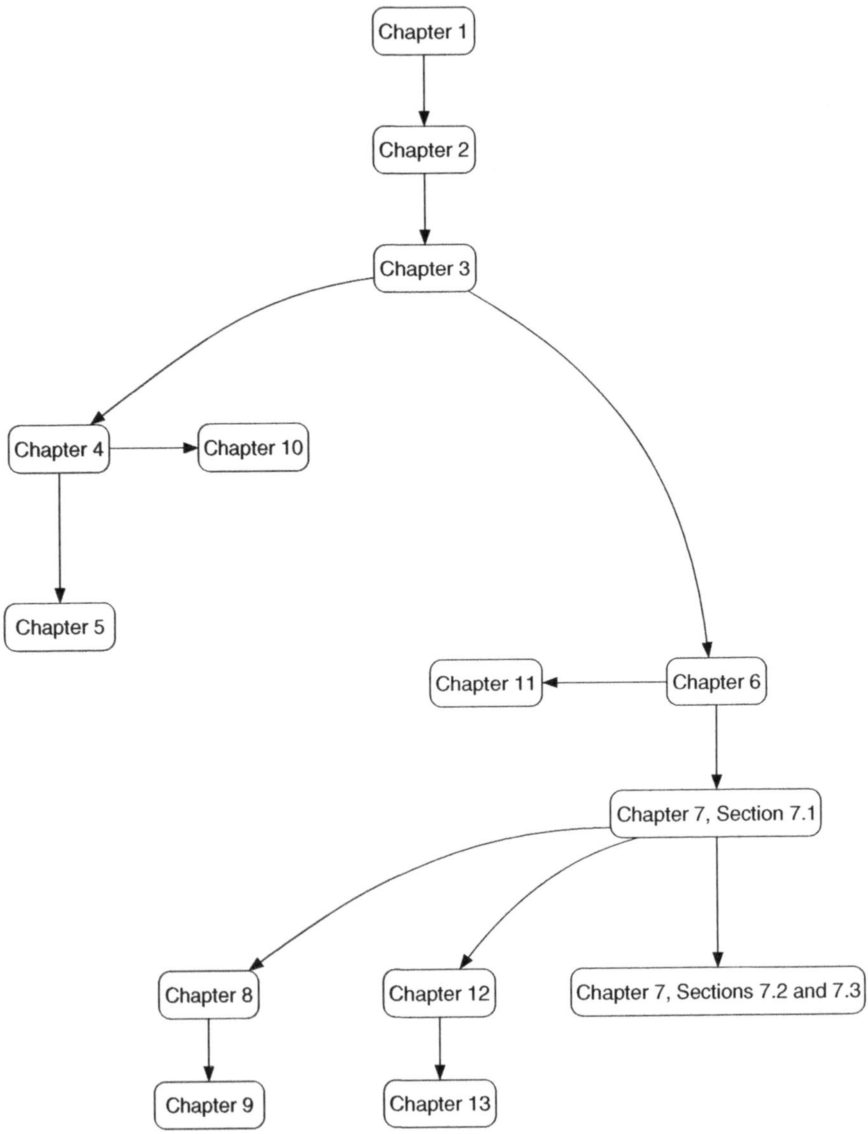

Figure I.1. Reading diagram.

line having C^∞-paths and satisfying the nondegeneracy condition $\mathrm{Var}(X(t)) > 0$ for every $t \in [0, T]$, all the moments of crossings of any level are finite.

Rice formulas for random fields are considered in Chapter 6. Proofs are new and self-contained except for the aforementioned co-area formula. In all cases, formulas for the moments of weighted (or "marked") crossings are stated and proved. They are used in the sequel for various applications and, moreover, are important by themselves.

Chapter 4 consists of two parts. In Sections 4.1 to 4.3, $\mathcal{X}$ is a Gaussian process defined on a bounded interval of the real line, and some initial estimates for $\mathrm{P}(M > u)$ are given, based on computations of the first two moments of the number of crossings. In Sections 4.4 and 4.5, two statistical applications are considered: the first to genomics and the second to statistical inference on mixtures of populations. The common feature of these two applications is that the relevant statistic for hypothesis testing is the maximum of a certain Gaussian process, so that the calculation of its distribution appears to be naturally related to the methods in the earlier sections.

Chapter 5 establishes a bridge between the distribution of the maximum on an interval of a one-parameter process and the factorial moments of up-crossings of the paths. The main result is the general formula (5.2), which expresses the tail of the distribution of the maximum as the sum of a series (the Rice series) defined in terms of certain factorial moments of the up-crossings of the given process. Rice series have been used for a long time with the aim of computing the distribution of the maximum of some special one-parameter Gaussian processes: as, for example, in the work of Miroshin (1974). The main point in Theorems 5.1, 5.6, and 5.7 is that they provide general sufficient conditions to compute or approximate the distribution of the maximum. Even though some of the theoretical results are valid for non-Gaussian processes, if one wishes to apply them in specific cases, it becomes difficult to compute the factorial moments of up-crossings for non-Gaussian processes. An interesting feature of the Rice series is its enveloping property: replacing the total sum of the series by partial sums gives upper and lower bounds for the distribution of the maximum, and a fortiori, the error when one replaces the total sum by a partial sum is bounded by the absolute value of the last term computed. This allows one to calculate the distribution of the maximum with some efficiency. We have included a comparison with the computation based on Monte Carlo simulation of the paths of the process. However, in various situations, more efficient methods exist; they are considered in Chapter 9.

In Section 7.1 we prove a general formula for the density of the probability distribution of the maximum which is valid for a large class of random fields. This is used in Section 7.2 to give strong results on the regularity of the distribution of a one-parameter Gaussian process; as an example, if the paths are of class C^∞ and the joint law of the process and its derivatives is nondegenerate (in the sense specified in the text), the distribution of the maximum is also of class C^∞. When it comes to random fields, the situation is more complicated and the known results are essentially weaker, as one can see in Section 7.3.

Chapters 4 and 5, as well as 8 and 9, point toward improving computation of the distribution of the maximum on the basis of special properties of the process, such as the regularity of the paths and the domain. In Chapter 8 one profits from the implicit formula for the density of the distribution of the maximum that has been proved in Chapter 7 to study second-order approximation of the tails of the distribution as the level u tends to $+\infty$, as done in Adler and Taylor's recent book (2007) by other methods. The direct method employed here is also suited to obtaining nonasymptotic results.

Chapter 10 contains a short account of limit theorems when the time domain grows to infinity, including the Volkonskii–Rozanov theorem on the asymptotic Poisson character of the stream of up-crossings for one-parameter stationary Gaussian processes under an appropriate normalization. This implies that the distribution of the maximum, after re-scaling, converges to a Gumbel distribution. Section 10.2 establishes a central limit theorem for nonlinear functionals of Gaussian process, which applies to the limit behavior of the number of crossings.

In Chapter 11 we describe the modeling of the surface of the sea using Gaussian random fields. Some geometric characteristics of waves, such as length of crests and velocities of contours, are introduced. The Rice formula is used to deduce from the directional spectrum of the sea, some properties of the distribution of these characteristics. Some non-Gaussian generalizations are proposed in Section 11.5.

In Chapter 12 we study random systems of equations over the reals, having more than one unknown. For polynomial systems, the first significant results on the number of roots were published in the 1990s, starting with the Shub–Smale theorem (1993), which gives a simple, elegant formula for the expectation of the number of roots. This is a fascinating subject in which the main questions remain unanswered. What can one say about the probability distribution of the number of solutions, and how does it depend on the probability law of the coefficients? What is the probability of having no solution? What can one say about the distribution of the roots in space, and how can this help to solve the system numerically? How do all these things behave as the number of unknowns grows indefinitely? What about the same questions for undetermined systems? And so on.

Answering some of these natural questions would imply at the same time making progress in key problems in numerical analysis and algorithmic complexity as well as in other areas. The content of the chapter is extremely modest with respect to the aforementioned questions, and it is our hope that it may stimulate other people to work on the numerous problems arising in this field.

Chapter 13 is on condition numbers. Roughly speaking, the condition number of a problem measures the difficulty of the problem to be solved by any algorithm. Of course, there are many measures of this sort. The standard procedure is to give a metric on the space of problems and define the condition number of a problem as the inverse of its distance to the set of ill-posed problems, possibly with some additional normalization. In this chapter we consider the simplest situation, in which a "problem" is a square system of linear equations and the set of ill-posed problems is the set of systems in which the matrix of coefficients is noninvertible.

The condition number turns out to be related to the singular values of the matrix of coefficients. The role of condition numbers in numerical linear algebra has been well known since the 1940s, when they were introduced by Von Neumann and Goldstine (1947) and Turing (1948) (see also Smale, 1997).

Condition numbers appear in the estimation of the complexity of algorithms in a natural way. When a problem is chosen at random, the condition number becomes a random variable. Of course, its probability distribution will depend on the underlying probability law on the set of problems. In the linear case, computing the probability distribution of the condition number becomes a problem on the spectrum of random matrices. The importance of studying the distribution of condition numbers of random matrices has been put forward by Smale (1985). The first precise result is due to Edelman (1988), who computed the equivalent of the expected value of the logarithm of the condition number when the matrix size n tends to infinity.

The methods we present to study the probability distribution of the condition number of square random matrices again rely on random fields and Rice formula. They are not optimal in some standard cases but in some others, produce at present the best known bounds. We do not consider the similar and much more difficult questions for nonlinear systems. The reader can consult the book by Blum et al. (1998), which includes a basic presentation of this subject, or Cucker et al. (2008).

CHAPTER 1

CLASSICAL RESULTS ON THE REGULARITY OF PATHS

This initial chapter contains a number of elements that are used repeatedly in the book and constitute necessary background. We will need to study the paths of random processes and fields; the analytical properties of these functions play a relevant role. This raises a certain number of basic questions, such as whether the paths belong to a certain regularity class of functions, what one can say about their global or local extrema and about local inversion, and so on. A typical situation is that the available knowledge on the random function is given by its probability law, so one is willing to know what one can deduce from this probability law about these kinds of properties of paths. Generally speaking, the result one can expect is the existence of a *version* of the random function having good analytical properties. A version is a random function which, at each parameter value, coincides almost surely with the one given. These are the contents of Section 1.4, which includes the classical theorems due to Kolmogorov and the results of Bulinskaya and Ylvisaker about the existence of critical points or local extrema having given values. The essence of all this has been well known for a long time, and in some cases proofs are only sketched. In other cases we give full proofs and some refinements that will be necessary for further use.

As for the earlier sections, Section 1.1 contains starting notational conventions and a statement of the Kolmogorov extension theorem of measure theory, and Sections 1.2 and 1.3 provide a quick overview of the Gaussian distribution and some connected results. Even though this is completely elementary, we call the reader's attention to Proposition 1.2, the Gaussian regression formula, which will appear now and again in the book and can be considered as the basis of calculations using the Gaussian distribution.

1.1. KOLMOGOROV'S EXTENSION THEOREM

Let $(\Omega, \mathcal{A}, \mathrm{P})$ be a probability space and $(F, \mathcal{F})$ a measurable space. For any measurable function

$$Y : (\Omega, \mathcal{A}) \to (F, \mathcal{F}),$$

that is, a random variable with values in F, the image measure

$$Q(A) = \mathrm{P}(Y^{-1}(A)) \qquad A \in \mathcal{F}$$

is called the *distribution* of Y.

Except for explicit statements to the contrary, we assume that probability spaces are complete; that is, every subset of a set that has zero probability is measurable. Let us recall that if $(F, \mathcal{F}, \mu)$ is a measure space, one can always define its completion $(F, \mathcal{F}_1, \mu_1)$ by setting

$$\mathcal{F}_1 = \{A : \exists B, C, A = B \bigtriangleup C, \text{such that } B \in \mathcal{F}, C \subset D \in \mathcal{F}, \mu(D) = 0\}, \tag{1.1}$$

and for $A \in \mathcal{F}_1$, $\mu_1(A) = \mu(B)$, whenever A admits the representation in (1.1). One can check that $(F, \mathcal{F}_1, \mu_1)$ is a complete measure space and μ_1 an extension of μ.

A *real-valued stochastic process indexed by the set* T is a collection of random variables $\{X(t) : t \in T\}$ defined on a probability space $(\Omega, \mathcal{A}, \mathrm{P})$. In what follows we assume that the process is *bi-measurable*. This means that we have a σ-algebra $\mathcal{T}$ of subsets of T and a Borel-measurable function of the pair (t, ω) to the reals:

$$X : (T \times \Omega, \mathcal{T} \times \mathcal{A}) \to (\mathbb{R}, \mathcal{B}_{\mathbb{R}})$$

($\mathcal{B}_{\mathbb{R}}$ denotes the Borel σ-algebra in $\mathbb{R}$), so that

$$X(t)(\omega) = X(t, \omega).$$

Let T be a set and $\mathbb{R}^T = \{g : T \to \mathbb{R}\}$ the set of real-valued functions defined on T. (In what follows in this section, one may replace $\mathbb{R}$ by $\mathbb{R}^d$, $d > 1$.) For $n = 1, 2, \ldots$, $t_1, t_2, \ldots, t_n$, n distinct elements of T, and $B_1, B_2, \ldots, B_n$ Borel sets in $\mathbb{R}$, we denote

$$C(t_1, t_2, \ldots, t_n; B_1, B_2, \ldots, B_n) = \{g \in \mathbb{R}^T : \; g(t_j) \in B_j, \; j = 1, 2, \ldots, n\}$$

and $\mathcal{C}$ the family of all sets of the form $C(t_1, t_2, \ldots, t_n; B_1, B_2, \ldots, B_n)$. These are usually called *cylinder sets depending on* $t_1, t_2, \ldots, t_n$. The smallest σ-algebra of parts of $\mathbb{R}^T$ containing $\mathcal{C}$ will be called the *Borel σ-algebra of* $\mathbb{R}^T$ and denoted by $\sigma(\mathcal{C})$.

Consider now a family of probability measures

$$\{P_{t_1,t_2,\ldots,t_n}\}_{t_1,t_2,\ldots,t_n \in T;\ n=1,2,\ldots} \tag{1.2}$$

as follows: For each $n = 1, 2, \ldots$ and each n-tuple $t_1, t_2, \ldots, t_n$ of distinct elements of T, $P_{t_1,t_2,\ldots,t_n}$ is a probability measure on the Borel sets of the product space $X_{t_1} \times X_{t_2} \times \cdots \times X_{t_n}$, where $X_t = \mathbb{R}$ for each $t \in T$ (so that this product space is canonically identified as $\mathbb{R}^n$).

We say that the probability measures (1.2) satisfy the *consistency condition* if for any choice of $n = 1, 2, \ldots$ and distinct $t_1, \ldots, t_n, t_{n+1} \in T$, we have

$$P_{t_1,\ldots,t_n,t_{n+1}}(B \times \mathbb{R}) = P_{t_1,\ldots,t_n}(B)$$

for any Borel set B in $X_{t_1} \times \cdots \times X_{t_n}$. The following is the basic Kolmogorov extension theorem, which we state but do not prove here.

Theorem 1.1 (Kolmogorov). $\{P_{t_1,t_2,\ldots,t_n}\}_{t_1,t_2,\ldots,t_n \in T;\ n=1,2,\ldots}$ *satisfy the consistency condition if and only if there exists one and only one probability measure* P *on* $\sigma(\mathcal{C})$ *such that*

$$P(C(t_1, \ldots, t_n; B_1, \ldots, B_n)) = P_{t_1,\ldots,t_n}(B_1 \times \cdots \times B_n) \tag{1.3}$$

for any choice of $n = 1, 2, \ldots$, *distinct* $t_1, \ldots, t_n \in T$ *and* B_j *Borel sets in* X_{t_j}, $j = 1, \ldots, n$.

It is clear that if there exists a probability measure P on $\sigma(\mathcal{C})$ satisfying (1.3), the consistency conditions must hold since

$$C(t_1, \ldots, t_n, t_{n+1}; B_1, \ldots, B_n, X_{t_{n+1}}) = C(t_1, \ldots, t_n; B_1, \ldots, B_n).$$

So the problem is how to prove the converse. This can be done in two steps: (1) define P on the family of cylinders $\mathcal{C}$ using (1.3) and show that the definition is unambiguous (note that each cylinder has more than one representation); and (2) apply Caratheodory's theorem on an extension of measures to prove that this P can be extended in a unique form to $\sigma(\mathcal{C})$.

Remarks

1. Theorem 1.1 is interesting when T is an infinite set. The purpose is to be able to measure the probability of sets of functions from T to $\mathbb{R}$ (i.e., subsets of $\mathbb{R}^T$) which cannot be defined by means of a finite number of coordinates, which amounts to looking only at the values of the functions at a finite number of t-values.

Notice that in the case of cylinders, if one wants to know whether a given function $g : T \to \mathbb{R}$ belongs to $C(t_1, \ldots, t_n; B_1, \ldots, B_n)$, it suffices to look at the values of g at the finite set of points $t_1, \ldots, t_n$ and check if $g(t_j) \in B_j$ for $j = 1, \ldots, n$. However, if one takes, for example, $T = \mathbb{Z}$ (the integers) and considers the sets of functions

$$A = \{g : g : T \to \mathbb{R}, \lim_{t \to +\infty} g(t) \text{ exists and is finite}\}$$

or

$$B = \{g : g : T \to \mathbb{R}, \sup_{t \in T} |g(t)| \leq 1\},$$

it is clear that these sets are in $\sigma(\mathcal{C})$ but are not cylinders (they "depend on an infinite number of coordinates").

2. In general, $\sigma(\mathcal{C})$ is strictly smaller than the family of all subsets of $\mathbb{R}^T$. To see this, one can check that

$$\sigma(\mathcal{C}) = \{A \subset \mathbb{R}^T : \exists T_A \subset T, T_A \text{ countable and } B_A \text{ a Borel set in } \mathbb{R}^{T_A}, \text{ such that } g \in A \text{ if and only if } g/T_A \in B_A\}. \quad (1.4)$$

The proof of (1.4) follows immediately from the fact that the right-hand side is a σ-algebra containing $\mathcal{C}$. Equation (1.4) says that a subset of $\mathbb{R}^T$ is a Borel set if and only if it "depends only on a countable set of parameter values." Hence, if T is uncountable, the set

$$\{g \in \mathbb{R}^T : g \text{ is a bounded function}\}$$

or

$$\{g \in \mathbb{R}^T : g \text{ is a bounded function}, |g(t)| \leq 1 \text{ for all } t \in T\}$$

does not belong to $\sigma(\mathcal{C})$. Another simple example is the following: If $T = [0, 1]$, then

$$\{g \in \mathbb{R}^T : g \text{ is a continuous function}\}$$

is not a Borel set in $\mathbb{R}^T$, since it is obvious that there does not exist a countable subset of $[0, 1]$ having the determining property in (1.4). These examples lead to the notion of *separable process* that we introduce later.

3. In the special case when $\Omega = \mathbb{R}^T$, $\mathcal{A} = \sigma(\mathcal{C})$, and $X(t)(\omega) = \omega(t)$, $\{X(t) : t \in T\}$ is called a *canonical process*.

4. We say that the stochastic process $\{Y(t) : t \in T\}$ is a version of the process $\{X(t) : t \in T\}$ if $P(X(t) = Y(t)) = 1$ for each $t \in T$.

1.2. REMINDER ON THE NORMAL DISTRIBUTION

Let μ be a probability measure on the Borel subsets of $\mathbb{R}^d$. Its Fourier transform $\widehat{\mu} : \mathbb{R}^d \to \mathbb{C}$ is defined as

$$\widehat{\mu}(z) = \int_{\mathbb{R}^d} \exp(i\langle z, x\rangle)\mu(dx),$$

where $\langle \cdot, \cdot \rangle$ denotes the usual scalar product in $\mathbb{R}^d$.

We use *Bochner's theorem* (see, e.g., Feller, 1966): $\widehat{\mu}$ is the Fourier transform of a Borel probability measure on $\mathbb{R}^d$ if and only if the following three conditions hold true:

1. $\widehat{\mu}(0) = 1$.
2. $\widehat{\mu}$ is continuous.
3. $\widehat{\mu}$ is positive semidefinite; that is, for any $n = 1, 2, \ldots$ and any choice of the complex numbers $c_1, \ldots, c_n$ and of the points $z_1, \ldots, z_n$, one has

$$\sum_{j,k=1}^{n} \widehat{\mu}(z_j - z_k)c_j\bar{c}_k \geq 0.$$

The random vector ξ with values in $\mathbb{R}^d$ is said to have the *normal distribution*, or the *Gaussian distribution*, with parameters (m, Σ)" [$m \in \mathbb{R}^d$ and Σ a $d \times d$ positive semidefinite matrix] if the Fourier transform of the probability distribution μ_ξ of ξ is equal to

$$\widehat{\mu_\xi}(z) = \exp\left[i\langle m, z\rangle - \tfrac{1}{2}\langle z, \Sigma z\rangle\right] \qquad z \in \mathbb{R}^d.$$

When $m = 0$ and $\Sigma = I_d =$ identity $d \times d$ matrix, the distribution of ξ is called *standard normal in* $\mathbb{R}^d$. For $d = 1$ we use the notation

$$\varphi(x) = \frac{1}{\sqrt{2\pi}} e^{-(1/2)x^2} \quad \text{and} \quad \Phi(x) = \int_{-\infty}^{x} \varphi(y)\, dy$$

for the density and the cumulative distribution function of a standard normal random variable, respectively.

If Σ is nonsingular, μ_ξ is said to be nondegenerate and one can verify that it has a density with respect to Lebesgue measure given by

$$\mu_\xi(dx) = \frac{1}{(2\pi)^{d/2}(\det(\Sigma))^{1/2}} \exp\left[-\frac{1}{2}(x - m)^{\mathrm{T}}\Sigma^{-1}(x - m)\right] dx$$

x^{T} denotes the transpose of x. One can check that

$$m = \mathrm{E}(\xi), \qquad \Sigma = \mathrm{Var}(\xi) = \mathrm{E}((\xi - m)(\xi - m)^{\mathrm{T}}),$$

so m and Σ are, respectively, the mean and the variance of ξ.

From the definition above it follows that if the random vector ξ with values in $\mathbb{R}^d$ has a normal distribution with parameters m and Σ, A is a real matrix with n rows and d columns, and b is a nonrandom element of $\mathbb{R}^n$, then the random vector $A\xi + b$ with values in $\mathbb{R}^n$ has a normal distribution with parameters $(Am + b, A\Sigma A^{\mathrm{T}})$. In particular, if Σ is nonsingular, the coordinates of the random vector $\Sigma^{-1/2}(\xi - m)$ are independent random variables with standard normal distribution on the real line.

Assume now that we have a pair ξ and η of random vectors in $\mathbb{R}^d$ and $\mathbb{R}^{d'}$, respectively, having finite moments of order 2. We define the $d \times d'$ covariance matrix as

$$\mathrm{Cov}(\xi, \eta) = \mathrm{E}((\xi - \mathrm{E}(\xi)(\eta - \mathrm{E}(\eta)^{\mathrm{T}}).$$

It follows that if the distribution of the random vector (ξ, η) in $\mathbb{R}^{d+d'}$ is normal and $\mathrm{Cov}(\xi, \eta) = 0$, the random vectors ξ and η are independent. A consequence of this is the following useful formula, which is standard in statistics and gives a version of the conditional expectation of a function of ξ given the value of η.

Proposition 1.2. *Let ξ and η be two random vectors with values in $\mathbb{R}^d$ and $\mathbb{R}^{d'}$, respectively, and assume that the distribution of (ξ, η) in $\mathbb{R}^{d+d'}$ is normal and* $\mathrm{Var}(\eta)$ *is nonsingular. Then, for any bounded function $f : \mathbb{R}^d \to \mathbb{R}$, we have*

$$\mathrm{E}(f(\xi)|\eta = y) = \mathrm{E}(f(\zeta + Cy)) \tag{1.5}$$

for almost every y, where

$$C = \mathrm{Cov}(\xi, \eta)[\mathrm{Var}(\eta)]^{-1} \tag{1.6}$$

and ζ is a random vector with values in $\mathbb{R}^d$, having a normal distribution with parameters

$$\big(\mathrm{E}(\xi) - C\mathrm{E}(\eta), \mathrm{Var}(\xi) - \mathrm{Cov}(\xi, \eta)[\mathrm{Var}(\eta)]^{-1}[\mathrm{Cov}(\xi, \eta)]^{\mathrm{T}}\big). \tag{1.7}$$

Proof. The proof consists of choosing the matrix C so that the random vector

$$\zeta = \xi - C\eta$$

becomes independent of η. For this purpose, we need the fact that

$$\mathrm{Cov}(\xi - C\eta, \eta) = 0,$$

and this leads to the value of C given by (1.6). The parameters (1.7) follow immediately. □

In what follows, we call the version of the conditional expectation given by formula (1.5), *Gaussian regression*. To close this brief list of basic properties, we mention that a useful property of the Gaussian distribution is stability under passage to the limit (see Exercise 1.5).

Let $r : T \times T \to \mathbb{R}$ be a positive semidefinite function and $m : T \to \mathbb{R}$ a function. In this more general context, that r is a positive semidefinite function, means that for any $n = 1, 2, \ldots$ and any choice of distinct $t_1, \ldots, t_n \in T$, the matrix $((r(t_j, t_k)))_{j,k=1,\ldots,n}$ is positive semidefinite. [This is consistent with the previous definition, which corresponds to saying that $r(s,t) = \widehat{\mu}(s-t)$, $s, t \in \mathbb{R}^d$ is positive semidefinite.]

Take now for $P_{t_1,\ldots,t_n}$ the Gaussian probability measure in $\mathbb{R}^n$ with mean

$$m_{t_1,\ldots,t_n} := (m(t_1), \ldots, m(t_n))^{\mathrm{T}}$$

and variance matrix

$$\Sigma_{t_1,\ldots,t_n} := ((r(t_j, t_k)))_{j,k=1,\ldots,n}.$$

It is easily verified that the set of probability measures $\{P_{t_1,\ldots,t_n}\}$ verifies the consistency condition, so that Kolmogorov's theorem applies and there exists a unique probability measure P on the measurable space $(\mathbb{R}^T, \sigma(\mathcal{C}))$, which restricted to the cylinder sets depending on $t_1, \ldots, t_n$ is $P_{t_1,\ldots,t_n}$ for any choice of distinct parameter values $t_1, \ldots, t_n$. P is called the *Gaussian measure generated by the pair* (m, r). If $\{X(t) : t \in T\}$ is a real-valued stochastic process with distribution P, one verifies that:

- For any choice of distinct parameter values $t_1, \ldots, t_n$, the joint distribution of the random variables $X(t_1), \ldots, X(t_n)$ is Gaussian with mean $m_{t_1,\ldots,t_n}$ and variance $\Sigma_{t_1,\ldots,t_n}$.
- $E(X(t)) = m(t)$ for $t \in T$.
- $\text{Cov}(X(s), X(t)) = E((X(s) - m(s))(X(t) - m(t))) = r(s,t)$ for $s, t \in T$.

A class of examples that appears frequently in applications is the d-parameter real-valued Gaussian processes, which are centered and stationary, which means that

$$T = \mathbb{R}^d, \qquad m(t) = 0, \qquad r(s,t) = \Gamma(t-s).$$

A general definition of strictly stationary processes is given in Section 10.2.

If the function Γ is continuous, $\Gamma(0) \neq 0$, one can write

$$\Gamma(\tau) = \int_{\mathbb{R}^d} \exp(i\langle \tau, x\rangle)\mu(dx),$$

where μ is a Borel measure on $\mathbb{R}^d$ with total mass equal to $\Gamma(0)$. μ is called the *spectral measure* of the process. We usually assume that $\Gamma(0) = 1$: that is, that μ is a probability measure which is obtained simply by replacing the original process $\{X(t) : t \in \mathbb{R}^d\}$ by the process $\{X(t)/(\Gamma(0))^{1/2} : t \in \mathbb{R}^d\}$.

Example 1.1 (Trigonometric Polynomials). *An important example of stationary Gaussian processes is the following. Suppose that μ is a purely atomic probability symmetric measure on the real line; that is, there exists a sequence $\{x_n\}_{n=1,2,\dots}$ of positive real numbers such that*

$$\mu(\{x_n\}) = \mu(\{-x_n\}) = \tfrac{1}{2}c_n \text{ for } n = 1, 2, \dots ; \qquad \mu(\{0\}) = c_0; \qquad \sum_{n=0}^{\infty} c_n = 1.$$

Then a centered Gaussian process having μ as its spectral measure is

$$X(t) = c_0^{1/2}\xi_0 + \sum_{n=1}^{\infty} c_n^{1/2}(\xi_n \cos tx_n + \xi_{-n} \sin tx_n) \qquad t \in \mathbb{R}, \tag{1.8}$$

where the $\{\xi_n\}_{n\in Z}$ is a sequence of independent identically distributed random variables, each having a standard normal distribution. In fact, the series in (1.8) converges in $L^2(\Omega, \mathcal{F}, P)$ and

$$\mathrm{E}(X(t)) = 0 \quad \text{and} \quad \mathrm{E}(X(s)X(t)) = c_0 + \sum_{n=1}^{\infty} c_n \cos[(t-s)x_n] = \widehat{\mu}(t-s).$$

We use the notation

$$\lambda_k := \int_{\mathbb{R}} x^k \mu(dx) \qquad k = 0, 1, 2, \dots \tag{1.9}$$

whenever the integral exists. λ_k is the kth *spectral moment* of the process.

An extension of the preceding class of examples is the following. Let $(T, \mathcal{T}, \rho)$ be a measure space, $H = L^2_{\mathbb{R}}(T, \mathcal{T}, \rho)$ the Hilbert space of real-valued square-integrable functions on it, and $\{\varphi_n(t)\}_{n=1,2,\dots}$ an orthonormal sequence in H. We assume that each function $\varphi_n : T \to \mathbb{R}$ is bounded and denote $M_n = \sup_{t\in T}|\varphi_n(t)|$. In addition, let $\{c_n\}_{n=1,2,\dots}$ be a sequence of positive numbers such that

$$\sum_{n=1}^{\infty} c_n < \infty, \qquad \sum_{n=1}^{\infty} c_n M_n^2 < \infty$$

and $\{\xi_n\}_{n=1,2,\dots}$ a sequence of independent identically distributed (i.i.d.) random variables, each with standard normal distribution in $\mathbb{R}$.

Then the stochastic process

$$X(t) = \sum_{n=1}^{\infty} c_n^{1/2}\xi_n\varphi_n(t) \tag{1.10}$$

is Gaussian, centered with covariance

$$r(s,t) = \mathrm{E}(X(s)X(t)\} = \sum_{n=1}^{\infty} c_n \varphi_n(s)\varphi_n(t).$$

Formulas (1.8) and (1.10) are simple cases of spectral representations of Gaussian processes, which is an important subject for both theoretical purposes and for applications. A compact presentation of this subject, including the Karhunen–Loève representation and the connection with reproducing kernel Hilbert spaces, may be found in Fernique's lecture notes (1974).

1.3. 0–1 LAW FOR GAUSSIAN PROCESSES

We will prove a 0–1 law for Gaussian processes in this section without attempting full generality. This will be sufficient for our requirements in what follows. For a more general treatment, see Fernique (1974).

Definition 1.3. *Let $\mathcal{X} = \{X(t) : t \in T\}$ and $\mathcal{Y} = \{Y(t) : t \in S\}$ be real-valued stochastic processes defined on some probability space $(\Omega, \mathcal{A}, P)$. $\mathcal{X}$ and $\mathcal{Y}$ are said to be* independent *if for any choice of the parameter values $t_1, \dots, t_n \in T; s_1, \dots, s_m \in S$, $n, m \geq 1$, the random vectors*

$$(X(t_1), \dots, X(t_n)), \ (Y(s_1), \dots, Y(s_m))$$

are independent.

Proposition 1.4. *Let the processes $\mathcal{X}$ and $\mathcal{Y}$ be independent and E (respectively, F) belong to the σ-algebra generated by the cylinders in $\mathbb{R}^T$ (respectively, $\mathbb{R}^S$). Then*

$$\mathrm{P}(X(\cdot) \in E, Y(\cdot) \in F) = \mathrm{P}(X(\cdot) \in E)\mathrm{P}(Y(\cdot) \in F). \tag{1.11}$$

Proof. Equation (1.11) holds true for cylinders. Uniqueness in the extension theorem provides the result. □

Theorem 1.5 (0–1 Law for Gaussian Processes). *Let $\mathcal{X} = \{X(t) : t \in T\}$ be a real-valued centered Gaussian process defined on some probability space $(\Omega, \mathcal{A}, \mathrm{P})$ and $(\mathrm{E}, \mathcal{E})$ a measurable space, where* E *is a linear subspace of $\mathbb{R}^T$ and the σ-algebra $\mathcal{E}$ has the property that for any choice of the scalars $a, b \in \mathbb{R}$, the function $(x, y) \rightsquigarrow ax + by$ defined on* E × E *is measurable with respect to the product σ-algebra. We assume that the function $X : \Omega \to \mathrm{E}$ defined as $X(\omega) = X(\cdot, \omega)$ is measurable $(\Omega, \mathcal{A}) \to (\mathrm{E}, \mathcal{E})$. Then, if L is a measurable subspace of* E*, one has*

$$\mathrm{P}(X(\cdot) \in L) = 0 \quad or \quad 1.$$

Proof. Let $\{X^{(1)}(t) : t \in T\}$ and $\{X^{(2)}(t) : t \in T\}$ be two independent processes each having the same distribution as that of the given process $\{X(t) : t \in T\}$. For each λ, $0 < \lambda < \pi/2$, consider a new pair of stochastic processes, defined for $t \in T$ by

$$\begin{aligned} Z_\lambda^{(1)}(t) &= X^{(1)}(t)\cos\lambda + X^{(2)}(t)\sin\lambda \\ Z_\lambda^{(2)}(t) &= -X^{(1)}(t)\sin\lambda + X^{(2)}(t)\cos\lambda. \end{aligned} \tag{1.12}$$

Each of the processes $Z_\lambda^{(i)}(t)(i = 1, 2)$ has the same distribution as $\mathcal{X}$.

In fact, $\mathrm{E}(Z_\lambda^{(1)}(t)) = 0$ and since $\mathrm{E}(X^{(1)}(s)X^{(2)}(t)) = 0$, we have $\mathrm{E}(Z_\lambda^{(1)}(s) Z_\lambda^{(1)}(t)) = \cos^2\lambda\ \mathrm{E}(X^{(1)}(s)X^{(1)}(t)) + \sin^2\lambda\ \mathrm{E}(X^{(2)}(s)X^{(2)}(t)) = \mathrm{E}(X(s)X(t))$.

A similar computation holds for $Z_\lambda^{(2)}$.

Also, the processes $Z_\lambda^{(1)}$ and $Z_\lambda^{(2)}$ are independent. To prove this, note that for any choice of $t_1, \ldots, t_n; s_1, \ldots, s_m, n, m \geq 1$, the random vectors

$$(Z_\lambda^{(1)}(t_1), \ldots, Z_\lambda^{(1)}(t_n)), (Z_\lambda^{(2)}(s_1), \ldots, Z_\lambda^{(2)}(s_m))$$

have a joint Gaussian distribution, so it suffices to show that

$$\mathrm{E}(Z_\lambda^{(1)}(t)Z_\lambda^{(2)}(s)) = 0$$

for any choice of $s, t \in T$ to conclude that they are independent. This is easily checked.

Now, if we put $q = \mathrm{P}(X(\cdot) \in L)$, independence implies that for any λ,

$$q(1-q) = \mathrm{P}(E_\lambda) \qquad \text{where} \quad E_\lambda = \{Z_\lambda^{(1)} \in L, Z_\lambda^{(2)} \notin L\}.$$

If $\lambda, \lambda' \in (0, \pi/2)$, $\lambda \neq \lambda'$, the events E_λ and $E_{\lambda'}$ are disjoint. In fact, the matrix

$$\begin{pmatrix} \cos\lambda & \sin\lambda \\ \cos\lambda' & \sin\lambda' \end{pmatrix}$$

is nonsingular and (1.12) implies that if at the same time $Z_\lambda^{(1)} \in L$, $Z_{\lambda'}^{(1)} \in L$, then $X^{(1)}(\cdot), X^{(2)}(\cdot) \in L$ also, since $X^{(1)}(\cdot), X^{(2)}(\cdot)$ are linear combinations of $Z_\lambda^{(1)}$ and $Z_{\lambda'}^{(1)}$. Hence, $Z_\lambda^{(2)}, Z_{\lambda'}^{(2)} \in L$ and E_λ, $E_{\lambda'}$ cannot occur simultaneously. To finish, the only way in which we can have an infinite family $\{E_\lambda\}_{0<\lambda<\pi/2}$ of pairwise disjoint events with equal probability is for this probability to be zero. That is, $q(1-q) = 0$, so that $q = 0$ or 1. □

In case the parameter set T is countable, the above shows directly that any measurable linear subspace of $\mathbb{R}^T$ has probability 0 or 1 under a centered Gaussian law. If T is a σ-compact topological space, E the set of real-valued continuous functions defined on T, and $\mathcal{E}$ the σ-algebra generated by the topology of uniform convergence on compact sets, one can conclude, for example, that the subspace of E of bounded functions has probability 0 or 1 under a centered Gaussian measure. The theorem can be applied in a variety of situations similar to standard function spaces. For example, put a measure on the space (E, $\mathcal{E}$) and take for L an L^p of this measure space.

1.4. REGULARITY OF PATHS

1.4.1. Conditions for Continuity of Paths

Theorem 1.6 (Kolmogorov). *Let $\mathcal{Y} = \{Y(t) : t \in [0,1]\}$ be a real-valued stochastic process that satisfies the condition*

(K) *For each pair t and $t + h \in [0,1]$,*

$$\mathrm{P}\{|Y(t+h) - Y(t)| \geq \alpha(h)\} \leq \beta(h),$$

where α and β are even real-valued functions defined on $[-1,1]$, increasing on $[0,1]$, that verify

$$\sum_{n=1}^{\infty} \alpha(2^{-n}) < \infty, \qquad \sum_{n=1}^{\infty} 2^n \beta(2^{-n}) < \infty.$$

Then there exists a version $\mathcal{X} = \{X(t) : t \in T\}$ of the process $\mathcal{Y}$ such that the paths $t \rightsquigarrow X(t)$ are continuous on $[0,1]$.

Proof. For $n = 1, 2, \ldots; k = 0, 1, \ldots, 2^n - 1$, let

$$E_{k,n} = \left\{\left|Y\left(\frac{k+1}{2^n}\right) - Y\left(\frac{k}{2^n}\right)\right| \geq \alpha(2^{-n})\right\}, \qquad E_n = \bigcup_{k=0}^{2^n-1} E_{k,n}.$$

From the hypothesis, $\mathrm{P}(E_n) \leq 2^n \beta(2^{-n})$, so that $\sum_{n=1}^{\infty} \mathrm{P}(E_n) < \infty$. The Borel–Cantelli lemma implies that $\mathrm{P}(\limsup_{n\to\infty} E_n) = 0$, where

$$\limsup_{n\to\infty} E_n = \{\omega : \omega \text{ belongs to infinitely many } E_n\text{'s}\}.$$

In other words, if $\omega \notin \limsup_{n\to\infty} E_n$, one can find $n_0(\omega)$ such that if $n \geq n_0(\omega)$, one has

$$\left|Y\left(\frac{k+1}{2^n}\right) - Y\left(\frac{k}{2^n}\right)\right| < \alpha(2^{-n}) \qquad \text{for all } k = 0, 1, \ldots, 2^n - 1.$$

Denote by $Y^{(n)}$ the function whose graph is the polygonal with vertices $(k/2^n, Y(k/2^n))$, $k = 0, 1, \ldots, 2^n$; that is, if $k/2^n \le t \le (k+1)/2^n$, one has

$$Y^{(n)}(t) = (k + 1 - 2^n t)Y\left(\frac{k}{2^n}\right) + (2^n t - k)Y\left(\frac{k+1}{2^n}\right).$$

The function $t \rightsquigarrow Y^{(n)}(t)$ is continuous. Now, if $\omega \notin \limsup_{n\to\infty} E_n$, one easily checks that there exists some integer $n_0(\omega)$ such that

$$\left\| Y^{(n+1)} - Y^{(n)} \right\|_\infty \le \alpha\left(2^{-(n+1)}\right) \qquad \text{for } n + 1 \ge n_0(\omega)$$

(here $\|\cdot\|_\infty$ denotes the sup norm on [0, 1]). Since $\sum_{n=1}^{\infty} \alpha(2^{-(n+1)}) < \infty$ by the hypothesis, the sequence of functions $\{Y^{(n)}\}$ converges uniformly on [0, 1] to a continuous limit function that we denote $X(t)$, $t \in [0, 1]$.

We set $X(t) \equiv 0$ when $\omega \in \limsup_{n\to\infty} E_n$. To finish the proof, it suffices to show that for each $t \in [0, 1]$, $\mathrm{P}(X(t) = Y(t)) = 1$.

- If t is a dyadic point, say $t = k/2^n$, then given the definition of the sequence of functions $Y^{(n)}$, it is clear that $Y^{(m)}(t) = Y(t)$ for $m \ge n$. Hence, for $\omega \notin \limsup_{n\to\infty} E_n$, one has $X(t) = \lim_{m\to\infty} Y^{(m)}(t) = Y(t)$. The result follows from $\mathrm{P}((\limsup_{n\to\infty} E_n)^C) = 1$ (A^C is the complement of the set A).
- If t is not a dyadic point, for each n, $n = 1, 2, \ldots$, let k_n be an integer such that $|t - k_n/2^n| \le 2^{-n}$, $k_n/2^n \in [0, 1]$. Set

  $$F_n = \left\{ \left| Y(t) - X\left(\frac{k_n}{2^n}\right) \right| \ge \alpha(2^{-n}) \right\}.$$

  We have the inequalities

  $$\mathrm{P}(F_n) \le \mathrm{P}\left(\left| Y(t) - X\left(\frac{k_n}{2^n}\right) \right| \ge \alpha\left(\left| t - \frac{k_n}{2^n} \right| \right) \right) \le \beta\left(\left| t - \frac{k_n}{2^n} \right| \right) \le \beta(2^{-n}),$$

  and a new application of the Borel–Cantelli lemma gives $\mathrm{P}(\limsup_{n\to\infty} F_n) = 0$. So if $\omega \notin [\limsup_{n\to\infty} E_n] \cup [\limsup_{n\to\infty} F_n]$, we have at the same time, $X(k_n/2^n)(\omega) \to X(t)(\omega)$ as $n \to \infty$ because $t \rightsquigarrow X(t)$ is continuous, and $X(k_n/2^n)(\omega) \to Y(t)(\omega)$ because $|Y(t) - X(k_n/2^n)| < \alpha(2^{-n})$ for $n \ge n_1(\omega)$ for some integer $n_1(\omega)$.

This proves that $X(t)(\omega) = Y(t)(\omega)$ for almost every ω. □

Corollary 1.7 *Assume that the process $\mathcal{Y} = \{Y(t) : t \in [0, 1]\}$ satisfies one of the following conditions for $t, t + h \in [0, 1]$:*

(a)
$$\mathrm{E}(|Y(t+h) - Y(t)|^p) \le \frac{K|h|}{|\log|h||^{1+r}}, \tag{1.13}$$

where p,r, and K are positive constants, $p < r$.

(b) *$\mathcal{Y}$ is Gaussian, $m(t) := \mathrm{E}(Y(t))$ is continuous, and*

$$\mathrm{Var}(Y(t+h) - Y(t)) \le \frac{C}{|\log|h||^a} \tag{1.14}$$

for all t, sufficiently small h, C some positive constant, and $a > 3$.

Then the conclusion of Theorem 1.6 holds.

Proof

(a) Set

$$\alpha(h) = \frac{1}{|\log|h||^b} \qquad 1 < b < \frac{r}{p}$$
$$\beta(h) = \frac{|h|}{|\log|h||^{1+r-bp}}$$

and check condition (K) using a Markov inequality.

(b) Since the expectation is continuous, it can be subtracted from $Y(t)$, so that we may assume that $\mathcal{Y}$ is centered. To apply Theorem 1.6, take

$$\alpha(h) = \frac{1}{|\log|h||^b} \text{ with } 1 < b < (a-1)/2 \text{ and } \beta(h) = \exp\left[-\frac{1}{4C}|\log|h||^{a-2b}\right].$$

Then

$$\mathrm{P}(|Y(t+h) - Y(t)| \ge \alpha(h)) = \mathrm{P}\left(|\xi| \ge \frac{\alpha(h)}{\sqrt{\mathrm{Var}(Y(t+h) - Y(t))}}\right),$$

where ξ stands for standard normal variable. We use the following usual bound for Gaussian tails, valid for $u > 0$:

$$\mathrm{P}(|\xi| \ge u) = 2\mathrm{P}(\xi \ge u) = \sqrt{\frac{2}{\pi}} \int_u^{+\infty} e^{-(1/2)x^2} dx \le \sqrt{\frac{2}{\pi}} \frac{1}{u} e^{-(1/2)u^2}.$$

With the foregoing choice of $\alpha(\cdot)$ and $\beta(\cdot)$, if $|h|$ is small enough, one has $\alpha(h)/\sqrt{\mathrm{Var}(Y(t+h) - Y(t))} > 1$ and

$$\mathrm{P}(|Y_{t+h} - Y(t)| \ge \alpha(h)) \le (\text{const})\, \beta(h).$$

where (const) denotes a generic constant that may vary from line to line. On the other hand, $\sum_1^\infty \alpha(2^{-n}) < \infty$ and $\sum_1^\infty 2^n \beta(2^{-n}) < \infty$ are easily verified. □

Some Examples

1. *Gaussian stationary processes.* Let $\{Y(t) : t \in \mathbb{R}\}$ be a real-valued Gaussian centered stationary process with covariance $\Gamma(\tau) = \mathrm{E}(Y(t)\, Y(t+\tau))$. Then condition (1.14) is equivalent to

$$\Gamma(0) - \Gamma(\tau) \leq \frac{C}{|\log|\tau||^a}$$

for sufficiently small $|\tau|$, with the same meaning for C and a.

2. *Wiener process.* Take $T = \mathbb{R}^+$. The function $r(s,t) = s \wedge t$ is positive semidefinite. In fact, if $0 \leq s_1 < \cdots < s_n$ and $x_1, \ldots, x_n \in \mathbb{R}$, one has

$$\sum_{j,k=1}^{n} (s_j \wedge s_k) x_j x_k = \sum_{k=1}^{n} (s_k - s_{k-1})(x_k + \cdots + x_n)^2 \geq 0, \tag{1.15}$$

where we have set $s_0 = 0$.

Then, according to Kolmogorov's extension theorem, there exists a centered Gaussian process $\{Y(t) : t \in \mathbb{R}^+\}$ such that $\mathrm{E}(Y(s)Y(t)) = s \wedge t$ for $s, t \geq 0$. One easily checks that this process satisfies the hypothesis in Corollary 1.7(b), since the random variable $Y(t+h) - Y(t)$, $h \geq 0$ has the normal distribution $N(0, h)$ because of the simple computation

$$\mathrm{E}([Y(t+h) - Y(t)]^2) = t + h - 2t + t = h.$$

It follows from Corollary 1.7(b) that this process has a continuous version on every interval of the form $[n, n+1]$. The reader will verify that one can also find a version with continuous paths defined on all $\mathbb{R}^+$. This version, called the *Wiener process*, is denoted $\{W(t) : t \in \mathbb{R}^+\}$.

3. *Ito integrals.* Let $\{W(t) : t \geq 0\}$ be a Wiener process on a probability space $(\Omega, \mathcal{A}, \mathrm{P})$. We define the *filtration* $\{\mathcal{F}_t : t \geq 0\}$ as $\mathcal{F}_t = \tilde{\sigma}\{W(s) : s \leq t\}$, where the notation means the σ-algebra generated by the set of random variables $\{W(s) : s \leq t\}$ (i.e., the smallest σ-algebra with respect to which these random variables are all measurable) completed with respect to the probability measure P.

Let $\{a_t : t \geq 0\}$ be a stochastic process adapted to the filtration $\{\mathcal{F}_t : t \geq 0\}$. This means that a_t is $\mathcal{F}_t$-measurable for each $t \geq 0$. For simplicity we assume that $\{a_t : t \geq 0\}$ is uniformly locally bounded in the sense that for each $T > 0$ there exists a constant C_T such that $|a_t(\omega)| \leq C_T$ for every ω and all $t \in [0, T]$. For each $t > 0$, one can define the stochastic Ito integral

$$Y(t) = \int_0^t a_s \, dW(s)$$

as the limit in $L^2 = L^2(\Omega, \mathcal{A}, \mathrm{P})$ of the Riemann sums

$$S_Q = \sum_{j=0}^{m-1} \widetilde{a}_{t_j} (W(t_{j+1}) - W(t_j))$$

when $N_Q = \sup\{(t_{j+1} - t_j) : 0 \le j \le m-1\}$ tends to 0. Here Q denotes the partition $0 = t_0 < t_1 < \cdots < t_m = t$ of the interval $[0, t]$ and $\{\widetilde{a}_t : t \ge 0\}$ an adapted stochastic process, bounded by the same constant as $\{a_t : t \ge 0\}$ and such that

$$\sum_{j=0}^{m-1} \widetilde{a}_{t_j} \mathbf{1}_{\{t_j \le s < t_{j+1}\}}$$

tends to $\{a_t : 0 \le s \le t\}$ in the space $L^2([0, t] \times \Omega, \lambda \times \mathrm{P})$ as $N_Q \to 0$. λ is a Lebesgue measure on the line.

Of course, the statements above should be proved to be able to define $Y(t)$ in this way (see, e.g., McKean, 1969). Our aim here is to prove that the process $\{Y(t) : t \ge 0\}$ thus defined has a version with continuous paths. With no loss of generality, we assume that t varies on the interval $[0, 1]$ and apply Corollary 1.7(a) with $p = 4$.

We will prove that

$$\mathrm{E}((Y(t+h) - Y(t))^4) \le (\mathrm{const})h^2.$$

For this, it is sufficient to see that if Q is a partition of the interval $[t, t+h]$, $h > 0$,

$$\mathrm{E}(S_Q^4) \le (\mathrm{const})h^2, \tag{1.16}$$

where (const) does not depend on t, h, and Q, and then apply Fatou's lemma when $N_Q \to 0$.

Let us compute the left-hand side of (1.16). Set $\Delta_j = W(t_{j+1}) - W(t_j)$. We have

$$\mathrm{E}(S_Q^4) = \sum_{j_1, j_2, j_3, j_4 = 0}^{m-1} \mathrm{E}(\widetilde{a}_{t_{j_1}} \widetilde{a}_{t_{j_2}} \widetilde{a}_{t_{j_3}} \widetilde{a}_{t_{j_4}} \Delta_{j_1} \Delta_{j_2} \Delta_{j_3} \Delta_{j_4}). \tag{1.17}$$

If one of the indices, say j_4, satisfies $j_4 > j_1, j_2, j_3$, the corresponding term becomes

$$\begin{aligned}\mathrm{E}\left(\prod_{h=1}^{4} (\widetilde{a}_{t_{j_h}} \Delta_{j_h})\right) &= \mathrm{E}\left(\mathrm{E}\left(\prod_{h=1}^{4} (\widetilde{a}_{t_{j_h}} \Delta_{j_h}) | \mathcal{F}_{t_{j_4}}\right)\right) \\ &= \mathrm{E}\left(\prod_{h=1}^{3} (\widetilde{a}_{t_{j_h}} \Delta_{j_h}) \widetilde{a}_{t_{j_4}} \mathrm{E}(\Delta_{j_4} | \mathcal{F}_{t_{j_4}})\right) = 0\end{aligned}$$

since

$$\mathrm{E}(\Delta_j | \mathcal{F}_{t_j}) = \mathrm{E}(\Delta_j) = 0 \quad \text{and} \quad \prod_{h=1}^{3} (\widetilde{a}_{t_{j_h}} \Delta_{j_h}) \widetilde{a}_{t_{j_4}} \ \text{is} \mathcal{F}_{t_{j_4}} - \text{measurable}.$$

In a similar way, if $j_4 < j_1 = j_2 = j_3$ (and similarly, if any one of the indices is strictly smaller than the others and these are all equal), the corresponding term vanishes since in this case

$$\mathrm{E}\left(\prod_{h=1}^{4}\left(\widetilde{a}_{t_{j_h}}\Delta_{j_h}\right)\right) = \mathrm{E}\left(\mathrm{E}\left(\left(\widetilde{a}_{t_{j_1}}\Delta_{j_1}\right)^3 \widetilde{a}_{t_{j_4}}\Delta_{j_4}|\mathcal{F}_{t_{j_1}}\right)\right)$$
$$= \mathrm{E}\left(\widetilde{a}^3_{t_{j_1}}\widetilde{a}_{t_{j_4}}\Delta_{j_4}\mathrm{E}\left(\Delta^3_{j_1}|\mathcal{F}_{t_{j_1}}\right)\right) = 0$$

because

$$\mathrm{E}\left(\Delta_j^3|\mathcal{F}_{t_j}\right) = \mathrm{E}\left(\Delta_j^3\right) = 0.$$

The terms with $j_1 = j_2 = j_3 = j_4$ give the sum

$$\sum_{j=0}^{m-1}\mathrm{E}\left(\left(\widetilde{a}_{t_j}\Delta_j\right)^4\right) \le C_1^4 \sum_{j=0}^{m-1} 3\,(t_{j+1} - t_j)^2 \le 3\,C_1^4\,h^2.$$

Finally, we have the sum of the terms corresponding to 4-tuples of indices j_1, j_2, j_3, and j_4 such that for some permutation (i_1, i_2, i_3, i_4) of $(1, 2, 3, 4)$, one has $j_{i_1}, j_{i_2} < j_{i_3} = j_{i_4}$. This is

$$6\sum_{j_3=1}^{m-1}\ \sum_{0\le j_1, j_2 < j_3}\mathrm{E}\left(\widetilde{a}_{t_{j_1}}\widetilde{a}_{t_{j_2}}\widetilde{a}^2_{t_{j_3}}\Delta_{j_1}\Delta_{j_2}\Delta^2_{j_3}\right).$$

Conditioning on $\mathcal{F}_{t_{j_3}}$ in each term yields for this sum

$$6\sum_{j_3=1}^{m-1}\ \sum_{0\le j_1, j_2 < j_3}(t_{j_3+1} - t_{j_3})\mathrm{E}\left(\widetilde{a}_{t_{j_1}}\widetilde{a}_{t_{j_2}}\widetilde{a}^2_{t_{j_3}}\Delta_{j_1}\Delta_{j_2}\right)$$
$$= 6\mathrm{E}\left(\sum_{j_3=1}^{m-1}(t_{j_3+1} - t_{j_3})\widetilde{a}^2_{t_{j_3}}\left(\sum_{j=0}^{j_3-1}\widetilde{a}_{t_j}\Delta_j\right)^2\right)$$
$$\le 6\,C_1^2\sum_{j_3=1}^{m-1}(t_{j_3+1} - t_{j_3})\mathrm{E}\left(\left(\sum_{j=0}^{j_3-1}\widetilde{a}_{t_j}\Delta_j\right)^2\right)$$
$$= 6\,C_1^2\sum_{j_3=1}^{m-1}(t_{j_3+1} - t_{j_3})\sum_{j=0}^{j_3-1}\mathrm{E}\left(\widetilde{a}^2_{t_j}\right)(t_{j+1} - t_j) \le 3C_1^4\,h^2.$$

Using (1.17), one obtains (1.16), and hence the existence of a version of the Itô integral possessing continuous paths.

Separability. Next, we consider the separability of stochastic processes. The separability condition is shaped to avoid the measurability problems that we have already mentioned and to use, without further reference, versions of stochastic processes having good path properties. We begin with a definition.

Definition 1.8. *We say that a real-valued stochastic process $\{X(t) : t \in T\}$, T a topological space, is* separable *if there exists a fixed countable subset D of T such that with probability 1,*

$$\sup_{t \in V \cap D} X(t) = \sup_{t \in V} X(t) \quad \textit{and} \quad \inf_{t \in V \cap D} X(t) = \inf_{t \in V} X(t) \quad \textit{for all open sets } V.$$

A consequence of Theorem 1.6 is the following:

Proposition 1.9. *Let $\{Y(t) : t \in I\}$, I an interval in the line, be a separable random process that satisfies the hypotheses of Theorem 1.6. Then, almost surely (a.s.), its paths are continuous.*

Proof. Denote by D the countable set in the definition of separability. With no loss of generality, we may assume that D is dense in I. The theorem states that there exists a version $\{X(t) : t \in I\}$ that has continuous paths, so that

$$\mathrm{P}(X(t) = Y(t) \text{ for all } t \in D) = 1.$$

Let

$$E = \{X(t) = Y(t) \text{ for all } t \in D\}$$

and

$$F = \bigcap_{J \subset I, J=(r_1,r_2), r_1,r_2 \in Q} \left\{ \sup_{t \in J \cap D} Y(t) = \sup_{t \in J} Y(t) \quad \text{and} \quad \inf_{t \in J \cap D} Y(t) = \inf_{t \in J} Y(t) \right\}.$$

Since $\mathrm{P}(E \cap F) = 1$, it is sufficient to prove that if $\omega \in E \cap F$, then $X(s)(\omega) = Y(s)(\omega)$ for all $s \in I$.

So, let $\omega \in E \cap F$ and $s \in I$. For any $\varepsilon > 0$, choose $r_1, r_2 \in Q$ such that

$$s - \varepsilon < r_1 < s < r_2 < s + \varepsilon.$$

Then, setting $J = (r_1, r_2)$,

$$Y(s)(\omega) \leq \sup_{t \in J} Y(t)(\omega) = \sup_{t \in J \cap D} Y(t)(\omega) = \sup_{t \in J \cap D} X(t)(\omega) \leq \sup_{t \in J} X(t)(\omega).$$

Letting $\varepsilon \to 0$, it follows that

$$Y(s)(\omega) \leq \limsup_{t \to s} X(t)(\omega) = X(s)(\omega)$$

since $t \rightsquigarrow X(t)(\omega)$ is continuous.

In a similar way, one proves that $Y(s)(\omega) \geq X(s)(\omega)$. □

The separability condition is usually met when the paths have some minimal regularity (see Exercise 1.7). For example, if $\{X(t) : t \in \mathbb{R}\}$ is a real-valued process having a.s. *càd-làg paths* (i.e., paths that are right-continuous with left limits), it is separable. All processes considered in the sequel are separable.

Some Additional Remarks and References. A reference for Kolmogorov's extension theorem and the regularity of paths, at the level of generality we have considered here, is the book by Cramér and Leadbetter (1967), where the reader can find proofs that we have skipped as well as related results, examples, and details. For d-parameter Gaussian processes, a subject that we consider in more detail in Chapter 6, in the stationary case, necessary and sufficient conditions to have continuous paths are due to Fernique (see his St. Flour 1974 lecture notes) and to Talagrand (1987) in the general nonstationary case. In the Gaussian stationary case, Belayev (1961) has shown that either: with probability 1 the paths are continuous, or with probability 1 the supremum (respectively, the infimum) on every interval is $+\infty$ (respectively, $-\infty$). General references on Gaussian processes are the books by Adler (1990) and Lifshits (1995).

1.4.2. Sample Path Differentiability and Hölder Conditions

In this section we state some results, without detailed proofs. These follow the lines of the preceding section.

Theorem 1.10. *Let $\mathcal{Y} = \{Y(t) : t \in [0, 1]\}$ be a real-valued stochastic process that satisfies the hypotheses of Theorem 1.6 and additionally, for any triplet $t - h$, t, $t + h \in [0, 1]$, one has*

$$\mathrm{P}(|Y(t+h) + Y(t-h) - 2Y(t)| \geq \alpha_1(h)) \leq \beta_1(h),$$

where α_1 and β_1 are two even functions, increasing for $h > 0$ and such that

$$\sum_{n=1}^{\infty} 2^n \alpha_1(2^{-n}) < \infty, \qquad \sum_{n=1}^{\infty} 2^n \beta_1(2^{-n}) < \infty.$$

Then there exists a version $\mathcal{X} = \{X(t) : t \in T\}$ of the process $\mathcal{Y}$ such that almost surely the paths of $\mathcal{X}$ are of class C^1.

Sketch of the Proof. Consider the sequence $\{Y^{(n)}(t) : t \in [0, 1]\}_{n=1,2,\ldots}$ of polygonal processes introduced in the proof of Theorem 1.6. We know that a.s. this sequence converges uniformly to $\mathcal{X} = \{X(t) : t \in [0, 1]\}$, a continuous version of $\mathcal{Y}$. Define:

$$\widetilde{Y}^{(n)}(t) := Y^{(n)\prime}(t^-) \quad \text{for } 0 < t \leq 1 \text{ (left derivative)}$$
$$\widetilde{Y}^{(n)}(0) := Y^{(n)\prime}(0^+) \quad \text{(right derivative).}$$

One can show that the hypotheses imply:

1. Almost surely, as $n \to \infty$, $\widetilde{Y}^{(n)}(\cdot)$ converges uniformly on $[0, 1]$ to a function $\widetilde{X}(\cdot)$.
2. Almost surely, as $n \to \infty$, $\sup_{t\in[0,1]}|\widetilde{Y}^{(n)}(t^+) - \widetilde{Y}^{(n)}(t)| \to 0$.

To complete the proof, check that the function $t \rightsquigarrow \widetilde{X}(t)$ a.s. is continuous and coincides with the derivative of $X(t)$ at every $t \in [0, 1]$. □

Example 1.2 (Stationary Gaussian Processes). *Let $\mathcal{Y} = \{Y(t) : t \in \mathbb{R}\}$ be a centered stationary Gaussian process with covariance of the form*

$$\Gamma(\tau) = \mathrm{E}(Y(t)Y(t+\tau)) = \Gamma(0) - \frac{1}{2}\lambda_2\tau^2 + O\left(\frac{\tau^2}{|\log|\tau||^a}\right)$$

with $\lambda_2 > 0$, $a > 3$. Then there exists a version of $\mathcal{Y}$ with paths of class C^1. For the proof, apply Theorem 1.10.

A related result is the following. The proof is left to the reader.

Proposition 1.11 (Hölder Conditions). *Assume that*

$$\mathrm{E}(|Y(t+h) - Y(t)|^p) \leq K|h|^{1+r} \qquad \text{for } t, t+h \in [0, 1], \tag{1.18}$$

where K, p, and r are positive constants, $r \leq p$. Then there exists a version of the process $\mathcal{Y} = \{Y(t) : t \in [0, 1]\}$ with paths that satisfy a Hölder condition with exponent α for any α such that $0 < \alpha < r/p$.

Note that, for example, this proposition can be applied to the Wiener process (Brownian motion) with $r = (p-2)/2$, showing that it satisfies a Hölder condition for every $\alpha < \frac{1}{2}$.

1.4.3. Higher Derivatives

Let $\mathcal{X} = \{X(t) : t \in \mathbb{R}\}$ be a stochastic process and assume that for each $t \in \mathbb{R}$, one has $X(t) \in L^2(\Omega, \mathcal{A}, \mathrm{P})$.

Definition 1.12. *$\mathcal{X}$ is* differentiable in quadratic mean *(q.m.) if for all $t \in \mathbb{R}$,*

$$\frac{X(t+h) - X(t)}{h}$$

converges in quadratic mean as $h \to 0$ to some limit that will be denoted $X'(t)$.

The stability of Gaussian random variables under passage to the limit implies that the derivative in q.m. of a Gaussian process remains Gaussian.

Proposition 1.13. *Let $\mathcal{X} = \{X(t) : t \in \mathbb{R}\}$ be a stochastic process with mean $m(t)$ and covariance $r(s,t)$ and suppose that m is C^1 and that r is C^2. Then $\mathcal{X}$ is differentiable in the quadratic mean.*

Proof. We use the following result, which is easy to prove: The sequence $Z_1, \ldots, Z_n$ of real random variables converges in q.m. if and only if there exists a constant C such that $\mathrm{E}(Z_m Z_n) \to C$ as the pair (m,n) tends to infinity. Since $m(t)$ is differentiable, it can be substracted from $X(t)$ without changing its differentiability, so we can assume that the process is centered. Then for all real h and k,

$$\begin{aligned}
&\mathrm{E}\left(\frac{X(t+h) - X(t)}{h}\,\frac{X(t+k) - X(t)}{k}\right) \\
&\quad = \frac{1}{hk}\big[r(t+h, t+k) - r(t, t+k) - r(t, t+h) + r(t,t)\big] \\
&\quad\quad \to r_{11}(t,t) \quad \text{as } (k,h) \to (0,0),
\end{aligned}$$

where $r_{11}(s,t) := \partial^2 r(s,t)/\partial s \partial t$. This shows differentiability in q.m. □

We assume, using the remark in the proof above, that $\mathcal{X}$ is centered and satisfies the conditions of the proposition. It is easy to prove that

$$\mathrm{E}(X(s)X'(t)) = r_{01}(s,t) := \frac{\partial r}{\partial t}(s,t),$$

and similarly, that the covariance of $\mathcal{X}' = \{X'(t) : t \in \mathbb{R}\}$ is $r_{11}(s,t)$. Now let $\mathcal{X}$ be a Gaussian process and $\mathcal{X}'$ its derivative in quadratic mean. If this satisfies, for example, the criterion in Corollary 1.7(b), it admits a continuous version $\mathcal{Y}' = \{Y'(t) : Y'(t); t \in \mathbb{R}\}$. Set

$$Y(t) := X(0) + \int_0^t Y'(s)\, ds.$$

Clearly, $\mathcal{Y}$ has $\mathcal{C}^1$-paths and $\mathrm{E}(X(s), Y(s)) = r(s,0) + \int_0^s r_{01}(s,t)\,dt = r(s,s)$. In the same way, $\mathrm{E}(Y(s)^2) = r(s,s)$, so that $\mathrm{E}([X(s) - Y(s)]^2) = 0$. As a consequence, $\mathcal{X}$ admits a version with $\mathcal{C}^1$ paths.

Using this construction inductively, one can prove the following:

- Let $\mathcal{X}$ be a Gaussian process with mean $\mathcal{C}^k$ and covariance $\mathcal{C}^{2k}$ and such that its kth derivative in quadratic mean satisfies the weak condition of Corollary 1.7(b). Then $\mathcal{X}$ admits a version with paths of class $\mathcal{C}^k$.
- If $\mathcal{X}$ is a Gaussian process with mean of class $\mathcal{C}^\infty$ and covariance of class $\mathcal{C}^\infty$, $\mathcal{X}$ admits a version with paths of class $\mathcal{C}^\infty$.

In the converse direction, regularity of the paths implies regularity of the expectation and of the covariance function. For example, if $\mathcal{X}$ has continuous sample paths, the mean and the variance are continuous. In fact, if $t_n, n = 1, 2, \ldots$ converges to t, then $X(t_n)$ converges a.s. to $X(t)$, hence also in distribution. Using the form of the Fourier transform of the Gaussian distribution, one easily proves that this implies convergence of the mean and the variance. Since for Gaussian variables, all the moments are polynomial functions of the mean and the variance, they are also continuous. If the process has differentiable sample paths, in a similar way one shows the convergence

$$\frac{m(t+h) - m(t)}{h} \to \mathrm{E}(X'(t))$$

as $h \to 0$, showing that the mean is differentiable.

For the covariance, restricting ourselves to stationary Gaussian processes defined on the real line, without loss of generality we may assume that the process is centered. Put $\Gamma(t) = r(s, s+t)$. The convergence in distribution of $(X(h) - X(0))/h$ to $X'(0)$ plus the Gaussianity imply that $\mathrm{Var}\big((X(h) - X(0))/h\big)$ has a finite limit as $h \to 0$. On the other hand,

$$\mathrm{Var}\left(\frac{X(h) - X(0)}{h}\right) = 2\int_{-\infty}^{+\infty} \frac{1 - \cos hx}{h^2}\mu(dx),$$

where μ is the spectral measure.

Letting $h \to 0$ and applying Fatou's lemma, it follows that

$$\lambda_2 = \int_{-\infty}^{+\infty} x^2 \mu(dx) \le \liminf_{h\to 0} \mathrm{Var}\left(\frac{X(h) - X(0)}{h}\right) < \infty.$$

Using the result in Exercise 1.4, Γ is of class $\mathcal{C}^2$.

This argument can be used in a similar form to show that if the process has paths of class $\mathcal{C}^k$, the covariance is of class $\mathcal{C}^{2k}$. As a conclusion, roughly speaking, for Gaussian stationary processes, the order of differentiability of the sample paths is half of the order of differentiability of the covariance.

1.4.4. More General Tools

In this section we consider the case when the parameter of the process lies in $\mathbb{R}^d$ or, more generally, in some general metric space. We begin with an extension of Theorem 1.6.

Theorem 1.14. *Let $\mathcal{Y} = \{Y(t) : t \in [0, 1]^d\}$ be a real-valued random field that satisfies the condition*

> ($\mathrm{K_d}$) *For each pair $t, t+h \in [0, 1]^d$,*
>
> $$P\{|Y(t+h) - Y(t)| \geq \alpha(\bar{h})\} \leq \beta(\bar{h}),$$
>
> *where $h = (h_1, \ldots, h_d)$, $\bar{h} = \sup_{1 \leq i \leq d} |h_i|$, and α, β are even real-valued functions defined on $[-1, 1]$, increasing on $[0, 1]$, which verify*
>
> $$\sum_{n=1}^{\infty} \alpha(2^{-n}) < \infty, \qquad \sum_{n=1}^{\infty} 2^{dn} \beta(2^{-n}) < \infty.$$
>
> *Then there exists a version $\mathcal{X} = \{X(t) : t \in [0, 1]^d\}$ of the process $\mathcal{Y}$ such that the paths $t \rightsquigarrow X(t)$ are continuous on $[0, 1]^d$.*

Proof. The main change with respect to the proof of Theorem 1.6 is that we replace the polygonal approximation, adapted to one-variable functions by another interpolating procedure. Denote by $\mathcal{D}_n$ the set of dyadic points of order n in $[0, 1]^d$; that is,

$$\mathcal{D}_n = \left\{ t = (t_1, \ldots, t_d) : t_i = \frac{k_i}{2^n}, k_i \text{ integers }, 0 \leq k_i \leq 2^n, i = 1, \ldots, d \right\}.$$

Let $f : [0, 1]^d \to \mathbb{R}$ be a function. For each $n = 1, 2, \ldots$, one can construct a function $f^{(n)} : [0, 1]^d \to \mathbb{R}$ with the following properties:

- $f^{(n)}$ is continuous.
- $f^{(n)}(t) = f(t)$ for all $t \in \mathcal{D}_n$.
- $\|f^{(n+1)} - f^{(n)}\|_\infty = \max_{t \in \mathcal{D}_{n+1} \setminus \mathcal{D}_n} |f(t) - f^{(n)}(t)|$, where $\|\cdot\|_\infty$ denotes sup-norm on $[0, 1]^d$.

A way to define $f^{(n)}$ is the following: Let us consider a cube $\mathcal{C}_{t,n}$ of the nth-order partition of $[0, 1]^d$; that is,

$$\mathcal{C}_{t,n} = t + \left[0, \frac{1}{2^n}\right]^d,$$

where $t \in \mathcal{D}_n$ with the obvious notation for the sum. For each vertex τ, set

$$f^{(n)}(\tau) = f(\tau).$$

Now, for each permutation π of $\{1, 2, \dots, d\}$, let $\mathcal{S}_\pi$ be the simplex

$$\mathcal{S}_\pi = \left\{ t + s : s = (s_{\pi(1)}, \dots, s_{\pi(d)}), 0 \le s_{\pi(1)} \le \cdots \le s_{\pi(d)} \le \frac{1}{2^n} \right\}.$$

It is clear that $\mathcal{C}_{t,n}$ is the union of the $\mathcal{S}_\pi$'s over all permutations. In a unique way, extend $f^{(n)}$ to $\mathcal{S}_\pi$ as an affine function. It is then easy to verify the afore mentioned properties and that

$$\|f^{(n+1)} - f^{(n)}\|_\infty \le d \sup_{s,t \in \mathcal{D}_{n+1}, |t-s|=2^{-(n+1)}} |f(s) - f(t)|.$$

The remainder of the proof is essentially similar to that of Theorem 1.6. □

From this we deduce easily

Corollary 1.15. *Assume that the process $\mathcal{Y} = \{Y(t) : t \in [0,1]^d\}$ verifies one of two conditions:*

(a)

$$\mathrm{E}(|Y(t+h) - Y(t)|^p) \le \frac{K_d |h|^d}{|\log |h||^{1+r}}, \tag{1.19}$$

where p, r, and K are positive constants, $p < r$.

(b) *If $\mathcal{Y}$ is Gaussian, $m(t) = \mathrm{E}(Y(t))$ is continuous and*

$$\mathrm{Var}(Y(t+h) - Y(t)) \le \frac{C}{|\log |h||^a} \tag{1.20}$$

for all t and sufficiently small h and $a > 3$.

Then the process has a version with continuous paths.

Note that the case of processes with values in $\mathbb{R}^{d'}$ need not to be considered separately, since continuity can be addressed coordinate by coordinate. For Hölder regularity we have

Proposition 1.16. *Let $\mathcal{Y} = \{Y(t) : t \in [0,1]^d\}$ be a real-valued stochastic process with continuous paths such that for some $q > 1$, $\alpha > 0$,*

$$\mathrm{E}\big(|Y(s) - Y(t)|^q\big) \le (\mathrm{const})\|s - t\|^{d+\alpha}.$$

Then almost surely, $\mathcal{Y}$ has Hölder paths with exponent $\alpha/2q$.

Until now, we have deliberately chosen elementary methods that apply to general random processes, not necessarily Gaussian. In the Gaussian case, even when the parameter varies in a set that does not have a restricted geometric structure, the question of continuity can be addressed using specific methods. As we have remarked several times already, we only need to consider centered processes.

Let $\{X(t) : t \in T\}$ be a centered Gaussian process taking values in $\mathbb{R}$. We assume that T is some metric space with distance denoted by τ. On T we define the canonical distance d,

$$d(s,t) := \sqrt{\mathrm{E}(X(t) - X(s))^2}.$$

In fact, d is a pseudodistance because two distinct points can be at d distance zero. A first point is that when the covariance $r(s,t)$ function is τ-continuous, which is the only relevant case (otherwise there is no hope of having continuous paths), d-continuity and τ-continuity are equivalent. The reader is referred to Adler (1990) for complements and proofs.

Definition 1.17. *Let (T, d) be a metric space. For $\varepsilon > 0$ denote by $N(\varepsilon) = N(T, d, \varepsilon)$ the minimum number of closed balls of radius ε with which we can cover T (the value of N_ε can be $+\infty$).*

We have the following theorem:

Theorem 1.18 (Dudley, 1973). *A sufficient condition for $\{X(t) : t \in T\}$ to have continuous sample paths is*

$$\int_0^{+\infty} \left(\log(N(\varepsilon))\right)^{1/2} d\varepsilon < \infty.$$

$\log(N(\varepsilon))$ *is called the* entropy *of the set T.*

A very important fact is that this condition is necessary in some relevant cases:

Theorem 1.19 (Fernique, 1974). *Let $\{X(t) : t \in T\}$, T compact, a subset of $\mathbb{R}^d$, be a stationary Gaussian process. Then the following three statements are equivalent:*

- *Almost surely, $X(\cdot)$ is bounded.*
- *Almost surely, $X(\cdot)$ is continuous.*
- $\displaystyle\int_0^{+\infty} \left(\log(N(\varepsilon))\right)^{1/2} d\varepsilon < \infty.$

This condition can be compared with Kolmogorov's theorem. The reader can check that Theorem 1.19 permits us to weaken the condition of Corollary 1.7(b) to $a > 1$. On the other hand, one can construct counterexamples (i.e., processes not having continuous paths) such that (1.14) holds true with $a = 1$. This shows that the condition of Corollary 1.7(b) is nearly optimal and sufficient for most applications. When the Gaussian process is no longer stationary, M. Talagrand has given necessary and sufficient conditions for sample path continuity in terms of the existence of majorizing measures (see Talagrand, 1987).

The problem of differentiability can be addressed in the same manner as for $d = 1$. A sufficient condition for a Gaussian process to have a version with $\mathcal{C}^k$ sample paths is for its mean to be $\mathcal{C}^k$, its covariance $\mathcal{C}^{2k}$, and its kth derivative in quadratic mean to satisfy some of the criteria of continuity above.

1.4.5. Tangencies and Local Extrema

In this section we give two classical results that are used several times in the book. The first gives a simple sufficient condition for a one-parameter random process not to have a.s. critical points at a certain specified level. The second result states that under mild conditions, a Gaussian process defined on a quite general parameter set with probability 1 does not have local extrema at a given level. We will use systematically the following notation: If ξ is a random variable with values in $\mathbb{R}^d$ and its distribution has a density with respect to Lebesgue measure, this density is denoted as

$$p_\xi(x) \qquad x \in \mathbb{R}^d.$$

Proposition 1.20 (Bulinskaya, 1961). *Let $\{X(t) : t \in I\}$ be a stochastic process with paths of class C^1 defined on the interval I of the real line. Assume that for each $t \in I$, the random variable $X(t)$ has a density $p_{X(t)}(x)$ which is bounded as t varies in a compact subset of I and x in a neighborhood v of $u \in \mathbb{R}$. Then*

$$\mathrm{P}(T_u^X \neq \emptyset) = 0,$$

where $T_u^X = \{t : t \in I, X(t) = u, X'(t) = 0\}$ is the set of critical points with value u of the random path $X(\cdot)$.

Proof. It suffices to prove that $\mathrm{P}(T_u^X \cap J \neq \emptyset) = 0$ for any compact subinterval J of I. Let ℓ be the length of J and $t_0 < t_1 < \cdots < t_m$ be a uniform partition of J (i.e., $t_{j+1} - t_j = \ell/m$ for $j = 0, 1, \cdots, m-1$). Denote by $\omega_{X'}(\delta, J)$ the modulus of continuity X' on the interval J and $E_{\delta,\varepsilon}$ the event

$$E_{\delta,\varepsilon} = \{\omega_{X'}(\delta, J) \geq \varepsilon\}.$$

Let $\varepsilon > 0$ be given; choose $\delta > 0$ so that $P(E_{\delta,\varepsilon}) < \varepsilon$ and m so that $\ell/m < \delta$, and $[u - l/m, u + l/m] \subset v$. We have

$$P(T_u^X \cap J \neq \emptyset) \leq P(E_{\delta,\varepsilon}) + \sum_{j=0}^{m-1} P(\{T_u^X \cap [t_j, t_{j+1}] \neq \emptyset\} \cap E_{\delta,\varepsilon}^C)$$

$$< \varepsilon + \sum_{j=0}^{m-1} P\left(|X(t_j) - u| \leq \varepsilon \frac{\ell}{m}\right) = \varepsilon + \sum_{j=0}^{m-1} \int_{|x-u| \leq \varepsilon(\ell/m)} p_{X(t_j)}(x)\, dx.$$

If C is an upper bound for $p_{X(t)}(x)$, $t \in J$, $|x - u| \leq \varepsilon l/m$, we obtain

$$P(T_u^X \cap J \neq \emptyset) \leq \varepsilon + C\varepsilon\ell.$$

Since $\varepsilon > 0$ is arbitrary, the result follows. □

The second result is an extension of Ylvisaker's theorem, which has the following statement:

Theorem 1.21 (Ylvisaker, 1968). *Let $\{Z(t) : t \in T\}$ be a real-valued Gaussian process indexed on a compact separable topological space T having continuous paths and $\mathrm{Var}(Z(t)) > 0$ for all $t \in T$. Then, for fixed $u \in \mathbb{R}$, one has $P(E_u^Z \neq \emptyset) = 0$, where E_u^Z is the set of local extrema of $Z(\cdot)$ having value equal to u.*

The extension is the following:

Theorem 1.22 *Let $\{Z(t) : t \in T\}$ be a real-valued Gaussian process on some parameter set T and denote by $M^Z = \sup_{t\in T} Z(t)$ its supremum (which takes values in $\mathbb{R} \cup \{+\infty\}$). We assume that there exists a nonrandom countable set $\mathcal{D}$, $\mathcal{D} \subset T$, such that a.s. $M^Z = \sup_{t\in\mathcal{D}} Z(t)$. Assume further that there exist $\sigma_0^2 > 0, m_- > -\infty$ such that*

$$m(t) = E(Z(t)) \geq m_-$$

$$\sigma^2(t) = \mathrm{Var}(Z(t)) \geq \sigma_0^2 \qquad \textit{for every } t \in T.$$

Then the distribution of the random variable M^Z is the sum of an atom at $+\infty$ and a (possibly defective) probability measure on $\mathbb{R}$ which has a locally bounded density.

Proof. STEP 1. Suppose first that $\{X(t) : t \in T\}$ satisfies the hypotheses of the theorem, and, moreover,

$$\mathrm{Var}(X(t)) = 1, \qquad E(X(t)) \geq 0$$

for every $t \in T$. We prove that the supremum M^X has a density p_{M^X}, which satisfies the inequality

$$p_{M^X}(u) \leq \psi(u) := \frac{\exp(-u^2/2)}{\displaystyle\int_u^\infty \exp(-v^2/2)\, dv} \qquad \text{for every u} \in \mathbb{R}. \tag{1.21}$$

Let $\mathcal{D} = \{t_k\}_{k=1,2,\ldots}$. Almost surely, $M^X = \sup\{X(t_1)\ldots\ X(t_n)\ldots\}$. We set

$$M_n := \sup_{1 \leq k \leq n} X(t_k).$$

Since the joint distribution of $X(t_k), k = 1, \ldots, n$, is Gaussian, for any choice of $k, \ell = 1, \ldots, n; k \neq \ell$, the probability $\mathrm{P}\{X(t_k) = X(t_\ell)\}$ is equal to 0 or 1. Hence, possibly excluding some of these random variables, we may assume that these probabilities are all equal to 0 without changing the value of M_n on a set of probability 1. Then the distribution of the random variable M_n has a density $g_n(\cdot)$ that can be written as

$$g_n(x) = \sum_{k=1}^n \mathrm{P}(X(t_j) < x, j = 1, \ldots, n;\ j \neq k | X(t_k) = x)$$
$$\times \frac{e^{-(1/2)(x-m(t_k))^2}}{\sqrt{2\pi}} = \varphi(x) G_n(x),$$

where φ denotes the standard normal density and

$$G_n(x) = \sum_{k=1}^n \mathrm{P}(Y_j < x - m(t_j), j = 1, \ldots, n;\ j \neq k | Y_k$$
$$= x - m(t_k))e^{xm(t_k)-(1/2)m^2(t_k)} \tag{1.22}$$

with

$$Y_j = X(t_j) - m(t_j) \qquad j = 1, \ldots, n.$$

Let us prove that $x \rightsquigarrow G_n(x)$ is an increasing function.

Since $m(t) \geq 0$, it is sufficient that the conditional probability in each term of (1.22) be increasing as a function of x. Write the Gaussian regression

$$Y_j = Y_j - c_{jk} Y_k + c_{jk} Y_k \qquad \text{with} \quad c_{jk} = \mathrm{E}(Y_j Y_k),$$

where the random variables $Y_j - c_{jk} Y_k$ and Y_k are independent. Then the conditional probability becomes

$$\mathrm{P}(Y_j - c_{jk} Y_k < x - m(t_j) - c_{jk}(x - m(t_k)), j = 1, \ldots, n;\ j \neq k).$$

This probability increases with x because $1 - c_{jk} \geq 0$, due to the Cauchy–Schwarz inequality. Now, if $a, b \in \mathbb{R}$, $a < b$, since $M_n \uparrow M^X$,

$$\mathrm{P}\{a < M^X \leq b\} = \lim_{n\to\infty} \mathrm{P}(a < M_n \leq b).$$

Using the monotonicity of G_n, we obtain

$$G_n(b) \int_b^{+\infty} \varphi(x)\, dx \leq \int_b^{+\infty} G_n(x)\varphi(x)\, dx = \int_b^{+\infty} g_n(x)\, dx \leq 1,$$

so that

$$\begin{aligned}\mathrm{P}\{a < M_n \leq b\} &= \int_a^b g_n(x)\, dx \leq G_n(b) \int_a^b \varphi(x)\, dx \\ &\leq \int_a^b \varphi(x)\, dx \left(\int_b^{+\infty} \varphi(x)\, dx \right)^{-1}.\end{aligned}$$

This proves (1.21).

STEP 2. Now let Z satisfy the hypotheses of the theorem without assuming the added ones in step 1. For given $a, b \in \mathbb{R}$, $a < b$, choose $A \in \mathbb{R}^+$ so that $|a| < A$ and consider the process

$$X(t) = \frac{Z(t) - a}{\sigma(t)} + \frac{|m_-| + A}{\sigma_0}.$$

Clearly, for every $t \in T$,

$$\mathrm{E}\big(X(t)\big) = \frac{m(t) - a}{\sigma(t)} + \frac{|m_-| + A}{\sigma_0} \geq -\frac{|m_-| + |a|}{\sigma_0} + \frac{|m_-| + A}{\sigma_0} \geq 0$$

and

$$\mathrm{Var}\big(X(t)\big) = 1,$$

so that (1.21) holds for the process X.

On the other hand,

$$\{a < M^Z \leq b\} \subset \{\mu_1 < M^X \leq \mu_2\},$$

where

$$\mu_1 = \frac{|m_-| + A}{\sigma_0}, \qquad \mu_2 = \frac{|m_-| + A}{\sigma_0} + \frac{b - a}{\sigma_0}.$$

It follows that

$$\mathrm{P}\big\{a < M^Z \leq b\big\} \leq \int_{\mu_1}^{\mu_2} \psi(u)\, du = \int_a^b \frac{1}{\sigma_0} \psi\Big(\frac{v - a + |m_-| + A}{\sigma_0}\Big)\, dv,$$

which proves the statement. □

Theorem 1.21 follows directly from Theorem 1.22, since under the hypotheses of Theorem 1.21, we can write

$$\{E_u^X \neq \emptyset\} \subset \bigcup_{U \in \mathcal{F}} (\{M_U = u\} \cup \{m_U = u\}),$$

where M_U (respectively, m_u) is the maximum (respectively, the minimum) of the process on the set U and $\mathcal{F}$ denotes a countable family of open sets being a basis for the topology of T.

Remark. We come back in later chapters to the subject of the regularity properties of the probability distribution of the supremum of a Gaussian process.

EXERCISES

1.1. Let $T = \mathcal{N}$ be the set of natural numbers. Prove that the following sets belong to $\sigma(\mathcal{C})$.

- **(a)** c_0 (the set of real-valued sequences $\{a_n\}$ such that $a_n \to 0$). *Suggestion*: Note that $c_0 = \bigcap_{k=1}^{\infty} \bigcup_{m=1}^{\infty} \bigcap_{n \geq m} \{|a_n| < 1/k\}$.
- **(b)** ℓ^2 (the set of real-valued sequences $\{a_n\}$ such that $\sum_n |a_n|^2 < \infty$).
- **(c)** The set of real-valued sequences $\{a_n\}$ such that $\overline{\lim}_{n\to\infty} a_n \leq 1$.

1.2. Take $T = \mathbb{R}$, $\mathcal{T} = \mathcal{B}_{\mathbb{R}}$. Then if for each $\omega \in \Omega$ the function

$$t \rightsquigarrow X(t, \omega), \tag{1.23}$$

the *path corresponding to* ω, is a continuous function, the process is bi-measurable. In fact, check that

$$X(t, \omega) = \lim_{n \to +\infty} X^{(n)}(t, \omega),$$

where for $n = 1, 2, \ldots$, $X^{(n)}$is defined by

$$X^{(n)}(t, \omega) = \sum_{k=-\infty}^{k=+\infty} X_{k/2^n}(\omega) \mathbf{1}_{\{k/2^n \leq t < (k+1)/2^n\}},$$

which is obviously measurable as a function of the pair (t, ω). So the limit function X has the required property. If one replaces the continuity of the path (1.23) by some other regularity properties such as right continuity, bi-measurability follows in a similar way.

1.3. Let U be a random variable defined on some probability space $(\Omega, \mathcal{A}, \mathrm{P})$, having uniform distribution on the interval $[0, 1]$. Consider the two stochastic processes

$$Y(t) = \mathbb{1}_{t=U}$$
$$X(t) \equiv 0.$$

The process $Y(t)$ is sometimes called the *random parasite*.

(**a**) Prove that for all $t \in [0, 1]$, a.s. $X(t) = Y(t)$.

(**b**) Deduce that the processes $X(t)$ and $Y(t)$ have the same probability distribution P on $\mathbb{R}^{[0,1]}$ equipped with its Borel σ-algebra.

(**c**) Notice that for each ω in the probability space, $\sup_{t\in[0,1]} Y(t) = 1$ and $\sup_{t\in[0,1]} X(t) = 0$, so that the suprema of both processes are completely different. Is there a contradiction with the previous point?

1.4. Let μ be a Borel probability measure on the real line and Γ its Fourier transform; that is,

$$\Gamma(\tau) = \int_{\mathbb{R}} \exp(i\tau x)\mu(dx).$$

(**a**) Prove that if

$$\lambda_k = \int_{\mathbb{R}} |x|^k \mu(dx) < \infty$$

for some positive integer k, the covariance $\Gamma(\cdot)$ is of class C^k and

$$\Gamma^{(k)}(\tau) = \int_{\mathbb{R}} (ix)^k \exp(i\tau x)\mu(dx).$$

(**b**) Prove that if k is even, $k = 2p$, the reciprocal is true: If Γ is of class C^{2p}, then λ_{2p} is finite and

$$\Gamma(t) = 1 - \lambda_2\frac{t^2}{2!} + \lambda_4\frac{t^4}{4!} + \cdots + (-1)^{2p}\lambda_{2p}\frac{t^{2p}}{(2p)!} + o(t^{2p}).$$

Hint: Using induction on p and supposing that λ_k is infinite, then for every $A > 0$, one can find some $M > 0$ such that

$$\int_{M}^{M} x^k \mu(dx) \geq A.$$

Show that it implies that

$$(-1)^k \frac{k!}{t^k}\left[\Gamma(t) - \left(1 - \lambda_2 \frac{t^2}{2!} + \cdots + (-1)^{k-2}\lambda_{k-2}\frac{t^{k-2}}{(k-2)!}\right)\right]$$

has a limit, when t tends to zero, greater than A, which contradicts differentiability.

(c) When k is odd, the result is false [see Feller, 1966, Chap. XVII, example (c)].

1.5. Let $\{\xi_n\}_{n=1,2,\ldots}$ be a sequence of random vectors defined on some probability space taking values in $\mathbb{R}^d$, and assume that $\xi_n \to \xi$ in probability for some random vector ξ. Prove that if each ξ_n is Gaussian, ξ is also Gaussian.

1.6. Prove the following statements on the process defined by (1.10).

(a) For each $t \in T$ the series (1.10) converges a.s.

(b) Almost surely, the function $t \rightsquigarrow X(t)$ is in H and $\|X_{(\cdot)}\|_H^2 = \sum_{n=1}^{\infty} c_n \xi_n^2$.

(c) $\{\varphi_n\}_{n=1,2,\ldots}$ are eigenfunctions—with eigenvalues $\{c_n\}_{n=1,2,\ldots}$, respectively—of the linear operator $A : H \to H$ defined by

$$(Af)(s) = \int_T r(s,t) f(t) \rho(dt).$$

1.7. Let $\{X(t) : t \in T\}$ be a stochastic process defined on some separable topological space T.

(a) Prove that if $X(t)$ has continuous paths, it is separable.

(b) Let $T = \mathbb{R}$. Prove that if the paths of $X(t)$ are càd-làg, $X(t)$ is separable.

1.8. Let $\{X(t) : t \in \mathbb{R}^d\}$ be a separable stochastic process defined on some (complete) probability space $(\Omega, \mathcal{A}, \mathrm{P})$.

(a) Prove that the subset of Ω $\{X(\cdot)$ is continuous$\}$ is in $\mathcal{A}$.

(b) Prove that the conclusion in part (a) remains valid if one replaces "continuous" by "upper continuous", "lower continuous," or "continuous on the right" [a real-valued function f defined on $\mathbb{R}^d$ is said to be *continuous on the right* if for each t, $f(t)$ is equal to the limit of $f(s)$ when each coordinate of s tends to the corresponding coordinate of t on its right].

1.9. Show that in the case of the Wiener process, condition (1.18) holds for every $p \geq 2$, with $r = p/2 - 1$. Hence, the proposition implies that a.s., the paths of the Wiener process satisfy a Hölder condition with exponent α, for every $\alpha < \frac{1}{2}$.

1.10. (*Wiener integral*) Let $\{W_1(t) : t \geq 0\}$ and $\{W_2(t) : t \geq 0\}$ be two independent Wiener processes defined on some probability space $(\Omega, \mathcal{A}, \mathrm{P})$, and denote by $\{W(t) : t \in \mathbb{R}\}$ the process defined as

$$W(t) = W_1(t) \text{ if } t \geq 0 \quad \text{and} \quad W(t) = W_2(-t) \text{ if } t \leq 0.$$

$L^2(\mathbb{R}, \lambda)$ denotes the standard L^2-space of real-valued measurable functions on the real line with respect to Lebesgue measure and $L^2(\Omega, \mathcal{A}, \mathrm{P})$ the L^2 of the probability space. $\mathcal{C}_K^1(\mathbb{R})$ denotes the subspace of $L^2(\mathbb{R}, \lambda)$ of $\mathcal{C}^1$-functions with compact support. Define the function $I : \mathcal{C}_K^1(\mathbb{R}) \to L^2(\Omega, \mathcal{A}, \mathrm{P})$ as

$$I(f) = -\int_{\mathbb{R}} f'(t) W(t)\, dt \qquad (1.24)$$

for each nonrandom $f \in \mathcal{C}_K^1(\mathbb{R})$. Equation (1.24) is well defined for each $\omega \in \Omega$ since the integrand is a continuous function with compact support.

(**a**) Prove that I is an isometry, in the sense that $\int_{\mathbb{R}} f^2(t)\, dt = \mathrm{E}\big(I^2(f)\big)$.

(**b**) Show that for each f, $I(f)$ is a centered Gaussian random variable. Moreover, for any choice of $f_1, \ldots, f_p \in \mathcal{C}_K^1(\mathbb{R})$, the joint distribution of $(I(f_1), \ldots ., I(f_p))$ is centered Gaussian. Compute its covariance matrix.

(**c**) Prove that I admits a unique isometric extension $\tilde{I}$ to $L^2(\mathbb{R}, \lambda)$ such that:

(1) $\tilde{I}(f)$ is a centered Gaussian random variable with variance equal to $\int_{\mathbb{R}} f^2(t)\, dt$; similarly for joint distributions.

(2) $\displaystyle\int_{\mathbb{R}} f(t) g(t)\, dt = \mathrm{E}\big(\tilde{I}(f)\tilde{I}(g)\big)$.

Comment: $\tilde{I}(f)$ is called the *Wiener integral of* f.

1.11. (*Fractional Brownian motion*) Let H be a real number, $0 < H < 1$. We use the notation and definitions of Exercise 1.10.

(**a**) For $t \geq 0$, define the function $K_t : \mathbb{R} \to \mathbb{R}$:

$$K_t(u) = \left[(t-u)^{H-1/2} - (-u)^{H-1/2}\right] \mathbb{1}_{u<0} + (t-u)^{H-1/2} \mathbb{1}_{0<u<t}.$$

Prove that $K_t \in L^2(\mathbb{R}, \lambda)$.

(**b**) For $t \geq 0$, define the Wiener integral $\tilde{I}(K_t)$, and for $s, t \geq 0$, prove the formula

$$\mathrm{E}\big(\tilde{I}(K_s)\tilde{I}(K_t)\big) = \frac{C_H}{2}\left[s^{2H} + t^{2H} - |t-s|^{2H}\right],$$

where C_H is a positive constant depending only on H. Compute C_H.

(c) Prove that the stochastic process $\{C_H^{-1/2}\tilde{I}(K_t) : t \geq 0\}$ has a version with continuous paths. This normalized version with continuous paths is usually called the *fractional Brownian motion with Hurst exponent* H and is denoted $\{W_H(t) : t \geq 0\}$.

(d) Show that if $H = \frac{1}{2}$, then $\{W_H(t) : t \geq 0\}$ is the standard Wiener process.

(e) Prove that for any $\delta > 0$, almost surely the paths of the fractional Brownian motion with Hurst exponent H satisfy a Hölder condition with exponent $H - \delta$.

1.12. (*Local time*) Let $\{W(t) : t \geq 0\}$ be a Wiener process defined in a probability space $(\Omega, \mathcal{A}, \mathrm{P})$. For $u \in \mathbb{R}$, I an interval $I \subset [0, +\infty]$ and $\delta > 0$, define

$$\mu_\delta(u, I) = \frac{1}{2\delta}\int_I \mathbf{1}_{|W(t)-u|<\delta}\,dt = \frac{1}{2\delta}\lambda(\{t \in I : |W(t) - u| < \delta\}).$$

(a) Prove that for fixed u and I, $\mu_\delta(u, I)$ converges in $L^2(\Omega, \mathcal{A}, \mathrm{P})$ as $\delta \to 0$. Denote the limit by $\mu_0(u, I)$. *Hint:* Use Cauchy's criterion.

(b) Denote $Z(t) = \mu_0(u, [0, t])$. Prove that the random process $Z(t) : t \geq 0$ has a version with continuous paths. We call this version the *local time of the Wiener process at the level* u, and denote it by $L^W(u, t)$.

(c) For fixed u, $L^W(u, t)$ is a continuous increasing function of $t \geq 0$. Prove that a.s. it induces a measure on $\mathbb{R}^+$ that is singular with respect to Lebesgue measure; that is, its support is contained in a set of Lebesgue measure zero.

(d) Study the Hölder continuity properties of $L^W(u, t)$. For future reference, with a slight abuse of notation, we will write, for any interval $I = [t_1, t_2], 0 \leq t_1 \leq t_2$:

$$L^W(u, I) = L^W(u, t_2) - L^W(u, t_1).$$

CHAPTER 2

BASIC INEQUALITIES FOR GAUSSIAN PROCESSES

This chapter is about inequalities for the probability distribution of the supremum of Gaussian processes. Among the numerous results giving upper and lower bounds, we have chosen the ones we consider to be more useful for the subjects considered in this book: comparison inequalities, isoperimetric inequalities, and their applications to obtain bounds for the tails of the distribution of the supremum and its moments.

The results in this chapter are very general, in the sense that beyond Gaussianity and almost sure boundedness of the paths, we do not require the random function to satisfy other hypotheses. They are essential basic tools and, at the same time, provide bounds that may turn out to be rough when applied to special families of random functions. One of our purposes in subsequent chapters is to refine these inequalities under additional hypotheses, as explained in the Introduction.

A good part of the theory was already well established more than 30 years ago. However, some results have been improved significantly more recently. Two relevant examples of this evolution are the Li–Shao comparison inequality and the C. Borell proof of the Ehrhard conjecture, which we consider in Sections 2.1 and 2.2, respectively.

Level Sets and Extrema of Random Processes and Fields, By Jean-Marc Azaïs and Mario Wschebor

2.1. SLEPIAN INEQUALITIES

Lemma 2.1 (Li and Shao, 2002). *Let $X := (X_1, \dots, X_n)^{\mathrm{T}}$ and $Y = (Y_1, \dots, Y_n)^{\mathrm{T}}$ be two centered Gaussian random vectors in $\mathbb{R}^n$, $n \geq 2$. Denote*

$$\Sigma^X = ((r_{jk}^X))_{j,k=1,\dots,n} \text{ with } r_{jk}^X = \mathrm{E}(X_j X_k)$$

and use similar notation for Y. We will assume that $r_{jj}^X = r_{jj}^Y$ for all $j = 1, \dots, n$ (i.e., that the variances are, respectively, equal) and with no loss of generality for our purposes, that their common value is equal to 1. Then, for any choice of the real numbers $a_1, \dots, a_n$, one has

$$\begin{aligned} &\mathrm{P}\{X_1 \leq a_1, \dots, X_n \leq a_n\} - \mathrm{P}\{Y_1 \leq a_1, \dots, Y_n \leq a_n\} \\ &\leq \frac{1}{2\pi} \sum_{1 \leq i < j \leq n} \left(\arcsin r_{ij}^X - \arcsin r_{ij}^Y\right)^+ \exp\left(-\frac{a_i^2 + a_j^2}{2(1 + \rho_{ij})}\right), \end{aligned} \tag{2.1}$$

where $\rho_{ij} = \max(|r_{ij}^X|, |r_{ij}^Y|)$.

This lemma, known under the generic name *normal comparison lemma*, has quite a long history. As far as we know, its first version is due to Plackett (1954), who proved that if $r_{jk}^X \leq r_{jk}^y$ for all $1 \leq j < k \leq n$, then

$$\mathrm{P}\{X_1 \leq a_1, \dots, X_n \leq a_n\} \leq \mathrm{P}\{Y_1 \leq a_1, \dots, Y_n \leq a_n\}. \tag{2.2}$$

We call this original version the Plackett–Slepian comparison lemma. Further versions have been given by Slepian (1961), Berman (1964), and Leadbetter et al. (1983). The present statement, due to Li and Shao (2002), contains and refines the previous ones.

Proof. We introduce some additional notation. For $t \in [0, 1]$, let

$$\Sigma_t = (1 - t)\Sigma^X + t\Sigma^Y.$$

It is obvious that Σ_t is positive semidefinite. Let $Z = (Z_1, \dots, Z_n)^{\mathrm{T}}$ be a centered Gaussian random vector in $\mathbb{R}^n$ with covariance Σ_t, that is, $\mathrm{E}(Z Z^{\mathrm{T}}) = \Sigma_t$, and

$$F(t) = \mathrm{P}(Z_1 \leq a_1, \dots, Z_n \leq a_n).$$

Our aim is to give an upper bound for $F(0) - F(1)$, and for this purpose we consider the derivative $F'(t)$. Let us first notice that it is sufficient to prove the result when Σ_t is nonsingular for all $t \in [0, 1]$. In fact, if this has been proved, in the general case we proceed as follows.

Take n i.i.d. random variables $\xi_1, \dots, \xi_n$, each of them having a standard normal distribution. $\xi = (\xi_1, \dots, \xi_n)^T$ is also assumed to be independent of X and Y. For any $\varepsilon > 0$, the variance of the centered Gaussian vector

$$((1-t)X_1 + tY_1 + \varepsilon\xi_1, \dots, (1-t)X_n + tY_n + \varepsilon\xi_n)$$

is $\Sigma_t + \varepsilon I_n$, which is nonsingular for any $t \in [0, 1]$. Hence, we may apply the inequality (2.1) to the pair of random vectors $X_1 + \varepsilon\xi_1, \dots, X_n + \varepsilon\xi_n$ and $Y_1 + \varepsilon\xi_1, \dots, Y_n + \varepsilon\xi_n$. Then we pass to the limit as $\varepsilon \to 0$. This should be done carefully and is left to the reader.

So assume that Σ_t is nonsingular for all $t \in [0, 1]$. For $\Sigma = \left((r_{jk})\right)_{j,k=1,\dots,n}$ positive definite and nonsingular, we denote by $\varphi_\Sigma(x)$, $x = (x_1, \dots, x_n)^T \in \mathbb{R}^n$ the density of the centered normal distribution in $\mathbb{R}^n$ with covariance Σ. We have the identity

$$\frac{\partial \varphi_\Sigma}{\partial r_{jk}} = \frac{\partial^2 \varphi_\Sigma}{\partial x_j \partial x_k} \qquad (j,k = 1, \dots, n, \quad j < k). \tag{2.3}$$

To prove (2.3) we use the inversion formula for the Fourier transform and the form of the Fourier transform of the normal distribution in $\mathbb{R}^n$:

$$\varphi_\Sigma(x) = \frac{1}{(2\pi)^n} \int_{\mathbb{R}^n} \exp\left[-i\,\langle x, z\rangle - \frac{1}{2}\langle z, \Sigma z\rangle\right] dz.$$

In this equality we may differentiate under the integral sign either with respect to r_{jk} or with respect to x_j. This can be justified using dominated convergence, since the nonsingularity of Σ_t implies the existence of a positive constant c such that $\langle z, \Sigma z\rangle \ge c\,\|z\|^2$ for all $z \in \mathbb{R}^n$. Equation (2.3) follows.

Using this, we can compute the derivative $F'(t)$:

$$\begin{aligned}
F'(t) &= \frac{d}{dt} \int_{x_h \le a_h, h=1,\dots,n} \varphi_{\Sigma_t}(x)\, dx \\
&= \int_{x_h \le a_h, h=1,\dots,n} \left[\sum_{1\le j<k\le n} \frac{\partial \varphi_{\Sigma_t}}{\partial r^t_{jk}}(x) \frac{dr^t_{jk}}{dt}\right] dx \\
&= \int_{x_h \le a_h, h=1,\dots,n} \left[\sum_{1\le j<k\le n} \frac{\partial^2 \varphi_{\Sigma_t}(x)}{\partial x_j \partial x_k}(r^Y_{jk} - r^X_{jk})\right] dx \\
&= \sum_{1\le j<k\le n} (r^Y_{jk} - r^X_{jk}) \int_{\substack{x_h \le a_h, h=1,\dots,n \\ h\ne j,k}} \prod_{\substack{h=1 \\ h\ne j,k}}^{n} dx_h \int_{-\infty}^{a_j}\int_{-\infty}^{a_k} \frac{\partial^2 \varphi_{\Sigma_t}(x)}{\partial x_j \partial x_k} dx_j dx_k.
\end{aligned}$$

using (2.3). r^t_{jk} stands for $(1-t)r^X_{jk} + tr^Y_{jk}$, the element (j, k) of Σ_t, and dx stands for $dx_1, \dots, dx_n$.

In each term of this equality we integrate twice (first in x_k and second in x_j), obtaining

$$F'(t) = \sum_{1\le j<k\le n} (r^Y_{jk} - r^X_{jk}) \int_{\substack{x_h \le a_h, h=1,\ldots,n \\ h\ne j,k}} \varphi_{\Sigma_t}(\widetilde{x}_{j,k}) \prod_{\substack{h=1 \\ h\ne j,k}}^{n} dx_h,$$

where $\widetilde{x}_{j,k} = (\widetilde{x}_1, \ldots., \widetilde{x}_n)$, $\widetilde{x}_h = x_h$ for $h \ne j, k$, $\widetilde{x}_h = a_h$ for $h = j, k$. Now majorizing the integral above by the same integral but with integration over all $\mathbb{R}$ for x_h, $h \ne j, k$, we get

$$-F'(t) \le \sum_{1\le j<k\le n} (r^X_{jk} - r^Y_{jk})^+ \varphi(a_j, a_k; r^t_{j,k}),$$

where $\varphi(u, v; \rho)$ is the joint density at point (u, v) of two jointly Gaussian random variables with zero expectation, variance 1, and covariance ρ. Now standard algebra shows that

$$\varphi(a_j, a_k; r^t_{j,k}) \le \frac{1}{2\pi} \frac{1}{\sqrt{1-(r^t_{jk})^2}} \exp\left(-\frac{a_j^2 + a_k^2}{2(1+\rho_{jk})}\right).$$

As a consequence, we get

$$F(0) - F(1) \le \frac{1}{2\pi} \sum_{1\le j<k\le n} (r^X_{jk} - r^Y_{jk})^+ \exp\left(-\frac{a_j^2 + a_k^2}{2(1+\rho_{jk})}\right) \int_0^1 \frac{1}{\sqrt{1-(r^t_{jk})^2}} dt.$$

Now, due to the form of r^t_{jk}, changing variables in the integral, we have, whenever $r^Y_{jk} \le r^X_{jk}$,

$$(r^X_{jk} - r^Y_{jk}) \int_0^1 \frac{1}{\sqrt{1-(r^t_{jk})^2}} dt = \int_{r^Y_{jk}}^{r^X_{jk}} \frac{1}{\sqrt{1-w^2}} dw = \arcsin r^X_{jk} - \arcsin r^Y_{jk}.$$

□

Corollary 2.2. *Let $\{X(t) : t \in T\}$ and $\{Y(t) : t \in T\}$ be separable centered Gaussian processes with a.s. bounded paths, defined on a topological space T. Let us assume that*

$$\mathrm{E}(X(t)^2) = \mathrm{E}(Y(t)^2) \qquad \text{for all } t \in T$$

$$\mathrm{E}\left((Y(t) - Y(s))^2\right) \le \mathrm{E}\left((X(t) - X(s))^2\right) \qquad \text{for all } s, t \in T.$$

Then, for each $x \in \mathbb{R}$,

$$\mathrm{P}\{\sup_{t\in T} X(t) \le x\} \le \mathrm{P}\{\sup_{t\in T} Y(t) \le x\}.$$

We will say that $\sup_{t\in T} X(t)$ is stochastically greater or equal to $\sup_{t\in T} Y(t)$.

Proof. Because of the separability, it is enough to prove the result for finite T. This follows immediately from the Plackett–Slepian version of the normal comparison lemma. □

Example 2.1. *Let T be a positive number. Consider the three centered Gaussian stationary processes $X(t)$, $Y(t)$, and $Z(t)$, $t \in [0, T]$, with respective covariances*

$$\Gamma_X(t) := \exp(-t^2/2), \qquad \Gamma_Y(t) := \exp(-|t|), \qquad \Gamma_Z(t) := (1 - |t|)^+$$

(see Figure 2.1). $X(t)$ is the stationary process with Gaussian covariance, $Y(t)$ the Ornstein–Uhlenbeck process, and $Z(t)$ the Slepian process. It is easy to check that

$$\Gamma_Z(t) \le \Gamma_Y(t) \le \Gamma_X(t) \qquad \text{for } |t| \le 2.$$

So let $T \le 2$; then

$$\sup_{t\in T} X(t) \overset{s}{\le} \sup_{t\in T} Y(t) \overset{s}{\le} \sup_{t\in T} Z(t),$$

where $\overset{s}{\le}$ is the stochastic order.

We finish this section by stating without proof two related results having important applications.

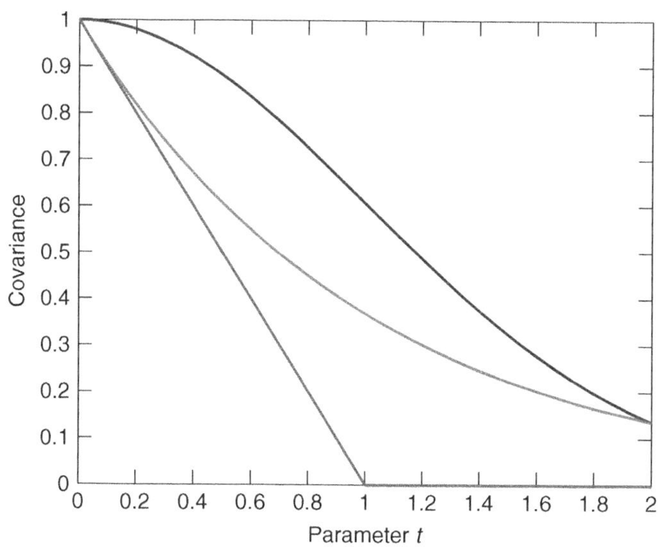

Figure 2.1. Representation of the three covariances.

Theorem 2.3 (Li and Shao, 2002). *Using the notations of Lemma 2.1, assume that $n \geq 3$ and that*

$$r_{ij}^X \geq r_{ij}^Y \geq 0 \qquad \text{for all } 1 \leq i, j \leq n.$$

Then for $a \leq 0$, one has

$$\begin{aligned}
\mathrm{P}\{Y_1 \leq a, \ldots, Y_n \leq a\} &\leq \mathrm{P}\{X_1 \leq a, \ldots, X_n \leq a\} \\
&\leq \mathrm{P}\{Y_1 \leq a, \ldots, Y_n \leq a\} \\
&\quad \times \exp\left[\sum_{1 \leq i < j \leq n} \log\left(\frac{\pi - 2\arcsin r_{ij}^Y}{\pi - 2\arcsin r_{ij}^X}\right) \exp\left(-\frac{a^2}{1 + r_{ij}^X}\right)\right].
\end{aligned}$$

Theorem 2.4 (Sudakov and Fernique). *Let $\{X(t) : t \in T\}$ and $\{Y(t) : t \in T\}$ be separable centered Gaussian processes with a.s. bounded paths, defined on a topological space T. Let us assume that*

$$\mathrm{E}\left((X(t) - X(s))^2\right) \leq \mathrm{E}\left((Y(t) - Y(s))^2\right) \qquad \text{for all } t \text{ and } s.$$

Then

$$\mathrm{E}(\sup_{t \in T} X(t)) \leq \mathrm{E}(\sup_{t \in T} Y(t)).$$

A proof of this theorem can be found in a book by Adler (1990). We will see later that under the conditions of this theorem, both expectations are finite (which is not evident at all!). Notice that under the (more restrictive) conditions of Corollary 2.2, the conclusion of Theorem 2.4 follows from it immediately, since for any integrable real-valued random variable η, one can express the expectation as

$$\mathrm{E}(\eta) = \int_0^{+\infty} \mathrm{P}(\eta > x)\, dx - \int_{-\infty}^0 \mathrm{P}(\eta < -x)\, dx.$$

Apply this to $\sup_{t \in T} X(t)$ and to $\sup_{t \in T} Y(t)$.

2.2. EHRHARD'S INEQUALITY

Theorem 2.5. *Let γ_n be the standard Gaussian probability measure on $\mathbb{R}^n$. Then for any pair A and B of Borel subsets of $\mathbb{R}^n$ and any λ, $0 < \lambda < 1$,*

$$\Phi^{-1}(\gamma_n(\lambda A + (1 - \lambda)B)) \geq \lambda \Phi^{-1}(\gamma_n(A)) + (1 - \lambda)\Phi^{-1}(\gamma_n(B)) \tag{2.4}$$

holds true.

This theorem was proved by Ehrhard (1983) for convex A and B and by Latala (1996) when at least one of the two sets is convex. The proof in its general form is due to Borell (2003) and is the following.

Proof. It suffices to prove (2.4) for compact A and B. Let $0 < \varepsilon < 1$ and $0 < \delta < \varepsilon$. We put $\alpha := 1 - \varepsilon + \delta$ so that $\delta < \alpha < 1$, and $A_\varepsilon := A + \overline{B}_n(0; \varepsilon)$, where $\overline{B}_n(0; \varepsilon)$ is the closed ball in $\mathbb{R}^n$ centered at the origin and having radius ε. Let $f_1 : \mathbb{R}^n \to \mathbb{R}$ be a C^∞-function such that $f_1|_A \equiv \alpha$ and $f_1|_{A^C} \equiv \delta$. We have $\delta \leq f_1(x) \leq \alpha$ for all $x \in \mathbb{R}^n$.

In a similar way, define f_2 by changing, in the definition of f_1, the set A by the set B, and f_3, by changing in the definition of f_1 the set A by the set $\lambda A_\varepsilon + (1 - \lambda) B_\varepsilon$ and the minimum value δ by

$$\kappa = \max\left(\Phi\left[\lambda\Phi^{-1}(\alpha) + (1-\lambda)\Phi^{-1}(\delta)\right], \Phi\left[\lambda\Phi^{-1}(\delta) + (1-\lambda)\Phi^{-1}(\alpha)\right]\right).$$

Notice that $\kappa \to 0$ as $\delta \to 0$. We will prove the inequality

$$\Phi^{-1}\left(\int_{\mathbb{R}^n} f_3\, d\gamma_n\right) \geq \lambda\Phi^{-1}\left(\int_{\mathbb{R}^n} f_1\, d\gamma_n\right) + (1-\lambda)\Phi^{-1}\left(\int_{\mathbb{R}^n} f_2\, d\gamma_n\right). \quad (2.5)$$

Inequality (2.4) follows from (2.5) by letting $\delta \to 0$ and $\varepsilon \to 0$, in this order.

Let us define, for $j = 1, 2, 3$ and $(t, x) \in [0, 1] \times \mathbb{R}^n$,

$$u_j(t, x) = \int_{\mathbb{R}^n} f_j(x + \sqrt{t}z)\gamma_n(dz).$$

Instead of (2.5) we will prove, in fact, the more general inequality

$$\Phi^{-1}(u_3(t, \lambda x + (1-\lambda)y) \geq \lambda\Phi^{-1}(u_1(t,x)) + (1-\lambda)\Phi^{-1}(u_2(t, y)). \quad (2.6)$$

Putting $t = 1$, $x = y = 0$ in (2.6), we obtain (2.5).

So, our aim is to prove that

$$C(t, x, y) \geq 0 \qquad \text{for all } t \in [0, 1],\ x, y \in \mathbb{R}^n, \quad (2.7)$$

where

$$C(t, x, y) = U_3(t, \lambda x + (1-\lambda)y) - \lambda U_1(t, x) + (1-\lambda)U_2(t, y) \quad (2.8)$$

with the notation $U_j = \Phi^{-1} \circ u_j$ $(j = 1, 2, 3)$.

An instant reflection shows that the definitions of the functions f_j and u_j imply that the functions U_j, as functions of the space variable x, are C^∞ and the partial derivatives of all orders are bounded for $t \geq 0, x \in \mathbb{R}^n$.

Let us check that (2.7) holds true when $t = 0$, in which case it becomes

$$f_3(\lambda x + (1-\lambda)y \geq \Phi\left[\lambda\Phi^{-1}(f_1(x)) + (1-\lambda)\Phi^{-1}(f_2(y))\right]. \quad (2.9)$$

If $x \notin A_\varepsilon$, the right-hand side of (2.9) is bounded above by $\Phi\left[\lambda\Phi^{-1}(\delta) + (1-\lambda)\Phi^{-1}(\alpha)\right] \leq \kappa$, and κ is a lower bound of f_3; similarly if $y \notin B_\varepsilon$. So it remains to prove (2.9) when $x \in A_\varepsilon$ and also $y \in B_\varepsilon$, in which case the left-hand side of (2.9) is equal to α and this is an upper bound for the right-hand side of (2.9).

To show that (2.7) also holds true for all $t \in [0, 1]$, Borell's proof uses a method that reminds us of the maximum principle for parabolic equations. We denote ∇ and Δ, respectively, the gradient and Laplace operators with respect to the space variables.

First, u_j $(j = 1, 2, 3)$ verifies the heat equation

$$\frac{\partial u_j}{\partial t} = \frac{1}{2}\Delta u_j \qquad \text{on} \quad [0, 1] \times \mathbb{R}^n,$$

and this implies, by a simple computation, that

$$\frac{\partial U_j}{\partial t} = \frac{1}{2}\Delta U_j - \frac{1}{2}U_j \left\|\nabla U_j\right\|^2 \qquad \text{on} \quad [0, 1] \times \mathbb{R}^n. \tag{2.10}$$

Using the identities (2.10) for $j = 1, 2, 3$, we can compute the value of the differential operator

$$\mathcal{L} = \frac{1}{2}\sum_{j=1}^{n}\left[\frac{\partial^2}{\partial x_j^2} + 2\frac{\partial^2}{\partial x_j \partial y_j} + \frac{\partial^2}{\partial y_j^2}\right]$$

on the function $C(t, \cdot, \cdot)$. We obtain

$$\mathcal{L}C(t, x, y) = \frac{\partial C}{\partial t}(t, x, y) + \frac{1}{2}\left\|(\nabla U_3)(t, \lambda x + (1-\lambda)y)\right\|^2 C(t, x, y) + b(t, x, y), \tag{2.11}$$

where

$$b(t, x, y) = U_1 \langle \nabla_x C, \nabla U_1 + \nabla U_3 \rangle + U_2 \left\langle \nabla_y C, \nabla U_2 + \nabla U_3 \right\rangle. \tag{2.12}$$

In (2.12), U_1 and ∇U_1 are computed at (t, x), U_2 and ∇U_2 at (t, y), U_3 and ∇U_3 at $(t, \lambda x + (1-\lambda)y)$, and ∇_x and ∇_y denote the gradient with respect to x and y. The computation of derivatives to check (2.11) is left to the reader.

Let us suppose that (2.7) does not hold and show that this leads to a contradiction. We prove first that

$$\liminf_{\|x\|+\|y\| \to +\infty} \inf_{0 \leq t \leq 1} C(t, x, y) \geq 0. \tag{2.13}$$

Since A and B are bounded, one can find $a > 0$ such that if $\|w\| \geq a$, then $f_1(w) = f_2(w) = \delta$, $f_3(w) = \kappa$. Note that

$$u_1(t,x) = \mathrm{E}\left(f_1\left(x + \sqrt{t}\xi\right)\right),$$

where ξ is standard normal in $\mathbb{R}^n$. Hence,

$$\begin{aligned} u_1(t,x) &= \mathrm{E}\left(f_1\left(x + \sqrt{t}\xi\right)\mathbf{1}_{\{\|x+\sqrt{t}\xi\|\geq a\}}\right) + \mathrm{E}\left(f_1\left(x + \sqrt{t}\xi\right)\mathbf{1}_{\{\|x+\sqrt{t}\xi\|< a\}}\right) \\ &= \delta + R, \end{aligned}$$

with $|R| \leq \mathrm{P}\left(\left\|x + \sqrt{t}\xi\right\| < a\right)$. Now choose x so that $\|x\| \geq 2a$, and we get for $0 \leq t \leq 1$,

$$\begin{aligned} |R| &\leq \mathrm{P}\left(\sqrt{t}\,\|\xi\| > \frac{\|x\|}{2}\right) \leq \mathrm{P}\left(\|\xi\| > \frac{\|x\|}{2}\right) \\ &= \sigma_{n-1}(S^{n-1}) \int_{\|x\|/2}^{+\infty} \rho^{n-1} \exp\left(-\frac{\rho^2}{2}\right) d\rho \leq C_1 \exp\left(-C_2\,\|x\|^2\right), \end{aligned}$$

where C_1 and C_2 are positive constants. This shows that as $\|x\| \to +\infty$, $u_1(t,x)$ converges to δ, uniformly on $t \in [0,1]$. A similar result holds for u_2 and u_3, in the latter case replacing δ by κ. Going back to the definitions of $C(t,x,y)$ and κ, (2.13) follows.

On the other hand, (2.13) implies that if $C(t,x,y)$ takes a negative value somewhere in $[0,1] \times \mathbb{R}^n \times \mathbb{R}^n$, it has a minimum, and since we also know that $C(0,x,y) \geq 0$ for any choice of x, y, choosing $\varepsilon > 0$ small enough, we can assure that the function $C(t,x,y) + \varepsilon t$ will also have a negative minimum at some point $(\bar{t}, \bar{x}, \bar{y})$ with $0 < \bar{t} \leq 1$. Clearly, this implies that

$$\nabla_x C(\bar{t},\bar{x},\bar{y}) = 0, \qquad \nabla_y C(\bar{t},\bar{x},\bar{y}) = 0, \qquad \frac{\partial C}{\partial t}(\bar{t},\bar{x},\bar{y}) \leq -\varepsilon.$$

Denote by M the $2n \times 2n$ matrix of second partial derivatives of the function $C(\bar{t},\cdot,\cdot)$ computed at the point $(\bar{x},\bar{y})$; that is, if we rename the vector $(x_1,\ldots,x_n,y_1,\ldots,y_n)$ as $(z_1,\ldots,z_{2n})$, we have

$$M = \left(\left(\left.\frac{\partial^2 C}{\partial z_h \partial z_k}\right|_{(t,x,y)=(\bar{t},\bar{x},\bar{y})}\right)\right)_{h,k=1,\ldots,2n}.$$

Since there is a minimum of $C(\bar{t},\cdot,\cdot)$ at the point $(\bar{x},\bar{y})$, M has to be semidefinite positive. One also has

$$\mathcal{L}C(\bar{t},\bar{x},\bar{y}) = \frac{1}{2}\sum_{j=1}^{n}\left\langle M,\theta_j,\theta_j\right\rangle,$$

where θ_j denotes the vector $(x_1, \ldots, x_n, y_1, \ldots, y_n)$ such that $x_k = y_k = \delta_{jk}$ $(k = 1, \ldots, n)$. So $\mathcal{L}C(\bar{t}, \bar{x}, \bar{y}) \geq 0$. However, putting $(t, x, y) = (\bar{t}, \bar{x}, \bar{y})$ on the right-hand side of (2.11), we see that it becomes strictly negative. This ends the proof. □

2.3. GAUSSIAN ISOPERIMETRIC INEQUALITY

One possible version of the classical isoperimetric inequality for Lebesgue measure in $\mathbb{R}^n$ states that if A is a Borel subset of $\mathbb{R}^n$ and the ball $B(0; r)$ has the same Lebesgue measure as A, then for any $t > 0$, $\lambda_n(A_t) \geq \lambda_n(B(0; r + t))$ holds true. The notation for A_t is the one introduced at the beginning of the proof of Theorem 2.5

In the mid-1970s, Borell (1975) and Sudakov and Tsirelson (1974) proved independently a similar property for Gaussian measures, which is the statement of the next theorem. The use of isoperimetric methods for Gaussian distributions seems to have started with a paper by Landau and Shepp (1970) in which they studied the tails of the distribution of the supremum of Gaussian processes. The results were improved by Marcus and Shepp (1972) in an article in which they gave what seems to be the first published proof of (2.33). An independent and purely probabilistic proof of (2.33) is given in Fernique's lecture notes (1974), along with other, connected results.

All this was well established around 1975 and will be sufficient for our uses in the next chapters. For the many interesting directions of the relationship between isoperimetry and Gaussian and related measures, the important references are the monographs by Ledoux (1996, 2001). A synthesis of known results and open problems on Gaussian and related inequalities and the relations with isoperimetry is given by Latala (2002).

Theorem 2.6. *Let A be a Borel subset of $\mathbb{R}^n$ and H a half-space in $\mathbb{R}^n$ such that $\gamma_n(A) = \gamma_n(H) = \Phi(a)$ for some $a \in \mathbb{R}$. Then*

$$\gamma_n(A_t) \geq \gamma_n(H_t) = \Phi(a + t) \qquad \text{for every } t > 0. \tag{2.14}$$

Proof. Let $0 < \lambda < 1$. Applying Theorem 2.5 yields

$$\begin{aligned}
\Phi^{-1}\left(\gamma_n\left(A_t\right)\right) &= \Phi^{-1}\left(\gamma_n\left(A + \overline{B_n(0;t)}\right)\right) \\
&= \Phi^{-1}\left[\gamma_n\left(\lambda\frac{1}{\lambda}A + (1-\lambda)\frac{1}{1-\lambda}\overline{B_n(0;t)}\right)\right] \\
&\geq \lambda\Phi^{-1}\left(\gamma_n\left(\frac{1}{\lambda}A\right)\right) + (1-\lambda)\Phi^{-1}\left(\gamma_n\left(\frac{1}{1-\lambda}\overline{B_n(0;t)}\right)\right).
\end{aligned} \tag{2.15}$$

Let $\lambda \uparrow 1$ in (2.15). The first term on the right-hand side tends to $\Phi^{-1}(\gamma_n(A)) = a$.. To compute the limit of the second term, put $r = 1/(1-\lambda) \to +\infty$ as $\lambda \uparrow 1$ and $\gamma_n\left(\overline{rB_n(0;t)}\right) = \Phi(y)$, so that $y \to +\infty$ as $r \to +\infty$. Using the standard formulas for the Gaussian distribution, we obtain

$$1 - \Phi(y) \approx \frac{1}{\sqrt{2\pi}} \frac{1}{y} e^{-(1/2)y^2}$$

$$1 - \gamma_n\left(\overline{rB_n(0;t)}\right) \approx \frac{1}{2^{(n/2)-1}\Gamma(n/2)} (rt)^{n-2} e^{-(1/2)r^2t^2}$$

since both left-hand members are equal, we conclude that $y/r \to t$. Summing up gives us

$$\Phi^{-1}(\gamma_n(A_t)) \geq a + t. \qquad \square$$

2.4. INEQUALITIES FOR THE TAILS OF THE DISTRIBUTION OF THE SUPREMUM

Let $\mathcal{X} = \{X(t) : t \in T\}$ be a real-valued centered Gaussian process, $M_T(\omega) = \sup_{t \in T} X(t)(\omega)$ [$M_T(\omega)$ may have the value $+\infty$]. We assume in this section that there exists a countable subset $\mathcal{D}$ of the parameter set T such that a.s. $M_T = \sup_{t \in \mathcal{D}} X(t)$. In particular, this condition holds true if $\mathcal{X}$ is a separable process. We denote $\sigma^2(t) = \mathrm{E}(X^2(t))$.

This section contains two theorems with general bounds for the probability distribution of M_T. The first is the Borell–Sudakov–Tsirelson inequality, which gives an exponential bound for $\mathrm{P}(|M_T - \mu(M_T)| > x)$, where $\mu(\xi)$ denotes a median of the distribution of the real-valued random variable ξ. The proof is a consequence of the isoperimetric inequality (2.14).

The second theorem is similar, but instead of the median, appears in the statement of the expectation $\mathrm{E}(M_T)$. We have included a proof due to Ibragimov et al. (1976), which is independent of the foregoing arguments. This proof is interesting by itself, since it is based on Ito's formula, so that it establishes a link between the theory of Gaussian processes and stochastic analysis.

In what follows, f is the function $f(x) = \sup_{1 \leq j \leq N} x^j$, where $x = (x^1, \ldots, x^N)^{\mathrm{T}}$ and

$$f_\varepsilon(x) = \int_{\mathbb{R}^N} \prod_{j=1}^{N} g_\varepsilon(x^j - y^j) f(y)\, dy \tag{2.16}$$

is a regularization of f by convolution. Here $g : \mathbb{R} \to \mathbb{R}^+$ is a function of class C^∞ with support in the interval $[-1, 1]$, $\int_{-1}^{+1} g(r)\, dr = 1$ and $g_\varepsilon(r) = (1/\varepsilon) g(r/\varepsilon)$, $0 < \varepsilon < 1$.

We start with the following lemma:

Lemma 2.7

1. $\sum_{j=1}^{N} \left| \partial f_\varepsilon(x)/\partial x_j \right| = 1 \;\; \forall x \in \mathbb{R}^N$.
2. *Let* $A = (a_{ij})_{i,j=1,\ldots,N}$ *be a real* $N \times N$ *matrix and set* $B = AA^T = (b_{ij})_{i,j=1,\ldots,N}$.
   The function $h_A(x) = f(Ax)$, $x \in \mathbb{R}^N$ *satisfies the Lipshitz condition*

$$|h_A(x) - h_A(y)| \leq \bar{b}\, \|x - y\| \;\forall\; x, y \in \mathbb{R}^N,$$

where $\bar{b}^2 = \sup\{b_{ii} : i = 1, ..., N\}$.

Proof. To prove (1), let us compute the partial derivatives of f_ε:

$$\begin{aligned}
\frac{\partial f_\varepsilon}{\partial x^j}(x) &= \int_{\mathbb{R}^{N-1}} \prod_{k=1, k\neq j}^{N} \left[g_\varepsilon(x^k - y^k)\, dy^k\right] \int_{\mathbb{R}} g_\varepsilon'(x^j - y^j) f(y)\, dy^j \\
&= \int_{\mathbb{R}^{N-1}} \prod_{k=1, k\neq j}^{N} \left[g_\varepsilon(x^k - y^k)\, dy^k\right] \int_{\mathbb{R}} g_\varepsilon(x^j - y^j) \mathbf{1}_{y^j > \sup_{k\neq j} y^k} dy^j \\
&= \int_{A_j} \prod_{j=1}^{N} g_\varepsilon(x^j - y^j)\, dy
\end{aligned}$$

with $A_j = \left\{y^j > \sup_{k\neq j} y^k\right\}$. The second equality above comes from integration by parts in the inner integral. Now it becomes plain that

$$\sum_{j=1}^{N} \left| \frac{\partial f_\varepsilon}{\partial x^j}(x) \right| = 1 \tag{2.17}$$

since the sets A_j $(j = 1, \ldots, N)$ are a partition of $\mathbb{R}^N$, modulo Lebesgue measure.

To prove (2), notice that it suffices to show the result for the function h_ε instead of h, where $h_\varepsilon(x) = f_\varepsilon(Ax)$ and then pass to the limit as $\varepsilon \downarrow 0$. This will be achieved if we prove

$$\|\nabla h_\varepsilon(x)\| \leq \bar{b} \;\; \forall\; x \in \mathbb{R}^N.$$

In fact,

$$\|\nabla h_\varepsilon(x)\|^2 = \sum_{i=1}^{N} \sum_{k,k'=1}^{N} \frac{\partial f_\varepsilon}{\partial x^k} a_{ik} \frac{\partial f_\varepsilon}{\partial x^{k'}} a_{ik'} \sum_{k,k'=1}^{N} \frac{\partial f_\varepsilon}{\partial x^k} b_{kk'} \frac{\partial f_\varepsilon}{\partial x^{k'}} \leq \bar{b}^2 \left[\sum_{j=1}^{N} \left| \frac{\partial f_\varepsilon}{\partial x^j} \right| \right]^2 = \bar{b}^2. \tag{2.18}$$

□

Let us now state and prove the first theorem.

Theorem 2.8. *Assume that* $\mathrm{P}(M_T < \infty) = 1$. *Then*

$$\sigma_T^2 = \sup_{t \in T} \sigma^2(t) < +\infty,$$

and for every $u > 0$,

$$\mathrm{P}(|M_T - \mu(M_T)| > u) \leq \exp\left(-\frac{1}{2}\frac{u^2}{\sigma_T^2}\right). \tag{2.19}$$

Remark. We will prove the stronger inequality

$$\mathrm{P}(|M_T - \mu(M_T)| > u) \leq 2[1 - \Phi(u/\sigma_T)].$$

Proof. Let us prove that $\sigma_T^2 < \infty$. In fact, if $\sigma_T^2 = \infty$ and $\{t_n\}$ is a sequence in T such that $\sigma^2(t_n) \to \infty$, it follows that for $u > 0$,

$$\begin{aligned}
\mathrm{P}(M_T > u) \geq \mathrm{P}(X(t_n) > u) &= \frac{1}{\sqrt{2\pi}\sigma(t_n)} \int_u^{+\infty} e^{-(1/2)[y^2/\sigma^2(t_n)]} dy \\
&= \frac{1}{\sqrt{2\pi}} \int_{u/\sigma(t_n)}^{+\infty} e^{-(1/2)y^2} dy \to \frac{1}{2}.
\end{aligned}$$

So $\mathrm{P}(M_T > u) \geq \frac{1}{2}$ for every $u > 0$. This implies that $\mathrm{P}(M_T = +\infty) \geq \frac{1}{2}$, which contradicts the hypothesis that $\mathrm{P}(M_T = +\infty) = 0$.

One can also assume that $\sigma_T^2 > 0$ since this only excludes the trivial case that a.s. $X_t = 0$ for every $t \in T$. On the other hand, due to the hypothesis that a.s. M_T is the supremum over a countable set of parameter values, a simple approximation argument shows that it is enough to prove (2.19) when the parameter T is finite: say that it consists of N points. So our aim is to prove the inequality

$$\mathrm{P}(|f(X) - \mu(f(X))| > u) \leq 2[1 - \Phi(u/\sigma)], \tag{2.20}$$

where $X = (X_1, \ldots, X_N)^{\mathrm{T}}$ is a centered Gaussian vector in $\mathbb{R}^N$, $\mathrm{Var}(X) = \mathrm{E}(XX^{\mathrm{T}}) = V = ((V_{ij}))_{i,j=1,\ldots,N}$ and $\sigma^2 = \sup_{1 \leq j \leq N} V_{jj}$.

Let $V^{1/2}$ be a square root of the variance matrix V; that is, $V^{1/2}(V^{1/2})^{\mathrm{T}} = V$. Then the random vectors X and $V^{1/2}\eta$, where η has a standard normal distribution in $\mathbb{R}^N$, have the same distribution, and our problem is to prove that

$$\mathrm{P}(|h(\eta) - \mu(h(\eta))| > u) \leq 2[1 - \Phi(u/\sigma)], \tag{2.21}$$

where $h(x) = f(V^{1/2}x)$.

We denote by $\bar{\mu}$ the median of the random variable $h(\eta)$. Notice that for any $x, y \in \mathbb{R}^N$, using Lemma 2.7, part (2) with h instead of h_A, we obtain

$$L_h = \sup\left\{\frac{|h(x) - h(y)|}{\|x - y\|} \ : \ x, y \in \mathbb{R}^N, x \neq y\right\} \leq \sigma.$$

Since $u > 0$, it follows that

$$\mathrm{P}(h(\eta) - \bar{\mu} > u\sigma) \leq \mathrm{P}(h(\eta) - \bar{\mu} > uL_h). \tag{2.22}$$

Define $A := \{x \in \mathbb{R}^N, h(x) \leq \bar{\mu}\}$. Then

$$A_u \subset \{y \in \mathbb{R}^N \ : \ h(y) \leq \bar{\mu} + uL_h\}.$$

In fact, according to the definition of A_u, if $w \in A_u$, one can write $w = x + z$, where $h(x) \leq \bar{\mu}$, $\|z\| \leq u$. So

$$h(w) \leq h(x) + uL_h.$$

It follows, using (2.22), that

$$\mathrm{P}(h(\eta) - \bar{\mu} > u\sigma) \leq \mathrm{P}(\eta \in A_u^C). \tag{2.23}$$

We now use the isoperimetric inequality (2.14). Since $\bar{\mu}$ is the median of the distribution of $h(\eta)$, one has $\mathrm{P}(\eta \in A) \geq \frac{1}{2}$, so that

$$\mathrm{P}(\eta \in A_u) \geq \Phi(u).$$

So (2.23) implies that

$$\mathrm{P}\left(h(\eta) - \bar{\mu} > u\sigma\right) \leq 1 - \Phi(u) \leq \tfrac{1}{2}e^{-(1/2)u^2}, \tag{2.24}$$

where checking the last inequality is an elementary computation.

A similar argument applies to $\mathrm{P}(h(\eta) - \bar{\mu} < -u\sigma)$. This proves (2.21) and finishes the proof of the theorem. □

Let us now turn to the second general inequality.

Theorem 2.9. *Assume that the process $\mathcal{X}$ satisfies the same hypotheses as in Theorem 2.8.*

Then:

1. $\mathrm{E}\left(|M_T|\right) < \infty$.
2. *For every $u > 0$, the inequality*

$$\mathrm{P}(|M_T - \mathrm{E}(M_T)| > u) \le 2\exp\left(-\frac{1}{2}\frac{u^2}{\sigma_T^2}\right). \tag{2.25}$$

Proof. We first prove the inequality for finite parameter set T, say having N points. The first part of the proof is exactly the same as in the preceding proof, and we use the same notations as above.

Let $\{W(t) : t \ge 0\}$ be a Wiener process in $\mathbb{R}^N$; that is,

$$W(t) = (W^1(t), \ldots, W^N(t))^{\mathrm{T}} \qquad t \ge 0,$$

where $W^1, \ldots, W^N$ are real-valued independent Wiener processes. We want to prove that

$$\mathrm{P}(|h(W(1)) - \mathrm{E}(h(W(1)))| > u) \le 2\exp\left(-\frac{1}{2}\frac{u^2}{\sigma^2}\right) \tag{2.26}$$

for any $u > 0$. It suffices to prove (2.26) for the smooth function h_ε. Then, passing to the limit as $\varepsilon \downarrow 0$ gives the result. In what follows, for brevity, we have put h instead of h_ε.

Consider the function $H : \mathbb{R}^N \times (0, 1) \to \mathbb{R}$, defined by means of

$$H(x,t) = \mathrm{E}(h(x + W(1-t))) = \int_{R^d} h(x+y)\, p_{1-t}(y)\, dy = (h * p_{1-t})(x),$$

where

$$p_t(y) = \frac{1}{(2\pi t)^{N/2}} \exp\left(-\frac{\|y\|^2}{2t}\right) \qquad t > 0$$

is the density of the random variable $W(t)$ in $\mathbb{R}^N$. One can easily check that

$$\frac{\partial p_t}{\partial t} = \frac{1}{2}\sum_{j=1}^{N} \frac{\partial^2 p_t}{(\partial x^j)^2}$$

$$\frac{\partial H}{\partial t} = -\frac{1}{2}\sum_{j=1}^{N} \frac{\partial^2 H}{(\partial x^j)^2} \tag{2.27}$$

and that the function H has the boundary values:

$$H(x,t) \to \begin{Bmatrix} h(y) & \text{as } t \uparrow 1 \text{ and } x \to y \\ \mathrm{E}(h(y + W(1))) & \text{as } t \downarrow 0 \text{ and } x \to y. \end{Bmatrix}$$

Let us apply Itô's formula, $0 < s < t < 1$:

$$H(W(t),t) - H(W(s),s) = \int_s^t \sum_{j=1}^N \left[\frac{\partial H}{\partial x^j}(W(u),u)dW^j(u) + \frac{1}{2}\frac{\partial^2 H}{\left(\partial x^j\right)^2}(W(u),u)du \right] + \int_s^t \frac{\partial H}{\partial t}(W(u),u)du$$
$$= \int_s^t \sum_{j=1}^N \frac{\partial H}{\partial x^j}(W(u),u)dW^j(u)$$

using (2.27). Now take limits as $t \uparrow 1$ and $s \downarrow 0$, obtaining

$$h(W(1)) - \mathrm{E}(h(W(1))) = Z(1), \tag{2.28}$$

where $\{Z(t) : t \geq 0\}$ is the martingale

$$Z(t) = \int_0^t \sum_{j=1}^N \frac{\partial H}{\partial x^j}(W(u),u)dW^j(u).$$

Let us prove that the quadratic variation of Z at the time 1 verifies

$$[Z]_1 \leq \sigma^2. \tag{2.29}$$

From the proof of Lemma 2.7, we know that

$$\sup_{y \in \mathbb{R}^N} \|\nabla h(y)\|^2 \leq \sigma^2.$$

So

$$\sum_{j=1}^N \left(\frac{\partial H}{\partial x^j}(x,u)\right)^2 = \sum_{j=1}^N \left[\mathrm{E}\left(\frac{\partial h}{\partial x^j}(x + W(1-u))\right)\right]^2$$
$$\leq \sum_{j=1}^N \mathrm{E}\left[\left(\frac{\partial h}{\partial x^j}(x + W(1-u))\right)^2\right] \leq \sigma^2$$

and we obtain the bound (2.29) since

$$[Z]_1 = \int_0^1 \sum_{j=1}^N \left(\frac{\partial H}{\partial x^j}(W(u),u)\right)^2 du \leq \sigma^2.$$

Now, for each $\theta \in \mathbb{R}$ we consider the exponential martingale (see McKean, 1969):

$$Y(t) = e^{\theta\, Z(t) - (1/2)\theta^2 [Z]_t}, \qquad 0 \le t \le 1,$$

which satisfies

$$\mathrm{E}\,(Y(t)) = 1 \qquad \text{for every } t,\ 0 \le t \le 1.$$

This, together with (2.29), implies that for every $\theta \in \mathbb{R}$, $\mathrm{E}(e^{\theta\, Z(1) - (1/2)\theta^2\sigma^2}) \le 1$, so that

$$\mathrm{E}\left(e^{\theta\, Z(1)}\right) \le e^{(1/2)\theta^2\sigma^2}. \tag{2.30}$$

Write the left-hand side of (2.26) for $u > 0$ as

$$\mathrm{P}(|Z(1)| > u) = \mathrm{P}(Z(1) > u) + \mathrm{P}(Z(1) < -u).$$

For the first term use (2.30) with $\theta = u/\sigma^2$. Then

$$\begin{aligned}\mathrm{P}(Z(1) > u) &= \mathrm{P}(e^{\theta Z(1)} > e^{\theta}) \le e^{-\theta u}\mathrm{E}\left(e^{\theta Z(1)}\right) \le \exp\left(-\theta u + \tfrac{1}{2}\theta^2\sigma^2\right)\\ &= \exp\left(-\frac{1}{2}u^2\sigma^2\right).\end{aligned}$$

A similar argument produces the same bound for the second term. This proves (2.26).

To finish we must show that the result holds for infinite T. Due to the hypothesis, it suffices to consider the case when T is countable. Put $T = \{t_n\}_{n=1,2,\ldots}$, $T_N = \{t_1, \ldots, t_N\}$, $N \ge 1$. Clearly,

$$M_{T_N} \uparrow M_T, \qquad \sigma^2_{T_N} \uparrow \sigma^2_T \quad \text{as } N \uparrow +\infty$$

and

$$0 \le M_{T_N} - X(t_1) \uparrow M_T - X(t_1) \quad \text{as } N \uparrow +\infty.$$

Beppo Levi's theorem implies that

$$\mathrm{E}\left(M_{T_N}\right) \uparrow \mathrm{E}\,(M_T) \qquad \text{as } N \uparrow +\infty.$$

Since we already know that (2.25) holds true for T_N instead of T, it will be sufficient to prove that $\mathrm{E}\,(M_T) < \infty$ to obtain (2.25), by letting $N \uparrow +\infty$.

Let us suppose that this were not true, that is, that $E(M_T) = +\infty$. Using the fact that a.s. the paths are bounded, choose x_0 large enough to have

$$\mathrm{P}(M_T < x_0) > \frac{3}{4} \quad \text{and} \quad \exp\left(-\frac{x_0^2}{2\sigma_T^2}\right) < \frac{1}{4}.$$

Now, if $\mathrm{E}(M_T) = +\infty$ using Beppo Levi's theorem, we can choose N large enough so that

$$\mathrm{E}\left(M_{T_N}\right) > 2\,x_0.$$

Then, if $\omega \in \{M_T < x_0\}$, one has $M_{T_N}(\omega) \leq M_T(\omega) < \mathrm{E}\left(M_{T_N}\right) - x_0$, which implies that $\left|M_{T_N}(\omega) - \mathrm{E}\left(M_{T_N}\right)\right| > x_0$. Hence,

$$\begin{aligned}\frac{3}{4} &< \mathrm{P}(M_T < x_0) \leq \mathrm{P}\left(\left|M_{T_N} - \mathrm{E}\left(M_{T_N}\right)\right| > x_0\right)\\ &\leq 2\,\exp\left(-\frac{x_0^2}{2\sigma_{T_N}^2}\right) \leq 2\,\exp\left(-\frac{x_0^2}{2\sigma_T^2}\right) < \frac{1}{2},\end{aligned}$$

which is a contradiction. This implies that $\mathrm{E}(M_T) < \infty$, and we are done. □

2.4.1. Some Derived Tail Inequalities

1. The same arguments show that Theorems 2.8 and 2.9 have unilateral versions: namely, for x greater than the mean $\mathrm{E}(M_T)$ or the median $\mu(M_T)$ of the process:

$$\mathrm{P}(M > u) \leq \exp\left(-\frac{(u - \mathrm{E}(M_T))^2}{2}\right) \tag{2.31}$$

and

$$\mathrm{P}(M > u) \leq \frac{1}{2}\exp\left(-\frac{(u - \mu(M_T))^2}{2}\right). \tag{2.32}$$

2. A weaker form of inequalities (2.31) and (2.32) is the following: Under the same hypotheses of Theorem 2.8 or 2.9, for each $\varepsilon > 0$ there exists a positive constant C_ε such that for all $u > 0$,

$$\mathrm{P}(|M_T| > u) \leq C_\varepsilon\,\exp\left(-\frac{1}{2}\frac{u^2}{\sigma_T^2 + \varepsilon}\right). \tag{2.33}$$

Inequality (2.33) is a consequence of (2.25) because $(u - E\{M_T\})/u \to 1$ as $u \to +\infty$. *Grosso modo*, this says that the tail of the distribution of the random variable M_T is bounded (except for a multiplicative constant) by the value of the centered normal density having variance larger than, and arbitrarily close to, σ_T^2.

The problem with this kind of inequality is that, in general, the constant C_ε can grow (and tend to infinity) as ε decreases to zero. Even for fixed ε, the usual situation is that, in general, one can have only rough bounds for C_ε, and this implies serious limitations for the use of these inequalities in statistics and in other fields. We return to this problem in some of the next chapters, with the aim of giving more accurate results on the value of the tails of the distribution of the supremum, at least for certain classes of Gaussian processes satisfying regularity assumptions.

3. Inequalities (2.31) and (2.25) show that one can do better than (2.33), since, for example, (2.25) has the form

$$\mathrm{P}(|M_T| > u) \leq C \exp\left(-\frac{u^2}{2\sigma_T^2} + C_1 u\right).$$

The difficulty in using this inequality is that the positive constants C and C_1 depend on $\mathrm{E}(M_T)$, which is finite but unknown. The problem of giving bounds on $\mathrm{E}(M_T)$ are addressed in Section 2.5.

4. Under the same hypotheses, if $\sigma^2 > \sigma_T^2$, one has

$$\mathrm{E}\left(\exp\left(\frac{M_T^2}{2\sigma^2}\right)\right) < \infty. \tag{2.34}$$

This is a direct consequence of (2.33) and implies that all moments of M_T are finite, since for any positive integer k,

$$\mathrm{E}\left(M_T^{2k}\right) \leq (2\sigma^2)^k k!\, \mathrm{E}\exp\left(\frac{M_T^2}{2\sigma^2}\right).$$

A straightforward consequence (which we have already proved by direct means in Chapter 1) is that if $X = \{X(t) : t \in T\}$ is a Gaussian process defined on a compact separable topological space such that almost surely the paths $t \rightsquigarrow X(t)$ are continuous, the mean $m(t)$ and the covariance function $r(s,t) := \mathrm{Cov}(X(s), X(t))$ are continuous functions. In fact, it suffices to note that each continuous path is bounded, so that the theorem can be applied and one can use Lebesgue dominated convergence to prove continuity.

5. To illustrate the fact that these inequalities are not usually significant from a numerical point of view, let us consider the simplest case, given by the Wiener process (Brownian motion). Let $\{W(t) : t \in [0,1]\}$ be the Brownian motion on the unit interval and M its maximum. It is well known (McKean, 1969) that the reflection principle implies that the distribution of M is that of the absolute value of a standard normal variable. It implies that

$$\mathrm{E}(M) = \sqrt{2/\pi} = 0.7989\ldots$$

and that the median $\mu(M)$ satisfies $\mu(M) = 0.675....$ If we apply Borell's type inequality to the Wiener process (Brownian motion) (with the advantage that the mean and the median are known, which is of course exceptional), we get

u	True Values of $\mathrm{P}(M_W > u)$	Borell's Mean	Borell's Median
2	0.045	0.4855	0.2077
3	0.0027	0.0885	0.0347
4	6.33×10^{-5}	5.93×10^{-3}	1.98×10^{-3}
5	5.73×10^{-7}	1.46×10^{-4}	4.32×10^{-5}

In this table we have taken unilateral versions of Borell's type inequality: namely, (2.31) and (2.32). The inequality with the median is sharper but even in this very favorable case, both inequalities are imprecise.

2.5. DUDLEY'S INEQUALITY

We now turn to obtaining a bound for $\mathrm{E}(M_T)$. A classical result in this direction is the next theorem. The proof is taken from Talagrand (1996).

Let $\{X(t) : t \in T\}$ be a stochastic process not necessarily Gaussian. As in Section 1.4.4, we define the canonical distance by $d(s,t) := \sqrt{\mathrm{E}\left((X(t) - X(s))^2\right)}$, identifying as usual points s, t when $d(s,t) = 0$. We define the covering number $N_\varepsilon := N(T, d, \varepsilon)$ as in Definition 1.17 using this metric.

Theorem 2.10 (Dudley). *With the preceding notations, assume that*

$$\mathrm{E}(X(t)) = 0 \qquad \textit{for every } t \in T$$

$$\mathrm{P}(|X(t) - X(s)| > u) \le 2\,e^{-(1/2)[u^2/d^2(s,t)]} \quad \textit{for all} \quad s, t \in T, \; u > 0. \tag{2.35}$$

Then

$$\mathrm{E}(\sup_{t \in T} X(t)) \le K \int_0^{+\infty} (\log N_\varepsilon)^{1/2}\, d\varepsilon, \tag{2.36}$$

where K is a universal constant.

Remarks. Let us make some comments on the statement before giving the proof.

1. It is clear from the definition that $0 < \varepsilon < \varepsilon'$ implies that $N_\varepsilon \ge N_{\varepsilon'}$. Hence if $N_{\varepsilon'} = +\infty$ for some $\varepsilon' > 0$, then $N_\varepsilon = +\infty$ for all $\varepsilon < \varepsilon'$ and the integral on the right-hand side of (2.36) is $+\infty$. In particular, this is the case when $\mathrm{diam}(T)$ (the diameter of T) is ∞. On the other hand, if $\mathrm{diam}(T) < \infty$ and $\varepsilon > \mathrm{diam}(T)$, then $N_\varepsilon = 1$; hence, the integral on the right-hand side of (2.36) is, in fact, an integral over a finite interval.

2. Condition (2.35) is easily verified for Gaussian processes. In fact, in this case, if $u > 0$,

$$P(|X(t) - X(s)| > u) = \sqrt{\frac{2}{\pi}} \int_{u/d(s,t)}^{+\infty} \exp\left(\frac{y^2}{2}\right) dy$$
$$\leq \sqrt{\frac{2}{\pi}} \frac{d(s,t)}{u} \exp\left(\frac{u^2}{2d^2(s,t)}\right).$$

If $d(s,t)/u \leq \sqrt{2\pi}$, (2.35) follows. If $d(s,t)/u > \sqrt{2\pi}$, then $2\exp(u^2/2d^2(s,t))$ $> 2\exp(-1/4\pi) > 1$ and (2.35) also holds true.

3. Under the general conditions of the theorem, it is necessary, to avoid measurability problems, to make precise the meaning of $\mathrm{E}\left(\sup_{t\in T} X(t)\right)$. This will be, by definition, equal to

$$\sup_{F\subset T, F \text{ finite}} \mathrm{E}\left(\sup_{t\in F} X(t)\right).$$

It is easy to see that if the conditions of Section 2.4 hold true, this coincides with the ordinary expectation of $\sup_{t\in T} X(t)$. This will be the situation in the processes we deal with in the remainder of the book.

So it will be sufficient to prove (2.36) when one replaces $\sup_{t\in T} X(t)$ on the left-hand side by $\sup_{t\in F} X(t)$, with F a finite subset of T.

Proof of Theorem 2.10. According to the remarks above, it suffices to consider the case $\mathrm{diam}(T) < \infty$, since otherwise the right-hand side of (2.36) is infinite. Let F be a finite subset of T. Choose any $t_0 \in F$ and fix it for the remainder of the proof. Since the process is centered, we have

$$\mathrm{E}\left(\sup_{t\in F} X(t)\right) = \mathrm{E}\left(\sup_{t\in F} (X(t) - X(t_0))\right)$$
$$= \int_0^{+\infty} P\left(\sup_{t\in F}(X(t) - X(t_0)) > x\right) dx \tag{2.37}$$

given that $\sup_{t\in F}(X(t) - X(t_0)) \geq 0$. Let j_0 be the (unique) integer such that

$$2^{-j_0} < \mathrm{diam}(T) \leq 2^{-j_0+1}.$$

We define the following sequence $\{E_j\}_{j=j_0, j_0+1,\ldots}$ of subsets of T: The first member of the sequence is $\{E_{j_0}\} = \{t_0\}$. For each integer $j \geq j_0 + 1$, take a set of $N_{2^{-j}}$ closed balls of radius 2^{-j} such that the union covers T (which exists, according to the definition of N_ε), and let E_j be the set of the centers of these balls. This implies that for each $t \in T$ and each $j \geq j_0 + 1$, one can define $\pi_j(t) \in E_j$ such that

$$d(t, \pi_j(t)) \leq 2^{-j}.$$

Also set $\pi_{j_0}(t) = t_0$ for every $t \in T$. Clearly,

$$d(\pi_j(t), \pi_{j-1}(t)) \leq d(\pi_j(t), t) + d(\pi_{j-1}(t), t) \leq 3.2^{-j} \qquad j \geq j_0 + 2$$

and

$$d(\pi_{j_0+1}(t), \pi_{j_0}(t)) \leq \text{diam}(T) \leq 2^{-j_0+1},$$

so that we have

$$d(\pi_j(t), \pi_{j-1}(t)) \leq 4.2^{-j} \qquad j \geq j_0 + 1. \tag{2.38}$$

Let us prove that for each $t \in T$ one has

$$\text{a.s.} \quad X(t) - X(t_0) = \sum_{j=j_0+1}^{\infty} \left(X(\pi_j(t)) - X(\pi_{j-1}(t))\right). \tag{2.39}$$

In fact, using the hypothesis and (2.38), for $\alpha_j > 0$,

$$\begin{aligned} \mathrm{P}\left(\left|X(\pi_j(t)) - X(\pi_{j-1}(t))\right| \geq \alpha_j\right) &\leq 2 \exp\left(-\frac{\alpha_j^2}{2d^2(\pi_j(t), \pi_{j-1}(t))}\right) \\ &\leq 2 \exp\left(-\frac{\alpha_j^2}{32 \times 2^{-2j}}\right). \end{aligned}$$

Taking $\alpha_j = 2^{-j/2}$ and applying the Borel–Cantellli lemma, it follows that a.s. the series in (2.39) converges. On the other hand,

$$\mathrm{E}\left[\left(X(\pi_j(t)) - X(t)\right)^2\right] = d^2(\pi_j(t), t) \to 0 \quad \text{as} \quad j \to \infty.$$

This proves (2.39).

It follows from (2.39) that if $\{a_j\}_{j=j_0, j_0+1, \ldots}$ is a sequence of positive numbers, and $u > 0$,

$$\begin{aligned} &\mathrm{P}\left(\sup_{t \in F}(X(t) - X(t_0)) > u \sum_{j=j_0+1}^{\infty} a_j\right) \\ &\leq \mathrm{P}\left(\exists t \in F \text{ and } \exists j > j_0 \text{ such that } X(\pi_j(t)) - X(\pi_{j-1}(t)) > u a_j\right). \end{aligned}$$

Now use the fact that there are at most $N_{2^{-j}} N_{2^{-(j-1)}}$ points in the product set $E_j \times E_{j-1}$, which implies that

$$P\left(\sup_{t\in F}(X(t) - X(t_0)) > u \sum_{j=j_0+1}^{\infty} a_j\right) \leq \sum_{j=j_0+1}^{\infty} N_{2^{-j}} N_{2^{-(j-1)}}\, 2 \exp\left(-\frac{u^2 a_j^2}{32 \times 2^{-2j}}\right). \quad (2.40)$$

Choosing

$$a_j = 4.\, 2^{-j+1/2} \left[\log\left(2^{j-j_0} N_{2^{-j}} N_{2^{-(j-1)}}\right)\right]^{1/2},$$

the expression on the right-hand side of (2.40) becomes, for $u \geq 1$,

$$2 \sum_{j=j_0+1}^{\infty} N_{2^{-j}} N_{2^{-(j-1)}} \left(2^{j-j_0} N_{2^{-j}} N_{2^{-(j-1)}}\right)^{-u^2} \leq 2 \sum_{j=j_0+1}^{\infty} 2^{-(j-j_0)u^2}$$
$$\leq 2\, 2^{-u^2} \sum_{k=0}^{\infty} 2^{-k} = 4\, .2^{-u^2}.$$

If we denote

$$S = \sum_{j=j_0+1}^{\infty} a_j,$$

then if $v/S \geq 1$, we get

$$P\left(\sup_{t\in F}(X(t) - X(t_0)) > v\right) \leq 4\, e^{-v^2/S^2},$$

which implies that

$$E\left[\sup_{t\in F}(X(t) - X(t_0))\right] \leq S + \int_S^{+\infty} P\left(\sup_{t\in F}(X(t) - X(t_0)) > v\right) dv$$
$$\leq S + 4 \int_S^{+\infty} e^{-v^2/S^2}\, dv \leq 2S \quad (2.41)$$

by a simple computation. On the other hand,

$$S = \sum_{j=j_0+1}^{\infty} a_j \leq \sum_{j=j_0+1}^{\infty} 4.2^{-j+1/2}\left[(j-j_0)\log 2 + 2(\log N_{2^{-j}})\right]^{1/2}$$

$$\leq \sum_{j=j_0+1}^{\infty} 4.2^{-j+1/2}\left[(j-j_0)^{1/2}(\log 2)^{1/2} + 2^{1/2}(\log N_{2^{-j}})^{1/2}\right] = T_1 + T_2.$$

For T_1 we have

$$T_1 \leq 4.2^{1/2}(\log 2)^{1/2} \sum_{j=j_0+1}^{\infty} 2^{-j}(j-j_0) = 16\sqrt{2\,(\log 2)}\,2^{-(j_0+1)}$$

$$\leq 16\sqrt{2} \sum_{j=j_0+1}^{\infty} 2^{-(j+1)}(\log N_{2^{-j}})^{1/2}$$

because $N_{2^{-j}} \geq 2$ for $j \geq j_0+1$ given that the definition of j_0 implies that one needs at least two balls of radius $2^{-(j_0+1)}$ to cover T.

As for T_2,

$$T_2 \leq 16 \sum_{j=j_0+1}^{\infty} 2^{-(j+1)}(\log N_{2^{-j}})^{1/2}.$$

Putting the two pieces together, we obtain

$$S \leq 16(1+\sqrt{2}) \sum_{j=j_0+1}^{\infty} 2^{-(j+1)}(\log N_{2^{-j}})^{1/2} \leq 16(1+\sqrt{2}) \times \int_0^{2^{-(j_0+1)}} (\log N_\varepsilon)^{1/2}\, d\varepsilon,$$

where the last inequality is a standard lower bound for the integral of the monotone decreasing function $\varepsilon \rightsquigarrow (\log N_\varepsilon)^{1/2}$ by Riemann's sums. This finishes the proof. □

EXERCISES

2.1. Give a direct geometric proof of the Plackett–Slepian lemma without using Fourier transform methods.
Hint: Prove the lemma for $n = 2$ by means of a comparison of measures in the plane. For general n, it suffices to prove that $\mathrm{P}(X_1 \leq a_1, \ldots, X_n \leq a_n)$

increases, in the broad sense, if one of the covariances r_{jk} $(j \neq k)$ increases, say r_{12}. For that purpose, write $X = A\,\xi$, A a nonrandom supertriangular matrix and ξ standard normal in $\mathbb{R}^n$. Then reduce the problem to dimension 2 by means of conditioning on the values of $\xi_3, \ldots, \xi_n$.

2.2. Prove that a direct consequence of the normal comparison lemma is that if $X_1, \ldots, X_n$ are standard normal variables with $\mathrm{Cov}(X_i, X_j) = r_{ij}$, then for any real numbers $u_1, \ldots, u_n$,

$$P\left|\left(\bigcap_{j=1}^{n}\{X_j \leq u_j\}\right) - \prod_{j=1}^{n} P\{X_j \leq u_j\}\right|$$
$$\leq \frac{1}{4} \sum_{1 \leq j < k \leq n} |r_{jk}| \exp\left(-\frac{u_j^2 + u_k^2}{2(1 + |r_{jk}|)}\right).$$

2.3. Give an example showing that in the Plackett–Slepian version of the normal comparison lemma, one cannot withdraw the equality-of-variances condition.

CHAPTER 3

CROSSINGS AND RICE FORMULAS FOR ONE-DIMENSIONAL PARAMETER PROCESSES

3.1. RICE FORMULAS

Let $f : I \to \mathbb{R}$ be a real-valued function defined on an interval I of the real line. We denote

$$C_u(f, I) := \{t \in I : f(t) = u\}$$
$$N_u(f, I) := \#C_u(f; I).$$

$C_u(f, I)$ is the set of roots of the equation $f(t) = u$ in the interval I and $N_u(f, I)$ is the number of these roots, which may be finite or infinite. We usually replace $C_u(f, I)$ by C_u [respectively, $N_u(f, I)$ by N_u] in case there is no doubt about the function f and the interval I.

In a similar way, if f is differentiable, we define

$$U_u(f, I) := \#\left\{t \in I : f(t) = u, f'(t) > 0\right\}$$
$$D_u(f, I) := \#\left\{t \in I : f(t) = u, f'(t) < 0\right\}.$$

N_u (respectively, U_u, D_u) will be called the *number of crossings* (respectively, *up-crossings* and *down-crossings*) of the level u by the function f on the interval I.

Level Sets and Extrema of Random Processes and Fields, By Jean-Marc Azaïs and Mario Wschebor

Our interest will be focused on $N_u(X, I), U_u(X, I), D_u(X, I)$ when $X(\cdot)$ is a path of a stochastic process. Even though these random variables are important in a large variety of problems, their probability distributions are unknown except for a small number of trivial cases. The Rice formulas that we study in this chapter provide certain expressions, having the form of integral formulas, for the moments of $N_u(X, I), U_u(X, I), D_u(X, I)$ and also some related random variables.

Rice formulas for one-parameter stochastic processes have long been used in various contexts, such as telecommunications and signal processing (Rice, 1944, 1945), ocean waves (Longuett-Higgins, 1957, 1962a,b), and random mechanics (Krée and Soize, 1983).

Rigorous results and a systematic treatment of the subject in the case of Gaussian processes came in the 1960s with the works of Ito (1964), Cramér and Leadbetter (1965), and Belayev (1966), among others. A landmark in the subject was the book by Cramér and Leadbetter (1967). The simple proof we have included below for general (not necessarily stationary) Gaussian processes with C^1-paths is given here for the first time. Formulas for wider classes of processes can be found, for example, in a book by Adler (1981) and articles by Marcus (1977) and Wschebor (1985). The proof of Theorem 3.4, which contains the Rice formula for general processes, not necessarily Gaussian, is an adaptation of Wschebor's proof.

We will say that the real-valued function f defined on the interval $I = [t_1, t_2]$ satisfies hypothesis H_{1u} if:

- f is a function of class C^1.
- $f(t_1) \neq u, \ f(t_2) \neq u$.
- $\{t : t \in I, \ f(t) = u, \ f'(t) = 0\} = \emptyset$.

Lemma 3.1 (Kac's Counting Formula). *If f satisfies H_{1u}, then*

$$N_u(f, I) = \lim_{\delta \to 0} \frac{1}{2\delta} \int_I \mathbf{1}_{\{|f(t)-u|<\delta\}} \left|f'(t)\right| dt. \tag{3.1}$$

Proof. The hypothesis H_{1u} implies that $N_u(f, I)$ is finite, say $N_u(f, I) = n$. If $n = 0$, the result is obvious, since the integrand on the right-hand side of (3.1) is identically zero if δ is small enough. If $n \geq 1$, set $C_u(f, I) = \{s_1, \ldots, s_n\}$. Since $f'(s_j) \neq 0$ for every $j = 1, \ldots, n$, if $\delta > 0$ is small enough the inverse image of the interval $(u - \delta, u + \delta)$ by the function f is the union of exactly n pairwise disjoint intervals $J_1, \ldots, J_n$, which contain, respectively, the points $s_1, \ldots, s_n$. The restriction of f to each of the intervals J_k $(k = 1, \ldots, n)$ is a diffeomorphism and one easily checks, changing variables, that

$$\int_{J_k} \left|f'(t)\right| dt = 2\delta$$

for each k. So if $\delta > 0$ is small enough,

$$\frac{1}{2\delta}\int_I \mathbb{I}_{\{|f(t)-u|<\delta\}}\,|f'(t)|\,dt = \frac{1}{2\delta}\sum_{k=1}^{n}\int_{J_k}|f'(t)|\,dt = n,$$

and we are done. □

Remarks on the Lemma. Lemma 3.1 holds true for polygonal f even though these are not C^1. More precisely, let

$$t_1 = \tau_0 < \tau_1 < \cdots < \tau_m = t_2$$

be a partition of the interval $[t_1, t_2]$ and f a function having the polygonal graph with vertices $(\tau_i, f(\tau_i))$, $i = 0, 1, \ldots, m$. Then, if $f(\tau_i) \neq u$ for $i = 0, 1, \ldots, m$, formula (3.1) holds true.

The proof is immediate, since formula (3.1) is satisfied for each partition interval and, under these hypotheses, is additive as a function of I. Moreover, notice that if f is such a polygonal function, the expression

$$\frac{1}{2\delta}\int_I \mathbb{I}_{\{|f(t)-u|<\delta\}}\,|f'(t)|\,dt$$

on the right-hand side of (3.1) is bounded by m. This is again simple, since the integral on each partition interval is bounded by 1 if it contains a crossing point and by $\frac{1}{2}$ if it does not.

Some Basic Ideas. An informal presentation of the Rice formula for the expectation $\mathrm{E}\big(N_u(X, I)\big)$, where X is a stochastic process, can be the following: Replace the function f in (3.1) by the random path $X(\cdot)$, and take expectations on both sides. Then

$$\begin{aligned}
\mathrm{E}\big(N_u(X, I)\big) &= \lim_{\delta\to 0}\mathrm{E}\left(\frac{1}{2\delta}\int_I \mathbb{I}_{\{|X(t)-u|<\delta\}}\,|X'(t)|\,dt\right)\\
&= \int_I dt \lim_{\delta\to 0}\frac{1}{2\delta}\int_{u-\delta}^{u+\delta}\mathrm{E}\big(|X'(t)|\big|X(t)=x\big)p_{X(t)}(x)\,dx\\
&= \int_I \mathrm{E}\big(|X'(t)|\big|X(t)=u\big)p_{X(t)}(u)\,dt.
\end{aligned}$$

We will pay attention in this chapter to the justification of these equalities as concerns passages to the limit. It is easy to prove weak forms of the Rice formula, such as equality for almost every level u (see Exercise 3.8) or giving upper bounds for $\mathrm{E}\big(N_u(X, I)\big)$ (see Exercise 3.9).

However, a formula for almost every u is not satisfactory for a number of uses. For example, if one is willing to compute the moments of the number of critical points or the number of local maxima of a random function, one has to count the number of points at which the derivative is equal to zero, and a formula of this kind, valid for almost every u, is uninteresting, since one needs it for $u = 0$. So it is worthwhile to expend some energy to prove an exact formula for each level u. In all cases, some hypotheses on the processes will be necessary (see Exercise 3.3, a simple counterexample in which the formula fails to hold true).

When the process $\{X(t) : t \in \mathbb{R}\}$ is Gaussian centered stationary with variance 1, the Rice formula for the expectation takes the simple form

$$\mathrm{E}\big(N_u(X, I)\big) = \frac{\sqrt{\lambda_2}}{\pi} e^{-u^2/2} |I|, \tag{3.2}$$

where $|I|$ denotes the length of the interval I. This formula is due to S. O. Rice, who stated it for the first time in a series of pioneering papers pointing to electrical engineering applications (1944, 1945).

3.1.1. Gaussian Case

We start with a statement and proof for Gaussian processes.

Theorem 3.2 (Gaussian Rice Formula). *Let $\mathcal{X} = \{X(t) : t \in I\}$, I an interval on the real line, be a Gaussian process having C^1-paths. Let k be a positive integer. We assume that for every k pairwise distinct points $t_1, \ldots, t_k$ in I, the joint distribution of $X(t_1), \ldots, X(t_k)$ does not degenerate. Then*

$$\mathrm{E}\big(N_u^{[k]}\big) = \int_{I^k} \mathrm{E}\big(|X'(t_1) \ldots X'(t_k)| \big| X(t_1) = u \cdots X(t_k) = u\big)$$
$$p_{X(t_1),\ldots,X(t_k)}(u, \ldots, u)\, dt_1 \cdots dt_k, \tag{3.3}$$

where

$$N_u := N_u(X, I)$$

$$m^{[k]} := \begin{cases} m(m-1)\cdots(m-k+1) & \text{if } m, k \text{ are positive integers, } m \geq k \\ 0 & \text{otherwise.} \end{cases}$$

Proof

STEP 1. Let $k = 1$. With no loss of generality, we assume that $I = [0, 1]$. Define $X^{(n)}(t)$ as the dyadic polygonal approximation of $X(t)$. As in the proof of Theorem 1.6, $X^{(n)}(t) - X(t)$ tends uniformly to zero and is bounded by the random variable $2 \sup_{t \in I} |X(t)|$, which has finite moments of all orders, because of the results of Chapter 2. Using dominated convergence, it follows that $\mathrm{Var}(X^{(n)}(t))$ converges uniformly for $t \in I$ to $\mathrm{Var}(X(t))$. So for n large enough, $\mathrm{Var}(X^{(n)}(t)) \geq b$ for some $b > 0$ and all $t \in [0, 1]$. (The reader might show this using the more elementary arguments of Section 1.2.)

For such an n, a.s. the process $X^{(n)}(t)$ does not take the value u at the partition points $j \cdot 2^{-n}$ $(j = 0, 1, \ldots, 2^n)$, since the random variable $X(t)$ has a density for each $t \in I$. So using the remarks after Lemma 3.1, we obtain

$$N_u(X^{(n)}, I) = \lim_{\delta \to 0} \frac{1}{2\delta} \int_I \mathbb{1}_{\{|X^{(n)}(t)-u|<\delta\}} \left|X^{(n)\prime}(t)\right| dt \text{ a.s.} \tag{3.4}$$

and the expression next to the limit on the right-hand side of (3.4) is bounded by 2^n.

Applying dominated convergence as $\delta \to 0$, for fixed n,

$$\begin{aligned} \mathrm{E}\big(N_u(X^{(n)}, I)\big) &= \lim_{\delta \to 0} \frac{1}{2\delta} \int_I \mathrm{E}\big(\mathbb{1}_{\{|X^{(n)}(t)-u|<\delta\}} |X^{(n)\prime}(t)|\big)\, dt \\ &= \lim_{\delta \to 0} \int_I dt \frac{1}{2\delta} \int_{u-\delta}^{u+\delta} \mathrm{E}\big(|X^{(n)\prime}(t)| \big| X^{(n)}(t) = x\big) \\ &\quad \times p_{X^{(n)}(t)}(x)\, dx, \end{aligned} \tag{3.5}$$

where the conditional expectation is the one defined by means of Gaussian regression.

Since the process has continuous sample paths, its expectation $m(t)$ and covariance $r(s,t)$ are continuous (see Section 1.4.3). On the other hand, the regression formulas show that $\mathrm{E}(|X^{(n)\prime}(t)| \big| X^{(n)}(t) = x) p_{X^{(n)}(t)}(x)$ is a continuous function of the pair (t, x), and thus it is bounded for $t \in I$ and x in a neighborhood of u. This implies that we may pass the limit sign inside the integral on the right-hand side of (3.5), so that

$$\mathrm{E}\big(N_u(X^{(n)}, I)\big) = \int_I \mathrm{E}\big(|X^{(n)\prime}(t)| \big| X^{(n)}(t) = u\big) p_{X^{(n)}(t)}(u)\, dt. \tag{3.6}$$

To finish the proof, let us take limits on both sides as $n \to +\infty$ in (3.6). By Ylvisaker's theorem (Theorem 1.21), with probability 1 there exists no local extrema at the level u. An instant reflection shows that this implies that a.s.,

$$N_u(X^{(n)}, I) \uparrow N_u(X; I),$$

so that the left-hand side of (3.6) tends to $\mathrm{E}\big(N_u(X, I)\big)$, using monotone convergence. On the other hand, as already mentioned, as $n \to +\infty$, the expectation and variance matrix of the pair $(X^{(n)}(t), X^{(n)\prime}(t))$ converge uniformly to those of $(X(t), X'(t))$, and this implies the convergence of the right-hand side of (3.6) to the corresponding expression for $X(t)$. This finishes the proof for $k = 1$.

STEP 2. For $k > 1$, let us denote by

$$C_u^k := C_u \times \cdots \times C_u \qquad \text{with } C_u = C_u(X, I)$$

the cartesian product of C_u k times by itself and

$$\mu(J) = \#(C_u^k \cap J),$$

the number of points of C_u^k belonging to J for each Borel subset J of I^k.

Let $D_k(I)$, the diagonal set of the cube I^k, be defined as

$$D_k(I) = \{(t_1, \ldots, t_k) : t_j \in I \text{ for } j = 1, \ldots, k$$
$$\text{and there exist } j, j', j \neq j' \text{ such that } t_j = t_{j'}\}. \tag{3.7}$$

It is easy to check that

$$N_u^{[k]} = \mu(I^k \backslash D_k(I)),$$

so it suffices to prove that

$$\mathrm{E}(\mu(J)) = \int_J A_{t_1,\ldots,t_k}(u, \ldots, u)\, dt_1 \cdots dt_k, \tag{3.8}$$

where

$$A_{t_1,\ldots,t_k}(u_1, \ldots, u_k) := \mathrm{E}\big(|X'(t_1) \ldots X'(t_k)| / X(t_1) = u_1, \ldots, X(t_k) = u_k\big)$$
$$p_{X(t_1),\ldots,X(t_k)}(u_1, \ldots, u_k) \tag{3.9}$$

for every compact rectangle $J = J_1 \times \cdots \times J_k$ contained in $I^k \backslash D_k(I)$ (which amounts to saying that the closed intervals $J_1, \ldots, J_k$ are pairwise disjoint). In fact, if this is proved, the two Borel measures

$$J \rightsquigarrow \mathrm{E}(\mu(J))$$
$$J \rightsquigarrow \int_J A_{t_1,\ldots,t_k}(u, \ldots, u)\, dt_1 \cdots dt_k$$

coincide on these rectangles, hence on all Borel subsets of $I^k \setminus D_k(I)$. So,

$$\mathrm{E}\big(N_u^{[k]}\big) = \int_{I^k \backslash D_k(I)} A_{t_1,\ldots,t_k}(u, \ldots, u)\, dt_1 \cdots dt_k.$$

This proves (3.3), since $D_k(I)$ has Lebesgue measure zero.

To end, let us turn to the proof of (3.8). We use the same arguments as in step 1: First, we prove the equality for the polygonal approximation using Kac's formula, and second, a similar domination argument makes it possible to pass to the limit as one refines the partition. □

Remark. A by-product of the Rice formula for $k = 1$ in the Gaussian case is that under the conditions of the theorem, $\mathrm{E}(N_u)$ is finite. This follows from the fact that the right-hand side of (3.3) is finite when $k = 1$ since it is the integral of a bounded function on a bounded interval. For $k > 1$, both sides in (3.3) can be infinite.

3.1.2. Non-Gaussian Case

For general processes, Lemma 3.1 will still be useful to get an upper bound for $\mathrm{E}(N_u(X, I))$ via Fatou's lemma. The next result will be helpful in the opposite direction.

Lemma 3.3. *Let f be a function that satisfies H_{1u}, and let $0 < \varepsilon < \delta < (t_2 - t_1)/2$. Let ψ be a real-valued function of one real variable, of class C^1, with support contained in $[-1, 1]$, $\psi(s) \geq 0$, $\int_{\mathbb{R}} \psi(s)\, ds = 1$. We define $\psi_\varepsilon(s) = (1/\varepsilon)\psi(s/\varepsilon)$, and for each locally integrable function g,*

$$g_\varepsilon(t) = (\psi_\varepsilon * g)(t) = \int_{\mathbb{R}} \psi_\varepsilon(t-s) g(s)\, ds,$$

the convolution of g with the approximation of unity ψ_ε. Then

$$N_u(f; I) \geq \int_{I_{-\delta}} \left| g'_\varepsilon(t) \right| dt, \tag{3.10}$$

where

$$g(t) = \mathbf{1}_{(u,+\infty)}(f(t)) \quad \text{and} \quad I_{-\delta} = [t_1 + \delta, t_2 - \delta].$$

Proof. By a duality argument, it suffices to show that

$$N_u(f; I) \geq \int_{I_{-\delta}} v(t)\, g'_\varepsilon(t)\, dt \tag{3.11}$$

for every C^1-function v with support in $I_{-\delta}$ and such that $\|v\|_\infty \leq 1$.

Let $\{h_m\}_{m=1,2,\ldots}$ be a sequence of C^1-functions that approximate the step function $\mathbf{1}_{x>u}$. More precisely, h_m is monotone increasing in the broad sense and $h_m(x) = 0$ for $x \leq u$, $h_m(x) = 1$ for $x \geq u + 1/m$.

Applying dominated convergence, we obtain

$$\begin{aligned}\int_{I_{-\delta}} v(t) g'_\varepsilon(t)\, dt &= \int_{I_{-\delta}} v(t)\, dt \int_{\mathbb{R}} \psi'_\varepsilon(t-s) \mathbf{1}_{(u,+\infty)}(f(s))\, ds \\ &= \lim_{m\to\infty} \int_{I_{-\delta}} v(t)\, dt \int_{\mathbb{R}} \psi'_\varepsilon(t-s) h_m(f(s))\, ds.\end{aligned}$$

Integrate by parts, use Fubini's theorem, and observe that if $t \in I_{-\delta}$ and $t - s$ is in the support of ψ_ε, s must be in I:

$$\int_{I_{-\delta}} v(t) g'_\varepsilon(t)\, dt = \lim_{m\to\infty} \int_{I_{-\delta}} v(t)\, dt \int_{\mathbb{R}} \psi_\varepsilon(t-s) h'_m(f(s)) f'(s)\, ds$$

$$= \lim_{m\to\infty} \int_{\mathbb{R}} h'_m(f(s))f'(s)\,ds \int_{I_{-\delta}} v(t)\psi_\varepsilon(t-s)\,dt$$

$$\le \overline{\lim_{m\to\infty}} \int_I h'_m(f(s))\,\left|f'(s)\right|\,ds.$$

The hypothesis H_{1u}, plus an argument similar to the one in Lemma 3.1, show that for m large enough,

$$\int_I h'_m(f(s))\,\left|f'(s)\right|\,ds = N_u(f;I).$$

This shows (3.11). □

Next, we are going to impose a certain number of hypotheses on the stochastic process $\mathcal{X} = \{X(t) : t \in I\}$ for which we will state and prove the Rice formulas. They are the following:

(A1) The paths of $\mathcal{X}$ are of class C^1.

(A2k) Let k be a positive integer. For any choice of the k-tuples $(t_1, \ldots, t_k)$, $(t'_1, \ldots, t'_k) \in I^k \backslash D_k(I)$, where $D_k(I)$ is the diagonal set defined in (3.7), the random vector $(X(t_1), \ldots, X(t_k), X'(t'_1), \ldots, X'(t'_k))$ has a density in $\mathbb{R}^{2k}$ denoted by

$$p_{X(t_1),\ldots,X(t_k),X'(t'_1),\ldots,X'(t'_k)}(x_1, \ldots, x_k, x'_1, \ldots, x'_k). \tag{3.12}$$

We also define

$$I_k(x_1, \ldots, x_k) := \int_{I^k} A_{t_1,\ldots,t_k}(x_1, \ldots, x_k)\,dt_1 \cdots dt_k,$$

where $A_{t_1,\ldots,t_k}(x_1, \ldots, x_k)$ has already been defined in the proof of the Rice formula in the Gaussian case. Notice that in the general case, this function is only defined for almost every point $(x_1, \ldots, x_k)$. We are assuming that it has a continuous version, and it is this version that appears in what follows.

These integrals may have the value $+\infty$, but are always well defined. We will assume that the density (3.12) is a continuous function of $(x_1, \ldots, x_k)$ at the point $(u, \ldots, u)$ (the other variables remaining constant) and of $(t_1, \ldots, t_k)$ in $I^k \backslash D_k(I)$ (the other variables remaining constant). We also assume that $p_{X(t)}(x)$ is continuous for $t \in I$ and x in a neighborhood of u.

(A3k) The function $(t_1, \ldots, t_k, x_1, \ldots, x_k) \rightsquigarrow A_{t_1,\ldots,t_k}(x_1, \ldots, x_k)$ is assumed to be continuous for $(t_1, \ldots, t_k)$ in $I^k \backslash D_k(I)$ and $x_1, \ldots, x_k$ in a neighborhood of $(u, \ldots, u)$.

(A4k)

$$\int_{R^3} |x_1'|^{k-1} |x_2' - x_3'|$$
$$p_{X(t_1),\ldots,X(t_k),X'(t_1'),X'(t_2'),X'(t_1)}(x_1,\ldots,x_k,x_1',x_2',x_3')\,dx_1'\,dx_2'\,dx_3'$$

tends to zero as $t_2' - t_1 \to 0$ uniformly for $(t_1,\ldots,t_k)$ in a compact subset of $I^k \backslash D_k(I)$ and $x_1,\ldots,x_k$ in a neighborhood of $(u,\ldots,u)$.

Theorem 3.4 (Rice's Formula). *If $\mathcal{X}$ satisfies* (A1), (A2*k*), (A3*k*), *and* (A4*k*), *then*

$$\mathrm{E}\big(N_u^{[k]}\big) = \int_{I^k} \mathrm{E}\big(|X'(t_1)\cdots X'(t_k)|\big|X(t_1) = u\cdots X(t_k) = u\big)$$
$$\times\, p_{X(t_1),\ldots,X(t_k)}(u,\ldots,u)\,dt_1\cdots dt_k. \tag{3.13}$$

Proof. Using the same arguments as in the proof of Theorem 3.2, it is sufficient to prove that

$$\mathrm{E}\big(\mu(J)\big) = \int_J A_{t_1,\ldots,t_k}(u,\ldots,u)\,dt_1\cdots dt_k \tag{3.14}$$

for every compact rectangle $J = J_1 \times \cdots \times J_k$ contained in $I^k \backslash D_k(I)$ [for $k = 1$, we put $D_k(I) = \emptyset$].

First we use Lemma 3.1. It is easy to check that a.s. the paths of the process satisfy hypothesis H_{1u}.

$$\mu(J) = \prod_{i=1}^{k} N_u(X, J_i) = \lim_{\delta\to 0}\frac{1}{(2\delta)^k}\prod_{i=1}^{k}\left[\int_{J_i} \mathbb{I}_{\{|X(t_i)-u|<\delta\}}\,|X'(t_i)|\,dt_i\right].$$

By Fatou's lemma, the definition of $A_{t_1,\ldots,t_k}(x_1,\ldots,x_k)$, and hypothesis (A3k), we obtain

$$\mathrm{E}\big(\mu(J)\big) \le \liminf_{\delta\to 0}\frac{1}{(2\delta)^k}\int_J dt_1\cdots dt_k$$
$$\int_{u-\delta}^{u+\delta}\cdots\int_{u-\delta}^{u+\delta} A_{t_1,\ldots,t_k}(x_1,\ldots,x_k)\,dx_1\cdots dx_k$$
$$= \int_J A_{t_1,\ldots,t_k}(u,\ldots,u)\,dt_1\cdots dt_k. \tag{3.15}$$

The converse inequality is somewhat more complicated. We apply Lemma 3.3 to each of the intervals $J_1, \ldots, J_k$. We have

$$\mathrm{E}\big(\mu(J)\big) = \mathrm{E}\left(\prod_{i=1}^{k} N_u(X, J_i)\right) \geq \mathrm{E}\left(\prod_{i=1}^{k}\left[\int_{(J_i)_{-\delta}} |g'_\varepsilon(t_i)|\, dt_i\right]\right),$$

where $g(t) = 1_{(u,+\infty)}(X(t))$.

Define the sequence of functions $\{h_m\}_{m=1,2,\ldots}$ as in the proof of Lemma 3.3. Dominated convergence, Fubini's theorem, and integration by parts provide

$$\mathrm{E}\big(\mu(J)\big) \geq \lim_{m\to\infty} \int_{J_{-\delta}} dt_1 \cdots dt_k \; E\left(\prod_{i=1}^{k}\left|\int_{\mathbb{R}} \psi_\varepsilon(t_i - s_i) h'_m(X(s_i)) X'(s_i)\, ds_i\right|\right) \tag{3.16}$$

with $J_{-\delta} = (J_1)_{-\delta} \times \cdots \times (J_k)_{-\delta}$.

To obtain a lower bound for the mathematical expectation on the right-hand side of the last inequality, we use

$$\prod_{i=1}^{k} a_i \geq \prod_{i=1}^{k} b_i - \sum_{i=1}^{k} b_1 \cdots b_{i-1} c_i a_{i+1}, \ldots, a_k, \tag{3.17}$$

which holds true whenever $a_i, b_i, c_i \geq 0$ and $a_i \geq b_i - c_i$ for $i = 1, \ldots, k$. One can check (3.17) by induction; this is left to the reader.

We apply (3.17) with

$$\begin{aligned} a_i &= \left|\int_{\mathbb{R}} \psi_\varepsilon(t_i - s_i) h'_m(X(s_i)) X'(s_i)\, ds_i\right| \\ b_i &= \int_{\mathbb{R}} \psi_\varepsilon(t_i - s_i) h'_m(X(s_i))\; \left|X'(t_i)\right| ds_i \\ c_i &= \int_{\mathbb{R}} \psi_\varepsilon(t_i - s_i) h'_m(X(s_i))\; \left|X'(s_i) - X'(t_i)\right| ds_i. \end{aligned}$$

For the remaining part, choose $\varepsilon > 0$ small enough so that it is sufficient to consider the k-tuple $(s_1, \ldots, s_k)$ in the integral in (3.16) as varying in a compact subset of $I^k \backslash D_k(I)$ outside which the integrand is equal to zero. This can be done given that the distance between J and the diagonal $D_k(I)$ is strictly positive and the support of ψ_ε is contained in $[-\varepsilon, \varepsilon]$.

Consider the expectation of the first term of (3.17):

$$\begin{aligned} \mathrm{E}\left(\prod_{i=1}^{k} b_i\right) = \int_{\mathbb{R}^k} \left[\prod_{i=1}^{k} \psi_\varepsilon(t_i - s_i)\, ds_i\right] \int_{\mathbb{R}^k \times \mathbb{R}^k} \prod_{i=1}^{k} \left[h'_m(x_i)\, |x'_i|\right] \\ \cdot\, p_{X(s_1),\ldots,X(s_k),X'(t_1),\ldots,X'(t_k)}(x_1, \ldots, x_k, x'_1, \ldots, x'_k) \\ \times\, dx_1 \cdots dx_k\, dx'_1 \cdots dx'_k. \end{aligned}$$

Let $m \to \infty$ and $\varepsilon \downarrow 0$ (in this order). Using hypotheses (A2k) and (A3k) and Fatou's lemma, we get

$$\liminf_{\varepsilon \downarrow 0} \liminf_{m \to \infty} \mathrm{E}\left(\prod_{i=1}^{k} b_i\right) \geq A_{t_1,\ldots,t_k}(u, \ldots, u). \tag{3.18}$$

We now consider the expectation of each term of the sum on the right-hand side of (3.17):

$$\begin{aligned}
&\mathrm{E}\,(b_1 \ldots b_{i-1} c_i a_{i+1}, \ldots, a_k) \\
&\leq \mathrm{E}\Bigg(\int_{\mathbb{R}^k}\left[\prod_{h=1}^{k} \psi_\varepsilon(t_h - s_h) h'_m(X(s_h))\, ds_h\right] \\
&\quad \times \left[\prod_{h=1}^{i-1} |X'(t_h)|\right]\left[\prod_{h=i+1}^{k} |X'(s_h)|\right] |X'(s_i) - X'(t_i)|\Bigg) \\
&= \int_{\mathbb{R}^k} \prod_{h=1}^{k} [\psi_\varepsilon(t_h - s_h)\, ds_h] \int_{\mathbb{R}^k \times \mathbb{R}^{k+1}} \\
&\quad \times \left[\prod_{h=1}^{k} h'_m(x_h)\, dx_h\right]\left[\prod_{h=1, h\neq i}^{k} |x'_h|\right] |x'_i - y'_i| \\
&\quad \cdot p_{X(s_1),\ldots,X(s_k),X'(t_1),\ldots,X'(t_{i-1}),X'(s_i),X'(t_i),X'(s_{i+1}),\ldots,X'(s_k)} \\
&\qquad (x_1, \ldots, x_k, x'_1, \ldots, x'_{i-1}, x'_i, y'_i, x'_{i+1}, \ldots, x'_k)\, dx'_1 \cdots dx'_k\, dy'_i.
\end{aligned}$$

We use the trivial bound

$$\prod_{\ell=1, \ell\neq i}^{k} |x'_\ell| \leq \sum_{\ell=1, \ell\neq i}^{k} |x'_\ell|^{k-1}$$

and integrate in the variables x'_h $(h = 1, \ldots, k; h \neq i, \ell)$. We obtain

$$\begin{aligned}
&\mathrm{E}(b_1 \ldots b_{i-1} c_i a_{i+1}, \ldots, a_k) \\
&\leq \sum_{\ell=1, \ell\neq i}^{k} \int_{\mathbb{R}^k} \prod_{h=1}^{k} [\psi_\varepsilon(t_h - s_h)\, ds_h] \int_{\mathbb{R}^k} \left[\prod_{h=1}^{k} h'_m(x_h)\, dx_h\right] \int_{\mathbb{R}^3} |x'_\ell|^{k-1} |x'_i - y'_i| \\
&\quad \cdot p_{X(s_1),\ldots,X(s_k),X'(\tau_\ell),X'(s_i),X'(t_i)}(x_1, \ldots, x_k, x'_\ell, x'_i, y'_i)\, dx'_\ell\, dx'_i\, dy'_i,
\end{aligned} \tag{3.19}$$

where we have set

$$\tau_\ell = \begin{cases} t_\ell & \text{when } \ell = 1, \ldots, i-1 \\ s_\ell & \text{when } \ell = i+1, \ldots, k. \end{cases}$$

Now, if we choose m large enough, since the integrand on the right-hand side of (3.19) is zero when $|x_h - u| \geq 1/m$ for some $h = 1, \ldots, k$, we can use hypothesis (A4k), and the inner integral in (3.19) is uniformly small if $|s_i - t_i|$ is small. This shows that the right-hand side in (3.19) tends to zero as $\varepsilon \downarrow 0$, and based on (3.16), (3.17), and (3.18), we obtain

$$\mathrm{E}\left(\mu(J)\right) \geq \int_{J_{-\delta}} A_{t_1,\ldots,t_k}(u, \ldots, u)\, dt_1 \cdots dt_k.$$

The converse inequality to (3.15) follows by making $\delta \downarrow 0$. □

3.2. VARIANTS AND EXAMPLES

1. *Another form of Rice's formulas.* Under the hypotheses of Theorem 3.4, one can write a Rice formula for the k-factorial moment of crossings in the form

$$\mathrm{E}\left(N_u(N_u - 1) \cdots (N_u - k + 1)\right) = \int_{I^k} dt_1 \cdots dt_k \int_{\mathbb{R}^k} \left[\prod_{i=1}^{k} |x_i'|\right] \cdot$$

$$\cdot\, p_{X(t_1),\ldots,X(t_k),X'(t_1),\ldots,X'(t_k)}(u, \ldots, u, x_1', \ldots, x_k')\, dx_1' \cdots dx_k'.$$

2. *Factorial and ordinary moments.* A simple remark is that the ordinary moments of the random variable N_u [i.e., $\mathrm{E}(N_u^k)$] are linear combinations of the factorial moments $\mathrm{E}\big(N_u^{[j]}\big)$, $j = 1, \ldots, k$ with fixed integer coefficients depending only on k and j, and conversely. Hence, one can express $\mathrm{E}\big(N_u^k\big)$ as linear combinations of multiple integrals of the form that appear in Rice's formulas.

3. *First moments.* In the case $k = 1$ we have the formula

$$\mathrm{E}\big(N_u\big) = \int_I dt \int_{\mathbb{R}} |x'|\, p_{X(t),X'(t)}(u, x')\, dx'. \tag{3.20}$$

Essentially the same proof that we have given above for the Rice formula in the general non-Gaussian case works when $k = 1$ under slightly weaker hypotheses [and easier to check, especially (A4k)]. The reader may check that (3.20) holds true if:

(i) $(t, x) \rightsquigarrow p_{X(t)}(x)$ is continuous for $t \in I$, x in a neighborhood of u.
(ii) $(t, x, x') \rightsquigarrow p_{X(t),X'(t)}(x, x')$ is continuous for $t \in I$ x in a neighborhood of u and $x' \in \mathbb{R}$.
(iii) $\mathrm{E}\big(\omega_{X'}(I, \delta)\big) \to 0$ as $\delta \to 0$, where $\omega_{X'}(I, \delta)$ denotes the modulus of continuity of $X'(\cdot)$.

4. *Gaussian stationary processes*. This is an important case, in which we mention the simple classical formula (3.2). Suppose that the process $\mathcal{X}$ is centered Gaussian stationary with C^1-paths and covariance $\Gamma(t) = \mathrm{E}(X(s)X(s+t))$ normalized by $\Gamma(0) = 1$. It is clear that $\Gamma'(0) = \mathrm{E}(X(t)X'(t)) = 0$, since Γ has a maximum at the point $t = 0$. Given that the joint distribution of $X(t)$ and $X'(t)$ is Gaussian, this implies that for each t, $X(t)$ and $X'(t)$ are independent random variables. Hence

$$p_{X(t),X'(t)}(u,x') = p_{X(t)}(u)p_{X'(t)}(x') = \frac{1}{\sqrt{2\pi}}e^{-(1/2)u^2}\frac{1}{\sqrt{2\pi\lambda_2}}e^{-(1/2)(x'^2/\lambda_2)}$$

[notice that $\lambda_2 = -\Gamma''(0) = \mathrm{E}\left([X'(t)]^2\right)$]. Substituting into (3.20), we get (3.2). Formula (3.2) remains valid if we only require the Gaussian centered stationary process to have continuous paths (see Exercise 3.2).

5. *General Gaussian processes*. Verifying hypotheses (A1), (A2k), (A3k), and (A4k) for non-Gaussian processes can be a nontrivial task. For Gaussian processes, this approach is tractable, as shown by the next proposition, which we include to see how the verification of the general hypotheses can be performed in this case. Of course, this has a limited interest, since the direct approach for Gaussian processes, as we have seen, is simpler and permits us to deduce Rice formulas under weaker conditions.

Proposition 3.5. *If $\mathcal{X}$ is a real-valued centered Gaussian process defined on a compact interval I of the real line, has C^1-paths, and the densities in* (A1), (A2*k*), (A3*k*), *and* (A4*k*) *do not degenerate for a given k, then* (A1), (A2*k*), (A3*k*), *and* (A*k*4) *are verified.*

Proof. Let us recall that the functions $(s,t) \rightsquigarrow \mathrm{E}(X(s)X(t))$, $(s,t) \rightsquigarrow \mathrm{E}\,(X(s)X'(t))$, and $(s,t) \rightsquigarrow E(X'(s)X'(t))$ are continuous, so that the densities, since they do not degenerate, are also continuous. We have

$$A_{t_1,\ldots,t_k}(x_1,\ldots,x_k) = \mathrm{E}\left\{\prod_{i=1}^{k}\left|X'(t_i)\right| / X(t_1) = x_1,\ldots,X(t_k) = x_k\right\}$$
$$\times p_{X(t_1),\ldots,X(t_k)}(x_1,\ldots,x_k) \tag{3.21}$$

and the expression in hypothesis (A4k) is

$$\mathrm{E}\left\{\left|X'(t_1')\right|^{k-1}\left|X'(t_2') - X'(t_1)\right| / X(t_1) = x_1,\ldots,X(t_k) = x_k\right\}$$
$$\times p_{X(t_1),\ldots,X(t_k)}(x_1,\ldots,x_k). \tag{3.22}$$

If $(t_1,\ldots,t_k)$ varies in a compact subset of $I^k \backslash D_k(I)$ and $(x_1,\ldots,x_k)$ in a neighborhood of u, the density $p_{X(t_1),\ldots,X(t_k)}(x_1,\ldots,x_k)$ is continuous and bounded.

We want to get rid of the conditional expectation in both expressions (3.21) and (3.22). For this purpose we use linear regression (see Chapter 1) and write

$$X'(t) = X'(t) - \sum_{h=1}^{k} c_h(t)X(t_h) + \sum_{h=1}^{k} c_h(t)X(t_h)$$

and choose $c_h(t)$ ($h = 1, \ldots, k$) in such a way that

$$Y(t) = X'(t) - \sum_{h=1}^{k} c_h(t)X(t_h)$$

will be orthogonal to the components of the random vector $(X(t_1), \ldots, X(t_k))$.

Denote by $\Gamma(s,t)$ the covariance of the given process $\mathcal{X}$, $\Sigma = ((\Gamma(t_h, t_\ell))_{h,\ell=1,\ldots,k}$, and $\gamma(t) = (\Gamma_1(t,t_1), \ldots, \Gamma_1(t,t_k))^{\mathrm{T}}$, where we have used Γ_1 for the partial derivative of Γ with respect to the first variable.

The orthogonality condition is

$$\gamma(t) = \sum c(t) \qquad \text{with } c(t) = (c_1(t), \ldots, c_k(t))^{\mathrm{T}},$$

so that $c(t) = \sum^{-1} \gamma(t)$. The nondegeneracy hypothesis implies that the function

$$(t, t_1, \ldots, t_k) \rightsquigarrow c(t) \text{ is continuous for } (t, t_1, \ldots, t_k) \in I \times I^k \backslash D_k(I).$$

The conditional expectations in (3.21) and (3.22) become, respectively, the unconditional expectations

$$\mathrm{E}\left(\prod_{i=1}^{k}\left|Y(t_i) + \sum_{h=1}^{k} c_h(t_i)x_h\right|\right)$$

$$\mathrm{E}\left(\left|Y(t_1') + \sum_{h=1}^{k} c_h(t_1')x_h\right|^{k-1} \left|Y(t_2') + \sum_{h=1}^{k} c_h(t_2')x_h - Y(t_1) - \sum_{h=1}^{k} c_h(t_1)x_h\right|\right).$$

To verify hypotheses (A3k) and (A4k) and finish the proof, one can now pass to the limit under the expectation, using dominated convergence, due to the integrability of the moments of the supremum of Gaussian process. This is left to the reader. □

6. *Stationary Gaussian processes, nondegeneracy condition*. Let us consider a Gaussian process defined on an interval of the real line, and $t_1, \ldots, t_n$, n distinct parameter values. It is in general nontrivial to check whether the random variables $X(t_1), \ldots, X(t_n)$ have a nondegenerate joint distribution. Similar difficulties appear if one is willing to prove nondegeneracy of the joint distribution of the process and its derivative.

However, in the stationary case, one can give the following sufficient condition for nondegeneracy to hold: Support of the spectral measure μ^X of the process has some accumulation point. In particular, this happens if μ^X is not purely atomic, that is, if there does not exist a countable subset A of the reals such that $\mu^X(A^C) = 0$ (see Exercises 3.4 and 3.5).

7. *Finiteness of moments of crossings*. It may happen that Rice formula (3.13) holds true but that both sides are infinite. In a certain number of applications, one wants to know whether $\mathrm{E}(N_u^k)$ is finite but is not much interested in its value. On the other hand, the standard situation is that to compute the right-hand side of (3.13) or even to obtain good upper bounds for it can be a very complicated or actually intractable problem.

From a numerical point of view, the general question of efficient procedures to compute approximately the moments of the number of crossings remains wide open. We will come back to this subject in Chapters 4 and 5, adding some more or less recent results. This will be done in the context of relating crossings to the distribution of the maximum of a stochastic process, even though it has an independent interest.

Finiteness of moments of crossings of Gaussian processes has been considered by Belayev (1966), Miroshin (1977), and Cuzick (1975). For stationary Gaussian processes, the sufficient condition for finiteness of the variance of $N_0(X, I)$ in Exercise 3.6 is in Cramér and Leadbetter's book (1967), where an explicit formula for the variance is also given. Geman (1972) proved that this condition is also necessary for finiteness at the level $u = 0$. In a recent paper, Kratz and León (2006) proved that the same condition is necessary and sufficient for any level u and also for the number of crossings with some differentiable curves.

For non-Gaussian processes, sufficient conditions have been given by Besson and Wschebor (1983). The next theorem gives sufficient conditions, which are reasonably easy to check in specific cases, to be able to assure the finiteness of $\mathrm{E}(N_u^k)$. It is taken from Nualart and Wschebor (1991).

Theorem 3.6. *Let m be a positive real number. Consider a real-valued stochastic process $\mathcal{X} = \{X(t) : t \in I\}$ defined on a compact interval I of the real line, with paths of class C^{p+1}, $p > 2m$. We assume also that for each $t \in I$ the random variable $X(t)$ has a density and that for some $\eta > 0$,*

$$C = \sup\{p_{X(t)}(x) : t \in I,\ x \in [u - \eta, u + \eta]\} < \infty.$$

Then

$$\mathrm{E}(N_u^m) \le C_{p,m}\big[1 + C + \mathrm{E}(\|X^{(p+1)}\|_\infty)\big], \tag{3.23}$$

where $C_{p,m}$ is a constant depending only on p, m, and the length of the interval I.

Proof. With no loss of generality, we assume that $I = [0, 1]$. Using a standard bound for the expectation of nonnegative random variables,

$$\mathrm{E}\left(N_u^m\right) \leq \sum_{k=0}^{\infty} \mathrm{P}(N_u^m \geq k) \leq C_{p,m} + C_{p,m} \sum_{k>p^m} \mathrm{P}(N_u^m \geq (p+1)^m k)$$
$$= C_{p,m} + C_{p,m} \sum_{k>p^m} \mathrm{P}(N_u \geq (p+1)k^{1/m}).$$

Our aim is to give an upper bound for this sum.

We have the inclusion of events (use Rolle's theorem; here $|J|$ denotes the length of the interval J):

$$\begin{aligned} &\left\{N_u(X, I) \geq (p+1)k^{1/m}\right\} \\ &\subset \left\{\exists \text{ an interval } J \subset I, |J| = k^{-1/m}, N_u(X, J) \geq (p+1)\right\} \\ &\subset \left\{ \begin{array}{c} \exists \text{ an interval } J \subset I, |J| = k^{-1/m}, \text{ and points } \tau_1, \ldots, \tau_p \in J \\ \text{such that } X^{(j)}(\tau_j) = 0 \text{ for } j = 1, \ldots, p \end{array} \right\}. \end{aligned} \tag{3.24}$$

Let $\{\varepsilon_k\}_{k=1,2,\ldots}$ be a sequence of positive real numbers (that we will choose afterward) and denote for $k = 1, 2, \ldots$:

$$A_k = \left\{N_u(X, I) \geq (p+1)k^{1/m}\right\} \cap \left\{\omega_{X^{(p)}}(I, k^{-1/m}) < \varepsilon_k\right\}$$

[as usual, $\omega_f(I, \delta)$ denotes the continuity modulus].

Let us consider the random open set in I:

$$V_k = \left\{t : \left|X'(t)\right| < \varepsilon_k\, k^{-(p-1)/m}\right\}.$$

We show that

$$A_k \subset \{N_u(X, V_k) \geq (p+1)\}.$$

To prove this, it suffices to prove that if A_k occurs, the random interval J that appears in (3.24) is contained in V_k, since in that case, the number of roots of the equation $X(t) = u$ that belong to V_k will be greater than or equal to $N_u(X, J) \geq p + 1$.

Suppose that A_k occurs and that $t \in J$. Then

$$\left|X^{(p)}(t)\right| = \left|X^{(p)}(t) - X^{(p)}(\tau_p)\right| < \varepsilon_k$$

given that $X^{(p)}(\tau_p) = 0$, given the definition of A_k, and given that $|J| = k^{-1/m}$. Similarly, using the mean value theorem yields

$$\left|X^{(p-1)}(t)\right| = \left|X^{(p-1)}(t) - X^{(p-1)}(\tau_{p-1})\right| < \varepsilon_k\, k^{-1/m}.$$

In the same way, we can increase step by step the order of the derivative and obtain

$$\left|X'(t)\right| < \varepsilon_k\, k^{-(p-1)/m},$$

which shows that $t \in V_k$. It follows that

$$\mathrm{P}(N_u \geq (p+1)k^{1/m}) \leq \mathrm{P}(\omega_{X^{(p)}}(I, k^{-1/m}) \geq \varepsilon_k) + \mathrm{P}(A_k). \tag{3.25}$$

The first term in (3.25) is bounded by

$$\mathrm{P}\big(\|X^{(p+1)}\|_\infty \geq k^{1/m}\varepsilon_k\big) \leq k^{-1/m}\varepsilon_k^{-1}\mathrm{E}\big(\|X^{(p+1)}\|_\infty\big).$$

As for the second term in (3.25),

$$\mathrm{P}(A_k) \leq \mathrm{P}(N_u(X, V_k)) \geq (p+1)) \leq \frac{1}{p+1}\mathrm{E}\left(N_u(X, V_k)\right).$$

One can check as an exercise that Lemma 3.1 holds true, *mutatis mutandis*, whenever the set in which the function is defined is an open set in the real line (as is the case for V_k) instead of an interval. Hence, a.s.

$$N_u(X, V_k) = \lim_{\delta\to 0}\frac{1}{2\delta}\int_{V_k} \mathbf{I}_{\{|X(t)-u|<\delta\}} \left|X'(t)\right| dt,$$

and applying Fatou's lemma and the definition of the set V_k we obtain

$$\begin{aligned}
\mathrm{E}\left(N_u(X, V_k)\right) &\leq \liminf_{\delta\to 0}\frac{1}{2\delta}\mathrm{E}\left\{\int_{V_k} \mathbf{I}_{\{|X(t)-u|<\delta\}} \left|X'(t)\right| dt\right\} \\
&\leq \liminf_{\delta\to 0}\frac{1}{2\delta}\mathrm{E}\left\{\int_{I} \mathbf{I}_{\{|X(t)-u|<\delta\}}\, \varepsilon_k\, k^{-(p-1)/m} dt\right\} \\
&\leq \varepsilon_k\, k^{-(p-1)/m} \liminf_{\delta\to 0}\frac{1}{2\delta}\int_I dt \int_{u-\delta}^{u+\delta} p_{X(t)}(x)\,dx \leq C\varepsilon_k k^{-(p-1)/m}.
\end{aligned}$$

Substituting in (3.25) and then in the upper bound for $\mathrm{E}(N_u^m)$, we obtain

$$\mathrm{E}(N_u^m) \leq C_{p,m} + C_{p,m}\left[C\sum_{k=1}^{\infty}\varepsilon_k k^{-(p-1)/m} + \mathrm{E}\big(\|X^{(p+1)}\|_\infty\big)\sum_{k=1}^{\infty} k^{-1/m}\varepsilon_k^{-1}\right].$$

Choosing

$$\varepsilon_k = k^{\beta - 1/m} \qquad \text{with} \quad 1 < \beta < \frac{p}{m} - 1,$$

which is possible since $p/m > 2$, the two series converge and we have the statement of the theorem, with a new constant $C_{p,m}$. □

Corollary 3.7. *If $\mathcal{X}$ is Gaussian with $\mathcal{C}^\infty$-paths and* $\mathrm{Var}(X(t)) \geq a > 0$ *for $t \in I, x \in \mathbb{R}$, then*

$$\mathrm{E}\big(N_u^m\big) < \infty$$

for every $u \in \mathbb{R}$ and every $m = 1, 2, \ldots$.

Proof. Using the results in Chapter 2, we know that $\mathrm{E}\big(\big\|X^{(p+1)}\big\|_\infty\big) < \infty$ for every $p = 1, 2, \ldots$ and $p_{X(t)}(x) \leq 1/\sqrt{2\pi a}$ for $t \in I$, $x \in \mathbb{R}$. □

8. *Variations on Rice formulas*. In applications one frequently needs a certain number of variants of formula (3.13). We give some examples here.

(a) If instead of all crossings we consider only up-crossings or down-crossings, under the same hypotheses as in Theorem 3.4, we obtain the following formulas:

$$\begin{aligned}\mathrm{E}(U_u^{[k]}) &= \mathrm{E}(U_u(U_u - 1)\cdots(U_u - k + 1)) \\ &= \int_{I^k} \mathrm{E}\left(\prod_{i=1}^k X'^{+}(t_i)/X(t_1) = u, \ldots, X(t_k) = u\right) \\ &\quad \times p_{X(t_1),\ldots,X(t_k)}(u, \ldots, u)\, dt_1 \ldots dt_k\end{aligned} \tag{3.26}$$

$$\begin{aligned}\mathrm{E}(D_u^{[k]}) &= \mathrm{E}(D_u(D_u - 1)\cdots(D_u - k + 1)) \\ &= \int_{I^k} \mathrm{E}\left(\prod_{i=1}^k X'^{-}(t_i)/X(t_1) = u, \ldots, X(t_k) = u\right) \\ &\quad \times p_{X(t_1),\ldots,X(t_k)}(u, \ldots, u)\, dt_1 \cdots dt_k.\end{aligned} \tag{3.27}$$

The proofs are exactly the same.

(b) If instead of counting crossings we count *marked crossings*, that is, points t such that $X(t) = u$ and in which some other event is happening, we obtain various Rice-type formulas. For example, let $\{Y(t) : t \in I\}$ be a second stochastic process, a and b extended real numbers (they may take the values $-\infty$ or $+\infty$), $a < b$, and define

$$N_u(X, Y; I, a, b) = \#\{t : t \in I,\ X(t) = u,\ a < Y(t) \leq b\}.$$

Then

$$\mathrm{E}\big(N_u(X, Y; I, a, b)\big) = \int_I dt \int_a^b dy \int_{\mathbb{R}} |x'|\ p_{X(t),X'(t),Y(t)}(u, x', y)\, dx'. \tag{3.28}$$

Formulas similar to (3.13) can be written for the factorial moments of $N_u(X, Y; I, a, b)$. We leave it to the reader to establish hypotheses and give proofs. These follow the same lines as in Theorem 3.4.
A typical application is the computation of moments of:

$$M(X, I, a, b) = \#\{t : t \in I,\ X(\cdot) \text{ has a local maximum at } t,\\ a < X(t) \le b\},$$

where $a < b$. Under general conditions (the statement of which is left to the reader), a.s.,

$$M(X, I, a, b) = \#\left\{t : t \in I,\ X'(t) = 0,\ \ X''(t) < 0,\ a < X(t) \le b\right\}.$$

This means that a.s. $M(X, I, a, b)$ is the number of down-crossings of the level 0 by the stochastic process $\left\{X'(t) : t \in I\right\}$ in which the process $X(t)$ itself takes values between a and b. So we may apply (3.28) with X' instead of X, X instead of Y, and down-crossings instead of crossings, to get

$$\mathrm{E}\big(M(X, I, a, b)\big) = \int_I dt \int_a^b dx \int_{-\infty}^0 \left|x''\right|\, p_{X(t),X'(t),X''(t)}(x, 0, x'')\, dx''.$$

Again, similar formulas hold true for higher factorial moments, under appropriate hypotheses on the process.

(c) If ξ is a bounded random variable (one can relax this condition), and if the stochastic process $\{X(t) : t \in I\}$ satisfies the hypotheses of Theorem 3.4, one has the more general equality

$$\mathrm{E}(\xi\, N_u^{[k]}) = \int_{I^k} \mathrm{E}\left(\xi \prod_{i=1}^{k} |X'(t_i)| \,\Big|\, X(t_1) = \cdots = X(t_k) = u\right)\\ \times p_{X(t_1),\ldots,X(t_k)}(u, \ldots, u)\, dt_1 \cdots dt_k. \qquad (3.29)$$

The proof is left to the reader.

EXERCISES

3.1. (*Kac counting formula*) Prove the following inequality, related to Lemma 3.1. Assume that:

- $f : I \to \mathbb{R}$, $I = [t_1, t_2]$ is an absolutely continuous function.
- $f(t_1) \neq u$, $f(t_2) \neq u$.

Then

$$N_u(f; I) \le \liminf_{\delta \to 0} \frac{1}{2\delta} \int_I \mathbf{I}_{\{|f(t)-u|<\delta\}} \left|f'(t)\right| dt.$$

3.2. Prove that formula (3.2) is always true in the following sense: Let $\{X(t) : t \in I\}$ be a centered Gaussian stationary process with continuous paths, defined on a compact interval I of the real line, normalized by $r(0) = 1$. Then, if λ_2 is finite, (3.2) holds true, and if λ_2 is infinite, $\mathrm{E}\{N_u\} = +\infty$. (This means that the remaining hypotheses are not necessary in this case.)

3.3. (*A simple example in which the Rice formula does not hold*) Let $X(t) = \xi t$, $t \in [-1, 1]$, where ξ is a standard normal random variable. Show that $\mathrm{E}(N_u^X)$ can be computed by means of the Rice formula if $u \neq 0$, but that the formula fails to hold for $u = 0$.

3.4. **(a)** Prove that if the process $\{X(t) : t \in \mathbb{R}\}$ is Gaussian stationary, has C^1-paths, and the support of the spectral measure has an accumulation point, the set of random variables $X(t_1), \ldots, X(t_n)$ has a joint non-degenerate distribution for any choice of the distinct parameter values $t_1, \ldots, t_n$.
Hint: With no loss of generality, one can assume that the process is centered. Denote $Y = (X(t_1), \ldots, X(t_n))^{\mathrm{T}}$. The variance of the Gaussian vector Y is $\Lambda = \mathrm{E}(YY^{\mathrm{T}})$. The aim is to prove that the quadratic form $F(z) = z^{\mathrm{T}}\Lambda z$, $z \in \mathbb{R}^n$ is positive definite. To prove it, show that $F(z)$ can be written in terms of the spectral measure μ of the process X by means of the formula

$$F(z) = \int_{\mathbb{R}} \left| \sum_{k=1}^{n} e^{it_k x} z_k \right|^2 \mu(dx).$$

Conclude that $F(z) > 0$ whenever $z \neq 0$, using the fact that the function $x \rightsquigarrow \sum_{k=1}^{n} e^{it_k x} z_k$ is analytic.

(b) Deduce that under the hypotheses in part (a), Rice formulas can be applied on any compact interval I and for any $k = 1, 2, \ldots$.

3.5. Assume that the process $\{X(t) : t \in \mathbb{R}\}$ verifies the hypotheses of Exercise 3.4(a) and moreover, that its paths are C^k-functions, k an integer, $k \ge 1$. Then, for any choice of distinct parameter values $t_1, \ldots, t_n$, the joint distribution of the random variables

$$X(t_1), \ldots, X(t_n), X'(t_1), \ldots, X'(t_n), \ldots, X^{(k)}(t_1), \ldots, X^{(k)}(t_n)$$

does not degenerate. *Hint*: Use the same method as in Exercise 3.4.

3.6. Let $\{X(t) : t \in \mathbb{R}\}$ be a centered Gaussian stationary process. Assume the Geman condition:

(G2) $\Gamma(t) = \mathrm{E}(X(s)X(s+t)) \neq \pm 1$ for $t > 0$.
(G2) $\Gamma(\tau) = 1 - \lambda_2 \tau^2/2 + \theta(\tau)$ with

$$\int \frac{\theta'(\tau)}{\tau^2} d\tau \text{ converging at } \tau = 0^+.$$

Prove that this condition is sufficient to have

$$\mathrm{E}\left\{[N_0(X, I)]^2\right\} < \infty$$

for any bounded interval I.

3.7. (**a**) Let $f : J \to \mathbb{R}$ be a C^1-function, J an interval in the reals, and δ a positive number. Prove that

$$\frac{1}{2\delta} \int_J \mathbb{1}_{\{|f(t)-u|\}} |f'(t)| dt \leq N_0(f', J) + 1.$$

(**b**) Let $\{X(t) : t \in I\}$, I a compact interval in the reals, be a stochastic process with C^1-paths. Let k be an integer, $k \geq 1$, and $u \in \mathbb{R}$. Assume that the function $A_{t_1,\ldots,t_k}(x_1, \ldots, x_k)$ is a continuous function of its $2k$ arguments when $(t_1, \ldots, t_k) \in I^k \backslash D_k(I)$ and $x_1, \ldots, x_k$ are in some neighborhood of u. Prove that if

$$\mathrm{E}\Big(\big[N_0(X', I)\big]^k\Big) < \infty,$$

then Rice formula (3.3) for the factorial moments of $N_0(X, I)$ holds true.

3.8. (**a**) (*The Banach formula*) Let f be a continuous function $\mathbb{R} \to \mathbb{R}$. Then the total variation $\mathrm{TV}(f,I)$ of f over an interval I is defined as

$$\mathrm{TV}(f, I) = \sup \sum_{k=0}^{m-1} |f(t_{k+1}) - f(t_k|,$$

where $I = [a, b]$, $a = t_0 < t_1 < \cdots < t_m = b$ is a partition of I and the sup is taken over all possible partitions of I. Prove that

$$\int_{-\infty}^{+\infty} N_u(f, I)\, du = \mathrm{TV}(f, I). \tag{3.30}$$

Both sides can be finite or infinite.
Hint: For each partition $t_0 < t_1 < \cdots < t_m$ of the interval I, put $L_k(u) = 1$ if $N_u(f, [t_k, t_{k+1}]) \geq 1$ and $L_k(u) = 0$; otherwise,

$k = 0, 1, \ldots, m-1$. Show that

$$\int_{-\infty}^{+\infty} \sum_{k=0}^{m-1} L_k(u)du = \sum_{k=0}^{m-1}(M_k - m_k),$$

where M_k (respectively, m_k) is the maximum (respectively, minimum) of the function f on t_k, t_{k+1}. Use this equality to prove (3.30).

(b) Assume furthermore that f is also absolutely continuous. Prove that for any bounded Borel-measurable function $g : \mathbb{R} \to \mathbb{R}$, one has

$$\int_{-\infty}^{+\infty} N_u(f, I)g(u)\, du = \int_I |f'(t)|g(f(t))\, dt. \tag{3.31}$$

(c) Prove that if f is absolutely continuous, for every bounded Borel-measurable function $h(t, u)$,

$$\int_{\mathbb{R}} \sum_{t \in I : f(t) = u} h(t, u)du = \int_I |f'(t)|h((t, f(t))\, dt. \tag{3.32}$$

(d) Let $\{X(t) : t \in I\}$ be a real-valued stochastic process with absolutely continuous paths, I a compact interval in the real line. Assume that for each $t \in I$, the distribution of the random variable $X(t)$ has a density $p_{X(t)}(\cdot)$, and the conditional expectation $\mathrm{E}\big(|X'(t)|\big|X(t) = u\big)$ is well defined. Prove that

$$\mathrm{E}(N_u(X, I)) = \int_I \mathrm{E}\big(|X'(t)|\big|X(t) = u\big)p_{X(t)}(u)\, dt \tag{3.33}$$

for almost every $u \in \mathbb{R}$.. *Hint*: Let $g : \mathbb{R} \to \mathbb{R}$ be continuous with compact support. Apply (3.31), replacing f by $X(\cdot)$ and take expectations in both sides.

3.9. (*Upper-bound part of the Rice formula*) Using Fatou's lemma, show that if $\{X(t) : t \in I\}$ is a process with C^1-paths and such that:

(a) For every $t \in I$, $X(t)$ admits a density $p_{X(t)}$.
(b) The conditional expectation $\mathrm{E}\big(|X'(t)|\big|X(t)\big)$ is well defined.
(c) The function

$$\int_I \mathrm{E}\big(|X'(t)|\big|X(t) = x\big)p_{X(t)}(x)\, dt$$

is continuous as a function of x at the point $x = u$. Then

$$\mathrm{E}\big(N_u(X; I)\big) \le \int_I \mathrm{E}\big(|X'(t)|\big|X(t) = u\big)p_{X(t)}(u)\, dt.$$

3.10. (*Rice formulas for χ^2-processes*) Let $\{X(t) : t \in I\}$ be a centered Gaussian process with $\mathcal{C}^1$-paths and values in $\mathbb{R}^d$. Assume that for each $t \in I$, Var $(X(t) = Id$; then the process $Y(t) := \|X(t)\|^2$ is called a χ^2 *process*. Adapt the proof of Theorem 3.2 to this process.

3.11. Let $\psi : \mathbb{R} \to \mathbb{R}$ be a $\mathcal{C}^1$-function taking nonnegative values, with support contained in the interval $[-1, 1]$, such that $\int_{-\infty}^{+\infty} \psi(t)\,dt = 1$. For $\varepsilon > 0$ we set

$$\psi_\varepsilon(t) := \frac{1}{\varepsilon}\psi\left(\frac{t}{\varepsilon}\right).$$

If $f : \mathbb{R} \to \mathbb{R}$ is any locally integrable function, we define the regularized function f^ε:

$$f^\varepsilon(t) := \int_{\mathbb{R}} f(t-s)\psi_\varepsilon(s)\,ds.$$

Let $\{W(t) : t \in \mathbb{R}\}$ be a Wiener process defined on a probability space $(\Omega, \mathcal{A}, \mathrm{P})$ (see Exercise 1.10 for a definition on then whole line).

(**a**) Prove that for each $t \in \mathbb{R}$, the distribution of the random variable $[W^\varepsilon]'(t)$ is centered normal with variance $\|\psi\|_2^2/\varepsilon$, where $\|\psi\|_2$ is the L^2-norm of the function ψ.

(**b**) Prove that for each $u \in \mathbb{R}$ and $I = [t_1, t_2], 0 < t_1 < t_2$,

$$\lim_{\varepsilon \to 0} \sqrt{\frac{\pi\varepsilon}{2}}\frac{1}{\|\psi\|_2} N_u(W^\varepsilon, I) = L^W(u, I), \tag{3.34}$$

where convergence in (3.34) takes place in $L^2(\Omega, \mathcal{A}, \mathrm{P})$.
Hint: Use the definition of the local time $L^W(u, I)$ of the Wiener process given in Exercise 1.12 and use the Rice formula to estimate

$$\mathrm{E}\left(\left[\sqrt{\frac{\pi\varepsilon}{2}}\frac{1}{\|\psi\|_2} N_u(W^\varepsilon, I) - \frac{1}{2\delta}\int_I \mathbb{1}_{|W(t)-u|<\delta}\,dt\right]^2\right).$$

(**c**) Prove that convergence in (3.34) holds true in $L^p(\Omega, \mathcal{A}, \mathrm{P})$ for any $p > 0$.

3.12. Let $\{X(t) : t \in R\}$ be a one-parameter centered Gaussian stationary process with covariance function Γ, $\Gamma(0) = 1$ and finite fourth spectral moment λ_4 (see Chapter 1 for the notation). We denote by $M_{u_1,u_2}(I)$ the number of local maxima of the process in the interval I, with values lying in the interval $[u_1, u_2]$.

(**a**) Use the Rice formula to express the expectation

$$\mathrm{E}(M_{u_1,u_2}(I)) = |I| \int_{u_1}^{u_2} f(x)\,dx,$$

where f is a certain function and $|I|$ denotes the length of the interval I.

(**b**) The function

$$g(x) = \frac{f(x)}{\int_{-\infty}^{+\infty} f(x)dx}$$

is called the *density of the values of the local maxima per unit length*. Give a heuristic interpretation of the function g and verify that under the hypotheses noted above, one has

$$g(x) = \frac{1}{\sqrt{2\pi}} \left[a \exp\left(-\frac{x^2}{2a^2}\right) + \sqrt{1-a^2}x \right.$$
$$\left. \times \exp\left(-\frac{x^2}{2}\right) \int_{-\infty}^{x\sqrt{1-a^2}/a} \exp\left(-\frac{u^2}{2}\right) du \right],$$

where

$$a = \sqrt{\frac{\lambda_4 - \lambda_2^2}{\lambda_4}}.$$

(**c**) Conclude from part (b) that if $a \to 0$, then

$$g(x) \to x \exp\left(-\frac{x^2}{2}\right) \text{ if } x > 0, \quad \text{and} \quad g(x) \to 0 \text{ if } x < 0.$$

(This is the Rayleigh density of the values of the maxima for "narrow-spectrum" Gaussian processes.)

Remark. Exercise 3.11 is the beginning of the far-reaching subject of the approximation of local time by means of functionals defined on some smoothing of the paths, which has applications in statistical inference of continuous parameter random processes. In this exercise, only the Wiener process and so-called first-order approximations are considered, with the aim of utilizing Rice formulas in the computations required in the proof. For related work on this subject, the interested reader can see, for example, Azaïs (1989), Wschebor (1992), Azaïs and Wschebor (1996, 1997a), Jacod (1998, 2000), Perera and Wschebor (1998, 2002), Berzin and León (2005), and references therein. A review article on this subject is that of Wschebor (2006).

CHAPTER 4

SOME STATISTICAL APPLICATIONS

This chapter contains two independent subjects. In the first part we use Rice formulas to obtain bounds for the tails of the distribution of the maximum of one-parameter Gaussian processes having regular paths. We also include some results on the asymptotic behavior of the tails of the supremum on a fixed interval when the level tends to infinity. In the second part of the chapter we provide a quite detailed account of two examples of statistical applications of the distribution of the maximum, one to genetics and the other to the study of mixtures of distributions. These two examples share the common trait that statistical inference is performed in the presence of nuisance parameters, which are not identifiable under the null hypothesis.

In standard classical situations in hypothesis testing, the critical region consists of the event that some function of the empirical observations (the *test statistic*) falls into a subset of its value space. Computation of the probabilities that are relevant for the test, for large samples, follows from some weak limit theorem, allowing us to obtain the asymptotic law of the test statistic as the sample size goes to infinity. Typically, this will be a normal distribution or some function of the normal distribution. In certain more complicated situations, as happens when nuisance parameters are present and are not identifiable under the null hypothesis, it turns out that a reasonable choice consists of using as a test statistic the supremum of a process indexed by the possible values of the nuisance parameter. Then, when passing to the limit as the sample size grows, instead of limiting distributions of finite-dimensional-valued random variables, we have to deal with

Level Sets and Extrema of Random Processes and Fields, By Jean-Marc Azaïs and Mario Wschebor

a limit process, typically Gaussian; and the relevant random variable to compute probabilities becomes its supremum.

The literature on this subject has been growing during the recent years, including applications to biology, econometrics, and in general, stochastic models which include nuisance parameters, of which hidden Markov chains have become a quite popular example. The interested reader can see, for example, Andrews and Ploberger (1994), Hansen (1996), Dacunha-Castelle and Gassiat (1997, 1999), Gassiat (2002), Azaïs et al. (2006, 2008), and references therein. The first example in this chapter is extracted from Azaïs and Cierco-Ayrolles (2002) and the second from Delmas (2001, 2003a).

The theory includes two parts: first, limit theorems that allow us to find the asymptotic behavior of certain stochastic processes, and second, computing (or obtaining bounds) for the distribution of the supremum of the limiting process or its absolute value. For the second part, a common practice is to in simulate the paths and approximate the tails of the distribution of its supremum using Monte Carlo. This is not what we will be doing here. Our aim is a better understanding of the behavior of this distribution, and for that purpose we use the results of the first part of this chapter. Since these concern only one-parameter Gaussian processes, the models in the examples of this chapter have a one-dimensional nuisance parameter, and the asymptotic law of the relevant processes is Gaussian.

4.1. ELEMENTARY BOUNDS FOR P{$M > u$}

In this chapter, if not stated otherwise, $\mathcal{X} = \{X(t) : t \in \mathbb{R}\}$ is a real-valued centered Gaussian process with continuously differentiable sample paths and covariance function $r(s,t)$. Let us recall that for $T > 0$, we denote $M_T = \sup_{t\in[0,T]} X(t)$. We begin with some bounds for the tails of the distribution of M_T which are a first approximation to the relationship between Rice formulas and the distribution of the maximum, a subject that you will come across frequently in this book. These bounds are based on variants of the following simple inequality:

We use throughout the notation $\nu_k := \mathrm{E}(U_u^{[k]})$ and $\widetilde{\nu}_k := \mathrm{E}(U_u^{[k]}\mathbb{I}_{X(0)<u})$ for $k = 1, 2, \ldots$, where $U_u := U_u^X([0,T])$ is the number of up-crossings on the interval $[0,T]$. Then

$$\mathrm{P}\{M_T > u\} \le \mathrm{P}\{X(0) > u\} + \mathrm{P}\{U_u > 0\} \le \mathrm{P}\{X(0) > u\} + \nu_1. \tag{4.1}$$

In the next proposition we make precise the bounds under certain hypotheses for the process.

Proposition 4.1

(a) *Assume that* $\mathrm{Var}(X(t)) \equiv 1$. *Then*

$$P\{M_T > u\} \le \frac{e^{-u^2/2}}{2\pi} \int_0^T \sqrt{r_{11}(t,t)}\, dt + 1 - \Phi(u). \tag{4.2}$$

(b) *Assume in addition that the covariance function is of class C^4, that $r(s,t) \neq \pm 1$ for $s, t \in [0, T]$, $s \neq t$, and that $r_{11}(s,s) > 0$ for all $s \in [0, T]$. Then, if $u > 0$,*

$$P\{M_T > u\} = \frac{e^{-u^2/2}}{2\pi} \int_0^T \sqrt{r_{11}(t,t)}\, dt + 1 - \Phi(u) + O\left(\phi(u(1+\delta))\right)$$

for some positive real number δ.

Remarks

1. The bound (4.2) is sometimes called the *Davies bound* (Davies, 1977).

2. Proposition 4.1(b) was originally proved by Piterbarg in 1981 [see also his book (1996a)] for centered stationary Gaussian processes. In this case, under the hypotheses,

- $\lambda_8 < \infty$.
- The joint distribution of $X(s), X'(s), X''(s), X(t), X'(t), X''(t)$ does not degenerate for distinct $s, t \in [0, T]$.
- $\Gamma'(t) < 0$ for $0 < t \leq T$.

Azaïs et al. (2001) have been able to describe more precisely the complementary term in the asymptotic expansion given in part (b). The result is that as $u \to +\infty$,

$$\begin{aligned} \mathrm{P}(M_T > u) &= 1 - \Phi(u) + \sqrt{\frac{\lambda_2}{2\pi}} T\phi(u) \\ &\quad - \frac{3\sqrt{3}(\lambda_4 - \lambda_2^2)^{9/2}}{2\pi\lambda_2^{9/2}(\lambda_2\lambda_6 - \lambda_4^2)} \frac{T}{u^5} \phi\left(\sqrt{\frac{\lambda_4}{\lambda_4 - \lambda_2^2}}\, u\right) [1 + o(1)]. \end{aligned}$$

This asymptotic behavior does appear in Piterbarg's 1981 article, but only for sufficiently small T. It appears to be the only result known at present that gives a precise description of the second-order term for the asymptotics of $\mathrm{P}(M_T > u)$ as $u \to +\infty$. We are not going to prove it here, since it requires quite long and complicated computations. We refer the interested reader to the two papers noted above

Proof of Proposition 4.1

(a) It is clear that

$$\{M_T > u\} = \{X(0) \geq u\} \cup \{X(0) < u, U_u > 0\} \text{ a.s.},$$

where the convention is that if A and B are two events in a probability space, "$A = B$ a.s." means that $\mathrm{P}\{A \Delta B\} = 0$. Using Markov's inequality, it follows that

$$\mathrm{P}\{M_T > u\} \leq 1 - \Phi(u) + \mathrm{E}(U_u).$$

Now we apply the Rice formula (Theorem 3.2) to compute $\mathrm{E}(U_u)$:

$$\mathrm{E}(U_u) = \int_0^T \mathrm{E}(X'^{+}(t)\big|X(t) = u) p_{X(t)}(u)\, dt$$
$$= \phi(u) \int_0^T \mathrm{E}(X'^{+}(t))\, dt = \frac{\phi(u)}{\sqrt{2\pi}} \int_0^T \sqrt{r_{11}(t,t)}\, dt$$

since $X(t)$ and $X'(t)$ are independent. This proves part (a).

(b) We have the lower bound:

$$\mathrm{P}\{M_T > u\} = 1 - \Phi(u) + \mathrm{P}\{U_u > 0\} - \mathrm{P}\{U_u > 0;\, X(0) > u\}$$
$$\geq 1 - \Phi(u) + \nu_1 - \frac{\nu_2}{2} - \mathrm{P}\{U_u > 0, X(0) > u\}. \tag{4.3}$$

The result will be obtained as soon as we prove that the last two terms on the right-hand side are $O\,(\phi(u(1+\delta)))$, $\delta > 0$. We have

$$\mathrm{P}\{X(0) > u, U_u > 0\} \leq \mathrm{P}\{X(0) > u, X(T) > u\}$$
$$+\ \mathrm{P}\{X(0) > u, X(T) < u, U_u > 0\}$$
$$\leq \mathrm{P}\{X(0) > u, X(T) > u\} + \mathrm{P}\{D_u > 1\}. \tag{4.4}$$

Note that since $r(0,T) \neq \pm 1$,

$$\mathrm{P}\{X(0) > u;\, X(T) > u\} \leq \mathrm{P}\{X(0) + X(T) > 2u\}$$
$$= \int_{2u}^{+\infty} \frac{1}{2\sqrt{\pi}} \frac{1}{\sqrt{1 + r(0,T)}} \exp\left[-\frac{1}{4(1 + r(0,T))} x^2\right] dx = O\,(\phi(u(1+\delta)))\,.$$

Let us look at the second term on the right-hand side of inequality (4.4). It is clear that

$$\mathbb{I}_{D_u > 1} \leq \tfrac{1}{2} D_u(D_u - 1),$$

which implies that

$$\mathrm{P}\{D_u > 1\} \leq \tfrac{1}{2}\mathrm{E}\,(D_u(D_u - 1) = \tfrac{1}{2}\mathrm{E}\,(U_u(U_u - 1) = \tfrac{1}{2}\nu_2,$$

where the penultimate equality follows reversing the time in the interval $[0, T]$, that is, changing t into $T - t$.

So our aim is to show that $\nu_2 = O\,(\phi(u(1+\delta)))$. We have

$$\nu_2 = \iint_0^T A_{s,t}^{+}(u,u)\, ds\, dt. \tag{4.5}$$

where

$$A^+_{s,t}(u,u) := \mathrm{E}\left(X'^+(s)X'^+(t)|X(s)=X(t)=u\right) p_{X(s),X(t)}(u,u)$$

$$p_{X(s),X(t)}(u,u) = \frac{1}{2\pi}\left(1-r^2(s,t)\right)^{-1/2}\exp\left(\frac{-u^2}{1+r(s,t)}\right).$$

Conditionally on $\mathcal{C} := \{X(s)=X(t)=u\}$, the random variables $X'(s)$ and $X'(t)$ have a joint Gaussian distribution with expectations and variances given by the following formulas, which can be obtained using regression formulas:

$$\mathrm{E}(X'(s)\big|\mathcal{C}) = \frac{r_{10}(s,t)}{1+r(s,t)}u$$

$$\mathrm{E}(X'(t)\big|\mathcal{C}) = \frac{r_{10}(t,s)}{1+r(s,t)}u$$

$$\mathrm{Var}(X'(s)\big|\mathcal{C}) = r_{11}(s,s) - \frac{r_{10}^2(s,t)}{1-r^2(s,t)}$$

$$\mathrm{Var}(X'(t)\big|\mathcal{C}) = r_{11}(t,t) - \frac{r_{10}^2(t,s)}{1-r^2(s,t)}.$$

Our hypotheses imply that

$$\begin{aligned}\mathrm{E}\left(X'^+(s)X'^+(t)\big|\mathcal{C}\right) &\le \frac{1}{2}\mathrm{E}\left((X'(s))^2+(X'(t))^2\big|\mathcal{C}\right)\\ &\le \frac{1}{2}\left(r_{11}(s,s)+r_{11}(t,t)\right) + \frac{u^2}{2}\frac{r_{10}^2(s,t)+r_{10}^2(t,s)}{[1+r(s,t)]^2}.\end{aligned}$$

So for fixed $\gamma > 0$, one has

$$\begin{aligned}&\iint_{s,t\in[0,T]:|s-t|\ge\gamma} A^+_{s,t}(u,u)\,ds\,dt\\ &\quad\le (L_1u^2+L_2)\times\iint_{s,t\in[0,T]:|s-t|\ge\gamma}\exp\left(-L_3\frac{u^2}{2}\right)ds\,dt,\end{aligned}\tag{4.6}$$

where L_1, L_2, and L_3 are positive constants, $L_3 > 1$, since $1+r(s,t) < 2$ for all $s,t\in[0,T]$, $s\ne t$. This shows that

$$\iint_{s,t\in[0,T]:|s-t|\ge\gamma} A^+_{s,t}(u,u)\,ds\,dt = O\left(\phi(u(1+\delta))\right).$$

Let us now look at the double integral near the diagonal $\{s,t\in[0,T]: s=t\}$. We take into account that $\mathrm{Var}(X(t))$ is constant and that $r_{10}(t,t)=0$. A Taylor expansion in the expressions for the conditional expectations as s, t approach the same value t^* permits us to show that

$$\mathrm{E}(X'(s)\big|\mathcal{C}) = \left[r_{11}(t^*, t^*) + A(t-s)\right]\frac{(t-s)u}{2} \tag{4.7}$$

$$\mathrm{E}(X'(t)\big|\mathcal{C}) = \left[r_{11}(t^*, t^*) + B(t-s)\right]\frac{(s-t)u}{2}, \tag{4.8}$$

where A and B are bounded functions of the pair s and t.

A similar expansion for the conditional variances shows that

$$\mathrm{Var}(X'(s)\big|\mathcal{C}), \mathrm{Var}(X'(t)\big|\mathcal{C}) \leq L(s-t)^2$$

for some positive constant L. So

$$\begin{aligned}
&\iint_{s,t\in[0,T]:|s-t|<\gamma} A^+_{s,t}(u,u)\,ds\,dt \\
&= \iint_{s,t\in[0,T]:|s-t|<\gamma} \mathrm{E}\left(X'^+(s)X'^+(t)\big|\mathcal{C}\right) p_{X(s),X(t)}(u,u)\,ds\,dt \\
&\leq \iint_{s,t\in[0,T]:|s-t|<\gamma} \left[\mathrm{E}((X'^+(s))^2\big|\mathcal{C})\mathrm{E}((X'^+(t))^2\big|\mathcal{C})\right]^{1/2} \\
&\quad \frac{1}{2\pi\sqrt{1-r^2(s,t)}}\exp\left(-\frac{u^2}{1+r(s,t)}\right)\,ds\,dt.
\end{aligned} \tag{4.9}$$

To bound the conditional expectations in the integrand we use the following inequalities, which the reader can easily check. Let Z be a real-valued random variable with normal distribution having parameters μ and σ^2. Then

$$\mathrm{E}\left((Z^+)^2\right) \leq \sigma^2 + \mu^2 \quad \text{if } \mu > 0 \tag{4.10}$$

$$\leq (\sigma^2+\mu^2)\left[1-\Phi\left(-\frac{\mu}{\sigma}\right)\right] + \mu\sigma\phi\left(\frac{\mu}{\sigma}\right) \quad \text{if } \mu < 0. \tag{4.11}$$

Using the expressions for the conditional expectations, one can see that if $|t-s|$ is sufficiently small, there exists a positive constant D such that

$$\frac{\left|\mathrm{E}\left(X'^+(s)\big|\mathcal{C}\right)\right|}{\sqrt{\mathrm{Var}\left(X'^+(s)\big|\mathcal{C}\right)}} \geq Du.$$

A similar inequality holds for t instead of s.

Now, from the expansions of the conditional expectations, it follows that if $|s-t|$ is small enough, $s \neq t$, $\mathrm{E}(X'(s)\big|\mathcal{C})$ and $\mathrm{E}(X'(t)\big|\mathcal{C})$ have opposite signs, so that we can apply to each of them one of the inequalities (4.10) and (4.11). It follows that

$$\mathrm{E}((X'^{+}(s))^2|\mathcal{C})\mathrm{E}((X'^{+}(t))^2|\mathcal{C}) \le K_1|s-t|^4\left[1-\Phi(-K_2u)+\exp\left(-\frac{1}{2}K_3u^2\right)\right].$$

A Taylor expansion around a point in the diagonal shows that if $|s-t|$ is small enough,

$$1-r(s,t) \ge K_4(s-t)^2$$

for some positive constant K_4. It follows that if γ is small enough, one has

$$\iint_{s,t\in[0,T]:|s-t|<\gamma} A^{+}_{s,t}(u,u)\,ds\,dt \le K_5\exp(-K_6u^2)\exp\left(-\frac{u^2}{2}\right)\iint_0^T |s-t|\;ds\;dt,$$

where K_5 and K_6 are new positive constants. It is clear that the right-hand side is $O\left(\phi(u(1+\delta))\right)$, $\delta>0$. □

One can obtain the same type of asymptotic expansion as that given in Proposition 4.1 in an easier way when the process is also stationary. This is the next proposition; the hypotheses are less demanding and the result for the error term is weaker.

Proposition 4.2 (Stationary Processes). *Let $\{X(t) : t \in [0,T]\}$ be a centered stationary Gaussian process. We assume that*

(G1) $\Gamma(t) = \mathrm{E}(X(s)X(s+t)) \ne \pm 1$ for $0 < t \le T$.

(G2) The Geman condition: The integral

$$\int \frac{\theta'(\tau)}{\tau^2}d\tau \text{ converges at } \tau = 0^+,$$

where $\theta(\tau)$ is defined by means of $\Gamma(\tau) = 1-\lambda_2\tau^2/2+\theta(\tau)$.

Then, as $u \to +\infty$,

$$\mathrm{P}\{M_T > u\} = \sqrt{\frac{\lambda_2}{2\pi}}T\phi(u)\left[1+o(1)\right]. \tag{4.12}$$

Remark. Conditions (G1) and (G2) are, as already mentioned, necessary and sufficient for the finiteness of the second moment of the crossings (Kratz and León, 2006).

Proof. As in the proof of Proposition 4.1, we have

$$\nu_2 := \mathrm{E}\left(U_u^{[2]}\right) = \int_0^T 2(T-\tau)\mathrm{E}\left(X'^{+}(0)X'^{+}(\tau)\middle|X(0)=X(\tau)=u\right)$$
$$\cdot\frac{1}{2\pi}\frac{1}{\sqrt{1-\Gamma^2(\tau)}}\exp\left(-\frac{u^2}{1+\Gamma(\tau)}\right)d\tau \tag{4.13}$$

and

$$\mathrm{E}\left(X'^{+}(0)\middle|X(0)=X(\tau)=u\right) = \frac{-\Gamma'(\tau)u}{1+\Gamma(\tau)} = -\mathrm{E}\left(X'^{+}(\tau)\middle|X(0)=X(\tau)=u\right).$$

A standard regression shows that

$$\sigma^2(\tau) := \mathrm{Var}\left(X'(0)\middle|X(0)=X(\tau)=u\right)$$
$$= \mathrm{Var}\left(X'_\tau\middle|X(0)=X(\tau)=u\right) = \frac{\lambda_2(1-\Gamma^2(\tau)-\Gamma'^2(\tau)}{1-\Gamma^2(\tau)}.$$

Using inequality $a^+b^+ \le (a+b)^2/4$, we get

$$\nu_2 \le T\int_0^T \frac{\sigma^2(\tau)}{\sqrt{1-\Gamma^2(\tau)}}\frac{1}{2\pi}\exp\left(-\frac{u^2}{1+\Gamma(\tau)}\right)d\tau. \tag{4.14}$$

Since $\theta(\tau), \theta'(\tau)$, and $\theta''(\tau) \ge 0$, an elementary expansion shows that

$$\sigma^2(\tau)\left(1-\Gamma^2(\tau)\right) = \lambda_2\left(1-\Gamma^2(\tau)\right) - \Gamma'^2(\tau) \le 2\lambda_2\tau\theta'(\tau),\ 1-\Gamma^2(\tau)\approx\lambda_2\tau^2,$$

so that

$$\nu_2 \le T(\text{const})\int_0^T \frac{\theta'(\tau)}{\tau^2}\exp\left(-\frac{u^2}{1+\Gamma(\tau)}\right)d\tau = o\left(\phi(u)\right),$$

based on hypotheses (G1) and (G2). □

4.2. MORE DETAILED COMPUTATION OF THE FIRST TWO MOMENTS

We return to the lower bound for $\mathrm{P}\{M_T > u\}$ for stationary centered Gaussian processes. In what follows we also assume that the distribution of the triplet $(X(s), X(t), X'(s))$ does not degenerate for $s \neq t$. In the remainder of the chapter we use the following inequality, which is a slight modification of (4.3). The proof is immediate and is left to the reader.

$$\mathrm{P}(X(0)>u) + \widetilde{\nu}_1 - \frac{\nu_2}{2} \le \mathrm{P}(M_T > u) \le \mathrm{P}(X(0)>u) + \widetilde{\nu}_1. \tag{4.15}$$

Our goal is to give as simply as possible, formulas for the quantities involved in (4.15). Let us introduce or recall some notation that will be used in the remainder of this section. We set:

- $\overline{\nu}_1 := \nu_1 - \widetilde{\nu}_1$. For large values of u, $\overline{\nu}_1$ and $\widetilde{\nu}_1$ are worth being distinguished, since they tend to zero at different exponential rates as $u \to +\infty$.
- $\mu(t) = \mathrm{E}(X'(0) \big| X(0) = X(t) = u) = -\dfrac{\Gamma'(t)}{1+\Gamma(t)} u.$
- $\sigma^2(t) = \mathrm{Var}(X'(0) \big| X(0) = X(t) = u) = \lambda_2 - \dfrac{\Gamma'^2(t)}{1 - \Gamma^2(t)}.$
- $\rho(t) = \mathrm{Cor}\left(X'(0), X'(t) \big| X(0) = X(t) = u\right)$
  $= \dfrac{-\Gamma''(t)\left(1-\Gamma^2(t)\right) - \Gamma(t)\Gamma'^2(t)}{\lambda_2\left(1-\Gamma^2(t)\right) - \Gamma'^2(t)}$
- $k(t) = \sqrt{\dfrac{1+\rho(t)}{1-\rho(t)}}.$
- $b(t) = \dfrac{\mu(t)}{\sigma(t)}.$

In what follows, the variable t will be omitted whenever there is no confusion, so that we will be writing $\Gamma, \Gamma', \mu, \sigma, \rho, k, b$ instead of $\Gamma(t), \Gamma'(t), \mu(t), \sigma(t), \rho(t), k(t), b(t)$.

Lemma 4.3. *Let (X, Y) be a random vector in $\mathbb{R}^2$ having centered normal distribution with variance matrix $\begin{pmatrix} 1 & \rho \\ \rho & 1 \end{pmatrix}$, $|\rho| \neq 1$. Then $\forall a \in \mathbb{R}^+$:*

$$\mathrm{P}(X > a, Y > -a) = \frac{1}{\pi} \arctan \sqrt{\frac{1+\rho}{1-\rho}} - 2 \int_0^a \phi(x) \left[\Phi\left(\sqrt{\frac{1+\rho}{1-\rho}} x \right) - \frac{1}{2} \right] dx$$

$$= 2 \int_a^{+\infty} \left[\Phi\left(\sqrt{\frac{1+\rho}{1-\rho}} x \right) - \frac{1}{2} \right] \phi(x)\, dx.$$

Proof. We first give an integral expression for $\mathrm{P}(X > a, Y > a)$. Set $\rho = \cos\theta$, $\theta \in (0, \pi)$, and use the orthogonal decomposition $Y = \rho X + \sqrt{1-\rho^2} Z$. Then

$$\{Y > a\} = \left\{ Z > \frac{a - \rho X}{\sqrt{1-\rho^2}} \right\}.$$

Thus,

$$\mathrm{P}\{X > a, Y > a\} = \int_a^{+\infty} \phi(x) \left[1 - \Phi\left(\frac{a - \rho x}{\sqrt{1-\rho^2}} \right) \right] dx = \iint_D \phi(x)\phi(z)\, dx\, dz,$$

where $\mathcal{D}$ is the domain located between the two half straight lines starting from the point $\left(a, a\sqrt{(1-\rho)/(1+\rho)}\right)$ and with angles $\theta - \pi/2$ and $\pi/2$.

Using the symmetry with respect to the straight line with angle $\theta/2$ passing through the origin, we get

$$\mathrm{P}\{X > a, Y > a\} = 2\int_a^{+\infty} \phi(x)\left[1 - \Phi\left(\sqrt{\frac{1-\rho}{1+\rho}}x\right)\right] dx. \tag{4.16}$$

Now

$$\mathrm{P}\{X > a, Y > -a\} = 1 - \Phi(a) - \mathrm{P}\{X > a, (-Y) > a\}.$$

Applying (4.16) to the pair or random variables $(X, -Y)$ yields

$$\begin{aligned}\mathrm{P}(X > a, Y > -a) &= 1 - \Phi(a) - 2\int_a^{+\infty} \phi(x)\left[1 - \Phi\left(\sqrt{\frac{1+\rho}{1-\rho}}x\right)\right] dx \\ &= 2\int_a^{+\infty}\left[\Phi\left(\sqrt{\frac{1+\rho}{1-\rho}}x\right) - \frac{1}{2}\right]\phi(x)\, dx.\end{aligned}$$

Now, using polar coordinates, it is easy to establish that

$$\int_0^{+\infty}\left[\Phi(kx) - \frac{1}{2}\right]\phi(x)\, dx = \frac{1}{2\pi}\arctan k,$$

which proves the lemma. □

Proposition 4.4. *Let $\{X(t) : t \in \mathbb{R}\}$ be a centered stationary Gaussian process satisfying the conditions preceding Lemma 4.3. Then*

(i) $$\overline{\nu}_1 = \phi(u)\int_0^T \left(\sqrt{\frac{\lambda_2}{2\pi}}\left[1 - \Phi\left(\sqrt{\frac{1-}{\Gamma}}1 + \Gamma\frac{\sqrt{\lambda_2}}{\sigma}u\right)\right]\right. \\ \left. + \phi\left(\sqrt{\frac{1-\Gamma}{1+\Gamma}}u\right)\left[1 - \Phi(b)\right]\frac{\Gamma'}{\sqrt{1-\Gamma^2}}\right) dt.$$

(ii) $$\nu_2 = \int_0^T 2(T-t)\frac{1}{\sqrt{1-\Gamma^2(t)}}\phi^2\left(\frac{u}{\sqrt{1+\Gamma(t)}}\right)\left[T_1(t) + T_2(t) + T_3(t)\right] dt$$

with

$$T_1(t) = \sigma^2(t)\sqrt{1-\rho^2(t)}\phi\left(b(t)\right)\phi\left(k(t)b(t)\right),$$

$$T_2(t) = 2(\sigma^2(t)\rho(t) - \mu^2(t))\int_{b(t)}^{+\infty}\left[\Phi(k(t)x) - \tfrac{1}{2}\right]\phi(x)\, dx,$$

$$T_3(t) = 2\mu(t)\sigma(t)\left[\Phi(k(t)b(t)) - \tfrac{1}{2}\right]\phi(b(t)).$$

(iii) *A second expression for $T_2(t)$ is*

$$T_2(t) = (\sigma^2(t)\rho(t) - \mu^2(t))\left[\frac{1}{\pi}\arctan k(t) - 2\int_0^{b(t)}\left[\Phi(k(t)x) - \frac{1}{2}\right]\phi(x)\,dx\right]. \tag{4.17}$$

Remarks

1. The formula in Lemma 4.3 is analogous to (2.10.4) of Cramér and Leadbetter (1967, p. 27); that is:

$$P\{X > a, Y > -a\} = \Phi(a)[1 - \Phi(a)] + \int_0^{\rho} \frac{1}{2\pi\sqrt{1 - z^2}} \exp\left(-\frac{a^2}{1 - z}\right) dz.$$

The formula here is easier to prove and more adapted to numerical applications because as $t \to 0$, $\rho(t) \to -1$, and the integrand in Cramér and Leadbetter's formula tends to infinity.

2. These formulas allow us to compute ν_2 at the cost of a double integral with finite limits. This implies a significant reduction of complexity with respect to the original form. The form (4.17) is more adapted to effective computation because it involves an integral on a bounded interval.

Proof of Proposition 4.4

(i) Conditionally on $\{X(0) = x, X(t) = u\}$, $X'(t)$ is Gaussian with:

- Mean $m(t) = \dfrac{\Gamma'(t)(x - \Gamma(t)u)}{1 - \Gamma^2(t)}$.
- Variance $\sigma^2(t)$; see the proof of Proposition 4.2. □

If Z is a Gaussian random variable with mean m and variance σ^2, then

$$\mathrm{E}\left(Z^+\right) = \sigma\phi\left(\frac{m}{\sigma}\right) + m\Phi\left(\frac{m}{\sigma}\right).$$

These two remarks yield $\overline{\nu}_1(u, T) = I_1 + I_2$, with:

- $I_1 = \displaystyle\int_0^T dt \int_u^{+\infty} \sigma\phi\left(\frac{\Gamma'(x - ru)}{(1 - \Gamma^2)\sigma}\right) p_{X(0),X(t)}(x, u)\,dx$
- $I_2 = \displaystyle\int_0^T dt \int_u^{+\infty} \frac{\Gamma'(x - ru)}{1 - \Gamma^2}\Phi\left(\frac{\Gamma'(x - ru)}{(1 - r^2)\sigma}\right) p_{X(0),X(t)}(x, u)\,dx.$

I_1 can be written as

$$I_1 = \phi(u)\int_0^T \frac{\sigma^2}{\sqrt{2\pi\lambda_2}}\left[1 - \Phi\left(\frac{\sqrt{\lambda_2}}{\sigma}\sqrt{\frac{1 - \Gamma}{1 + \Gamma}}u\right)\right] dt.$$

Integrating by parts in I_2 leads to

$$I_2 = \phi(u) \int_0^T \frac{\Gamma'}{\sqrt{1-\Gamma^2}} \phi\left(\sqrt{\frac{1-\Gamma}{1+\Gamma}}u\right) [1 - \Phi(b)]$$
$$+ \frac{\Gamma'^2}{\sqrt{2\pi\lambda_2(1-\Gamma^2)}} \left[1 - \Phi\left(\frac{\sqrt{\lambda_2}}{\sigma}\sqrt{\frac{1-\Gamma}{1+\Gamma}}u\right)\right] dt.$$

Since $\sigma^2 + \Gamma'^2/(1-\Gamma^2) = \lambda_2$, we obtain

$$\overline{\nu_1} = \sqrt{\frac{\lambda_2}{2\pi}} \phi(u) \int_0^T \left[1 - \Phi\left(\frac{\sqrt{\lambda_2}}{\sigma}\sqrt{\frac{1-\Gamma}{1+\Gamma}}u\right)\right] dt$$
$$+ \phi(u) \int_0^T \frac{\Gamma'}{\sqrt{1-\Gamma^2}} \phi\left(\sqrt{\frac{1-\Gamma}{1+\Gamma}}u\right) [1 - \Phi(b)]\, dt.$$

(ii) We set:

- $v(x, y) = \dfrac{(x-b)^2 - 2\rho(x-b)(y+b) + (y+b)^2}{2(1-\rho^2)}$.
- For $(i, j) \in \{(0, 0); (1, 0); (0, 1); (1, 1); (2, 0); (0, 2)\}$,

$$J_{ij} = \int_0^{+\infty} \int_0^{+\infty} \frac{x^i y^j}{2\pi\sqrt{1-\rho^2}} \exp(-v(x, y))\, dy\, dx.$$

We have

$$J_{10} - \rho J_{01} - (1+\rho)bJ_{00} = \left(1-\rho^2\right) \int_0^{+\infty} \left(\int_0^{+\infty} \frac{\partial}{\partial x} v(x, y) \frac{\exp(-v(x, y))}{2\pi\sqrt{1-\rho^2}} dx\right) dy$$
$$= \left(1-\rho^2\right) [1 - \Phi(kb)]\phi(b). \tag{4.18}$$

Replacing x by y and b by $-b$ in (4.18) yields

$$J_{01} - \rho J_{10} + (1+\rho)bJ_{00} = \left(1-\rho^2\right) \Phi(kb)\, \phi(b). \tag{4.19}$$

In the same way, multiplying the integrand by y, we get

$$J_{11} - \rho J_{02} - (1+\rho)bJ_{01} = \left(1-\rho^2\right)^{3/2} [\phi(kb) - kb[1 - \Phi(kb)]\, \phi(b). \tag{4.20}$$

Now, multiplying the integrand by x leads to

$$J_{11} - \rho J_{20} + (1+\rho)bJ_{10} = \left(1-\rho^2\right)^{3/2} [\phi(kb) + kb\Phi(kb)]\, \phi(b). \tag{4.21}$$

So

$$J_{20} - \rho J_{11} - (1+\rho) b J_{10} = (1-\rho^2) \int_0^{+\infty} \int_0^{+\infty} x \frac{\partial}{\partial x} v(x,y) \frac{\exp(-v(x,y))}{2\pi\sqrt{1-\rho^2}}\, dx\, dy.$$

Then, integrating by parts yields

$$J_{20} - \rho J_{11} - (1+\rho) b J_{10} = (1-\rho^2) J_{00}. \tag{4.22}$$

Multiplying equation (4.22) by ρ and adding (4.21) gives

$$J_{11} = -b J_{10} + \rho J_{00} + \sqrt{1-\rho^2}\,[\phi(kb) + kb\Phi(kb)]\,\phi(b).$$

Multiplying equation (4.19) by ρ and adding equation (4.18) yields

$$J_{10} = b J_{00} + [1 - \Phi(kb) + \rho\Phi(kb)]\,\phi(b).$$

Using Lemma 4.3 gives us $J_{00} = 2\int_b^{+\infty} [\Phi(kx) - \frac{1}{2}]\phi(x)\, dx$. Gathering the various pieces, we have

$$J_{11} = J_{11}(b,\rho) = \sqrt{1-\rho^2}\, \phi^2 \frac{b}{\sqrt{1-\rho}} \phi(b) + 2(\rho - b^2)$$

$$\int_b^{+\infty} \left[\Phi(kx) - \frac{1}{2}\right] \phi(x)\, dx + 2b\left[\Phi(kb) - \frac{1}{2}\phi(b)\right].$$

The final result is obtained taking into account that

$$\mathrm{E}\left((X_0')^+ (X_t')^+ | X_0 = X_t = u \right) = \sigma^2(t) J_{11}(b(t), \rho(t)).$$

(iii) Expression (4.17) follows from the second expression of J_{00}. □

The numerical computation of ν_2 has some difficulties related to the behavior of the integrand near the diagonal. Since $\nu_2 = \iint_0^T A_{s,t}^+(u,u)\, ds\, dt$, one can use the next proposition to describe the function $A_{s,t}^+(u,u)$ when $|t-s|$ is small. For the proof, which we are not going to give here, one can use Maple to compute its Taylor expansion as a function of $t-s$ in a neighborhood of 0.

Proposition 4.5 (Azaïs et al., 1999). *Let the Gaussian stationary process $\{X(t) : t \in \mathbb{R}\}$ satisfy the hypotheses presented earlier in this section. Assume, moreover, that λ_8 is finite.*

(a) *As $t \to 0$:*

$$A_{0,t}^+(u,u) = \frac{1}{1296} \frac{(\lambda_2\lambda_6 - \lambda_4)^{3/2}}{\left(\lambda_4 - \lambda_2^2\right)^{1/2} \pi^2 \lambda_2^2} \exp\left(-\frac{1}{2}\frac{\lambda_4}{\lambda_4 - \lambda_2^2} u^2\right) t^4 + O(t^5).$$

(b) *There exists $T_0 > 0$ such that for every $T, 0 < T < T_0$,*

$$\overline{\nu}_1 = \frac{27}{4\sqrt{\pi}} \frac{\left(\lambda_4 - \lambda_2^2\right)^{11/2}}{\lambda_2^5 \left(\lambda_2\lambda_6 - \lambda_4^2\right)^{3/2}} \phi\left(\sqrt{\frac{\lambda_4}{\lambda_4 - \lambda_2^2}} u\right) u^{-6} \left(1 + O\left(\frac{1}{u}\right)\right)$$

$$\nu_2 = \frac{3\sqrt{3}T}{\pi} \frac{\left(\lambda_4 - \lambda_2^2\right)^{9/2}}{\lambda_2^{9/2} \left(\lambda_2\lambda_6 - \lambda_4^2\right)} \phi\left(\sqrt{\frac{\lambda_4}{\lambda_4 - \lambda_2^2}} u\right) u^{-5} \left(1 + O\left(\frac{1}{u}\right)\right)$$

as $u \to +\infty$.

We consider the same question on the behavior of the integrand $A^+_{s,t}(u, u)$ near the diagonal for nonstationary Gaussian processes in Chapter 5.

4.3. MAXIMUM OF THE ABSOLUTE VALUE

We set $M_T^* = \sup_{t\in[0,T]} |X(t)|$. The following inequality for $P\{M_T^* > u\}$ is elementary, the proof being left to the reader:

$$\begin{aligned} &P(|X(0)| > u) + E\left(U_u \mathbb{1}_{|X(0)|<u}\right) + E\left(D_{-u} \mathbb{1}_{|X(0)|<u}\right) - \tfrac{1}{2}E\left((U_u + D_{-u})^{[2]}\right) \\ &\quad \le P(M_T^* > u) \le P(|X(0)| > u) + E\left(U_u \mathbb{1}_{|X(0)|<u}\right) + E\left(D_{-u} \mathbb{1}_{|X(0)|<u}\right). \end{aligned} \tag{4.23}$$

Delmas (2001) has provided tables with formulas to compute the terms appearing in (4.15) and (4.23) as well as extensions to nonstationary processes. We refer to it for a comprehensive list of useful formulas. Here, we only describe the result for the distribution of M_T^* in the case of centered stationary Gaussian processes, normalized by $\Gamma(0) = 1$. The terms are included in Table 4.1.

For $u > 0$,

$$P(|X(0)| > u) = 2[1 - \Phi(u)], \qquad U_u \stackrel{D}{=} D_{-u},$$

and for $b > 0$ we use notation already defined:

$$\sigma^2 J_{11}(b, \rho) := T_1(b, \rho, \sigma^2) + T_2(b, \rho, \sigma^2) + T_3(b, \rho, \sigma^2)$$

with

$$T_1(b, \rho, \sigma^2) := \sigma^2\sqrt{1 - \rho^2}\phi(b)\phi(kb)$$

$$T_2(b, \rho, \sigma^2) := (\sigma^2\rho - \sigma^2 b^2)\left[\frac{1}{\pi}\arctan k - 2\int_0^b \phi(x)\Psi(kx)\,dx\right]$$

$$T_3(b, \rho, \sigma^2) := 2b\sigma^2\left[\Phi(kb) - \tfrac{1}{2}\right]\phi(b).$$

TABLE 4.1. Terms for the Centered Stationary Case

$\mathrm{E}[U_u^X[0,T]\mathbf{I}_{\{X(0)>u\}}]$	$F(\Gamma,\Gamma') := \phi(u)\int_0^T \frac{\sqrt{\lambda_2}}{\sqrt{2\pi}}\left(1-\Phi\left[\frac{u}{\sigma}\sqrt{\lambda_2}\sqrt{\frac{1-\Gamma}{1+\Gamma}}\right]\right) + \frac{\Gamma'}{\sqrt{1-\Gamma^2}}\phi\left(u\sqrt{\frac{1-\Gamma}{1+\Gamma}}\right)\left(1-\Phi\left[-\frac{u\Gamma'}{\sigma(1+\Gamma)}\right]\right)dt$
$\mathrm{E}[U_u^X[0,T]\mathbf{I}_{\{X(0)\le u\}}]$	$\frac{T\sqrt{\lambda_2}}{2\pi}\exp\left(-\frac{u^2}{2}\right) - F(\Gamma,\Gamma')$
$\mathrm{E}[U_u^X[0,T]\mathbf{I}_{\{\lvert X(0)\rvert\le u\}}]$	$\frac{T\sqrt{\lambda_2}}{2\pi}\exp\left(-\frac{u^2}{2}\right) - F(\Gamma,\Gamma') - F(-\Gamma,-\Gamma')$
$\mathrm{E}[U_u^X[0,T]^{[2]}]$	$2\int_0^T (T-t)\sigma^2 J_{11}\left(-\frac{\Gamma' u}{\sigma(1+\Gamma)},\rho\right)\frac{\phi^2(u/\sqrt{1+\Gamma})}{\sqrt{1-\Gamma^2}}dt$
$\mathrm{E}[U_u^X[0,T]D_{-u}^X[0,T]]$	$2\int_0^T (T-t)\sigma^2 J_{11}\left(-\frac{\Gamma' u}{\sigma(1-\Gamma)},-\rho\right)\frac{\phi^2(u/\sqrt{1-\Gamma})}{\sqrt{1-\Gamma^2}}dt$

4.4. APPLICATION TO QUANTITATIVE GENE DETECTION

We study a backcross population: A × (A × B), where A and B are purely homozygous lines and we address the problem of detecting a gene influencing some quantitative trait (i.e., which is able to be measured) on a given chromosome. The trait is observed on n individuals and we denote by $Y_k, k = 1, \dots, n$ the observations, which we will assume to be independent.

The mechanism of genetics, or more precisely of meiosis, implies that among the two chromosomes of each individual, one is inherited purely from A while the other (the "recombined" one) consists of parts that originate from A and parts that originate from B, due to crossing-overs. Using the Haldane (1919) distance and modeling, each chromosome will be represented by a segment $[0, L]$. The distance on $[0, L]$ is called the *genetic distance* (which is measured in morgans).

Now, let us describe the mathematical model that we will be using. We assume that the "recombined" chromosome starts at the left endpoint $t = 0$ with probability $\frac{1}{2}$ with some part originated from A or B and then switches from one value to another at every location of a crossing-over. We model the crossing-over points by a standard Poisson process, independent of the fact that the chromosome starts with A or B.

The influence of the putative gene [often called QTL (*quantitative trait locus*) by geneticists] on the quantitative trait is represented by a classical linear model:

$$Y_k = \mu + G_k(t_0)\,a/2 + \varepsilon_k \qquad k = 1, \dots, n, \tag{4.24}$$

where:

- μ is the general mean.
- t_0 is the location of the gene on the chromosome.
- $G_k(t)$ is the genotypic composition of the individual k at location t on the chromosome, $t \in [0, L]$. In a backcross crossing scheme it can only take two values (AB or AA), denoted $+1$ and -1.
- $\varepsilon_k, k = 1, \ldots, n$ are independent errors that are assumed to be i.i.d. with zero mean and finite fourth moment. We denote by σ^2 their common variance.

Formula (4.24) implies that the gene effect is a. Our problem is to test the null hypothesis of absence of gene influence; that is, $a = 0$.

The main statistical difficulties come from the following facts:

- The putative location t_0 of the gene is unknown.
- The functions $G_k(t)$ cannot be fully observed. However, some genetic markers are available, allowing us to determine the genetic composition at some fixed locations $t_1, \ldots, t_M$ on $[0, L]$.

Since the number of genetic markers available on a given species becomes larger and larger and since considering the limit model where the number of markers tends to infinity permits us to construct a test that is free of the markers' positions, we will assume that the number M_n of markers as well as their locations depend on the number of observations n. More precisely, we now use a local asymptotic framework in which:

- The number n of individuals observed tends to infinity.
- The number of genetic markers M_n tends to infinity with n, their locations being denoted by $t_{i,n}; i = 1, \ldots, M_n$.
- The size a of the gene effect is small, satisfying the contiguity condition $a = \delta n^{-1/2}$, where δ is a constant.
- The observation is

$$\{(Y_k^n, G_k(t_{1,n}), \ldots, G_k(t_{M_n,n})), k = 1, \ldots, n\}.$$

Notice that since we have set $a = \delta n^{-1/2}$, the distribution of the observation depends on n, so that the observations are denoted by Y_k^n instead of Y_k.

If the true position t_0 of the gene were known and would coincide with a marker location, a natural test would be to make a *comparison of means*, computing the statistics:

$$S_n(t_0) := \frac{\sum_{k=1}^n Y_k^n \mathbb{I}_{[G_k(t_0)=1]}}{\sum_{k=1}^n \mathbb{I}_{[G_k(t_0)=1]}} - \frac{\sum_{k=1}^n Y_k^n \mathbb{I}_{[G_k(t_0)=-1]}}{\sum_{k=1}^n \mathbb{I}_{[G_k(t_0)=-1]}}. \tag{4.25}$$

In case of Gaussian observations, the test based on these statistics is equivalent to the likelihood ratio test.

Since t_0 is unknown, the calculation of the quantity given by formula (4.25) should be performed at each location $t \in [0, L]$. To do this, we compute S_n at points $t_{1,n}, \ldots, t_{M_n,n}$, make a linear interpolation between two consecutive marker positions, and extend by a constant to the left of the first marker position and similarly to the right of the last one, so that the function thus obtained is continuous. We denote by $\{S_n(t) : t \in [0, L]\}$ the random process obtained in this form, which has continuous polygonal paths.

Our aim is to find a normalization of the process S_n in order to get weak convergence. Before that, let us make a short diversion to give an overview on this subject, which includes some general and useful results.

4.4.1. Weak Convergence of Stochastic Processes

Let us consider real-valued processes defined on [0, 1] (extensions to the whole line do not present essential difficulties). We need the following function spaces:

- $C = C([0, 1])$, the space of continuous functions $[0, 1] \to \mathbb{R}$, equipped with the topology of uniform convergence generated by the distance

$$d_u(f, g) = \|f - g\|_\infty := \sup_{t \in [0,1]} |f(t) - g(t)|.$$

- $D = D([0, 1])$, the set of càd-làg functions (functions that are right-continuous and have left-hand limits at each point), equipped with the Skorohod topology generated by the distance

$$d_s(f, g) := \inf_h \{\sup\{\|h - \mathrm{Id}\|_\infty, \|f - g \circ h\|_\infty\}\},$$

  where the infimum is taken over all strictly increasing continuous mappings $h : [0, 1] \to [0, 1]$, $h(0) = 0$, $h(1) = 1$, Id is the identity mapping (i.e., $\mathrm{Id}(x) = x$ for all $x \in [0, 1]$), and "$\circ$" denotes composition.

C is a Polish space (i.e., a complete separable metric space). D is not complete, but with a natural change of the metric, which does not modify the topology, it becomes complete [see Billingsley (1999) for details]. A possible modification consists of replacing d_s by d_s^0 defined by

$$d_s^0(f, g) := \inf_h \{\sup\{\|h\|^o, \|f - g \circ h\|_\infty\}\},$$

where the inf is over the same class of functions as above, and

$$\|h\|^o = \sup_{s<t} \left| \log \frac{h(t) - h(s)}{t - s} \right|.$$

Without further reference, we will usually denote by the same letters C and D the analogous spaces of functions defined on some interval $[t_1, t_2]$ other than $[0, 1]$.

Next we give two definitions that are basic in what follows: weak convergence and tightness.

Definition 4.6. *Let E be a Polish space and $\mathcal{E}$ the σ-algebra of subsets of E generated by the open sets. The sequence of probability measures $\{\mathrm{P}_n\}_{n=1,2,\ldots}$ on $(E, \mathcal{E})$ is said to* converge *weakly to the measure* P *defined on the same measurable space (this is denoted by $\mathrm{P}_n \Rightarrow \mathrm{P}$ as $n \to +\infty$) if for every continuous and bounded function $f : E \to \mathbb{R}$,*

$$\int_E f\,d\mathrm{P}_n \to \int_E f\,d\mathrm{P} \quad \textit{as } n \to +\infty.$$

Definition 4.7. *Let E be a Polish space. A collection of probability measures $\mathcal{F}$ on $(E, \mathcal{E})$ is said to be* tight *if for any $\varepsilon > 0$ there exists K_ε, a compact subset of E, such that for all $\mathrm{P} \in \mathcal{F}$, $\mathrm{P}(K_\varepsilon^C) \leq \varepsilon$.*

A sequence of random variables $\{Y_n\}_{n=1,2,\ldots}$ with values in E is said to *converge weakly* to the random variable Y if

$$\mathrm{P}^{Y_n} \Rightarrow \mathrm{P}^Y \quad \text{as } n \to +\infty,$$

that is, if the sequence of distributions converges weakly to the distribution of Y. P^{Y_n} and P^Y are the image measure of Y_n and Y, respectively, as defined in Section 1.1. In that case, we write $Y_n \Rightarrow Y$.

The next three theorems contain the results that we will be using in the statistical applications in this chapter. For proofs, examples, and a general understanding of the subject, the reader is referred to Billingsley's classical book (1999).

Theorem 4.8 (Prohorov). *Let E be a Polish space. From every tight sequence of probability measures on $(E, \mathcal{E})$, one can extract a weakly convergent subsequence. If a sequence of probability measures on $(E, \mathcal{E})$ is tight, it is weakly convergent if and only if all weakly convergent subsequences have the same limit.*

Theorem 4.9. *Let $\{X_n\}_{n=1,2,\ldots}$ be a sequence of random variables with values in the spaces C or D. $X_n(t)$ denotes the value of X_n at $t \in [0, 1]$. Then $X_n \Rightarrow X$ as $n \to +\infty$, if and only if:*

- *$\{X_n\}_{n=1,2,\ldots}$ is tight (in the corresponding space C or D).*
- *For any choice of $k = 1, 2, \ldots,$ and distinct $t_1, \ldots, t_k \in [0, 1]$, the random vector $(X_n(t_1), \ldots, X_n(t_k))$ converges weakly to $(X(t_1), \ldots, X(t_k))$ in $\mathbb{R}^k$ as $n \to +\infty$.*

When convergence takes place, if $g : C \to \mathbb{R}$ (respectively, $g : D \to \mathbb{R}$) is a continuous function with respect to the topology of C (respectively, D), the sequence of real-valued random variables $\{g(X_n)\}_{n=1,2,...}$ converges in distribution to the random variable $g(X)$. Moreover, if $X_n \Rightarrow X$ in D and X has continuous paths, weak convergence holds true in C (see Exercise 4.3).

The statement above contains the standard procedure, in a large set of statistical problems, to prove the existence of weak limits of stochastic processes and to compute the limit distributions whenever they exist. One has to check tightness and finite-dimensional weak convergence. If possible, one also wants to identify the limit measure (the distribution of X), and if the "observable" quantity in which one is interested in is $g(X_n)$, this also allows to find its limit distribution.

The next theorem gives a sufficient condition based on upper bounds for moments of increments, to verify tightness and prove weak convergence.

Theorem 4.10. *Let $\{X_n(t) : t \in [0, 1]\}, n = 1, 2, \ldots$ be a sequence of random processes and $\{X(t) : t \in [0, 1]$ a process with sample paths in C (respectively, D) satisfying:*

(1) *For any choice of $k = 1, 2, \ldots,$ and distinct $t_1, \ldots, t_k \in [0, 1]$, the sequence of random vectors $(X_n(t_1), \ldots, X_n(t_k))$ converges weakly to $(X(t_1), \ldots, X(t_k))$ in $\mathbb{R}^k$ as $n \to +\infty$.*

(2) *If the sample paths are in C, there exist three positive constants, α, β, and γ such that for all $s, t \in [0, 1]$,*

$$\mathrm{E}|X_n(s) - X_n(t)|^\alpha \le \beta |s - t|^{1+\gamma}.$$

If the sample paths are in D, there exist three positive constants, α, β, andγ such that for all $t_1, t, t_2 \in [0, 1], t_1 \le t \le t_2$,

$$\mathrm{E}\left(|X_n(t_1) - X_n(t)|^\alpha |X_n(t_2) - X_n(t)|^\alpha\right) \le \beta |t_2 - t_1|^{1+\gamma}. \tag{4.26}$$

Then $X_n \Rightarrow X$ as $n \to +\infty$ in C (respectively, in D).

In addition to the tightness of the family of probability distributions $X_n(0)$, property (2) is a sufficient condition for the tightness of the sequence $\{X_n\}_{n=1,2,...}$.

We will also use the *Skorohod embedding* (see, e.g., Dudley, 1989 p. 324), which states that if X_n and X are random variables taking values in a separable metric space (E, d) and if X_n converges weakly to X, we can construct a representation of these variables, say $\{Y_n\}_{n=1,2,...}, Y$, defined on some new probability space with values in E, so that X_n and Y_n have the same distribution for every $n = 1, 2, \ldots$ (as well as Y and X) and $d(Y_n, Y) \to 0$ a.s. as $n \to +\infty$.

4.4.2. Weak Convergence of the Detection Process

We go back to the genetic model.

Theorem 4.11

(a) $\text{Var}\,(S_n(t)) \approx 4\sigma^2/n$.

(b) *Since* $\text{Var}(Y_k^n) \approx \sigma^2$, *we can estimate the parameter* σ^2 *by the empirical variance:* $\hat{\sigma}_n^2$. *We define the normalized process*

$$X_n(t) := \frac{\sqrt{n}}{2\hat{\sigma}_n} S_n(t). \tag{4.27}$$

Assume that

$$\max_{i=0,\dots,M_n} \left(t_{i+1,n} - t_{i,n}\right) \to 0 \text{ as } n \to +\infty,$$

where we have used the convention $t_{0,n} = 0$; $t_{M_n+1,n} = L$.

Then the normalized process $X_n(t)$ *converges weakly in the space of continuous functions on* $[0, L]$ *to a Gaussian process* $\mathcal{X} = \{X(t) : t \in [0, L]\}$ *with*

$$\begin{aligned} \text{E}\,(X(t)) &= \frac{\delta}{2\sigma} \exp(-2|t_0 - t|) \\ \text{Cov}\,(X(t), X(t+h)) &= \exp(-2|h|). \end{aligned} \tag{4.28}$$

$\mathcal{X}$ *is an Ornstein–Uhlenbeck process with a change of scale and a deterministic drift (see Exercise 4.6).*

The proof of Theorem 4.11 is based on the following lemma.

Lemma 4.12. *Let* $\{\eta_k\}_{k=1,2,\dots}$ *be centered random variables with common variance* σ^2 *and finite fourth-order moment* μ_4. *Suppose that the collection of random variables and processes* $\eta_1, G_1(\cdot), \dots, \eta_n, G_n(\cdot), \dots$ *are independent. Here* $G_k(t), t \in [0, L]$ *is the genotypic composition of the* kth *individual, which follows the switching Poisson model that has been described above. Then, as* $n \to +\infty$:

(a) *The processes*

$$Z_n(t) := n^{-1/2} \sum_{k=1}^{n} \eta_k G_k(t)$$

converge weakly in space D *to a stationary Gaussian process with zero mean and covariance* $h \rightsquigarrow \sigma^2 \exp(-2|h|)$.

(b) *The processes*

$$\tilde{Z}_n(t) := n^{-1} \sum_{k=1}^{n} G_k(t_0) G_k(t)$$

converge uniformly (for $t \in [0, L]$*) in probability to the function* $\exp(-2|t_0 - t|)$.

Proof

(a) Independence implies that

$$\mathrm{E}\,(\eta_k G_k(t)) = 0.$$

If Z is a random variable having the Poisson distribution with parameter λ, one easily checks that

$$\mathrm{P}(Z \text{ even }) - \mathrm{P}(Z \text{ odd}) = \exp(-2\lambda).$$

This implies that

$$\mathrm{E}\left(G_k(t)G_k(t')\right) = \exp(-2|t-t'|), \qquad \mathrm{E}\left(\eta_k^2 G_k(t)G_k(t')\right) = \sigma^2 \exp(-2|t-t'|).$$

To prove part (a), we use Theorem 4.10. The convergence of the finite-dimensional distributions follows from a standard application of the multivariate central limit theorem and is left to the reader.

We now prove the moment condition (4.26). Let $t_1 < t < t_2$ be in $[0, L]$:

$$\begin{aligned}
&\mathrm{E}\Big[\,(Z_n(t) - Z_n(t_1))^2\,(Z_n(t) - Z_n(t_2))^2\,\Big] \\
&\quad = \frac{1}{n^2} \sum_{1 \le k_1, k_2, k_3, k_4 \le n} \mathrm{E}\left(\eta_{k_1} \cdots \eta_{k_4} \left(G_{k_1}(t) - G_{k_1}(t_1)\right)\left(G_{k_2}(t) - G_{k_2}(t_1)\right)\right. \\
&\qquad \left.\left(G_{k_3}(t) - G_{k_3}(t_2)\right)\left(G_{k_4}(t) - G_{k_4}(t_2)\right)\right). \qquad (4.29)
\end{aligned}$$

The independence implies that as soon as one index k_i is different from the other three, the expectation in the corresponding term of the sum vanishes. Hence,

$$\begin{aligned}
&\mathrm{E}\Big[\,(Z_n(t) - Z_n(t_1))^2\,(Z_n(t) - Z_n(t_2))^2\,\Big] \\
&\quad = \frac{1}{n^2} \sum_{1 \le k_1 \ne k_2 \le n} \mathrm{E}\left(\eta_{k_1}^2 \eta_{k_2}^2 \left(G_{k_1}(t) - G_{k_1}(t_1)\right)^2 \left(G_{k_2}(t) - G_{k_2}(t_2)\right)^2\right) \\
&\qquad + \frac{1}{n^2} \sum_{1 \le k_1 \ne k_2 \le n} \mathrm{E}\left(\eta_{k_1}^2 \eta_{k_2}^2 \left(G_{k_1}(t) - G_{k_1}(t_1)\right)\left(G_{k_1}(t) - G_{k_1}(t_2)\right)\right. \\
&\qquad \left.\left(G_{k_2}(t) - G_{k_2}(t_1)\right)\left(G_{k_2}(t) - G_{k_2}(t_2)\right)\right) \\
&\qquad + \frac{1}{n^2} \sum_{1 \le k \le n} \mathrm{E}\Big(\eta_k^4\,(G_k(t) - G_k(t_1))^2\,(G_k(t) - G_k(t_2))^2 \\
&\quad \le (\mathrm{const})\sigma^4 |t - t_1||t - t_2| + (\mathrm{const})\mu_4 |t - t_1||t - t_2| \le (\mathrm{const})(t_2 - t_1)^2,
\end{aligned}$$

(4.30)

where we have used the following facts:

- Almost surely, $(G_k(t) - G_k(t_1))(G_k(t) - G_k(t_2))$ vanishes unless the number of occurrences in intervals $[t_1, t]$ and $[t, t_2]$ are both odd. In that case it takes the value -4. In conclusion, its expectation is nonpositive.
- $(G_k(t) - G_k(t_2))^2$ depends only on the parity of the number of occurrences of the Poisson process on $[t, t_2]$, and is independent of $(G_k(t) - G_k(t_1))$.
- $\mathrm{E}\left(G_k(t) - G_k(t')\right)^2 = 2\left(1 - \exp(-2|t - t'|)\right) \leq (\text{const})|t - t'|$.

The inequality (4.30) implies (4.26). This proves part (a).

(b) We write

$$\begin{aligned}\tilde{Z}_n(t) &:= n^{-1}\sum_{k=1}^{n} G_k(t_0)G_k(t)\\ &= n^{-1}\sum_{k=1}^{n} (G_k(t_0)G_k(t) - \exp(-2|t - t_0|)) + \exp(-2|t - t_0|)\\ &= n^{-1}\sum_{k=1}^{n} T_k(t) + \exp(-2|t - t_0|) = \tilde{Z}_{n,1}(t) + \exp(-2|t - t_0|)\,. \quad (4.31)\end{aligned}$$

Part (b) follows from the result in Exercise 4.4 if we can prove that $\tilde{Z}_{n,1}$ tends weakly to zero in the space D as $n \to +\infty$. Because of the strong law of large numbers, for each t, $\tilde{Z}_{n,1}(t)$ a.s. tends to zero. This obviously implies that a.s. the k-tuple $\left(\tilde{Z}_{n,1}(t_1), \ldots, \tilde{Z}_{n,1}(t_k)\right)$ converges to $(0, \ldots, 0)$ as $n \to +\infty$.

So, to apply Theorem 4.10 it suffices to check (4.26). This can be done in much the same way that we did with formula (4.30) except that (1) the normalizing constant is now n^{-4}, (2) $G_k(t)$ is replaced by $T_k(t)$, and (3) the variables η_k are absent. To conclude, we have to check that

- $$\begin{aligned}\mathrm{E}\left(T_k(t) - T_k(t')\right)^2 &\leq \mathrm{E}\left(G_k(t_0)\left(G_k(t) - G_k(t')\right)\right)^2\\ &= \mathrm{E}\left(G_k(t) - G_k(t')\right)^2 \leq (\text{const})|t - t'|.\end{aligned}$$
- $$\begin{aligned}&\mathrm{E}\left((T_k(t) - T_k(t_1))(T_k(t) - T_k(t_2))\right)\\ &= \mathrm{Cov}\big(G_k(t_0)(G_k(t) - G_k(t_1)),\, G_k(t_0)(G_k(t) - G_k(t_2))\big)\\ &= \mathrm{E}\left((G_k(t) - G_k(t_1))(G_k(t) - G_k(t_2))\right)\\ &\quad - \left(e^{-2|t_0 - t|} - e^{-2|t_0 - t_1|}\right)\left(e^{-2|t_0 - t|} - e^{-2|t_0 - t_2|}\right)\\ &\leq -\left(e^{-2|t_0 - t|} - e^{-2|t_0 - t_1|}\right)\left(e^{-2|t_0 - t|} - e^{-2|t_0 - t_2|}\right) \leq (\text{const})(t_1 - t_2)^2.\end{aligned}$$
- If Z and T are two random variables:

$$\mathrm{E}\left[(Z - \mathrm{E}(Z))^2 (T - \mathrm{E}(T))^2\right] \leq 4\mathrm{E}(Z^2T^2) + 12\mathrm{E}(Z^2)\mathrm{E}(T^2).$$

Applying this with $Z - \mathrm{E}(Z) = T_k(t) - T_k(t_1)$; $T - \mathrm{E}(T) = T_k(t) - T_k(t_2)$, we get

$$\begin{aligned}
&\mathrm{E}\big[\,(T_k(t) - T_k(t_1))^2\,(T_k(t) - T_k(t_2))^2\,\big] \\
&\quad \le 16\mathrm{E}\,(G_k(t) - G_k(t_1))^2\,\mathrm{E}\,(G_k(t) - G_k(t_2))^2 \\
&\quad \le (\mathrm{const})(t_1 - t_2)^2.
\end{aligned}$$

Summing up, we have

$$\mathrm{E}\left[\left(\tilde{Z}_{n,1}(t_1) - \tilde{Z}_{n,1}(t)\right)^2\left(\tilde{Z}_{n,1}(t_2) - \tilde{Z}_{n,1}(t)\right)^2\right] \le (\mathrm{const})(t_1 - t_2)^2 n^{-2}.$$

□

Proof of Theorem 4.11

STEP 1: CONVERGENCE IN the CASE OF FULL GENETIC INFORMATION. In a first step, we assume that the genetic information $G_k(t)$ is available at every location of the chromosome, so that no interpolation is needed. We define several auxiliary random processes:

$$X_{n,1}(t) := \frac{\sqrt{n}}{2\sigma} S_n(t),$$

(It is clear that if the process $\{X_{n,1}(t) : t \in [0, L]\}$ converges weakly, the same holds true for the process $\{X_n(t) : t \in [0, L]\}$, with the same limit.)

$$X_{n,2}(t) := \frac{1}{\hat{\sigma}_n\sqrt{n}} \sum_{k=1}^{n} \left(Y_k^n - \bar{Y}_n\right) G_k(t),$$

which can actually be computed ($\bar{Y}_n$ is the mean of the sample), and

$$X_{n,3}(t) := \frac{1}{\sigma\sqrt{n}} \sum_{k=1}^{n} (Y_k^n - \mu) G_k(t),$$

which cannot actually be computed.

The convergence in probability of $\hat{\sigma}_n^2$ implies that if $X_{n,3}(t)$ converges weakly, it is also the case for the process

$$X_{n,4}(t) := \frac{1}{\hat{\sigma}_n\sqrt{n}} \sum_{k=1}^{n} \left(Y_k^n - \mu\right) G_k(t).$$

We have

$$X_{n,4}(t) - X_{n,2}(t) = \frac{\bar{Y}_n - \mu}{\hat{\sigma}_n\sqrt{n}} \sum_{k=1}^{n} G_k(t).$$

Since under the conditions of Lemma 4.12 the law of the process $\{\sum_{k=1}^{n} G_k(t) : t \in [0, L]\}$ is the same as that of $\{\sum_{k=1}^{n} \eta_k G_k(t) : t \in [0, L]\}$, we can use part (a) of this lemma and Exercise 4.4 to deduce that the process $\{n^{-1/2} \sum_{k=1}^{n} G_k(t) : t \in [0, L]\}$ is stochastically uniformly bounded. Now apply the law of large numbers to the sequence of means $\{\bar{Y}_n\}_{n=1,2,\ldots}$ to deduce that

$$\sup_{t\in[0,L]} \left|X_{n,4}(t) - X_{n,2}(t)\right|$$

tends to zero in probability.

We see also that

$$\sup_{t\in[0,L]} \left| X_{n,3}(t) - X_{n,1}(t)\right| \Rightarrow 0. \tag{4.32}$$

In fact, some algebra permits us to check that

$$\begin{aligned}
X_{n,3}(t) - X_{n,1}(t) &= \frac{1}{\sigma\sqrt{n}} \sum_{k=1}^{n} (Y_k^n - \mu) G_k(t) \\
&\quad - \frac{\sqrt{n}}{2\sigma} \sum_{k=1}^{n} Y_k^n \left(\frac{\mathbf{I}_{G_k(t)=1}}{\nu_n(t)} - \frac{\mathbf{I}_{G_k(t)=-1}}{n - \nu_n(t)} \right) \\
&= \frac{\sqrt{n}}{2\sigma} \sum_{k=1}^{n} (Y_k^n - \mu) \left[\frac{2}{n} G_k(t) - \left(\frac{\mathbf{I}_{G_k(t)=1}}{\nu_n(t)} - \frac{\mathbf{I}_{G_k(t)=-1}}{n - \nu_n(t)} \right) \right] \\
&= \frac{f_n(t) - 1/2}{\sigma\sqrt{n} f_n(t)[1 - f_n(t)]} \sum_{k=1}^{n} \left[G_k(t_0) \frac{\delta}{\sqrt{n}} + \varepsilon_k \right] \\
&\quad \times \left[\frac{1}{2} + (1/2 - f_n(t)) G_k(t) \right],
\end{aligned}$$

where we have set

$$\nu_n(t) = \#\{k : 1 \le k \le n, G_k(t) = 1\}$$

$$f_n(t) = \frac{1}{n} \nu_n(t) = \frac{1}{2n} \sum_{k=1}^{n} G_k(t) + \frac{1}{2}.$$

Relation (4.32) follows using Lemma 4.12.

So weak convergence of $X_{n,3}$ implies convergence of the other processes. The discussion above also proves that the variance of $X_n(t)$ is equivalent to that of $X_{n,3}(t)$, which tends to 1, thus proving assertion (a) in the statement of the theorem.

The model implies that at every location t and for every individual k, $\mathrm{P}\{G_k(t) = 1\} = \frac{1}{2}$ and

$$\mathrm{E}\,(\varepsilon_k G_k(t)) = 0, \qquad \mathrm{Var}\,(\varepsilon_k G_k(t)) = \sigma^2.$$

To finish this part, let us turn to the process $\{X_{n,3}(t) : t \in [0, L]\}$. Set

$$X_{n,3}(t) := \frac{1}{\sigma\sqrt{n}} \sum_{k=1}^{n} Y_k^n G_k(t) = \frac{\delta}{2n\sigma} \sum_{k=1}^{n} G_k(t_0) G_k(t)$$
$$+ \frac{1}{\sigma\sqrt{n}} \sum_{k=1}^{n} \varepsilon_k G_k(t) = X_{n,5}(t) + X_{n,6}(t).$$

By Lemma 4.12, $X_{n,5}(t)$ converges uniformly to the function $(\delta/2\sigma)\exp(-2|t-t_0|)$ and $X_{n,6}$ converges weakly to the Ornstein–Uhlenbeck process with a scale change, having covariance (4.28).

STEP 2: CONVERGENCE IN THE CASE OF PARTIAL GENETIC INFORMATION. Using the Skorohod embedding technique, the weak convergence of the process $X_{n,3}(t)$ toward the limit process $X(t)$ can be represented by an a.s. convergence in some probability space. Since $X(t)$ has continuous sample paths, the convergence is also true in the uniform topology:

$$\|X_{n,3}(\cdot) - X(\cdot)\|_\infty \to 0 \text{ a.s.}$$

Now let $\mathcal{D}_n$ be the operator on D that consists of (a) discretization at the locations $d_{i,n}$, $i = 1, \ldots, M_n$ followed by (b) linear interpolation between two marker positions and extending by a constant before the first and after the last marker. This operator is a contraction for the uniform norm. We can deduce that

$$\begin{aligned} \|\mathcal{D}_n[X_{n,3}(\cdot)] - X(\cdot)\|_\infty &\le \|\mathcal{D}_n[X_{n,3}(\cdot)] - \mathcal{D}_n[X(\cdot)]\|_\infty + \|\mathcal{D}_n[X(\cdot)] - X(\cdot)\|_\infty \\ &\le \|X_{n,3}(\cdot) - X(\cdot)\|_\infty + \|\mathcal{D}_n[X(\cdot)] - X(\cdot)\|_\infty \to 0 \end{aligned} \tag{4.33}$$

as $n \to +\infty$, and we are done. □

4.4.3. Smoothing the Detection Test Process

In the remaining study of this example, we will combine the previous theoretical results with a series of practical recipes to be able to answer the relevant questions. The original problem is to test the null hypothesis $\delta = 0$ against $\delta \neq 0$. The classical approach would be to use the test statistic $T_n = \sup_{t\in[0,L]} |X_n(t)|$, which corresponds to a likelihood ratio test in the case of Gaussian observations. This is inconvenient for two reasons:

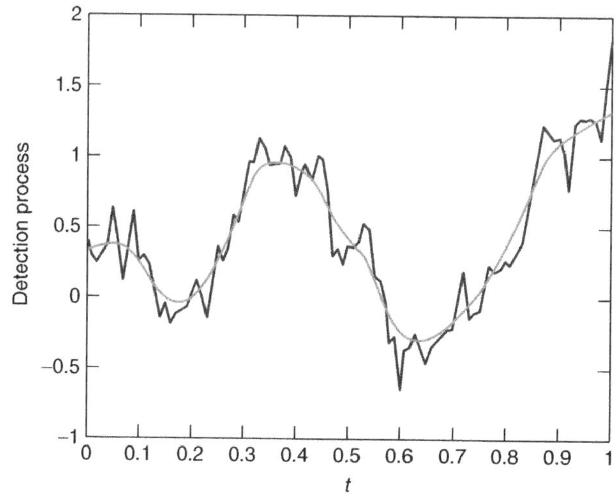

Figure 4.1. Realization of the detection process and its smoothing with $\varepsilon^2 = 10^{-3}$. There are 100 markers on a chromosome of size 1 morgan, and 500 individuals are observed.

1. The limit process has irregular sample paths (nondifferentiable), and the distribution of its supremum is known only when $\delta = 0$ (Delong, 1981) and for certain lengths of the observation interval. In the other cases, for $\delta = 0$ we can use asymptotic bounds for the maximum of *α-regular processes* that are due to Pickands (see, e.g., Leadbetter et al., 1983). But to compute the power of the test, that is, for $\delta \neq 0$, the only method available is Monte Carlo.
2. It does not take into account that the presence of a gene at t_0 modifies the expectation of the limit process in a neighborhood of t_0.

Given these two reasons, we will smooth the paths of the detection process $\{X_n(t) : t \in [0, L]\}$ by means of convolution with a regular kernel, which we take to be a centered Gaussian kernel having variance ε^2, which we denote ϕ_ε (Figure 4.1). Let $\{X_n^\varepsilon(t) : t \in [0, L]\}$ be the smoothed process, defined as $X_n^\varepsilon(t) = (X_n * \phi_\varepsilon)(t)$.

We consider the test statistic $T_n^\varepsilon = \sup_{t\in[0,L]} \left|X_n^\varepsilon(t)\right|$. The reader can check that the limit of $\left(X_n^\varepsilon(t)\right)_{t\in[0,L]}$ is the smoothed version of the previous limit process [or consult Billingsley's book (1999)] and compute the mean $m^\varepsilon(\cdot)$ and the covariance $\Gamma^\varepsilon(\cdot)$ of the new process:

$$m^\varepsilon(t) = \frac{\delta}{2\sigma}\left\{\exp[2(-t_0 + \varepsilon^2 + t)]\Phi\left(\frac{t_0 - 2\varepsilon^2 - t}{\varepsilon}\right)\right.$$
$$\left. + \exp[2(t_0 + \varepsilon^2 - t)]\left[1 - \Phi\left(\frac{t_0 + 2\varepsilon^2 - t}{\varepsilon}\right)\right]\right\} \qquad (4.34)$$

$$\Gamma^{\varepsilon}(t) = \exp(2(2\varepsilon^2 - t))\Phi\left(\frac{t - 4\varepsilon^2}{(2\varepsilon^2)^{1/2}}\right) + \exp(2(2\varepsilon^2 + t))\left[1 - \Phi\left(\frac{t + 4\varepsilon^2}{(2\varepsilon^2)^{1/2}}\right)\right].$$

To compute approximately the power of the test based on this process we use the basic inequalities (4.23). Of course, the use of crossings is feasible for the regularized process, but it is not for the original one, which has nondifferentiable paths. For determination of the threshold, it turns out (on the basis of numerical simulation) that the lower bound in it, which uses second-order factorial moment of crossings, is more accurate than the upper bound. So we determine thresholds using the lower bound.

4.4.4. Simulation Study

On the basis of a Monte Carlo experiment, one can evaluate the quality of the proposed method under a variety of conditions. Especially:

- The relationship between the value of the smoothing parameter and the validity of the asymptotic approximation for a reasonable number of markers and individuals in the sample
- The sharpness of the bounds given by the inequality (4.23) for various values of the smoothing parameter

Table 4.2 displays empirical levels for smoothed and unsmoothed procedures with thresholds computed under the asymptotic distribution.

- For the unsmoothed process ($\varepsilon = 0$), the threshold is calculated using Table II of Delong (1981). For this reason, the chromosome length, 0.98 Morgan, has been chosen to correspond to an entry of DeLong's table and to be close to lengths encountered for several vegetable species.
- For the smoothed process, we used the lower bound in inequality (4.23).

Simulations have been performed for two values of the smoothing parameter and three marker densities: a marker every icM with $i = 1, 2, 7$. The number of individuals is equal to 500; the crossing-overs are simulated according to a standard Poisson process; the simulation has 10^4 realizations; a 5% confidence interval for the empirical levels associated with the theoretical levels is indicated.

Table 4.2 presents the power associated with the detection test in the case of a gene of size $\delta = 6$ located at the position $t_0 = 0.4$. The length of the chromosome is 1 Morgan (M); calculations are made under the asymptotic distribution, using a test with nominal level equal to 5%.

- For the unsmoothed detection test process, the threshold is calculated via DeLong's table and the power using Monte Carlo with 10^4 simulations.

TABLE 4.2. Threshold and Empirical Level (%) of Test Using the Unsmoothed Detection Test Process ($\varepsilon = 0$) $(X_n(d))_{t\in[0,L]}$ and the Smoothed Detection Process $\left(X_n^{\varepsilon}(t)\right)_{d\in[0,L]}$[a]

	Nominal Level of the Test								
	10%			5%			1%		
5% confidence interval for the empirical level	9.41–10.59			4.57–5.43			0.80–1.19		
Threshold									
$\varepsilon = 0$	2.74			3.01			3.55		
$\varepsilon^2 = 10^{-2}$	2.019			2.276			2.785		
$\varepsilon^2 = 10^{-3}$	2.321			2.593			3.128		
Marker density	1 cM	2 cM	7 cM	1 cM	2 cM	7 cM	1 cM	2 cM	7 cM
Empirical level									
$\varepsilon = 0$	7.37	6.67	4.99	3.91	3.42	2.4	0.77	0.67	0.43
$\varepsilon^2 = 10^{-2}$	12.17	12.17	11.82	6.75	6.69	6.53	1.76	1.72	1.77
$\varepsilon^2 = 10^{-3}$	10.84	10.66	9.71	5.63	5.55	5.02	1.34	1.32	1.04

[a]The chromosome length is equal to 0.98 morgan, and the number of individuals is equal to 500. The second line of the table gives a confidence interval for the empirical proportion related to the nominal level over 10^4 simulations.

- For the smoothed process, the threshold is calculated as above using the lower bound in (4.23). The power of the test is calculated in three ways: (1) using the upper bound in (4.23), (2) using the lower bound in (4.23), and (3) by a Monte Carlo method.

Summing up, let us add some final comments on this study of the genetic model (4.24).

- Table 4.2 indicates clearly that the unsmoothed procedure is very conservative.
- The empirical level given by the smoothed procedure is close to the nominal value. For $\varepsilon^2 = 10^{-3}$, it is nearly inside the confidence interval.
- Table 4.3 shows clearly that smoothing at size $\varepsilon^2 = 10^{-2}$ instead of 10^{-3} does not imply a sizable loss of the power computed with the asymptotic distribution.
- It is also clear from Table 4.3 that at the sizes $\varepsilon^2 = 10^{-2}$ and 10^{-3}, the lower bound is almost exact.
- Use of the asymptotic test after smoothing with a window size $\varepsilon^2 = 10^{-3}$ and thresholds and powers computed by means of the lower bound in (4.23) has a number of advantages. The thresholds corresponding to levels α equal to 1%, 5%, and 10% and for certain chromosome lengths are given in Table 4.4. For other cases and power calculations, the S^+ program developed by Cierco-Ayrolles et al. (2003) can be used (see also Chapter 9).

TABLE 4.3. Power (%) Associated with the Detection Test[a]

	$\varepsilon^2 = 10^{-2}$	$\varepsilon^2 = 10^{-3}$	Unsmoothed Process
5% threshold	2.281	2.599	3.02
$P(\lvert Y(0)\rvert > u)$	15.70	9.81	
$E\left((U_u + D_{-u})\, \mathbf{I}_{\lvert Y(0)\rvert \le u}\right)$	55.57	72.30	
$\dfrac{E(U_u(U_u - 1))}{2}$	1.43	13.06	
$\dfrac{E(D_{-u}(D_{-u} - 1))}{2}$	7.84×10^{-6}	3.40×10^{-4}	
$E(U_u D_{-u})$	2.54×10^{-5}	4.10×10^{-3}	
Lower bound	69.84	69.05	
Upper bound	71.27	82.11	
Empirical power	71.37 ± 0.88	72.53 ± 0.87	68.99 ± 0.91

[a]Gene size $\delta = 6$, located at a distance $t_0 = 0.4$ from the origin of a chromosome of length 1 morgan. The value of σ is 1. In the smoothed procedure, the 5% level, the upper and lower bounds are calculated using the S^+ program mentioned previously. For the unsmoothed process, the 5% level is given by DeLong's table. The empirical powers are calculated over 10^4 simulations, and the corresponding 95% confidence intervals are given.

TABLE 4.4. Thresholds Calculated Using the Lower Bound in (4.23) for Different Values of Level α and Various Chromosome Lengths[a]

	Chromosome Length (Morgans)					
α Level	0.75	1	1.5	2	2.5	3
1%	3.059	3.133	3.239	3.315	3.375	3.423
5%	2.516	2.599	2.721	2.809	2.878	2.934
10%	2.239	2.328	2.458	2.553	2.626	2.687

[a]The smoothing parameter is $\varepsilon^2 = 10^{-3}$.

4.5. MIXTURES OF GAUSSIAN DISTRIBUTIONS

A classical problem in statistical inference is the one of deciding, on the basis of a sample, whether a population should be considered homogeneous or a mixture. We are going to address this question, but only when very simple possible models for the description of the population are present. This already leads to mathematical problems having a certain complexity and to the use of techniques that are related directly to our main subjects.

Our framework is restricted to Gaussian mixtures and we consider the following hypothetical testing situations:

1. The simple mixture model:

$$\begin{aligned} &H_0 : Y \sim N(0, 1) \\ &H_1 : Y \sim pN(0, 1) + (1 - p)N(\mu, 1) p \in [0, 1], \mu \in \mathcal{M} \subset \mathbb{R}. \end{aligned} \tag{4.35}$$

   We mean the following: Assume that we are measuring a certain magnitude Y on each individual in a population. Under the null hypothesis H_0, Y has a Normal (0,1) distribution for each individual. Under the alternative hypothesis H_1 each individual can be considered to have been chosen at random with probability p in a population in which the magnitude Y is normal (0,1) and with probability $1 - p$ in a population in which Y is normal $(\mu, 1)$. The purpose of the game is to make a decision about which is the underlying true situation, on the basis of the observation of a sample. The foregoing explanation applies to the other two cases that we describe next, *mutatis mutandis*.

2. The test of one population against two, variance known:

$$\begin{aligned} &H_0 : Y \sim N(\mu_0, 1) \mu_0 \in \mathcal{M} \\ &H_1 : Y \sim pN(\mu_1, 1) + (1 - p)N(\mu_2, 1) p \in [0, 1], \mu_1, \mu_2 \in \mathcal{M} \subset \mathbb{R}. \end{aligned} \tag{4.36}$$

3. The test of one population against two, variance unknown:

$$\begin{aligned} &H_0 : Y \sim N(\mu, \sigma^2) \mu \in \mathcal{M} \subset \mathbb{R}, \sigma^2 \in \Sigma \subset \mathbb{R}^+ \\ &H_1 : Y \sim pN(\mu_1, \sigma^2) + (1 - p)N(\mu_2, \sigma^2) p \in [0, 1], \\ &\quad \mu_1, \mu_2 \in \mathcal{M} \subset \mathbb{R}, \sigma^2 \in \Sigma \subset \mathbb{R}^+. \end{aligned} \tag{4.37}$$

These problems appear in many different types of applications and, of course, the statistical methods apply quite independent of the field generating the problem. However, to persist in biological questions considered in Section 4.4, let us again choose genetics to show a possible meaning of case 3.

Let us consider a quantitative trait on a given population: for example, the yield per unit surface in a plant-breeding experiment. A reasonable model consists of assuming that such a complex trait is influenced by a large number of genes, each with small effects. Assuming independence or almost-independence between the effects of the different genes, a heuristic application of the central limit theorem leads to a Gaussian distribution for the trait. That corresponds to the null hypothesis in formula (4.37).

Suppose now that a mutation appears in the population, introducing a new allele that alone has a nonnegligible effect on the trait. Let G be the new allelic form and g the old one and suppose that one form is dominant: for example, G (this means that Gg is equivalent to GG). Then the distribution of the trait in the population considered can be modeled by the general hypothesis in formula (4.37), with $1 - p$ being the frequency of individuals gg.

So, rejection of H_0 is associated with the detection of the existence of a new gene, and the purpose of the hypothesis testing is to take this decision on the basis of the observation of the value of the trait in a sample of the population. Of course, if H_0 is rejected, understanding the location of the gene will require, for example, some genetic marker information and the techniques of Section 4.4.

To perform a test of the type we described in (4.35), (4.36), or (4.37), there exist two main classical techniques:

1. A test based on moments: in a first approximation, expectation, variance, order-three moment
2. A test based on likelihood ratio

The asymptotic distribution of the likelihood ratio test was established by Ghosh and Sen (1985) under a strong separation hypothesis: for example, $|\mu_1 - \mu_2| > \varepsilon > 0$ for model (4.36). Without this hypothesis but still assuming that the set of the means $\mathcal{M}$ is compact and that the variance σ^2 is bounded away from zero, the asymptotic distribution has been studied by Dacunha-Castelle and Gassiat (1997, 1999) [see also Gassiat (2002) and Azaïs et al., (2008) for a simpler proof]. See also Azaïs et al. (2006) for further developments and a discussion of the behavior when $\mathcal{M}$ is not compact or is large in some sense.

On the other hand, moment-based tests do not demand compactness assumptions and have invariance properties. Since the distribution of the likelihood ratio test (LRT) is related to that of the maximum of a rather regular Gaussian process, we use a method based on Rice formulas to address the following problems:

Is the power of the LRT test much influenced by the size of the interval(s) in which the parameters are supposed to be?

Is it true, as generally believed but without proof, that the LRT test is more powerful than the moment tests? Notice that Azaïs et al. (2006) have proven that theoretically, the power of the LRT is smaller than the power of moment tests when the parameter set is very large.

Our aim here is to show how Rice formulas on crossings can be used to perform the computations required by the results below for hypothesis-testing problems on mixtures. We are not including proofs of the asymptotic statements, since they would lead us far away from the main subjects of this book. The interested reader can find them in the references above.

4.5.1. Simple Mixture

Theorem 4.13 (Asymptotic Distribution of the LRT). *Suppose that $\mathcal{M}$ is a bounded interval that contains zero and define the local asymptotic alternative:*

$$\mu = \mu_0 \in \mathcal{M}, \qquad 1 - p = \frac{\delta}{\sqrt{n}},$$

for fixed μ_0 and δ. Under this alternative the LRT of H_0 against H_1 has the distribution of the random variable

$$\frac{1}{2} \sup_{t \in \mathcal{M}} \{Z^2(t)\}, \tag{4.38}$$

where $Z(\cdot)$ is a Gaussian process with mean

$$m(t) = \frac{\delta(e^{t\mu_0} - 1)}{\sqrt{e^{t^2} - 1}}$$

and covariance function

$$r(s,t) = \frac{e^{st} - 1}{\sqrt{e^{s^2} - 1}\sqrt{e^{t^2} - 1}}.$$

Direct application of the regularity results of Chapter 1 shows that the process $Z(\cdot)$ has C^∞-paths on $(0, +\infty)$ and $(-\infty, 0)$, but has a discontinuity at $t = 0$, where it has right and left limits. Since

$$m(0^-) = -m(0^+) = -\delta\mu_0, \qquad r(0^-, 0^+) = -1,$$

it follows that a.s. $Z(0^-) = -Z(0^+)$.

We assume, for simplicity, that $\mathcal{M} = [-T, T]$ and set $M_T^* := \sup_{t \in [-T,T]} |Z(t)|$. Then we have the following inequalities, which are analogous to (4.23). Let

$$\xi := U_u[0,T] + D_{-u}[0,T] + D_u[-T,0] + U_{-u}[0,T].$$

Then

$$\mathrm{P}(|Z(0)| > u) + \mathrm{E}(\xi\, \mathbf{1}_{|Z(0)|<u}) - \tfrac{1}{2}\mathrm{E}(\xi^{[2]}) \le \mathrm{P}(M_T^* > u) \le \mathrm{P}(|Z(0)| > u) + \mathrm{E}(\xi\, \mathbf{1}_{|Z(0)|<u}) \tag{4.39}$$

and we can use the results in Section 4.2 to compute upper and lower bounds for $\mathrm{P}(M_T^* > u)$. The critical values deduced for the test are shown in Table 4.5. This table shows that the Rice method is very precise. It also shows that the critical values depend heavily on the size of the interval $\mathcal{M}$.

As for the power, we give some examples in Table 4.6. More examples may be found in articles by Delmas (2001, 2003a). In the table we can see that the power is affected by the size of $\mathcal{M}$. For example, for $\mu_0 = 1$, $\delta = 1$, the power varies from 16 to 23%.

TABLE 4.5. Critical Values or Thresholds for the LRT Test for a Simple Gaussian Mixture[a]

	Nominal Level of the Test		
Size, T	1%	5%	10%
0.5	3.6368	2.1743	1.5710
1	3.9015	2.3984	1.7707–1.7713
2	4.3432–4.3438	2.7908–2.7942	2.1300–2.1378
3	4.6784–4.6798	3.1002–3.1075	2.4209–2.4350
5	5.1357–5.1384	3.5353–3.5478	2.8390–2.8635
10	5.7903–5.7940	4.1736–4.1921	3.4641–3.5013
15	6.1837–6.1879	4.5621–4.5828	3.8480–3.8908
20	6.4657–6.4703	4.8414–4.8636	4.1253–4.1712
30	6.8658–6.8705	5.2392–5.2627	4.5212–4.5698
50	7.3725–7.3774	5.7443–5.7688	5.0245–5.0758

[a]Upper and lower bounds are given when they differ significantly.

TABLE 4.6. Power (%) of the LRT at the 5% Level as a Function of the Size of $\mathcal{M}$ and of δ and μ_0[a]

Size, T	Location, μ_0	$\delta = 1$	$\delta = 3$
0.5	0.25	5.79	11.84
	0.5	8.31	34.89
1	0.25	5.79	11.48
	0.5	8.40	34.59
	1	23.73–23.74	96.62–96.63
2	0.25	5.64–5.69	10.18–10.24
	1	22.93–23.09	95.99–96.19
	2	100	100
3	0.25	5.49–5.57	9.17–9.28
	1	20.21–20.50	94.76–95.17
	1.5	74.04–75.02	100
	3	100	100
5	0.25	5.31–5.44	8.02–8.20
	1	16.76–17.18	92.66–93.53
	1.5	68.34–69.99	99.99–100
	2.5	100	100

Source: From Delmas (2003a), with permission.
[a]Upper and lower bounds are given when they differ significantly.

Let us now compare the LRT test with two tests based on moments:

1. The $\bar{X}_n$ *test*, which is based on the fact the mean is zero under H_0 and that under the local alternative

$$\sqrt{n}\bar{X}_n \xrightarrow{\mathcal{D}} N(\delta\mu_0, 1) \quad \text{as } n \text{ tends to} + \infty. \tag{4.40}$$

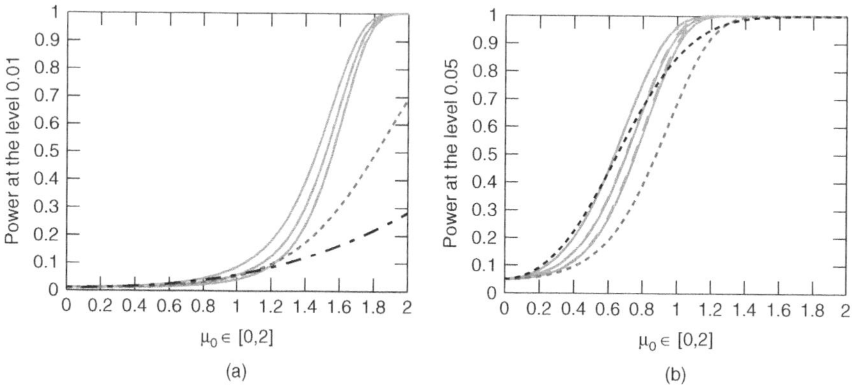

Figure 4.2. Variation of the power of the three tests as a function of μ_0. The power of the $\bar{X}_n$ test is represented in dashed-dotted line and that of the S_n^2 test in dotted line. For the power of the LRT the upper-bound is represented in dashed line and the lower-bound in solid line (they almost coincide). From top to bottom we find the cases $T = 2$, $T = 5$, and $T = 10$. In (a), $\delta^2 = 1$ and the level is 1%; in (b), $\delta^2 = 3$ and the level is 5%. (From Delmas, 2003a, with permission.)

2. The S_n^2 *test*, which is based on the fact the variance is 1 under H_0 and that under the local alternative

$$\sqrt{\frac{n}{2}}(S_n^2 - 1) \xrightarrow{\mathcal{D}} N\left(\frac{\delta\mu_0^2}{\sqrt{2}}, 1\right) \qquad \text{as } n \text{ tends to} + \infty. \tag{4.41}$$

Since the S_n^2 test has a one-sided rejection region while the $\bar{X}_n$ test has a two-sided rejection region, it is not straightforward to see which of the two tests is more powerful. Moreover, the answer depends on the level. A general fact is that for large μ_0, the S_n^2 test is more powerful. Comparisons of the three tests are presented in Figure 4.2.

The main point, which can be seen on Figure 4.2, is that *the likelihood ratio test is not uniformly most powerful*. Figure 4.2a is rather typical in the sense that the situations for which the LRT is not optimal (e.g., $T = 10$, $\mu_0 = 1$) correspond to very small power. They are actually uninteresting. Figure 4.2b corresponds to a deliberate choice of a situation where the LRT behaves badly. For example, for $\mu_0 = 0.6$ and $T = 10$ the lack of power of the LRT compared to that for the $\bar{X}_n$ test is important.

4.5.2. One Population Against Two, σ^2 Known

Theorem 4.14 (Asymptotic Distribution of the LRT). *Suppose that $\mathcal{M}$ is a bounded interval that will now be chosen of the form $\mathcal{M} = [0, T]$. We define the local asymptotic alternative in model by means of (4.36):*

$$1 - p = \frac{\delta^2}{\sqrt{n}}, \qquad \mu_1 - \mu_0 = \frac{\alpha}{\sqrt{n}}, \qquad \mu_2 = \mu_{2,0} \neq \mu_0 \in (0, T)$$

for fixed μ_0, $\mu_{2,0}$, and δ.

Then, under this alternative, the LRT of H_0 against H_1 has the limit distribution given by (4.36), where $Z(\cdot)$ is now a Gaussian process with mean

$$m(t) = \frac{\delta^2 (e^{(\mu_0 - \mu_{2,0})(\mu_0 - t)} - 1 - (\mu_0 - \mu_{2,0})(\mu_0 - t))}{\sqrt{e^{(\mu_0 - t)^2} - 1 - (\mu_0 - t)^2}}$$

and covariance function

$$r(s,t) = \frac{e^{(s-\mu_0)(t-\mu_0)} - 1 - (s - \mu_0)(t - \mu_0)}{\sqrt{e^{(s-\mu_0)^2} - 1 - (s - \mu_0)^2}\sqrt{e^{(t-\mu_0)^2} - 1 - (t - \mu_0)^2}}.$$

One should notice that these functions are of class C^∞ and so are the sample paths of the process. As a consequence, the Rice method applies directly. But a new problem arises since the null hypothesis is composite (it consists of more than one distribution) and the distribution of the LRT statistic under H_0 is not free from the parameter μ_0, which is unknown in practical applications. Table 4.7 illustrates the variation of the threshold as a function of μ_0 and T. We can clearly see in Table 4.7 that the value taken by μ_0 does not matter very much. So the LRT can be used in practical situations. As in Section 4.5.1, Figure 4.3 compares the power of the LRT test with the S_n^2 test, based on the fact that the variance is 1 under H_0. Of course, the $\bar{X}_n$ test cannot be performed since the expectation is not yet fixed under H_0. We observe roughly the same phenomenon as in the case of the simple mixture problem.

4.5.3. One Population Against Two, σ^2 Unknown

Theorem 4.15 (Asymptotic Distribution of the LRT). *Suppose that $\mathcal{M}$ is a bounded interval and that $\Sigma = [S_1, S_2]$, $0 < S_1 < S_2 < +\infty$ and define the local asymptotic alternative in model (4.37):*

$$1 - p = \frac{\delta^2}{\sqrt{n}}, \qquad \mu_1 - \mu_0 = \frac{\alpha}{\sqrt{n}}, \qquad \sigma^2 - \sigma_0^2 = \frac{\beta}{\sqrt{n}}, \qquad \mu_2 = \mu_{2,0} \neq \mu_0$$

for fixed μ_0, $\mu_{2,0}$, α, β, and δ. Then under this alternative the LRT of H_0 against H_1 has the limit distribution given by (4.38), where $Z(\cdot)$ is now a Gaussian process.

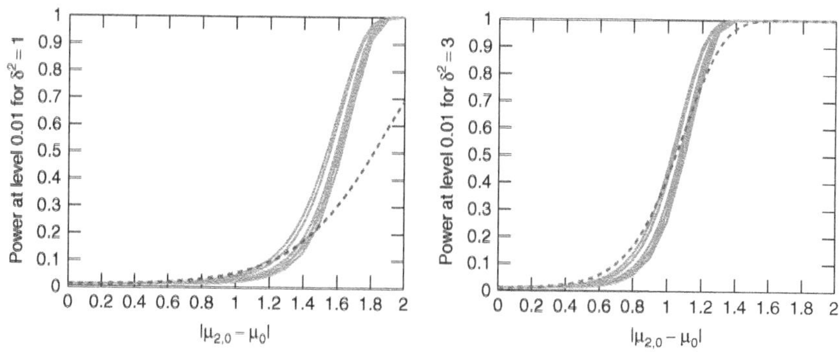

Figure 4.3. Variation of the power of the LRT and the S_n^2 test (at the level 1%) as a function of $\mu_{2,0} - \mu_0$. The power of the S_n^2 test is represented in dotted line for the S_n^2 test. For the power of the LRT the upper-bound is represented in dashed line and the lower-bound in solid line. From top to bottom we find the cases $T = 4$, $T = 10$, and $T15$. In (a), $\delta^2 = 1$ and on (b), $\delta^2 = 3$. Some lines are superposed because the upper-bound and the lower-bound are numerically equal. The difference between upper- and lower-bound is due to both the inequality (4.23) and to the variation of the nuisance parameter μ_0 inside $\mathcal{M}$. (From Delmas, 2003a, with permission.)

Define

$$\nu := \frac{t - \mu_0}{\sigma_0} \quad \nu' := \frac{s - \mu_0}{\sigma_0} \quad \nu_0 = \frac{\mu_{2,0} - \mu_0}{\sigma_0}$$

and

$$f(x, y) := \exp(xy) - 1 - xy - \frac{x^2 y^2}{2}$$

$Z(t)$ has expectation

$$m(t) = \delta^2 \frac{f(\nu, \nu_0)}{\sqrt{f(\nu, \nu)}}$$

and covariance function

$$r(s, t) = \frac{f(\nu, \nu')}{\sqrt{f(\nu, \nu)}\sqrt{f(\nu', \nu')}}.$$

The process $Z(\cdot)$ of Theorem 4.15 is, in fact, a random function of $\nu := (t - \mu_0)/\sigma_0$:

$$Z(t) = T\left(\frac{t - \mu_0}{\sigma_0}\right) \tag{4.42}$$

TABLE 4.7. Variation of the Threshold of the Test as a Function of the Level, the Size T of the Interval, and the Position μ_0[a]

		Level of the Test		
Size, T	Position, μ_0	1%	5%	10%
1	0.5	3.1524	1.6925	1.0991
	0.25	3.1561	1.6956	1.1018
	0+	3.1676	1.7049	1.1097
2	1	3.4921–3.4924	1.9779–1.9782	1.3487–1.3490
	0.5	3.5137–3.5139	1.9967–1.9970	1.3658–1.3661
	0	3.5712	2.0475–2.0477	1.4120–1.4121
4	2	4.0424–4.0439	2.4733–2.4772	1.8048–1.8106
	1	4.0904–4.0915	2.5189–2.5220	1.8486–1.8532
	0+	4.1876–4.1884	2.6108–2.6141	1.9362–1.9420
6	3	4.4663–4.4688	2.8738–2.8826	2.1869–2.2018
	1.5	4.4863–4.4882	2.8934–2.9009	2.2059–2.2194
	0+	4.5756–4.5776	2.9797–2.9879	2.2897–2.3048
10	5	5.0092–5.0126	3.3994–3.4137	2.6975–2.7252
	2.5	5.0100–5.0132	3.4001–3.4142	2.6981–2.7254
	0+	5.0746–5.0778	3.4636–3.4779	2.7608–2.7886

[a]Upper and lower bounds are given when they differ significantly.

TABLE 4.8. Upper Bound for the Threshold of the LRT Test as a Function of the Level and of the Set of Parameter After the Change of Variable (4.42)

	Level of the Test		
ν	1%	5%	10%
[-2;2]	4.1533	2.6209	1.9755
[-5;5]	5.0390	3.4514	2.7688
[-8;8]	5.5168	3.9183	3.2296
[-1;3]	4.2076	2.6687	2.0180
[-2.5;7.5]	5.0413	3.4532	2.7707
[-4;12]	5.5168	3.9183	3.2296
[0;4]	4.3092	2.7594	2.1002
[0;10]	5.1242	3.5330	2.8477
[0;16]	5.5708	3.9714	3.2814

and the process $T(\cdot)$ satisfies a.s. $T(0^-) = -T(0^+)$, as we found in the simple mixture model. The method based on crossings can be applied in much the same manner. Table 4.8 displays the variation of the threshold as a function of the level and of the size of the parameter set $\nu = (t - \mu_0)/\sigma_0$.

The intervals of variation for the parameter ν that are displayed in Table 4.8 correspond, for example, to $t \in \mathcal{M} = [5, 15]$ and several values of σ_0 and μ_0.

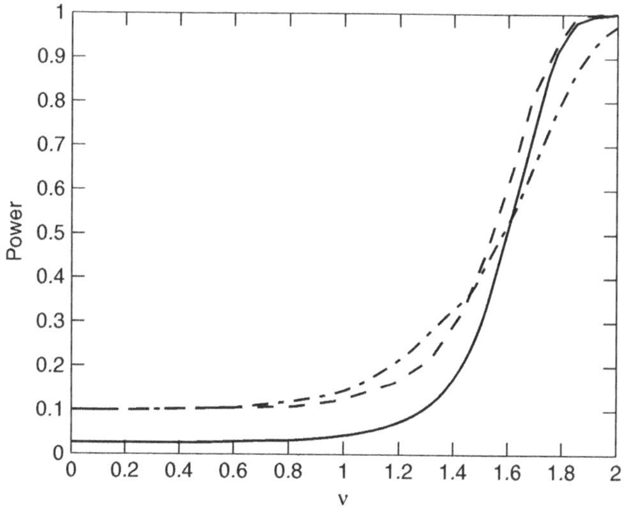

Figure 4.4. Variation of the power of the LRT and moment test at the level 10% as a function of $\nu = \frac{t-\mu_0}{\sigma_0}$. The power of the moment test is represented in dashed-dotted line. For the LRT upper-bound, in dotted line, corresponds to the best choice of the nuisance parameter μ_0/σ_0 and the lower-bound, in solid line, corresponds to the worst choice. (From Delmas, 2003a, with permission.)

The table is divided into three "great rows," corresponding to $\mu_0 = 10, 7$, and 5. Each great row is divided into three rows, corresponding to $\sigma_0 = 5/2$, 1, and 5/8.

The threshold now depends heavily on the form of the null hypothesis. If we know that (μ_0, σ_0) are close to some prior value, we can take the threshold corresponding to this prior value. In the others cases, a classical choice is to take the highest value of the threshold. This in general leads to an important loss of power. Another possibility would be to perform a "plug-in," that is, to take for μ_0 and σ_0^2 the values computed from an empirical estimation. The behavior of such a procedure does not seem to have been studied yet.

In this paragraph the LRT is compared with a moment test based on the difference between the rough estimator of variance S_n^2 and a robust estimator (Caussinus and Ruiz, 1995). Figure 4.4 displays the powers of the two tests.

Summing up, for the last two models, the distribution of the LRT is not free under H_0, since it depends on the position of the true mean μ_0 with respect of the interval in which the means are supposed to be. When σ^2 is known, this dependence is not heavy, so that LRT can be performed in practice without introducing a prior value or an estimation for μ_0. The situation is more complex when σ is unknown. In any case, LRT appears to be nonuniformly optimal, but in most relevant situations more powerful, than moment tests. It remains the best choice in practice.

EXERCISES

4.1. *(Exact formula for the distribution of M_T for the sine–cosine process)* This is the simplest periodic Gaussian process. Let the stochastic process $\mathcal{X} = \{X(t) : t \in \mathbb{R}\}$ be defined as

$$X(t) := \xi_1 \cos \omega t + \xi_2 \sin \omega t,$$

where ξ_1 and ξ_2 are two independent standard normal random variables and ω is a real number, $\omega \neq 0$.

(a) Show that we can write the process $\mathcal{X}$ as

$$X(t) = Z \cos(\omega(t + \theta)),$$

where Z and θ are independent random variables having, respectively, the distributions square root of a χ_2^2 for Z and uniform on $[0, 2\pi]$ for θ.

(b) Show that the covariance of $\mathcal{X}$ is $r(s,t) = \Gamma(t-s) = \cos(\omega(t-s))$, and its spectral measure is $\frac{1}{2}(\delta_\omega + \delta_{-\omega})$, where δ_x denotes the unit atom at point x. Prove that for $u > 0$:

$$\text{For } T \leq \frac{\pi}{\omega}: \quad \mathrm{P}\{M_T > u\} = 1 - \Phi(u) + \frac{T\omega}{2\pi} e^{-u^2/2} \tag{4.43}$$

$$\text{For } T \geq \frac{2\pi}{\omega}: \quad \mathrm{P}\{M_T > u\} = e^{-u^2/2} = \mathrm{P}\{|Z| > u\} \tag{4.44}$$

$$\text{For } \frac{\pi}{\omega} \leq T < \frac{2\pi}{\omega}: \quad \mathrm{P}\{M_T > u\} = 1 - \Phi(u) + \frac{T\omega}{2\pi} e^{-u^2/2} \tag{4.45}$$

$$- \int_{\pi}^{T/\omega} \frac{1}{2\pi} \times \exp\left[- \frac{u^2(1 - cos(t))}{\sin^2 t} \right] dt. \tag{4.46}$$

4.2. Let $\{X(t) : t \in \mathbb{R}\}$ be a centered stationary Gaussian process having covariance function $\Gamma(t-s) = \mathrm{E}(X(s)X(t))$, normalized by $\Gamma(0) = 1$. Assume that Γ satisfies hypothesis (A2) in Proposition 4.2 and denote, as usual, $\lambda_2 = -\Gamma''(0)$, the second spectral moment. Let $\tau = \inf\{t > 0 : \Gamma(t) = 1\}$. Exclude the trivial case in which $\lambda_2 = 0$. Let $M_T = \max_{t \in [0,T]} X(t)$.

(a) Prove that $\tau > 0$.

(b) Show that

$$\mathrm{P}(X(t) = X(t + \tau) \text{ for all } t \in \mathbb{R}) = 1$$

(i.e., the paths are a.s. periodic functions having period τ).

(c) Prove that if $T < \tau$, the conclusion of Proposition 4.2 holds true.

(d) Let $T \geq \tau$. Prove that as $u \to +\infty$,

$$\mathrm{P}(M_T > u) = \tau\sqrt{\frac{\lambda_2}{2\pi}}\phi(u) + O\left(\phi(u(1+\delta))\right)$$

for some $\delta > 0$.

Hint: Prove and use the following equalities: $M_T = M_\tau$, and $\mathrm{P}(M_\tau > u) = \mathrm{P}(X(t) > u \text{ for all } t \in [0, \tau]) + \mathrm{P}(U_u(X, [0, \tau]) \geq 1)$.

4.3. Suppose that f_n is a sequence of functions in D that converges to f for the Skorohod topology. Prove that if f is continuous, then in fact the convergence holds true for the uniform distance.

4.4. Let $\{Z_n\}_{n=1,2,\dots}$ be a sequence of random variables with values in the space D. Assume that $Z_n \Rightarrow 0$ as $n \to +\infty$ (which means that the distribution of Z_n tends to the unit measure at the identically zero function). Then, for each $\varepsilon > 0$,

$$\mathrm{P}\left(\sup_{t\in[0,1]} |Z_n(t)| \geq \varepsilon\right) \to 0$$

as $n \to +\infty$; that is, $\sup_{t\in[0,1]} |Z_n(t)|$ tends to zero in probability.

4.5. Prove relations (4.40) and (4.41). *Hint:* Use the central limit theorem under Lindeberg's condition.

4.6. Consider the function $\Gamma : \mathbb{R} \to \mathbb{R}^+$, defined as $\Gamma(t) = \exp(-|t|)$..

(a) Show that Γ is the covariance of a stationary centered Gaussian process $\{X(t) : t \in \mathbb{R}\}$ (the Ornstein–Uhlenbeck process)

(b) Compute the spectral density of the process.

(c) Study the Hölder properties of the paths.

(d) Show that the process is Markovian, that is, if $t_1 < t_2 < \cdots < t_k$, k a positive integer $k \geq 2$, then

$$\mathrm{P}\left(X(t_k) \in B \middle| X(t_1) = x_1, \dots, X(t_{k-1}) = x_{k-1}\right)$$
$$= \mathrm{P}\left(X(t_k) \in B \middle| X(t_{k-1}) = x_{k-1}\right)$$

for any Borel set B and any choice of $x_1, \dots, x_{k-1}$. (The conditional probability is to be interpreted in the sense of Gaussian regression.)

(e) Let $\{W(t) : t \geq 0\}$ be a Wiener process. Show that

$$X(t) = \exp(-t)W(\exp(2t)) \qquad t \in \mathbb{R}$$

is an Ornstein–Uhlenbeck process.

4.7. Prove the statements about the properties of the process $\{Z(t) : t \in \mathbb{R} \setminus \{0\}\}$ in Section 4.5.1.

CHAPTER 5

THE RICE SERIES

Let $\mathcal{X} = \{X(t) : t \in [0, T]\}$ be a one-parameter stochastic process with real values and let us denote by $M_T := \sup_{t\in T} X(t)$ its supremum. In this chapter we continue to study the distribution of the random variable M_T, that is, the function $F_{M_T}(u) := \mathrm{P}(M_T \le u)$, $u \in \mathbb{R}$, and we express this distribution by means of a series (the *Rice series*) whose terms contain the factorial moments of the number of up-crossings. The underlying ideas have been known for a long time (Rice, 1944, 1945; Slepian, 1962; Miroshin, 1974). The results in this chapter are taken from Azaïs and Wschebor (2002). We have included some numerical computations that have been performed with the help of A. Croquette and C. Delmas. The main result in this chapter is to prove the convergence of the Rice series in a general framework instead of considering only some particular processes. This provides a method that can be applied to a large class of stochastic processes.

A typical situation given by Theorem 5.6 states that if a stationary Gaussian process has a covariance with a Taylor expansion at zero that is absolutely convergent at $t = 2T$, then $F_{M_T}(u)$ can be computed by means of the Rice series. On the other hand, even though Theorems 5.1 and 5.7 do not refer specifically to Gaussian processes, for the time being we are able in practice to apply them to the numerical computation of $F_{M_T}(u)$ only in Gaussian cases.

Section 5.3 includes a comparison of the complexities of the computation of $F_{M_T}(u)$ using the Rice series versus the Monte Carlo method, in the case of a general class of stationary Gaussian processes. It shows that use of the Rice series is a priori better. More important is the fact that the Rice series is self-controlling for numerical errors. This implies that the a posteriori number of computations

Level Sets and Extrema of Random Processes and Fields, By Jean-Marc Azaïs and Mario Wschebor

can be much smaller than the number required in simulation. In fact, in relevant cases for standard bounds for the error, the actual computation is performed with a few terms of the Rice series.

As examples we give tables for $F_{M_T}(u)$ for a number of Gaussian processes. When the length of the interval T increases, one needs an increasing number of terms in the Rice series so as not to surpass a given bound for the error. For small values of T and large values of the level u, one can use Davies bound (4.2) or, more accurately, the first term in the Rice series, which is, in fact, the upper bound in inequality (4.15).

As T increases, for moderate values of u the Davies bound is far from the true value, and one requires the computation of several terms (see Figures 5.1 to 5.4). Numerical results are shown in the case of four Gaussian stationary processes for which no closed formula is known. The same examples are considered in Chapter 9 in relation to the record method.

One of the key points is the numerical computation of the factorial moments of up-crossings by means of Rice integral formulas. The main difficulty is the precise description of the behavior of the integrands appearing in these formulas near the diagonal, which is again a subject that is interesting on its own (see Belayev, 1966; Cuzick, 1975). Even though this is an old subject, it remains widely open. In Section 5.2 we give some partial answers that are helpful in improving the numerical methods and also have some theoretical interest.

The extension to processes with nonsmooth trajectories can be done by smoothing the paths by means of a deterministic device, applying the previous methods to the regularized process and estimating the error as a function of the smoothing bandwidth. Section 5.4 describes these type of results, which have been included even though they do not seem to have at present practical uses for actual computation of the distribution of the maximum.

5.1. THE RICE SERIES

We recall the following notation:

- $U_u = U_u(X, [0, T])$ is the number of up-crossings of the level u by the function $X(\cdot)$ on the interval $[0, T]$.
- $\widetilde{\nu}_m := \mathrm{E}(U_u^{[m]} \mathbf{I}_{\{X(0) \le u\}})(m = 1, 2, \ldots)$.
- $\nu_m := \mathrm{E}(U_u^{[m]})(m = 1, 2, \ldots)$.

$\widetilde{\nu}_m$ is the factorial moment of the number of up-crossings when starting below u at $t = 0$. The Rice formula to compute $\widetilde{\nu}_m$, whenever it holds true, is the following:

$$\widetilde{\nu}_m = \int_{[0,T]^m} dt_1 \cdots dt_m$$

$$\int_{-\infty}^{u} \mathrm{E}\left(X'^{+}(t_1) \cdots X'^{+}(t_m) \middle| X(0) = x, X(t_1) = \cdots = X(t_m) = u\right) \cdot p_{X(0),X(t_1),\ldots,X(t_m)}(x, u, \ldots, u)\, dx.$$

This section contains two principal results. The first is Theorem 5.1, which requires that the process have $\mathcal{C}^{\infty}$-paths and contain a general condition making it possible to compute $F_{M_T}(u)$ as the sum of a series. The second is Theorem 5.6, which illustrates the same situation for Gaussian stationary processes. As for Theorem 5.7, it contains upper and lower bounds on $F_{M_T}(u)$ for processes with $\mathcal{C}^k$-paths, verifying some additional conditions.

Theorem 5.1. *Assume that a.s. the paths of the stochastic process $\mathcal{X}$ are of class $\mathcal{C}^{\infty}$ and that the density $p_{X_{T/2}}(\cdot)$ is bounded by some constant D.*

(i) *If there exists a sequence of positive numbers $\{c_k\}_{k=1,2,\ldots}$ such that*

$$\gamma_k := \mathrm{P}\left(\|X^{(2k-1)}\|_\infty \geq c_k T^{-(2k-1)}\right) + \frac{D c_k}{2^{2k-1}\,(2k-1)!} = o\left(2^{-k}\right)(k \to \infty), \tag{5.1}$$

then

$$1 - F_{M_T}(u) = \mathrm{P}\left(X(0) > u\right) + \sum_{m=1}^{\infty} (-1)^{m+1} \,\frac{\widetilde{\nu}_m}{m!}. \tag{5.2}$$

(ii) *In formula (5.2) the error when one replaces the infinite sum by its m_0th partial sum is bounded by $\gamma^*_{m_0+1}$, where*

$$\gamma^*_m := \sup_{k \geq m}\left(2^{k+1}\gamma_k\right).$$

We call the series on the right-hand side of (5.2) the Rice series. For the proof we will assume, with no loss of generality, that $T = 1$. We start with the following lemma on the remainder for polynomial interpolation (Davis, 1975, Theorem 3.1.1). It is a standard tool in numerical analysis.

Lemma 5.2

(a) *Let $f : I \to \mathbb{R}$, $I = [0,1]$ be a function of class C^k, k a positive integer, $t_1, \ldots, t_k$, k points in I, and let $P(t)$ be the (unique) interpolation polynomial of degree $k-1$ such that $f(t_i) = P(t_i)$ for $i = 1, \ldots, k$, taking into account possible multiplicities. Then, for $t \in I$,*

$$f(t) - P(t) = \frac{1}{k!}(t-t_1)\cdots(t-t_k) f^{(k)}(\xi),$$

where

$$\min(t_1, \ldots, t_k, t) \leq \xi \leq \max(t_1, \ldots, t_k, t).$$

(b) *If f is of class C^k and has k zeros in I (taking into account possible multiplicities), then*

$$|f(1/2)| \leq \frac{1}{k!\,2^k}\|f^{(k)}\|_\infty.$$

Proof. Fix $t \in I, t \neq t_1, \ldots, t_k$ and define

$$F(v) = f(v) - P(v) - \frac{(v - t_1) \ldots (v - t_k)}{(t - t_1) \ldots (t - t_k)} \left[f(t) - P(t) \right].$$

Clearly, F has at least the $k + 1$ zeros $t_1, \ldots, t_k, t$, so that by Rolle's theorem, there exists ξ, $\min(t_1, \ldots, t_k, t) \leq \xi \leq \max(t_1, \ldots, t_k, t)$ such that $F^{(k)}(\xi) = 0$. This gives part (a). Part (b) is a simple consequence of (a), since in this case the interpolating polynomial vanishes. □

The next combinatorial lemma plays the central role in what follows. A proof is given in Lindgren (1972) similar to the one we include here.

Lemma 5.3. *Let ξ be a nonnegative integer-valued random variable having finite moments of all orders. Let $k, m, M (k \geq 0, m \geq 1, M \geq 1)$ be integers and denote*

$$p_k := \mathrm{P}(\xi = k), \qquad \mu_m := \mathrm{E}(\xi^{[m]}), \qquad S_M := \sum_{m=1}^{M} (-1)^{m+1} \frac{\mu_m}{m!}.$$

Then:

(i) *For each M:*

$$S_{2M} \leq \sum_{k=1}^{2M} p_k \leq \sum_{k=1}^{\infty} p_k \leq S_{2M+1}. \tag{5.3}$$

(ii) *The sequence $\{S_M\}_{M=1,2,\ldots}$ has a finite limit if and only if $\mu_m/m! \to 0$ as $m \to \infty$, and in that case,*

$$\mathrm{P}(\xi \geq 1) = \sum_{k=1}^{\infty} p_k = \sum_{m=1}^{\infty} (-1)^{m+1} \frac{\mu_m}{m!}. \tag{5.4}$$

Proof. Part (ii) is an immediate consequence of (i). As for (i), denote by $\binom{k}{m}$ the binomial numbers and write

$$S_M = \sum_{m=1}^{M} (-1)^{m+1} \sum_{k=m}^{\infty} \binom{k}{m} p_k = \sum_{k=1}^{\infty} p_k B_{k,M} \tag{5.5}$$

with

$$B_{k,M} := \sum_{m=1}^{k \wedge M} (-1)^{m+1} \binom{k}{m}. \tag{5.6}$$

It is clear that $B_{k,M} = 1$ if $k \leq M$.

If $k > M$, we have two cases:

1. $k \geq 2M$. Note that $\binom{k}{m}$ increases with m if $1 \leq m \leq k/2$. It follows that $B_{k,M} \geq k$ if M is odd and $B_{k,M} \leq -k/2$ if M is even, since $B_{k,M} \leq \binom{k}{1} - \binom{k}{2} \leq -k/2$, given that $k \geq 2M \geq 4$.
2. $M < k < 2M$. Check that in this case

$$B_{k,M} = 1 + (-1)^{k+1} \sum_{h=0}^{k-M-1} (-1)^{h+1} \binom{k}{h} = 1 + (-1)^{k+1}(B_{k,k-M-1} - 1)$$

with the convention $B_{k,0} = 0$.

Since $k > 2(k - M - 1)$, if $0 < k - M - 1 < k$, we can apply the first case and it turns out that

$$k - M - 1 \text{ odd} \Rightarrow B_{k,k-M-1} \geq k$$
$$k - M - 1 \text{ even} \Rightarrow B_{k,k-M-1} \leq -k/2.$$

Finally, if $k = M + 1$, $B_{k,M} = 2$ when M is odd and $B_{k,M} = 0$ if M is even.

Summing up the two cases, if $k > M$, we have $B_{k,M} > 1$ if M is odd and $B_{k,M} \leq 0$ if M is even, so that from

$$S_M = \sum_{k=1}^{M} p_k + \sum_{k=M+1}^{\infty} p_k B_{k,M},$$

one gets (i). This proves the lemma. □

Remark. A by-product of Lemma 5.3 that will be used in the sequel is the following: If in (5.4) one substitutes the infinite sum by the M-partial sum, the absolute value $\mu_{M+1}/(M + 1)!$ of the first neglected term is an upper bound for the error in the computation of $\mathrm{P}(\xi \geq 1)$.

Lemma 5.4. *With the same notations as in Lemma 5.3, we have the equality*

$$\mathrm{E}(\xi^{[m]}) = m \sum_{k=m}^{\infty} (k - 1)^{[m-1]} \mathrm{P}(\xi \geq k) \qquad m = 1, 2, \ldots.$$

Proof. Check the identity

$$j^{[m]} = m \sum_{k=m-1}^{j-1} (k)^{[m-1]}$$

for each pair of integers j and m. So

$$\mathrm{E}(\xi^{[m]}) = \sum_{j=m}^{\infty} j^{[m]}\mathrm{P}(\xi = j) = \sum_{j=m}^{\infty} \mathrm{P}(\xi = j) m \sum_{k=m}^{j} (k-1)^{[m-1]}$$

$$= m \sum_{k=m}^{\infty} (k-1)^{[m-1]} \mathrm{P}(\xi \geq k).$$

□

Lemma 5.5. *Suppose that a.s. the paths of the process $\mathcal{X}$ are of class C^{∞} and that the density $p_{X_{1/2}}(\cdot)$ is bounded by the constant D. Then for any sequence $\{c_k\}_{k=1,2,\ldots}$ of positive numbers, one has*

$$\mathrm{E}((U_u)^{[m]}) \leq m \sum_{k=m}^{\infty} (k-1)^{[m-1]} \left[\mathrm{P}\left(\|X^{(2k-1)}\|_{\infty} \geq c_k\right) + \frac{Dc_k}{2^{2k-1}(2k-1)!}\right]. \tag{5.7}$$

Proof. Because of Lemma 5.4, it is enough to prove that $\mathrm{P}(U_u \geq k)$ is bounded by the expression in brackets on the right-hand side of (5.7). We have

$$\mathrm{P}(U_u \geq k) \leq \mathrm{P}(\|X^{(2k-1)}\|_{\infty} \geq c_k) + \mathrm{P}(U_u \geq k, \|X^{(2k-1)}\|_{\infty} < c_k).$$

Because of Rolle's theorem,

$$\{U_u \geq k\} \subset \{N_u(X; I) \geq 2k-1\}.$$

Applying Lemma 5.2 to the function $X(\cdot) - u$ and replacing k in its statement by $2k-1$, we obtain

$$\{U_u \geq k, \|X^{(2k-1)}\|_{\infty} < c_k\} \subset \left\{|X_{1/2} - u| \leq \frac{c_k}{2^{2k-1}(2k-1)!}\right\}.$$

The remainder follows easily. □

Proof of Theorem 5.1. Using Lemma 5.5 and the hypothesis, we obtain

$$\frac{\nu_m}{m!} \leq \frac{1}{m!} \sum_{k=m}^{\infty} k^{[m]} \gamma_m^* 2^{-(k+1)} = \frac{\gamma_m^*}{m!} 2^{-(m+1)} \left[\left(\frac{1}{1-x}\right)^{(m)} \Bigg|_{x=1/2}\right] = \gamma_m^*.$$

Since $\widetilde{\nu}_m \leq \nu_m$, we can apply Lemma 5.3 to the random variable $\xi = U_u 1_{\{X_0 \leq u\}}$ and the result follows from $\gamma_m^* \to 0$. □

One can replace condition "$p_{X(T/2)}(x) \leq D$ for all x" by "$p_{X(T/2)}(x) \leq D$ for x in some neighborhood of u." In this case, the statement of Theorem 5.1 holds if in (ii) one adds that the error is bounded by $\gamma^*_{m_0+1}$ for m_0 large enough. The proof is similar.

Also, one can substitute the one-dimensional density $p_{X(T/2)})(\cdot)$ by $p_{X(t)}(\cdot)$ for some other $t \in [0, T]$, introducing into the bounds the corresponding modifications.

The application of Theorem 5.1 requires an adequate choice of the sequence $\{c_k, k = 1, 2, \ldots\}$, which depends on the available description of the process X. The entire procedure will have some practical interest for the computation of $\mathrm{P}(M > u)$ only if we get appropriate bounds for the quantities γ^*_m and the factorial moments $\widetilde{\nu}_m$ can actually be computed by means of Rice formulas (or by some other procedure). The next theorem shows how this can be done in the case of a general class of Gaussian stationary processes.

Theorem 5.6. *Let $\mathcal{X} = \{X(t) : t \in \mathbb{R}\}$ be Gaussian centered and stationary, with covariance Γ normalized by $\Gamma(0) = 1$. Assume that Γ has a Taylor expansion at the origin which is absolutely convergent at $t = 2T$. Then the conclusion of Theorem 5.1 holds true, so that the Rice series converges and $F_{M_T}(u)$ can be computed by means of (5.2).*

Proof. Again we assume, with no loss of generality, that $T = 1$. Notice that the hypothesis implies that the spectral moments λ_k exist and are finite for every $k = 0, 1, 2, \ldots$. We will obtain the result assuming that

$$H_1 : \lambda_{2k} \leq C_1 (k!)^2.$$

It is easy to verify that if Γ has a Taylor expansion at zero that is absolutely convergent at $t = 2$, then H_1 holds true. (In fact, both conditions are only slightly different, since H_1 implies that the Taylor expansion of Γ at zero is absolutely convergent in $\{|\}t| < 2$.) Let us check that the hypotheses of Theorem 5.1 are satisfied. First, $p_{X(1/2)}(x) \leq D = (2\pi)^{-1/2}$. Second, let us show a sequence $\{c_k\}$ that satisfies (5.1). We have

$$\begin{aligned} \mathrm{P}(\|X^{(2k-1)}\|_\infty \geq c_k) &\leq \mathrm{P}(|X^{(2k-1)}(0)| \geq c_k) + 2\mathrm{P}(U_{c_k}(X^{(2k-1)}, I) \geq 1) \quad (5.8)\\ &\leq \mathrm{P}(|Z| \geq c_k \lambda_{4k-2}^{-1/2}) + 2\mathrm{E}(U_{c_k}(X^{(2k-1)}, I)), \end{aligned}$$

where Z is standard normal. One easily checks that $\{X^{(2k-1)}(t) : t \in \mathbb{R}\}$ is a Gaussian stationary centered process with covariance function $-\Gamma^{(4k-2)}(\cdot)$. So we can use a Rice formula for the expectation of the number of up-crossings of a Gaussian stationary centered process to compute the second term on the right-hand side of (5.8).

Using the inequality $1 - \Phi(x) \leq (1/x)\phi(x)$ valid for $x > 0$, one gets

$$\mathrm{P}(\|X^{(2k-1)}\|_\infty \geq c_k) \leq \left[\sqrt{\frac{2}{\pi}}\frac{\lambda_{4k-2}^{1/2}}{c_k} + (1/\pi)\left(\frac{\lambda_{4k}}{\lambda_{4k-2}}\right)^{1/2}\right] \exp\left(-\frac{c_k^2}{2\lambda_{4k-2}}\right). \tag{5.9}$$

Choose

$$c_k := \begin{cases} (B_1 k \lambda_{4k-2})^{1/2} & \text{if } \dfrac{\lambda_{4k}}{\lambda_{4k-2}} \le B_1 k \\ (\lambda_{4k})^{1/2} & \text{if } \dfrac{\lambda_{4k}}{\lambda_{4k-2}} > B_1 k. \end{cases}$$

Using hypothesis (H_1), if $B_1 > 1$, we obtain

$$\mathrm{P}(\|X^{(2k-1)}\|_\infty \ge c_k) \le \left[\sqrt{\frac{2}{\pi}} + \frac{1}{\pi}(B_1 k)^{1/2}\right] e^{-B_1 k/2}.$$

Finally, choosing $B_1 := 4\log(2)$ yields

$$\gamma_k \le \sqrt{\frac{2}{\pi}}(1 + 2(C_1^{1/2} + 1)k)2^{-2k} \qquad k = 1, 2, \ldots,$$

so that (5.1) is satisfied. As a by-product, notice that

$$\gamma_m^* \le \sqrt{\frac{8}{\pi}}(1 + 2(C_1^{1/2} + 1)m)2^{-m} \qquad m = 1, 2, \ldots. \tag{5.10}$$

□

Remark. For Gaussian processes, if one is willing to use Rice formulas to compute the factorial moments $\widetilde{\nu}_m$, it is enough to verify that the distribution of $X(0), X(t_1), \ldots, X(t_m)$ is nondegenerate for any choice of nonzero distinct $t_1, \ldots, t_m \in I$. For stationary Gaussian processes, a simple sufficient condition of nondegeneracy on the spectral measure was given in Chapter 3 (see Exercises 3.4 and 3.5).

If instead of requiring the paths of the process $\mathcal{X}$ to be of class C^∞, one relaxes this condition up to a certain order of differentiability, one can still get upper and lower bounds for $\mathrm{P}(M > u)$, as stated in the next theorem.

Theorem 5.7. *Let $X = \{X(t) : t \in I\}$ be a real-valued stochastic process. Suppose that $p_{X(t)}(x)$ is bounded for $t \in I$, $x \in \mathbb{R}$ and that the paths of $\mathcal{X}$ are of class C^{p+1}. Then*

$$\textit{if} \quad 2K + 1 < p/2: \qquad \mathrm{P}(M > u) \le \mathrm{P}(X(0) > u) + \sum_{m=1}^{2K+1} (-1)^{m+1} \frac{\widetilde{\nu}_m}{m!}$$

and

$$\textit{if} \quad 2K < p/2: \qquad \mathrm{P}(M > u) \ge \mathrm{P}(X(0) > u) + \sum_{m=1}^{2K} (-1)^{m+1} \frac{\widetilde{\nu}_m}{m!}.$$

Notice that all the moments in the formulas above are finite.

The proof is a straightforward application of Lemma 5.3 and Theorem 3.6. When the level u is high, a first approximation is given by Proposition 4.1, which shows that only the *first* term in the Rice series takes part in the equivalent of $1 - F_{M_T}(u)$ as $u \to +\infty$.

5.2. COMPUTATION OF MOMENTS

An efficient numerical computation of the factorial moments of crossings is associated with a fine description of the behavior as the k-tuple $(t_1, \dots, t_k)$ approaches the diagonal $D_k(I)$ of the integrands

$$A^+_{t_1,\dots,t_k}(u,\dots,u) = \mathrm{E}\Big(X'^+(t_1)\cdots X'^+(t_k)\big|X(t_1) = \cdots = X(t_k) = u\Big) \times p_{X(t_1),\dots,X(t_k)}(u,\dots,u)$$

$$\widetilde{A}^+_{t_1,\dots,t_k}(u,\dots,u) = \int_{-\infty}^{u} \mathrm{E}\Big(X'^+(t_1)\cdots X'^+(t_k)\big|X(0) = x, X(t_1) = \cdots = X(t_k) = u\Big) p_{X(0),X(t_1),\dots,X(t_k)}(x,u,\dots,u)\,dx. \tag{5.11}$$

We recall that $A^+_{t_1,\dots,t_k}(u,\dots,u)$ and $\widetilde{A}^+_{t_1,\dots,t_k}(u,\dots,u)$ appear, respectively, in Rice formulas for the kth factorial moment of up-crossings and the kth factorial moment of up-crossings with the additional condition that $X(0) \le u$.

If the process is Gaussian stationary and satisfies a certain number of regularity conditions, we have seen in Proposition 4.5 that

$$A^+_{s,t}(u,u) \approx \frac{1}{1296} \frac{\left(\lambda_2\lambda_6 - \lambda_4^2\right)^{3/2}}{\left(\lambda_4 - \lambda_2^2\right)^{1/2} \pi^2 \lambda_2^2} \exp\left(-\frac{1}{2}\frac{\lambda_4}{\lambda_4 - \lambda_2^2}u^2\right)(t-s)^4 \tag{5.12}$$

as $t - s \to 0$.

Equation (5.12) can be extended to nonstationary Gaussian processes, obtaining an equivalence of the form

$$A^+_{s,t}(u,u) \approx J(\widetilde{t})(t-s)^4 \quad \text{as } s,t \to \widetilde{t}, \tag{5.13}$$

where $J(\widetilde{t})$ is a continuous nonzero function of $\widetilde{t}$ depending on u, which can be expressed in terms of the mean and covariance functions of the process and its derivatives. We give a proof of an equivalence of the form (5.13) in the next proposition.

One can profit from this equivalence to improve the numerical methods to compute $\widetilde{\nu}_2$ [the second factorial moment of the number of up-crossings with the restriction $X(0) \le u$]. Equivalence formulas such as (5.12) or (5.13) can be used to avoid numerical degeneracies near the diagonal $D_2(I)$. Notice that even in case the process $\mathcal{X}$ is stationary at the departure, under conditioning on $X(0)$, the process that must be taken into account in the computation of the factorial moments of up-crossings for the Rice series (5.2) will be nonstationary, so that equivalence (5.13) is the appropriate tool for our main purpose here.

Proposition 5.8. *Suppose that $\mathcal{X}$ is a Gaussian process with C^5-paths and that for each $t \in I$ the joint distribution of $X(t)$, $X'(t)$, $X^{(2)}(t)$, and $X^{(3)}(t)$ does not degenerate. Then (5.13) holds true.*

Proof. We give the general scheme of the proof and leave the detailed computations to the reader. Denote by $\xi = \binom{\xi_1}{\xi_2}$ a two-dimensional random vector having as probability distribution the conditional distribution of $\binom{X'(s)}{X'(t)}$ given $X(s) = X(t) = u$. One has

$$A^+_{s,t}(u, u) = \mathrm{E}\left(\xi_1^+\xi_2^+\right) p_{X(s),X(t)}(u, u). \tag{5.14}$$

Set $\tau = t - s$ and check the following Taylor expansions around point s:

$$\mathrm{E}\left(\xi_1\right) = m_1\tau + m_2\tau^2 + L_1\tau^3 \tag{5.15}$$

$$\mathrm{E}\left(\xi_2\right) = -m_1\tau + m_2'\tau^2 + L_2\tau^3 \tag{5.16}$$

$$\mathrm{Var}\left(\xi\right) = \begin{pmatrix} a\tau^2 + b\tau^3 + c\tau^4 + \rho_{11}\tau^5 & -a\tau^2 - \dfrac{b+b'}{2}\tau^3 + d\tau^4 + \rho_{12}\tau^5 \\ -a\tau^2 - \dfrac{b+b'}{2}\tau^3 + d\tau^4 + \rho_{12}\tau^5 & a\tau^2 + b'\tau^3 + c'\tau^4 + \rho_{22}\tau^5 \end{pmatrix}, \tag{5.17}$$

where m_1, m_2, m_2', a, b, c, d, a', b', c' are continuous functions of s, and L_1, L_2, ρ_{11}, ρ_{12}, ρ_{22} are bounded functions of s and t. Equations (5.15), (5.16), and (5.17) follow directly from the regression formulas of the pair $\binom{X'(s)}{X'(t)}$ on the condition $X(s) = X(t) = u$.

Notice that (as in Belayev, 1966, or Azaïs and Wschebor, 2002)

$$\begin{aligned} \mathrm{Var}(\xi_1) &= \frac{\det[\mathrm{Var}(X(s), X(t), X'(s))^{\mathrm{T}}]}{\det[\mathrm{Var}(X(s), X(t))^{\mathrm{T}}]} \\ &= \frac{\det[\mathrm{Var}(X(s), X'(s), X(t) - X(s) - (t-s)X'(s))^{\mathrm{T}}]}{\det[\mathrm{Var}(X(s), X(t) - X(s))^{\mathrm{T}}]}. \end{aligned}$$

Direct computation gives

$$\det[\mathrm{Var}(X(s), X(t))^{\mathrm{T}}] \approx \tau^2 \det[\mathrm{Var}(X(s), X'(s))^{\mathrm{T}}]$$

$$\mathrm{Var}(\xi_1) \approx \frac{1}{4}\frac{\det[\mathrm{Var}(X(s), X'(s), X^{(2)}(s))^{\mathrm{T}}]}{\det[\mathrm{Var}(X(s), X'(s))^{\mathrm{T}}]}\tau^2, \tag{5.18}$$

where $\approx$ denotes equivalence as $\tau \to 0$. So

$$a = \frac{1}{4}\frac{\det[\mathrm{Var}(X(s), X'(s), X^{(2)}(s))^{\mathrm{T}}]}{\det[\mathrm{Var}(X(s), X'(s))^{\mathrm{T}}]},$$

which is a continuous nonvanishing function for $s \in I$. Notice that the coefficient of τ^3 in the Taylor expansion of $\mathrm{Cov}(\xi_1, \xi_2)$ is equal to $-(b+b')/2$. This follows either by direct computation or by taking into account the fact that $\det[\mathrm{Var}(\xi)]$ is a symmetric function of the pair s, t. Set

$$\Delta(s,t) = \det[\mathrm{Var}(\xi)].$$

The behavior of $\Delta(s,t)$ as $s, t \to \widetilde{t}$ can be obtained from

$$\Delta(s,t) = \frac{\det\left[\mathrm{Var}(X(s), X(t), X'(s), X'(t))^{\mathrm{T}}\right]}{\det[\mathrm{Var}(X(s), X(t))^{\mathrm{T}}]}$$

by applying the more general Proposition 5.9, which provides an equivalent for the numerator [or use Lemma 4.3, p. 76 in Piterbarg (1996a), which is sufficient in the present case]. We get

$$\Delta(s,t) \approx \overline{\Delta}(\widetilde{t})\tau^6, \tag{5.19}$$

where

$$\overline{\Delta}(\widetilde{t}) = \frac{1}{144}\frac{\det\left[\mathrm{Var}(X(\widetilde{t}), X'(\widetilde{t}), X^{(2)}(\widetilde{t}), X^{(3)}(\widetilde{t}))^{\mathrm{T}}\right]}{\det[\mathrm{Var}(X(\widetilde{t}), X'(\widetilde{t}))^{\mathrm{T}}]}.$$

The nondegeneracy hypothesis implies that $\overline{\Delta}(\widetilde{t})$ is continuous and nonzero. One has

$$\mathrm{E}\left(\xi_1^+\xi_2^+\right) = \frac{1}{2\pi\,[\Delta(s,t)]^{1/2}}\int_0^{+\infty}\int_0^{+\infty} xy \exp\left[-\frac{1}{2\Delta(s,t)}F(x,y)\right]dx\,dy, \tag{5.20}$$

where

$$F(x,y) = \mathrm{Var}(\xi_2)(x - \mathrm{E}(\xi_1))^2 + \mathrm{Var}(\xi_1)(y - \mathrm{E}(\xi_2))^2$$
$$-2\,\mathrm{Cov}(\xi_1, \xi_2)(x - \mathrm{E}(\xi_1))(y - \mathrm{E}(\xi_2)).$$

Substituting the expansions (5.15), (5.16), and (5.17) in the integrand of (5.20) and making the change of variables $x = \tau^2 v$, $y = \tau^2 w$, we get, as $s, t \to \widetilde{t}$,

$$\mathrm{E}\left(\xi_1^+\xi_2^+\right) \approx \frac{\tau^5}{2\pi\left[\overline{\Delta}(\widetilde{t})\right]^{1/2}}\int_0^{+\infty}\int_0^{+\infty} vw \exp\left[-\frac{1}{2\overline{\Delta}(\widetilde{t})}\overline{F}(v,w)\right]dv\,dw. \tag{5.21}$$

$\overline{\Delta}(\widetilde{t})$ can also be expressed in terms of the functions a, b, c, d, a', b', c':

$$\overline{\Delta}(\widetilde{t}) = ac' + ca' + 2ad - \left(\frac{b-b'}{2}\right)^2$$

and

$$\overline{F}(v, w) = a\left(v - m_2 + w - m_2'\right)^2 + m_1^2(c + c' + 2d)$$
$$-m_1(b - b')(v + w - m_2 - m_2').$$

The functions $a, b, c, d, b', c', m_1, m_2$ which appear in these formulas are all evaluated at point $\widetilde{t}$. Substituting (5.21) and (5.18) into (5.14), one gets (5.13). □

For $k \geq 3$, the general behavior of the functions $\widetilde{A}_{t_1}, \ldots, t_k(u, \ldots, u)$ and $A^+_{t_1,\ldots,t_k}(u, \ldots, u)$ when $(t_1, \ldots, t_k)$ approaches the diagonal is not known. Even though Proposition 5.10 below contains some restrictions [it requires that $\mathrm{E}(X(t)) = 0$ and $u = 0$], it can be used to improve efficiency in the computation of the kth factorial moments by means of a Monte Carlo method through the use of importance sampling. More precisely, this proposition can be used when computing the integral of $A^+_{t_1,\ldots,t_k}(u, \ldots, u)$ over I^k in the following way: Instead of choosing at random the point $(t_1, t_2, \ldots, t_k)$ in the cube I^k with a uniform distribution, we should do it with a probability law having a density proportional to the function $\prod_{1\leq i<j\leq k}\left(t_j - t_i\right)^4$. For the proof of Proposition 5.10 we will use the following auxiliary proposition, which has its own interest.

Proposition 5.9. *Let $\mathcal{X} = \{X(t) : t \in I\}$ be a Gaussian process defined on the compact interval I of the real line, k an integer $k \geq 2$, and $t_1, \ldots, t_k \in I$. When the paths of the process $\mathcal{X}$ are of class C^l, we denote*

$$D_l(t) = \det\left[\mathrm{Var}\left(X(t), X'(t), \ldots, X^{(l)}(t)\right)^{\mathrm{T}}\right].$$

(i) *If the paths of process $\mathcal{X}$ are of class C^{k-1} and $t_1, t_2, \ldots, t_k \to t^*$, then*

$$\det\left[\mathrm{Var}(X(t_1), X(t_2), \ldots, X(t_k))^{\mathrm{T}}\right]$$
$$\approx \frac{1}{[2!\cdots(k-1)!]^2}\left[\prod_{1\leq i<j\leq k}(t_j - t_i)^2\right] D_{k-1}(t^*). \tag{5.22}$$

(ii) *If the paths of $\mathcal{X}$ are of class C^{2k-1} and $t_1, t_2, \ldots, t_k \to t^*$, then*

$$\Delta = \det\left[\mathrm{Var}(X(t_1), X'(t_1), \ldots, X(t_k), X'(t_k))^{\mathrm{T}}\right]$$
$$\approx \frac{1}{[2!\cdot 3!\cdots(2k-1)!]^2}\prod_{1\leq i<j\leq k}(t_j - t_i)^8 D_{2k-1}(t^*). \tag{5.23}$$

Proof. We prove (ii). The proof of (i) can be done along the same lines as that of (ii). It is, in fact, simpler and is left to the reader.

With no loss of generality, we may assume that $t_1, t_2, \ldots, t_k$ are pairwise different. Suppose that $f : I \to \mathbb{R}$ is a function of class C^{2m-1}, $1 \leq m \leq k$. We use the following notation for interpolating polynomials: $P_m(t; f)$ is a polynomial of degree $2m - 1$ such that

$$P_m(t_j; f) = f(t_j) \quad \text{and} \quad P'_m(t; f) = f'(t_j) \quad \text{for } j = 1, \ldots, m.$$

$Q_m(t; f)$ is a polynomial of degree $2m - 2$ such that

$$\begin{aligned} Q_m(t_j; f) &= f(t_j) \quad \text{for } j = 1, \ldots, m, \\ Q'_m(t; f) &= f'(t_j) \quad \text{for } j = 1, \ldots, m-1. \end{aligned}$$

From Lemma 5.2 we know that

$$f(t) - P_m(t; f) = \frac{1}{(2m)!}(t - t_1)^2 \cdots (t - t_m)^2 f^{(2m)}(\xi) \tag{5.24}$$

$$f(t) - Q_m(t; f) = \frac{1}{(2m-1)!}(t - t_1)^2 \cdots (t - t_{m-1})^2 (t - t_m) f^{(2m-1)}(\eta), \tag{5.25}$$

where

$$\begin{gathered} \xi = \xi(t_1, t_2, \ldots, t_m, t), \eta = \eta(t_1, t_2, \ldots, t_m, t) \\ \min(t_1, t_2, \ldots, t_m, t) \leq \xi, \eta \leq \max(t_1, t_2, \ldots, t_m, t). \end{gathered}$$

The function

$$g(t) = f^{(2m-1)}(\eta(t_1, t_2, \ldots, t_m, t)) = \frac{(2m-1)!\left[f(t) - Q_m(t; f)\right]}{(t - t_1)^2 \cdots (t - t_{m-1})^2 (t - t_m)}$$

is differentiable at the point $t = t_m$, and differentiating in (5.25) yields

$$\begin{aligned} f'(t_m) - Q'_m(t_m; f) = {} & \frac{1}{(2m-1)!}(t_m - t_1)^2 \cdots (t_m - t_{m-1})^2 \\ & \times f^{(2m-1)}(\eta(t_1, t_2, \ldots, t_m, t_m)). \end{aligned} \tag{5.26}$$

Set

$$\xi_m = \xi(t_1, t_2, \ldots, t_m, t_m), \; \eta_m = \eta(t_1, t_2, \ldots, t_m, t_m).$$

Since $P_m(t; f)$ is a linear functional of

$$\left(f(t_1), \ldots, f(t_m), f'(t_1), \ldots, f'(t_m)\right)$$

and $Q_m(t; f)$ is a linear functional of

$$\big(f(t_1), \ldots, f(t_m), f'(t_1), \ldots, f'(t_{m-1})\big)$$

with coefficients depending (in both cases) only on $t_1, t_2, \ldots, t_m, t$, it follows that

$$\begin{aligned}
\Delta &= \det\Big[\mathrm{Var}\big(X(t_1), X'(t_1), X(t_2) - P_1(t_2; X), X'(t_2) - Q_2'(t_2, X), \ldots, \\
&\qquad X(t_k) - P_{k-1}(t_k; X), X'(t_k) - Q_k'(t_k; X)\big)^{\mathrm{T}}\Big] \\
&= \det\Big[\mathrm{Var}\big(X(t_1), X'(t_1), \frac{1}{2!}(t_2 - t_1)^2 X^{(2)}(\xi_1), \frac{1}{3!}(t_2 - t_1)^2 X^{(3)}(\eta_2), \ldots, \\
&\qquad \frac{1}{(2k-2)!}(t_k - t_1)^2 \cdots (t_k - t_{k-1})^2 X^{(2k-2)}(\xi_{k-1}), \\
&\qquad \frac{1}{(2k-1)!}(t_k - t_1)^2 \cdots (t_k - t_{k-1})^2 X^{(2k-1)}(\eta_{k-1})\big)^{\mathrm{T}}\Big] \\
&= \frac{\widetilde{\Delta}}{[2! \cdots (2k-1)!]^2} \prod_{1 \le i < j \le k} (t_j - t_i)^8
\end{aligned}$$

with

$$\begin{aligned}
\widetilde{\Delta} &= \det\Big[\mathrm{Var}\big(X(t_1), X'(t_1), X^{(2)}(\xi_1), X^{(3)}(\eta_2), \ldots, X^{(2k-2)}(\xi_{k-1}), \\
&\qquad X^{(2k-1)}(\eta_{k-1})\big)^{\mathrm{T}}\Big] \to \det\Big[\mathrm{Var}\big(X(t^*), X'(t^*), \ldots, X^{(2k-1)}(t^*)\big)^{\mathrm{T}}\Big] \\
&= D_{2k-1}(t^*)
\end{aligned}$$

as $t_1, t_2, \ldots, t_k \to t^*$. This proves (5.23). □

Proposition 5.10. *Suppose that $\mathcal{X}$ is a centered Gaussian process with C^{2k-1}-paths and that for each pairwise distinct values of the parameter $t_1, t_2, \ldots, t_k \in I$, the joint distribution of $(X(t_h), X'(t_h), \ldots, X^{(2k-1)}(t_h),\ h = 1, 2, \ldots, k)$ is nondegenerate. Then, as $t_1, t_2, \ldots, t_k \to t^*$,*

$$A^+_{t_1,\ldots,t_k}(0, \ldots, 0) \approx J_k(t^*) \prod_{1 \le i < j \le k} (t_j - t_i)^4,$$

where $J_k(t)$ is a continuous nonzero function of t.

Proof. For k distinct values $t_1, t_2, \ldots, t_k$, let $Z = (Z_1, \ldots, Z_k)^{\mathrm{T}}$ be a random vector having the conditional distribution of $(X'(t_1), \ldots, X'(t_k))^{\mathrm{T}}$ given

$X(t_1) = X(t_2) = \cdots = X(t_k) = 0$. The (Gaussian) distribution of Z is centered and we denote its covariance matrix by Σ. Also set

$$\Sigma^{-1} = \frac{1}{\det(\Sigma)} \left(\sigma^{ij}\right)_{i,j=1,\dots,k},$$

σ^{ij} being the cofactor of the position (i, j) in the matrix Σ. Then one can write

$$A^{+}_{t_1,\dots,t_k}(0,\dots,0) = \mathrm{E}(Z_1^{+} \dots Z_k^{+}) p_{X(t_1),\dots,X(t_k)}(0,\dots,0) \tag{5.27}$$

and

$$A^{+}_{t_1,\dots,t_k}(0,\dots,0) = \frac{1}{(2\pi)^{k/2}(\det(\Sigma))^{1/2}} \int_{(R^{+})^k} x_1 \cdots x_k \times \exp\left(-\frac{F(x_1,\dots,x_k)}{2\det(\Sigma)}\right) dx_1 \cdots dx_k, \tag{5.28}$$

where

$$F(x_1,\dots,x_k) = \sum_{i,j=1}^{k} \sigma^{ij} x_i x_j.$$

Letting $t_1, t_2, \dots, t_k \to t^*$ and using (5.23) and (5.22), we get

$$\det(\Sigma) = \frac{\det[\mathrm{Var}(X(t_1), X'(t_1), \dots, X(t_k), X'(t_k))^{\mathrm{T}}]}{\det[\mathrm{Var}(X(t_1), \dots, X(t_k))^{\mathrm{T}}]} \approx \frac{1}{[k! \cdots (2k-1)!]^2} \left[\prod_{1\le i<j\le k} (t_j - t_i)^6\right] \frac{D_{2k-1}(t^*)}{D_{k-1}(t^*)}.$$

We now consider the behavior of the $\sigma^{ij}(i, j = 1, \dots, k)$. Let us first look at σ^{11}. Using the same method as above, now applied to the cofactor of the position $(1, 1)$ in Σ, one has

$$\sigma^{11} = \frac{\det\left[\mathrm{Var}(X(t_1), X(t_2), \dots, X(t_k), X'(t_2), \dots, X'(t_k))^{\mathrm{T}}\right]}{\det[\mathrm{Var}(X(t_1), \dots, X(t_k))^{\mathrm{T}}]}$$

$$\approx \frac{[2! \cdots (2k-2)!]^{-2} \left[\prod_{2\le i<j\le k}(t_j - t_i)^8\right] \left[\prod_{2\le h\le k}(t_1 - t_h)^4\right] D_{2k-2}(t^*)}{[2! \cdots (k-1)!]^{-2} \left[\prod_{1\le i<j\le k}(t_j - t_i)^2\right] D_{k-1}(t^*)}$$

$$= [k! \cdots (2k-2)!]^{-2} \left[\prod_{2\le i<j\le k} (t_j - t_i)^6\right] \left[\prod_{2\le h\le k} (t_1 - t_h)^2\right] \frac{D_{2k-2}(t^*)}{D_{k-1}(t^*)}.$$

A similar computation holds for $\sigma^{ii}, i = 2, \dots, k$.

Now consider σ^{12}. One has

$$\sigma^{12} = -\frac{\det\left[\mathrm{E}\left\{(X(t_1), X(t_2), \dots, X(t_k), X'(t_2), \dots, X'(t_k))^{\mathrm{T}}(X(t_1), X(t_2), \dots, X(t_k), X'(t_1), X'(t_3) \dots, X'(t_k))\right\}\right]}{\det[\mathrm{Var}(X(t_1), \dots, X(t_k))^{\mathrm{T}}]}$$

$$= \frac{\det\left[\mathrm{E}\left\{(X(t_2), X'(t_2), \dots, X(t_k), X'(t_k), X(t_1))^{\mathrm{T}}(X(t_1), X'(t_1), X(t_3), X'(t_3), \dots, X(t_k), X'(t_k), X(t_2))\right\}\right]}{\det[\mathrm{Var}(X(t_1), \dots, X(t_k))^{\mathrm{T}}]}$$

$$\approx \frac{1}{[k! \cdots (2k-2)!]^2}\left[\prod_{3 \le i < j \le k} (t_j - t_i)^6\right]\left[\prod_{3 \le h \le k} (t_1 - t_h)^4 (t_2 - t_h)^4\right] \times (t_2 - t_1)^2 \frac{D_{2k-2}(t^*)}{D_{k-1}(t^*)}.$$

A similar computation applies to all the cofactors $\sigma^{ij}, i \neq j$. In the integral in (5.28) perform the change of variables

$$x_j = \left[\prod_{i=1, i \neq j}^{i=k} (t_i - t_j)^2\right] y_j \qquad j = 1, \dots, k$$

and the integral becomes

$$\left[\prod_{1 \le i < j \le k} (t_j - t_i)^8\right] \int_{(R^+)^k} y_1 \cdots y_k \exp\left[-\frac{1}{2\det(\Sigma)} G(y_1, \dots, y_k)\right] dy_1 \cdots dy_k,$$

where

$$G(y_1, \dots, y_k) = \sum_{i,j=1}^{k} \sigma^{ij} \left[\prod_{h=1, h \neq i}^{h=k} (t_h - t_i)^2\right]\left[\prod_{h=1, h \neq j}^{h=k} (t_h - t_j)^2\right] y_i\, y_j,$$

so that as $t_1, t_2, \dots, t_k \to t^*$,

$$\frac{G(y_1, \dots, y_k)}{\det(\Sigma)} \approx [(2k-1)!]^2 \frac{D_{2k-2}(t^*)}{D_{2k-1}(t^*)} \left(\sum_{i=1}^{i=k} y_i\right)^2.$$

Now, passage to the limit under the integral sign in (5.28), which is easily justified by application of the Lebesgue theorem, leads to

$$\mathrm{E}\{Z_1^+ \cdots Z_k^+\} \approx \frac{1}{(2\pi)^{k/2}} k! \cdots (2k-1)! \left[\prod_{1 \le i < j \le k} |t_j - t_i|^5 \right] \times \left(\frac{D_{k-1}(t^*)}{D_{2k-1}(t^*)} \right)^{1/2} I_k(\alpha^*),$$

where $I_k(\alpha)$, $\alpha > 0$ is

$$I_k(\alpha) = \int_{(R^+)^k} y_1 \cdots y_k \exp\left[-\frac{\alpha}{2} \left(\sum_{i=1}^{i=k} y_i \right)^2 \right] dy_1 \cdots dy_k = \frac{1}{\alpha^k} I_k(1)$$

and

$$\alpha^* = [(2k-1)!]^2 \frac{D_{2k-2}(t^*)}{D_{2k-1}(t^*)}.$$

Substituting in (5.27), one obtains the result

$$J_k(t) = \frac{2! \cdots (2k-2)!}{[2\pi(2k-1)!]^{2k-1}} \frac{I_k(1)}{[D_{2k-1}(t)]^{1/2}} \left[\frac{D_{2k-1}(t)}{D_{2k-2}(t)} \right]^k.$$

This completes the proof. □

5.3. NUMERICAL ASPECTS OF THE RICE SERIES

Let us compare the numerical computation based on Theorem 5.1 with the Monte Carlo method based on the simulation of the paths. We do this for stationary Gaussian processes that satisfy the hypotheses of Theorem 5.6 and also the non-degeneracy condition, which ensures that one is able to compute the factorial moments of crossings by means of Rice formulas.

Suppose that we want to compute $\mathrm{P}(M > u)$ with an error bounded by δ, where $\delta > 0$ is a given positive number. To proceed by simulation, we discretize the paths by means of a uniform partition $\{t_j := j/n, j = 0, 1, \ldots, n\}$. Denote

$$M^{(n)} := \sup_{0 \le j \le n} X(t_j).$$

Using Taylor's formula at the time where the maximum M of $X(\cdot)$ occurs, one gets

$$0 \le M - M^{(n)} \le \frac{\|X''\|_\infty}{2n^2}.$$

It follows that

$$0 \le \mathrm{P}(M > u) - \mathrm{P}(M^{(n)} > u) = \mathrm{P}(M > u, M^{(n)} \le u) \le \mathrm{P}(u < M \le u + \|X''\|_\infty/(2n^2)).$$

Let us admit that the distribution of M has a locally bounded density (see Ylvisaker's Theorem 1.22). The above suggests that $n =$ (const) $\delta^{-1/2}$ points are required if one wants the error $\mathrm{P}(M > u) - \mathrm{P}(M^{(n)} > u)$ to be bounded by δ.

On the other hand, to estimate $\mathrm{P}(M^{(n)} > u)$ by Monte Carlo methods with a mean square error smaller than δ, we require the simulation of $N =$ (const) δ^{-2} Gaussian n-tuples $(X_{t_1}, \ldots, X_{t_n})$ from the distribution determined by the given stationary process. Performing each simulation demands (const) $n \log(n)$ elementary operations [see, e.g., Dietrich and Newsam (1997) for this computational point]. Summing up, the total mean number of elementary operations required to get a mean square error bounded by δ in the estimation of $\mathrm{P}(M > u)$ has the form (const) $\delta^{-5/2} \log(1/\delta)$.

Suppose now that we apply Theorem 5.1 to a Gaussian stationary centered process verifying the hypotheses of Theorem 5.6 and the nondegeneracy condition. The bound for γ_m^* in equation (5.10) implies that computing a partial sum with (const) $\log(1/\delta)$ terms assures that the tail in the Rice series is bounded by δ. If one computes each $\widetilde{\nu}_m$ by means of a Monte Carlo method for the multiple integrals appearing in the Rice formulas, the number of elementary operations for the entire procedure will have the form (const) $\delta^{-2} \log(1/\delta)$. Hence, this is better than simulation as δ tends to zero.

As usual, given $\delta > 0$, the value of the generic constants determines the comparison between these methods, and these are very difficult to estimate for a general class of processes. More important is the fact that the enveloping property of the Rice series implies that the actual number of terms required by the use of Theorem 5.1 can be *much* smaller than the one resulting from the a priori bound for γ_m^*. More precisely, suppose that we have obtained each numerical approximation $\widetilde{\nu}_m^*$ of $\widetilde{\nu}_m$ with a precision η:

$$|\widetilde{\nu}_m^* - \widetilde{\nu}_m| \le \eta,$$

and that we stop when

$$\frac{\widetilde{\nu}_{m_0+1}^*}{(m_0+1)!} \le \eta. \tag{5.29}$$

Then it follows that

$$\left| \sum_{m=1}^{\infty} (-1)^{m+1} \frac{\widetilde{\nu}_m}{m!} - \sum_{m=1}^{m_0} (-1)^{m+1} \frac{\widetilde{\nu}_m^*}{m!} \right| \le (e+1)\eta.$$

Putting $\eta = \delta/(e+1)$, we get the bound desired. In other words, one can profit from the successive numerical approximations of $\widetilde{\nu}_m$ to determine a new m_0, which in certain interesting examples turns out to be much smaller than the one deduced from the a priori bound on γ_m^*.

Next, we give the results of an evaluation of $P(M_T > u)$ using up to three terms in the Rice series in a certain number of typical cases. We compare these results with the classical evaluation given by Proposition 4.1 For fixed T and high-level u, this bound is sharp. But when both T and u are fixed, the situation becomes essentially different, and using more than one term in the Rice series yields a remarkable improvement in the computation. We consider several stationary centered Gaussian processes in the following table, where the covariances and corresponding spectral densities are indicated.

Process	Covariance	Spectral Density
X_1	$\Gamma_1(t) = \exp(-t^2/2)$	$f_1(x) = (2\pi)^{-1/2}\exp(-x^2/2)$
X_2	$\Gamma_2(t) = (\cosh(t))^{-1}$	$f_2(x) = \left(2\cosh((\pi x)/2)\right)^{-1}$
X_3	$\Gamma_3(t) = \left(3^{1/2}t\right)^{-1}\sin(3^{1/2}t)$	$f_3(x) = 12^{-1/2}\mathbf{1}_{\{-\sqrt{3}<x<\sqrt{3}\}}$
X_4	$\Gamma_4(t) = e^{-\lvert\sqrt{5}t\rvert}((\sqrt{5}/3)\lvert t\rvert^3$ $+2t^2 + \sqrt{5}\lvert t\rvert + 1)$	$f_4(x) = (10^4/\sqrt{5}\pi)(5+x^2)^{-4}$

In all cases, we set $\lambda_0 = \lambda_2 = 1$ to be able to compare the various results. Notice that Γ_1 and Γ_3 have analytic extensions to the entire plane, so that Theorem 5.6 applies to the processes X_1 and X_3. On the other hand, even though all spectral moments of the process X_2 are finite, Theorem 5.6 applies only for a length less than $\pi/4$ since the meromorphic extension of $\Gamma_2(\cdot)$ has poles at the points $i\pi/2 + k\pi i$, k an integer. With respect to $\Gamma_4(\cdot)$ notice that it is obtained as the convolution $\Gamma_5 * \Gamma_5 * \Gamma_5 * \Gamma_5$, where $\Gamma_5(t) := e^{-|t|}$ is the covariance of the Ornstein–Uhlenbeck process, plus a change of scale to get $\lambda_0 = \lambda_2 = 1$. The process X_4 has $\lambda_6 < \infty$ and $\lambda_8 = \infty$ and its paths are C^3. For the processes X_2 and X_4 we use Theorem 5.7 to compute $F(T, u)$.

Table 5.1 contains the results for $T = 1, 4, 6, 8, 10$ and the values $u = -2, -1, 0, 1, 2, 3$ using three terms of the Rice series. A single value is given when a precision of 10^{-2} is met; otherwise, the lower and upper bounds given by two or three terms of the Rice series, respectively, are displayed. The calculation uses a deterministic evaluation of the first two moments, $\widetilde{\nu}_1$ and $\widetilde{\nu}_2$, using a program written by Cierco-Ayrolles et al. (2003) and a Monte Carlo evaluation of $\widetilde{\nu}_3$. In fact, for simpler and faster calculation, ν_3 has been evaluated instead of $\widetilde{\nu}_3$, providing a slightly weaker bound.

In addition, Figures 5.1 to 5.4 show the behavior of four bounds: from the highest to the lowest:

- The Davies' bound (D), defined by Proposition 4.1.
- One, three, or two terms of the Rice series ($R1$, $R3$, $R2$ in the sequel): that is,

$$P(X(0) > u) + \sum_{m=1}^{K} (-1)^{m+1} \frac{\widetilde{\nu}_m}{m!}$$

with $K = 1, 3$, or 2.

TABLE 5.1. Values of P($M > u$) for Various Processes[a]

	Length of the Time Interval T				
u	1	4	6	8	10
−2	0.99	1.00	1.00	1.00	1.00
	0.99	1.00	1.00	1.00	1.00
	1.00	1.00	1.00	1.00	1.00
	0.99	1.00	1.00	1.00	1.00
−1	0.93	1.00	1.00	0.99–1.00	0.98–1.00
	0.93	0.99	1.00	0.99–1.00	0.98–1.00
	0.93	1.00	1.00	1.00	0.99
	0.93	1.00	1.00	0.99–1.00	0.98–1.00
0	0.65	0.90	0.95	0.95–0.99	0.90–1.00
	0.65	0.89	0.94–0.95	0.93–0.99	0.87–1.00
	0.656	0.919	0.97	0.98–0.99	0.92–1.00
	0.65	0.89	0.94–0.95	0.94–0.99	0.88–1.00
1	0.25	0.49	0.61	0.69–0.70	0.74–0.77
	0.25	0.48	0.58	0.66–0.68	0.70–0.76
	0.26	0.51	0.62	0.71	0.76–0.78
	0.25	0.48	0.59	0.67–0.69	0.72–0.77
2	0.04	0.11	0.15	0.18	0.22
	0.04	0.11	0.14	0.18	0.21
	0.04	0.11	0.15	0.19	0.22
	0.04	0.11	0.14	0.18	0.22
3	0.00	0.01	0.01	0.02	0.02
	0.00	0.01	0.01	0.02	0.02
	0.00	0.01	0.01	0.02	0.02
	0.00	0.01	0.01	0.02	0.02

Source: From Azaïs and Wschebor, 2002, with permission.
[a]Each cell contains, from top to bottom, values corresponding to stationary centered Gaussian processes with covariances $\Gamma_1, \Gamma_2, \Gamma_3$, and Γ_4, respectively. The calculation uses three terms of the Rice series for the upper bound and two terms for the lower bound. Both are rounded up to two decimals, and when they differ, both are displayed.

Notice that the bound D differs from $R1$ due to the difference between $\widetilde{\nu}_1$ and ν_1. These bounds are evaluated for $T = 4, 6, 8, 10, 15$ and also for $T = 20$ and $T = 40$ when they fall in the range $[0, 1]$. Between these values, ordinary spline interpolation has been performed.

We illustrate the complete detailed calculation in three cases. They correspond to zero and positive levels u. For u negative, it is easy to check that the Davies bound is often greater than 1, thus noninformative.

- For $u = 0$, $T = 6$, $\Gamma = \Gamma_1$, we have $\mathrm{P}(X(0) > u) = 0.5$, $\tilde{\nu}_1 = 0.955$, $\widetilde{\nu}_1 = 0.602$, $\widetilde{\nu}_2/2 = 0.150$, and $\tilde{\nu}_3/6 = 0.004$, so that

$$D = 1.455, \qquad R1 = 1.103, \qquad R3 = 0.957, \qquad R2 = 0.953$$

$R2$ and $R3$ give a rather good evaluation of the probability. The Davies bound gives no information.

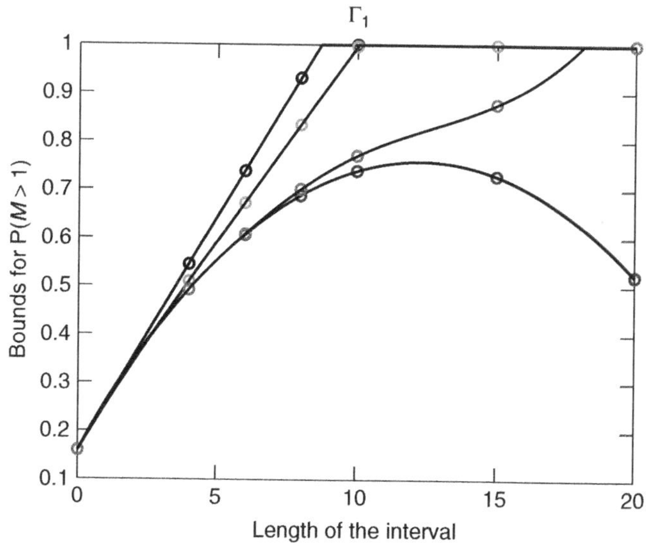

Figure 5.1. For the process with covariance Γ_1 and level $u = 1$, representation of the three upper bounds, D, $R1$, and $R3$, and the lower bound, $R2$ (from top to bottom), as a function of the length T of the interval. (From Azaïs and Wschebor, 2002, with permission.)

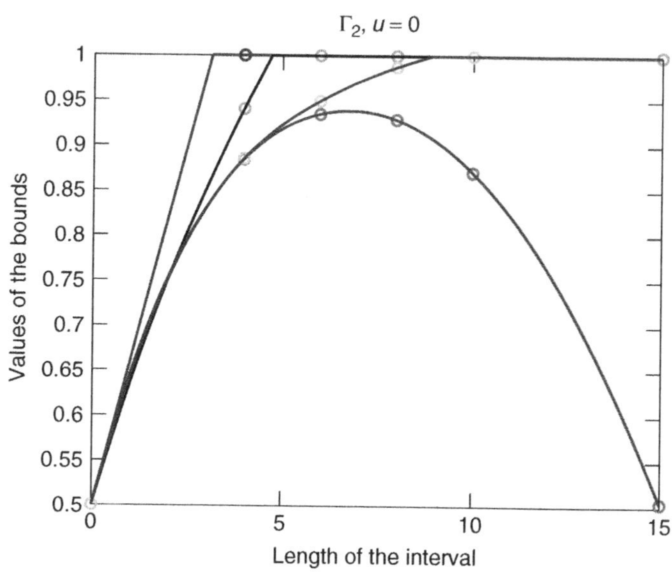

Figure 5.2. For the process with covariance Γ_2 and the level $u = 0$, representation of the three upper bounds, D, $R1$, and $R3$, and the lower bound, $R2$ (from top to bottom), as a function of the length T of the interval. (From Azaïs and Wschebor, 2002, with permission.)

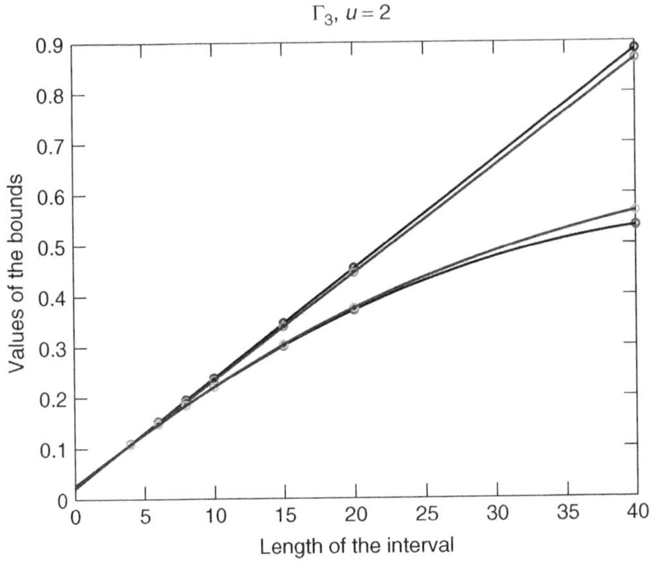

Figure 5.3. For the process with covariance Γ_3 and the level $u = 2$, representation of the three upper bounds, D, $R1$, and $R3$, and the lower bound, $R2$ (from top to bottom), as a function of the length T of the interval. (From Azaïs and Wschebor, 2002, with permission.)

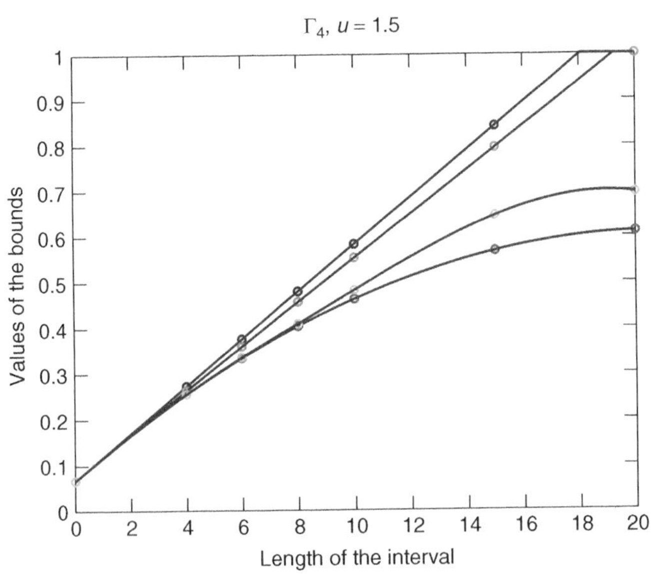

Figure 5.4. For the process with covariance Γ_4 and the level $u = 1.5$, representation of the three upper bounds, D, $R1$, and $R3$, and the lower bound, $R2$ (from top to bottom) as a function of the length T of the interval. (From Azaïs and Wschebor, 2002, with permission.)

- For $u = 1.5$, $T = 15$, $\Gamma = \Gamma_2$, we have $\mathrm{P}(X_0 > u) = 0.067$, $\nu_1 = 0.517$, $\widetilde{\nu}_1 = 0.488$, $\widetilde{\nu}_2/2 = 0.08$, and $\nu_3/6 = 0.013$, so that

$$D = 0.584, \quad R1 = 0.555, \quad R3 = 0.488, \quad R2 = 0.475$$

  In this case the Davies bound is not sharp and a very clear improvement is provided by $R2$ and $R3$.
- For $u = 2$, $T = 10$, $\Gamma = \Gamma_3$, we have $\mathrm{P}(X_0 > u) = 0.023$, $\tilde{\nu}_1 = 0.215$, $\nu_1 = 0.211$, $\widetilde{\nu}_2/2 = 0.014$, and $\nu_3/6 = 3.10^{-4}$, so that

$$D = 0.238, \quad R1 = 0.234, \quad R3 = 0.220, \quad R2 = 0.220$$

  In this case the Davies bound is rather sharp.

As a conclusion, these numerical results show that it is worth using several terms of the Rice series. In particular, the first three terms are relatively easy to compute and provide a good evaluation of the distribution of M under a rather broad set of conditions.

5.4. PROCESSES WITH CONTINUOUS PATHS

This section is devoted to a modification of Theorem 5.1 to include processes that do not have sufficiently differentiable paths. This is done using a regularization of the paths by convolution with a deterministic approximation of unity. For simplicity, we limit ourselves to the case of Gaussian kernels. Other kernels can be employed in a similar way.

Suppose that $\mathcal{X} = \{X(t) : t \in [0, 1]\}$ is a stochastic process with continuous paths. We let ε be a positive real number and define

$$X^{\varepsilon}(t) := (\phi_{\varepsilon} * X(\cdot))(t) = \int_{-\infty}^{+\infty} \phi_{\varepsilon}(t - s)X(s)\, ds, \tag{5.30}$$

where

$$\phi_{\varepsilon}(t) := (2\pi)^{-1/2}(\varepsilon)^{-1}e^{-t^2/2\varepsilon^2} \qquad t \in \mathbb{R},$$

and in (5.30) we have extended $X(\cdot)$ by $X(0)$ [respectively, $X(1)$] for $t \le 0$ (respectively, $t \ge 1$). Denote by $M^{\varepsilon}, \widetilde{\nu}_m^{\varepsilon}, \ldots$ the analog to $M, \widetilde{\nu}_m, \ldots$ for the process $\mathcal{X}^{\varepsilon} = \{X^{\varepsilon}(t), t \in [0, 1]\}$ instead of $\mathcal{X}$.

Theorem 5.11. *With the foregoing notation, suppose that the following conditions are satisfied:*

(a) $p_{X^{\varepsilon}(1/2)}(x)$ *is bounded by a constant D_1 for ε small enough.*
(b) $\mathrm{E}(\|X\|_{\infty}) < \infty$.
(c) *The distribution of M has no atoms. Then*

(i)

$$P(M > u) = P(X(0) > u) + \lim_{\varepsilon \to 0} \sum_{m=1}^{\infty} (-1)^{m+1} \frac{\widetilde{\nu}_m^{\varepsilon}}{m!}. \tag{5.31}$$

(ii) *In formula (5.31) when one replaces the limit by a given* ε $(0 < \varepsilon < \varepsilon_0 := e^{-2})$ *and the infinite sum by the* m_0 *partial sum, the error is bounded by*

$$[32 D_1 \mathrm{E}(\|X\|_\infty)]^{1/2}\, \Psi^{*,\varepsilon}_{m_0+1} + P(|X(0) - u| < \eta) + P(u < M \le u + \eta) + P(\omega_X(\delta(\varepsilon)) \ge \eta/2) + P\left(\|X\|_\infty > \frac{\sqrt{2\pi}\,\eta}{8\varepsilon} \right) \tag{5.32}$$

for each $\eta > 0$*, where*

$$\delta(\varepsilon) := \varepsilon (2 \log(1/\varepsilon))^{1/2}$$

$$\Psi^{*,\varepsilon}_m := \sup_{k \ge m} \left([(2k-1)!]^{1/2} \varepsilon^{2k-1} \right)^{-1/2}.$$

Note. If one wishes the bound for the error in formula (5.32) to be smaller than some positive number, proceed according to the following steps:

1. Choose $\eta > 0$ so that the second and third terms are small.
2. With that value of η, choose $\varepsilon > 0$ so that the fourth and fifth terms are small.
3. Choose m_0 large enough to make the first term small.

Proof. Consider the events

$$E_1 := \{|X(0) - u| < \eta\}, \quad E_2 := \{u < M \le u + \eta\},$$

$$E_3 := \{\omega_X(\delta(\varepsilon)) \ge \eta/2\}, \quad E_4 := \left\{ \|X\|_\infty > \frac{\sqrt{2\pi}\,\eta}{4\varepsilon} \right\}$$

$$E := E_1 \cup E_2 \cup E_3 \cup E_4.$$

Observe that if $\omega \notin E$ and $\varepsilon < \varepsilon_0$, then

$$|X^{\varepsilon}(t) - X(t)| \le \int_{-\infty}^{+\infty} \phi_\varepsilon(t-s) |X(s) - X(t)|\, ds \le \omega_X(\delta(\varepsilon)) + 2\|X\|_\infty \int_{|t-s| > \delta(\varepsilon)} \phi_\varepsilon(t-s)\, ds < \eta. \tag{5.33}$$

Using this relation, one gets

$$\begin{aligned}
P(M > u, X(0) \le u) &\le P(M > u + \eta, X(0) \le u - \eta, E^c) + P(E) \\
&\le P(M^\varepsilon > u, X^\varepsilon(0) < u) + P(E) \\
&\le P(U_u^\varepsilon \ge 1, X^\varepsilon(0) < u) + P(E).
\end{aligned}$$

Also,

$$\begin{aligned}
P(U_u^\varepsilon \ge 1, X^\varepsilon(0) \le u, E^c) &\le P(U_u^\varepsilon \ge 1, X^\varepsilon(0) \le u, X(0) \le u - \eta, E^c) \\
&\le P(M > u, X(0) \le u - \eta) \le P(M > u, X(0) \le u).
\end{aligned}$$

Summing up, we have

$$\begin{aligned}
P(X(0) > u) + P(U_u^\varepsilon \ge 1, X^\varepsilon(0) \le u) - P(E) &\le P(M > u) \\
&\le P(X(0) > u) \\
&\quad + P(U_u^\varepsilon \ge 1, X^\varepsilon(0) \le u) + P(E).
\end{aligned}$$

To compute $P(U_u^\varepsilon \ge 1, X^\varepsilon(0) \le u)$ we use the same method as in the proof of Theorem 5.1. For that purpose we need to show that the process X^ε satisfies the conditions for an appropriate choice of the sequence $\{c_k; k = 1, 2, \ldots\}$. Denoting by $\overline{H}_k(s)$ the kth modified Hermite polynomial (see Section 8.1), we have

$$\begin{aligned}
|X^{\varepsilon(k)}(t)| &\le \varepsilon^{-(k+1)} \|X\|_\infty \int_{-\infty}^{+\infty} |\phi^{(k)}((t-s)/\varepsilon)| \, ds \\
&= \varepsilon^{-k} \|X\|_\infty \int_{-\infty}^{+\infty} |\phi^{(k)}(u)| \, du \\
&= \varepsilon^{-k} \|X\|_\infty k! \int_{-\infty}^{+\infty} |\overline{H}_k(s)| \phi(s) \, ds \\
&\le \varepsilon^{-k} \|X\|_\infty k! \left(\int_{-\infty}^{+\infty} (\overline{H}_k(s))^2 \phi(s) \, ds \right)^{1/2} \\
&= \varepsilon^{-k} \|X\|_\infty (k!)^{1/2},
\end{aligned}$$

so

$$\begin{aligned}
\gamma_k^\varepsilon &= P(\|X^{\varepsilon(2k-1)}\|_\infty \ge c_k) + \frac{D_1 c_k}{2^{2k-1}(2k-1)!} \\
&\le \frac{((2k-1)!)^{1/2}}{\varepsilon^{2k-1} c_k} E(\|X\|_\infty) + \frac{D_1 c_k}{2^{2k-1}(2k-1)!}.
\end{aligned}$$

Choosing

$$c_k := \left[\frac{((2k-1)!)^{3/2} E(\|X\|_\infty)}{(\varepsilon/2)^{2k-1} D_1}\right]^{1/2},$$

we obtain

$$\gamma_k^\varepsilon \leq 2^{-k}\left[\frac{8D_1 E(\|X\|_\infty)}{\varepsilon^{2k-1}((2k-1)!)^{1/2}}\right]^{1/2}.$$

Hence,

$$\gamma_m^{\varepsilon *} = \sup_{k\geq m}(2^{k+1}\gamma_k^\varepsilon) \leq [32 D_1 E(\|X\|_\infty)]^{1/2}\,\Psi_m^{*,\varepsilon}.$$

The remainder follows as in the proof of Theorem 5.1. □

Remarks and Examples

1. Conditions (a), (b), and (c) in Theorem 5.11 are usually not trivial to check and the a priori estimation of the error can be a difficult problem. Moreover, when this can actually be done, the validity of Rice formulas and the feasibility of the method remains a problem if one is willing to use Theorem 5.11 as a tool for numerical computation. For a given error, a smaller ε implies a larger m_0, and the usefulness of Theorem 5.11 for numerical applications is still doubtful. The bound in (5.32) shows that a priori we require at least $m_0 \approx (1/2)\varepsilon^{-2}$ terms in the sum as $\varepsilon \to 0$.

2. Let $\mathcal{X}$ be a Gaussian process with continuous paths and

$$m(t) := \mathrm{E}(X(t)), \qquad \sigma^2(t) := \mathrm{Var}(X(t)) > 0$$

be the (continuous) mean and variance of $X(t)$. Condition (a) in Theorem 5.11 follows easily together with bounds on D_1 and $\mathrm{P}(E_1)$. Condition (b) is well known from the classical inequalities for Gaussian processes that we considered in Chapter 2. These inequalities also imply a priori bounds for $\mathrm{P}(E_4)$. Condition (c) follows from Ylvisaker's Theorem 1.22.

A priori bounds on $\mathrm{P}(E_2)$ follow from bounds on the density of the distribution of the random variable M, a subject that we consider again for certain classes of Gaussian processes in Chapter 7. $\mathrm{P}(E_3)$ can be bounded using classical methods to study the modulus of continuity of a stochastic process, as in Chapter 1.

3. Theorem 5.11 can be applied to one-dimensional diffusions satisfying certain assumptions. The reader who is not familiar with stochastic differential equations is referred to Ikeda and Watanabe's book (1981).

Let $\{X(t) : t \geq 0\}$ be the strong solution of the stochastic differential equation

$$dX(t) = \sigma(t, X(t))dW(t) + b(t, X(t))dt, \qquad X(0) = x_0,$$

where $\{W(t) : t \geq 0\}$ stands for the standard Wiener process; $\sigma, b : \mathbb{R}^+ \times \mathbb{R} \to \mathbb{R}$ are continuous; $\partial\sigma/\partial x$ and $\partial b/\partial x$ are continuous and bounded; and $x_0 \in \mathbb{R}$. We also assume that

$$\sigma(t, x) \geq \sigma_0 > 0, \qquad t \in \mathbb{R}^+, x \in \mathbb{R}.$$

The methods employed by Azaïs (1989) or Nualart and Wschebor (1991) (stochastic calculus and Malliavin calculus, respectively) allow us to prove that $p_{X^\varepsilon(t)}$ exists and is a bounded function for $t \in [\delta, 1]$ for each $\delta > 0$, $0 < \varepsilon < \varepsilon_0(\delta)$. Condition (b) is standard and well known. Condition (c) can be proved as in the article by Nualart and Vives (1988) using stochastic calculus of variations.

Hence, Theorem 5.11 can be used to obtain formula (5.31) for $\mathrm{P}(M_\delta > u)$, $M_\delta := \max_{\delta \leq t \leq 1} X(t)$ and bounds having the form (5.32) for the error. Adding an elementary bound on the local oscillation $\mathrm{P}(\max_{0 \leq t \leq \delta} |X(t) - x_0| \leq \eta)$, one is able to get $\mathrm{P}(M > u)$ with a controlled error. An obstacle to having an actual numerical computation for $\mathrm{P}(M > u)$ is the lack of a good description of the joint densities of $X^\varepsilon(t)$ and $X^{\varepsilon\prime}(t)$ at the k-tuple $(t_1, \ldots, t_k)$ to be used in Rice formulas. It appears that this problem did not have a satisfactory solution until now.

CHAPTER 6

RICE FORMULAS FOR RANDOM FIELDS

In this chapter we begin to study random fields, that is, random functions defined on multidimensional parameter sets. More precisely, the random fields that we consider throughout are defined on a probability space $(\Omega, \mathcal{A}, \mathrm{P})$ and have the form $\mathcal{X} = \{X(t) : t \in S\}$, where S is a subset of Euclidean space $\mathbb{R}^d$ and the function $X(\cdot)$ takes values in a Euclidean space $R^{d'}$, $d' \leq d$. We require the paths $t \rightsquigarrow X(t)$ to be smooth functions, and in some situations we also ask the domain S to have a geometric structure.

Our main interest lies in the random level sets $C_u(X, S) = \{t \in S : X(t) = u\}$ for each $u \in \mathbb{R}^{d'}$. We first consider the case in which, $d = d'$, in which, generally speaking, for each $u \in \mathbb{R}^{d'}$ the set $C_u(X, S)$ will be a locally finite random set and the main question is about the number of points belonging to it and lying in a subset T of S. We denote this random number by $N_u = N_u(X, T)$, as in the one-dimensional case. The first half of this chapter is devoted to proving Rice formulas for the moments of N_u.

When $d' < d$, the random level set $C_u(X, S)$ will, of course, have a more complicated geometry, and counting the number of points is no more interesting. Generally speaking, one expects the typical level set to be a $(d - d')$-dimensional differentiable manifold. We prove Rice formulas for the moments of the geometric measure of the level set.

These and related formulas have been used by Longuett-Higgins in the 1950s and 1960s (see, e.g., Longuett-Higgins, 1957). Systematic treatment seems to have begun with the book by Adler (1981), followed by papers by Aronowich and Adler (1985), Adler et al. (1993), Adler and Samorodnisky (1997), and

Level Sets and Extrema of Random Processes and Fields, By Jean-Marc Azaïs and Mario Wschebor

Adler (2000, 2004). A proof or Rice formula for the expectation of the geometric measure of level sets of real-valued random fields was given by Wschebor (1982) and followed by various extensions and higher moments (see Wschebor, 1983, 1985). See also the paper by Cabaña (1985), where proofs for Rice formulas are given. First moments for functionals describing the geometry of $C_u(X, S)$ may be found in a recent paper by Bürgisser et al. (2006).

The case $d' > d$ is uninteresting, since in a natural situation, for fixed u, $C_u(X, S)$ will be almost surely empty.

6.1. RANDOM FIELDS FROM $\mathbb{R}^d$ TO $\mathbb{R}^d$

Our next task is to prove Rice formulas for Gaussian random fields, that is, for the moments of the number of roots N_u. Our proof is self-contained and uses only elementary arguments. It is published here for the first time and follows the proof of Azaïs and Wschebor (2006). We also consider formulas for the moments of the total weight (as in Theorem 6.4) when random weights are put in each root.

6.1.1. Area Formula

Proposition 6.1 (Area Formula). *Let f be a C^1-function defined on an open subset U of $\mathbb{R}^d$ taking values in $\mathbb{R}^d$. Assume that the set of critical values of f has zero Lebesgue measure. Let g: $\mathbb{R}^d \to \mathbb{R}$ be continuous and bounded. Then*

$$\int_{\mathbb{R}^d} g(u) N_u(f, B)\, du = \int_B |\det(f'(t))| g(f(t))\, dt \tag{6.1}$$

for any Borel subset B of U whenever the integral in the right-hand side is well defined.

Remarks

1. The hypothesis that the set of critical values of f has zero Lebesgue measure (that will be a.s. satified in our case) is unnecessary, since it is implied by the fact that f is a C^1-function. (This is a special case of Sard's lemma.)

2. The result of Proposition 6.1 is true under the weaker hypothesis that the function f verifies a Lipschitz condition (see Federer, 1969, Theorem 3.2.5).

3. Using standard extension arguments, the continuous function g can be replaced by the indicator function of a Borel set T. Formula (6.1) can then be rewritten as

$$\int_{\mathbb{R}^d} \sum_{t \in f^{-1}(u)} h(t, u)\, du = \int_{\mathbb{R}^d} |\det(f'(t))| h((t, f(t))\, dt, \tag{6.2}$$

where h is the function $(t, u) \rightsquigarrow \mathbb{I}_{t \in T} g(u)$. Again by a standard approximation argument, (6.2) holds true for every bounded Borel-measurable function h such that the right-hand side of (6.2) is well defined.

Proof of Proposition 6.1. First notice that due to standard extension arguments, it suffices to prove (6.1) for nonnegative g and for T a compact parallelotope contained in U. Second, since T is a compact parallelotope, since f is C^1, the set $f(\partial T)$ of boundary values of f has Lebesgue measure zero.

Next, we define an auxiliary function $\delta(u)$ for $u \in \mathbb{R}^d$ in the following way:

- If u is neither a critical value nor a boundary value and $n := N_u(f, T)$ is nonzero, we denote by $x^{(1)}, \ldots, x^{(n)}$ the roots of $f(x) = u$ belonging to T. Using the local inversion theorem, we know that there exists some $\delta > 0$ and n neighborhoods $U_1, \ldots, U_n$ of of $x^{(1)}, \ldots, x^{(n)}$ such that:
  (1) f is a C^1-diffeomorphism $U_i \to B(u; \delta)$, the ball centered at u with radius δ.
  (2) $U_1, \ldots, U_n$ are pairwise disjoint and included in T.
  (3) If $t \notin \bigcup_{i=1}^n U_i$, then $f(t) \notin B(u; \delta)$.

  The compactness implies that n is finite. In this case we define

$$\delta(u) := \sup\{\delta > 0 : (1), (2), (3) \text{ hold true for all } \delta' \le \delta\}.$$

- If u is a critical value or a boundary value, we set $\delta(u) := 0$.
- If $N_u(f, T) = 0$, we set

$$\delta(u) := \sup\{\delta > 0 : f(T) \cap B(u; \delta) = \emptyset\}.$$

It is clear that in this case, $\delta(u) > 0$.

The function $\delta(u)$ is Lipschitz. In fact, let u be a value of f that is not a critical value or a boundary value. If u' belongs to $B(u; \delta(u))$, then $B(u'; \delta(u) - \|u' - u\|) \subset B(u; \delta(u))$, and as a consequence, $\delta(u') \ge \delta(u) - \|u' - u\|$. Exchanging the roles of u and u', we get

$$|\delta(u') - \delta(u)| \le \|u - u'\|.$$

The Lipschitz condition is easily checked in the other two cases.

Let $\mathcal{F}$ be a real-valued monotone continuous function defined on $\mathbb{R}^+$ such that

$$\mathcal{F} \equiv \begin{cases} 0 \text{ on } [0, \frac{1}{2}], \\ 1 \text{ on } [1 + \infty). \end{cases} \tag{6.3}$$

Let $\delta(u) > 0$ and $0 < \delta < \delta(u)$. Using the change-of-variable formula, we have

$$\int_T |\det(f'(t))| \mathbb{I}_{\|f(t)-u\|<\delta}\, dt = \sum_{i=1}^{n} \int_{U_i} |\det(f'(t))|\, dt = V(\delta)n,$$

where $V(\delta)$ is the volume of the ball with radius δ in $\mathbb{R}^d$. Thus, we have an exact counter for $N_u(f, T)$ when it is nonzero, which obviously also holds true when $N_u(f, T) = 0$ for $\delta < \delta(u)$.

Let $g : \mathbb{R}^d \to \mathbb{R}$ be continuous, bounded, and nonnegative and $\delta_0 > 0$. For every $\delta' < \delta_0/2$ we have

$$\int_{\mathbb{R}^d} g(u) N_u(f, T) \mathcal{F}\left(\frac{\delta(u)}{\delta_0}\right) du = \int_{\mathbb{R}^d} g(u) \mathcal{F}\left(\frac{\delta(u)}{\delta_0}\right) du \times \frac{1}{V(\delta')} \int_T |\det(f'(t))| \mathbb{I}_{\|f(t)-u\|<\delta'}\, dt.$$

Applying Fubini's theorem, we see that the expression above is equal to

$$A_{\delta_0,\delta'} := \int_T |\det(f'(t))|\, dt \frac{1}{V(\delta')} \int_{B(f(t);\delta')} \mathcal{F}\left(\frac{\delta(u)}{\delta_0}\right) g(u)\, du.$$

$A_{\delta_0,\delta'}$ does not, in fact, depend on δ', so it is equal to its limit as $\delta' \to 0$, which because of the continuity of the function $u \rightsquigarrow \mathcal{F}(\delta(u)/\delta_0)\, g(u)$ is equal to

$$\int_T |\det(f'(t))| \mathcal{F}\left(\frac{\delta(f(t))}{\delta_0}\right) g(f(t))\, dt.$$

Let δ_0 tend to zero and use monotone convergence. For the left-hand side, we take into account that the set of critical values and the set of boundary values have measure zero. For the right-hand side, we use the definition of $\mathcal{F}$, the fact that the boundary of T has Lebesgue measure zero, and the fact that the integrand is zero if t is a critical point of f. □

6.1.2. Rice Formulas for Gaussian Random Fields

Main Results

Theorem 6.2 (Rice Formula for the Expectation). *Let $Z : U \to \mathbb{R}^d$ be a random field, U an open subset of $\mathbb{R}^d$, and $u \in \mathbb{R}^d$ a fixed point. Assume that:*

(i) *Z is Gaussian.*

(ii) *Almost surely the function $t \rightsquigarrow Z(t)$ is of class C^1.*

(iii) *For each $t \in U$, $Z(t)$ has a nondegenerate distribution [i.e.,* $\text{Var}(Z(t)) \succ 0$].

(iv) $\mathrm{P}\{\exists t \in U, Z(t) = u, \det\left(Z'(t)\right) = 0\} = 0.$

Then for every Borel set B contained in U, one has

$$\mathrm{E}\left(N_u(Z, B)\right) = \int_B \mathrm{E}\left(|\det(Z'(t))| \big| Z(t) = u\right) p_{Z(t)}(u)\, dt. \tag{6.4}$$

If B is compact, both sides of (6.4) are finite.

Theorem 6.3 (Rice Formula for the kth Moment). *Let k, $k \geq 2$ be an integer. Assume the same hypotheses as in Theorem 6.2 except that (iii) is replaced by*

(iii′) For $t_1, \ldots, t_k \in U$ distinct values of the parameter, the distribution of

$$\left(Z(t_1), \ldots, Z(t_k)\right)$$

does not degenerate in $(\mathbb{R}^d)^k$.

Then for every Borel set B contained in U, one has

$$\begin{aligned} &\mathrm{E}\left[\left(N_u(Z, B)\right)\left(N_u(Z, B) - 1\right) \cdots \left(N_u(Z, B) - k + 1\right)\right] \\ &= \int_{B^k} \mathrm{E}\Big(\prod_{j=1}^{k} |\det\left(Z'(t_j)\right)| \big| Z(t_1) = \cdots = Z(t_k) = u\Big) \\ &\quad \times p_{Z(t_1),\ldots,Z(t_k)}(u, \ldots, u) dt_1 \cdots dt_k, \end{aligned} \tag{6.5}$$

where both sides may be infinite.

Remark. With the same proof as that of Theorem 6.3 and under the same conditions, we have for distinct $u_1, \ldots, u_k$,

$$\begin{aligned} &\mathrm{E}\left[\left(N_{u_1}(Z, B)\right)\left(N_{u_2}(Z, B)\right) \cdots \left(N_{u_k}(Z, B)\right)\right] \\ &= \int_{B^k} \mathrm{E}\Big(\prod_{j=1}^{k} |\det\left(Z'(t_j)\right)| \big| Z(t_1) = u_1, \ldots, Z(t_k) = u_k\Big) \\ &\quad \times p_{Z(t_1),\ldots,Z(t_k)}(u_1, \ldots, u_k)\, dt_1 \cdots dt_k. \end{aligned}$$

Theorem 6.4 (Expected Number of Weighted Roots). *Let Z be a random field that verifies the hypotheses of Theorem 6.2. Assume that for each $t \in U$ one has another random field, $Y^t : W \to \mathbb{R}^n$, where W is some topological space, verifying the following conditions:*

(a) *$Y^t(w)$ is a measurable function of (ω, t, w) and almost surely, $(t, w) \rightsquigarrow Y^t(w)$ is continuous.*

(b) *For each* $t \in U$ *the random process* $(s, w) \rightsquigarrow \big(Z(s), Y^t(w)\big)$ *defined on* $U \times W$ *is Gaussian.*

Moreover, assume that $g : U \times \mathcal{C}(W, \mathbb{R}^n) \to \mathbb{R}$ *is a bounded function, which is continuous when one puts on* $\mathcal{C}(W, \mathbb{R}^n)$ *the topology of uniform convergence on compact sets. Then, for each compact subset* I *of* U*, one has*

$$\mathrm{E}\left(\sum_{t \in I, Z(t)=u} g(t, Y^t)\right) = \int_I \mathrm{E}\big(|\det(Z'(t)|g(t, Y^t)\big|Z(t) = u\big)p_{Z(t)}(u)\,dt. \quad (6.6)$$

Proof of Theorem 6.2. Let $\mathcal{F} : \mathbb{R}^+ \to [0, 1]$ be the function defined in (6.3), For m and n positive integers and $x \geq 0$, define

$$F_m(x) := \mathcal{F}(mx), \qquad G_n(x) := 1 - \mathcal{F}(x/n). \quad (6.7)$$

A standard extension argument says that it is enough to prove the theorem when B is a compact rectangle included in U. So assume that B satisfies this condition. Let us introduce some more notation:

- $\Delta(t) := |\det(Z'(t))|(t \in U)$.
- For n and m positive integers and $u \in \mathbb{R}^d$:

$$C_u^m(B) := \sum_{s \in B : Z(s)=u} F_m(\Delta(s)). \quad (6.8)$$

$$Q_u^{n,m}(B) := C_u^m(B)G_n(C_u^m(B)). \quad (6.9)$$

In (6.8), when the summation index set is empty, we put $C_u^m(B) = 0$. Let $g : \mathbb{R}^d \to \mathbb{R}$ be continuous with compact support. We apply the area formula (6.2) for the function

$$h(t, u) = F_m(\Delta(t))G_n(C_u^m(B))g(u)\mathbb{I}_{t \in B}$$

to get

$$\int_{\mathbb{R}^d} g(u)Q_u^{n,m}(B)\,du = \int_B \Delta(t)F_m(\Delta(t))G_n(C_{Z(t)}^m(B))g(Z(t))\,dt.$$

Taking expectations in both sides provides

$$\int_{\mathbb{R}^d} g(u)\,\mathrm{E}(Q_u^{n,m}(B))\,du = \int_{\mathbb{R}^d} g(u)\,du$$

$$\int_B \mathrm{E}\big[\Delta(t)F_m(\Delta(t))G_n \times \ (C_u^m(B))\big|Z(t) = u\big]p_{Z(t)}(u)\,dt.$$

Since this equality holds for any g continuous with bounded support, it follows that

$$\mathrm{E}(Q_u^{n,m}(B)) = \int_B \mathrm{E}\big[\Delta(t)F_m(\Delta(t)G_n(C_u^m(B))\big|Z(t)=u\big]p_{Z(t)}(u)\,dt \tag{6.10}$$

for almost every $u \in \mathbb{R}^d$.

Notice that the hypotheses imply that if J is a subset of U, $\lambda_d(J)=0$, then $\mathrm{P}\{N_u(Z,J)=0\}=1$ for each $u \in \mathbb{R}^d$. In particular, almost surely, there are no roots of $Z(t)=u$ in the boundary of the rectangle B. Let us prove that the left-hand side of (6.10) is a continuous function of u. Fix $u \in \mathbb{R}^d$. Outside the compact set

$$\{t \in B : \Delta(t) \geq 1/2m\},$$

the contribution to the sum (6.8) defining $C_v^m(B)$ is zero for any $v \in \mathbb{R}^d$. Using the local inversion theorem, the number of points $t \in B$ such that $Z(t)=u$; $\Delta(t)\geq 1/2m$, say k, is finite. Almost surely, all these points are interior to B.

If k is nonzero, $Z(t)$ is locally invertible in k neighborhoods $V_1,\ldots,V_k \subset B$ around these k points. For v in some (random) neighborhood of u, there is exactly one root of $Z(s)=v$ in each $V_1,\ldots,V_k$ and the contribution to $C_v^m(B)$ of these points can be made arbitrarily close to the contribution corresponding to $v=u$. Outside the union of $V_1,\ldots,V_k$, $Z(t)-u$ is bounded away from zero in B, so that the contribution to $C_v^m(B)$ vanishes if v is sufficiently close to u. Therefore, the function $v \rightsquigarrow Q_v^{n,m}$ is a.s. continuous at $v=u$. On the other hand, it is obvious from its definition that $Q_v^{n,m}(B)\leq n$ and an application of the Lebesgue dominated convergence theorem implies the continuity of $\mathrm{E}(Q_u^{n,m}(B))$ as a function of u.

Let us now write the Gaussian regression formulas for fixed $t \in B$:

$$\begin{aligned} Z(s) &= a^t(s)Z(t) + Z^t(s) \\ Z'(s) &= (a^t)'(s)Z(t) + (Z^t)'(s), \end{aligned} \tag{6.11}$$

where the prime denotes the derivative with respect to s and the pair $\big(Z^t(s)$ and $(Z^t)'(s)\big)$ is independent from $Z(t)$ for all $s \in U$. Then we write the conditional expectation on the right-hand side of (6.10) as the unconditional expectation,

$$\mathrm{E}\big[\Delta_u^t(t)F_m(\Delta_u^t(t))G_n(\widetilde{C}_u^m(B))\big], \tag{6.12}$$

where we use the notation

$$\Delta_u^t(s) := |\det(Z_u^t)'(s)|$$
$$Z_u^t(s) := a^t(s)u + Z^t(s)$$
$$\widetilde{C}_u^m(B) := \sum_{s \in B, Z_u^t(s)=u} F_m\big(\Delta_u^t(s)\big).$$

Now, observe that (6.10) implies that for almost every $u \in \mathbb{R}^d$, one has the inequality

$$\mathrm{E}(Q_u^{n,m}(B)) \leq \int_B \mathrm{E}\big[\Delta(t)\big|Z(t) = u\big]p_{Z(t)}(u)\,dt, \tag{6.13}$$

which is in fact true for all $u \in \mathbb{R}^d$ since both sides are continuous functions of u.

The remainder of the proof consists of proving an inequality in the opposite sense. Let us fix n, m, u, and t. Let K be the compact set

$$K := \{s \in B : \Delta_u^t(s) \geq 1/4m\}.$$

If v varies in a sufficiently small (random) neighborhood of u, the points outside K do not contribute to the sum defining $\widetilde{C}_v^m(B)$.

Let k be the a.s. finite number of roots of $Z_u^t(s) = u$ lying in the set K. Assume that k does not vanish and denote these roots by $\bar{s}_1, \ldots, \bar{s}_k$. Consider the equation

$$Z_v^t(s) - v = 0 \tag{6.14}$$

in a neighborhood of each of the pairs $s = \bar{s}_i$, $v = u$. Applying the implicit function theorem, one can find k pairwise disjoint open sets sets $V_1, \ldots, V_k$ such that if v is sufficiently close to u, equation (6.14) has exactly one root $s_i = s_i(v)$ in V_i, $1 = 1, \ldots, k$. These roots vary continuously with v and $s_i(u) = \bar{s}_i$. On the other hand, for the compact set $K \backslash (V_1 \cup \ldots \cup V_k)$ the quantity $\|Z_u^t(s) - u\|$ is bounded away from zero, so $\|Z_v^t(s) - v\|$ does not vanish if v is sufficiently close to u. As a conclusion, we have

$$\limsup_{v \to u} \widetilde{C}_v^m(B) \leq \widetilde{C}_u^m(B),$$

where the inequality arises from the fact that some of the points $s_i(v)$ may not belong to B and hence don't contribute to the sum defining $\widetilde{C}_v^m(B)$. Now since (6.10) holds for u, almost everywhere (a.e.), one can find a sequence $\{u_N, N = 1, 2, \ldots\}$ converging to u such that (6.10) holds true for $u = u_N$ and all $N = 1, 2, \ldots$. Using the continuity (already proved) of the function

$u \rightsquigarrow \mathrm{E}(Q_u^{n,m}(B))$, Fatou's lemma, and the fact that G_n is nonincreasing, we have

$$\begin{aligned}\mathrm{E}\big(Q_u^{n,m}(B)\big) &= \lim_{N\to+\infty} \mathrm{E}\big(Q_{u_N}^{n,m}(B)\big)\\ &= \lim_{N\to+\infty} \int_B \mathrm{E}\big[\Delta_{u_N}^t(t) F_m(\Delta_{u_N}^t(t) G_n\big(\widetilde{C}_{u_N}^m(B)\big)\big] p_{Z(t)}(u_N)\, dt\\ &\geq \int_B \mathrm{E}\big[\Delta_u^t(t) F_m(\Delta_u^t(t) G_n\big(\widetilde{C}_u^m(B)\big)\big] p_{Z(t)}(u)\, dt.\end{aligned}$$

Since $\widetilde{C}_u^m(B)$ is a.s. finite, we can now pass to the limit as $n \to +\infty$, $m \to +\infty$ (in that order) and using Beppo–Levi's theorem, conclude the proof. □

Proof of Theorem 6.3. For each $\delta > 0$, define the domain

$$D_{k,\delta}(B) = \{(t_1, \ldots, t_k) \in B^k, \left\|t_i - t_j\right\| \geq \delta \text{ if } i \neq j, i, j = 1, \ldots, k\}$$

and the process $\widetilde{Z}$,

$$(t_1, \ldots, t_k) \in D_{k,\delta}(B) \rightsquigarrow \widetilde{Z}(t_1, \ldots, t_k) = \big(Z(t_1), \ldots, Z(t_k)\big).$$

It is clear that $\widetilde{Z}$ satisfies the hypotheses of Theorem 6.2 for every value $(u, \ldots, u) \in (\mathbb{R}^d)^k$. So

$$\begin{aligned}&\mathrm{E}\big[N_{(u,\ldots,u)}\big(\widetilde{Z}, D_{k,\delta}(B)\big)\big]\\ &= \int_{D_{k,\delta}(B)} \mathrm{E}\Big(\prod_{j=1}^k |\det\big(Z'(t_j)\big)| \big| Z(t_1) = \cdots = Z(t_k) = u\Big)\\ &\quad \times p_{Z(t_1),\ldots,Z(t_k)}(u, \ldots, u)\, dt_1 \cdots dt_k. \qquad (6.15)\end{aligned}$$

To finish, let $\delta \downarrow 0$, and take into account that $\big(N_u(Z, B)\big)\big(N_u(Z, B) - 1\big) \cdots \big(N_u(Z, B) - k + 1\big)$ is the monotone limit of $N_{(u,\ldots,u)}\big(\widetilde{Z}, \big(D_{k,\delta}(B)\big)$, and that the diagonal $D_k(B) = \big\{(t_1, \ldots, t_k) \in B^k, t_i = t_j \text{ for some pair } i, j, i \neq j\big\}$ has zero Lebesgue measure in $(\mathbb{R}^d)^k$. □

Proof of Theorem 6.4. The proof is essentially the same. It suffices to consider instead of $C_u^m(B)$ the quantity

$$C'^m_u(I) := \sum_{s \in I: Z(s) = u} \mathcal{F}_m(\Delta(s)) g_s(s, Y^s). \qquad (6.16)$$

□

Sufficient Conditions for Hypothesis* (iv) *in Theorem 6.2. These conditions are given by the following proposition:

Proposition 6.5. *Let $Z : U \to \mathbb{R}^d$, U a compact subset of $\mathbb{R}^d$ be a random field with paths of class C^1 and $u \in \mathbb{R}^d$. Assume that*

- *$p_{Z(t)}(x) \le C$ for all $t \in U$ and x in some neighborhood of u.*
- *At least one of the two following hypotheses is satisfied:*

  (a) a.s. $t \rightsquigarrow Z(t)$ is of class C^2.
  (b) $\alpha(\delta) = \sup_{t\in U, x\in V(u)} \mathrm{P}\{|\det(Z'(t))| < \delta \big| Z(t) = x\} \to 0$

  as $\delta \to 0$, where $V(u)$ is some neighborhood of u.

Then (iv) *holds true.*

Proof. Assume with no loss of generality that $I = [0,1]^d$ and that $u = 0$. Set $G_I = \{\exists t \in I, Z(t) = 0, \det\left(Z'(t)\right) = 0\}$.

UNDER CONDITION (a) (Cucker and Wschebor, 2003). For each integer N, consider I as a union of cubes of sides $1/N$ with sides parallel to the axis. We denote these cubes as $C_1, \ldots, C_{N^d}$. In a similar way, we consider each face at the boundary of the cube C_r as a union of $(d-1)$-dimensional cubes of sides $1/N^2$. We denote these cubes as D_{rs}, $s = 1, \ldots, 2dN^{d-1}$. In each D_{rs} fix a point τ_{rs}^*: for instance, the center.

We denote $Z = (Z_1, \ldots, Z_d)^{\mathrm{T}}$ and $t = (t_1, \ldots, t_d)^{\mathrm{T}}$. For a given $\eta > 0$, choose $B > 0$ large enough so that $\mathrm{P}\{F_B\} < \eta$, where F_B is the event

$$F_B = \left\{\left[\sup\left[\left|\frac{\partial Z_i}{\partial t_j}(t)\right|, \left|\frac{\partial^2 Z_i}{\partial t_j \partial t_h}(t)\right|\right] : i, j, h = 1, \ldots, d; t \in [0,1]^d\right] > B\right\}.$$

Clearly,

$$G_I = \bigcup_{r=1}^{r=N^d} \{\exists \tau_r \in C_r, v \in \mathbb{R}^d,$$

such that

$$Z(\tau_r) = 0, \|v\| = 1, Z'(\tau_r)v = 0\} = \bigcup_{r=1}^{r=N^d} G_r.$$

Assume that $G_r \cap F_B^C$ is nonempty. Denote $\tilde{\tau}_{rs}$ as an intersection point with the boundary of C_r, of the straight line through τ_r which is parallel to v. Consider the Taylor expansion of Z_i at point τ_r, evaluated at point $\tilde{\tau}_{rs}$:

$$Z_i(\tilde{\tau}_{rs}) = Z_i(\tau_r) + \sum_{j=1}^{d} \frac{\partial Z_i}{\partial t_j}(\tau_r)(\tilde{\tau}_{rs,j} - \tau_{r,j})$$

$$+ \sum_{j,h=1}^{d} \frac{\partial^2 Z_i}{\partial t_j \partial t_h}(\tau_r + \theta(\tilde{\tau}_{rs} - \tau_r))(\tilde{\tau}_{rs,j} - \tau_{r,j})(\tilde{\tau}_{rs,h} - \tau_{r,h})$$

with $0 < \theta < 1$. Since the first two terms in this sum are equal to zero, we deduce that $|Z_i(\tilde{\tau}_{rs})| \leq K_d B N^{-2}$ for all $i = 1, \ldots, d$, where K_d is a constant depending only on the dimension.

Since the diameter of each D_{rs} is bounded by a constant depending only on the dimension, times N^{-2}, it follows that $\|Z(\tau^*_{rs})\| \leq K N^{-2}$ for some constant K depending only on d and B. So

$$\mathrm{P}\{G_I\} \leq \mathrm{P}\{F_B\} + \mathrm{P}\{\exists r \leq N^d, s \leq N^{d-1}, \text{ s.t. } \|Z(\tau^*_{rs})\| \leq K N^{-2}\}$$

$$\leq \eta + \sum_{r=1}^{N^d} \sum_{s=1}^{2dN^{d-1}} \mathrm{P}\{\|Z(\tau^*_{rs})\| \leq K N^{-2}\}$$

$$\leq \eta + N^{2d-1} C K_1 N^{-2d},$$

where K_1 is a constant depending on d and B, using the hypotheses on the boundedness of the density. The remainder is plain. This proves (iv) under condition (a).

Under Condition (b) (Azaïs and Wschebor, 2005). Choose $\varepsilon > 0$ and $\eta > 0$; then there exists a positive number M such that

$$\mathrm{P}(E_M) = \mathrm{P}\left\{ \sup_{t \in I} \|Z'(t)\| > M \right\} \leq \varepsilon.$$

Denote by $\omega_{\det}$ the modulus of continuity of $|\det(X'(\cdot))|$ and choose N large enough so that

$$\mathrm{P}(F_{N,\eta}) = \mathrm{P}\left\{ \omega_{\det}\left(\frac{\sqrt{d}}{N} \right) \geq \eta \right\} \leq \varepsilon.$$

Consider the partition of I used in part (a) into N^d small cubes. Let τ^*_r be the center of C_r. Then

$$\mathrm{P}(G_I) \leq \mathrm{P}(E_M) + \mathrm{P}(F_{N,\eta}) + \sum_r \mathrm{P}\left(G_{C_r} \cap E^c_M \cap F^c_{N,\eta} \right). \tag{6.17}$$

When the event in the rth term occurs, we have

$$\left| Z_j(\tau^*_r) \right| \leq \frac{M}{N}\sqrt{d} \qquad j = 1, \ldots, d$$

and

$$\left|\det\left(Z'(\tau_r^*)\right)\right| < \eta.$$

So if N is chosen sufficiently large that $V(0)$ contains the ball centered at 0 with radius $M\sqrt{d}/N$, one has

$$\mathrm{P}(G_I) \leq 2\epsilon + N^d \left(\frac{2M}{N}\sqrt{d}\right)^d C\alpha(\eta).$$

Since ε and η are arbitrarily small, the result follows. □

6.1.3. Maxima and Critical Points on a Smooth Manifold

Let us write a Rice formula for the first moment in two special cases that will appear various times in the remainder of the book. These correspond to the number of local maxima and the number of critical points of a real-valued random field.

Assume that $\{X(t) : t \in W\}$ is a real-valued random field defined on the open subset W of $\mathbb{R}^d$ and such that $Z(t) = X'(t)$ satisfies the hypothesis of Theorem 6.2. Let S be a Borel subset of W and $u \in \mathbb{R}$. The following quantities are well defined and measurable: $M_{u,1}(X, S)$, the number of local maxima, and $M_{u,2}(X, S)$, the number of critical points of $X(\cdot)$ belonging to S in which the function $X(\cdot)$ takes a value bigger than u.

We also introduce the following notation: for each real symmetric matrix M, we set $\delta^1(M) := |\det(M)|\mathbf{1}_{M \prec 0}$ and $\delta^2(M) := |\det(M)|$. Then we have the following formulas for the expectation ($k = 1, 2$):

$$\mathrm{E}\left(M_{u,k}^X(S)\right) = \int_S ds \int_u^{+\infty} \mathrm{E}\left(\delta^k(X''(s))\big|X(s) = x, X'(s) = 0\right) \times p_{X(s),X'(s)}(x, 0)\, dx. \tag{6.18}$$

Similar expressions are obtained when extending the statements of Theorems 6.3 and 6.4 to this case. Let W be a C^2-manifold of dimension d. We suppose that W is orientable; that is, there exists an atlas $\left((U_i, \phi_i); i \in I\right)$ such that for any pair of intersecting charts (U_i, ϕ_i) and (U_j, ϕ_j), the Jacobian of the map $\phi_i \circ \phi_j^{-1}$ is positive.

We consider a Gaussian random field with real values and C^2-paths $X = \{X(t) : t \in W\}$ defined on the manifold W. In this section our aim is to write Rice formulas for this kind of random field under various geometric settings for W. More precisely, we consider three cases: first, when W is a manifold without any additional structure; second, when W has a Riemannian metric; and third, when it is embedded in Euclidean space. We use these formulas in the next chapters, but they are of interest in themselves [see Taylor and Adler (2003) for other details or similar results].

We denote the derivative along the manifold as $DX(t)$ to distinguish it from the free derivative in $\mathbb{R}^d$, and we assume that in every chart, the pair $X(t)$ and

$DX(t)$ has a nondegenerate joint distribution and that hypothesis (iv) of Theorem 6.2 is verified.

Abstract Manifold

Proposition 6.6. *For $k = 1, 2$, the quantity that is expressed in every chart ϕ with coordinates $s_1, \ldots, s_d$ as*

$$\int_u^{+\infty} dx \mathrm{E}\left(\delta^k(Y''(s)) \middle| Y(s) = x, Y'(s) = 0\right) p_{Y(s),Y'(s)}(x, o) \wedge_{i=1}^d ds_i, \qquad (6.19)$$

where $Y(s)$ is the process X written in the chart: $Y = X \circ \phi^{-1}$, defines a d-form Ω^k on the interior $\overset{\circ}{W}$ of W and for every Borel set $S \subset \dot{W}$,

$$\int_S d\Omega^k = \mathrm{E}\left(M_{u,k}^X(S)\right).$$

Proof. Note that a d-form is a measure on $\overset{\circ}{W}$ whose image in each chart is absolutely continuous with respect to Lebesgue measure $\wedge_{i=1}^d ds_i$. To prove that (6.19) defines a d-form, it is sufficient to prove that its density with respect to $\wedge_{i=1}^d ds_i$ satisfies locally the change-of-variable formula. Let (U_1, ϕ_1) and (U_2, ϕ_2) be two intersecting charts, and set

$$U_3 := U_1 \cap U_2, \qquad Y_1 := X \circ \phi_1^{-1}, \qquad Y_2 := X \circ \phi_2^{-1}, \qquad H := \phi_2 \circ \phi_1^{-1}.$$

Denote by s_i^1 and s_i^2, $i = i, \ldots, d$ the coordinates in each chart. We have

$$\frac{\partial Y_1}{\partial s_i^1} = \sum_{i'} \frac{\partial Y_2}{\partial s_{i'}^2} \frac{\partial H_{i'}}{\partial s_i^1}$$

$$\frac{\partial^2 Y_1}{\partial s_i^1 \partial s_j^1} = \sum_{i',j'} \frac{\partial^2 Y_2}{\partial s_{i'}^2 \partial s_{j'}^2} \frac{\partial H_{i'}}{\partial s_i^1} \frac{\partial H_{j'}}{\partial s_j^1} + \sum_{i'} \frac{\partial Y_2}{\partial s_{i'}^2} \frac{\partial^2 H_{i'}}{\partial s_i^1 \partial s_j^1}.$$

Thus at every point

$$Y_1'(s^1) = \left(H'(s^1)\right)^{\mathrm{T}} Y_2'(s^2),$$

$$p_{Y_1(s^1),Y_1'(s^1)}(x, 0) = p_{Y_2(s^2),Y_2'(s^2)}(x, 0) |\det(H'(s^1)|^{-1}$$

and at a singular point

$$Y_1''(s^1) = \left(H'(s^1)\right)^{\mathrm{T}} Y_2''(s^2) H'(s^1).$$

On the other hand, by the change-of-variable formula,

$$\wedge_{i=1}^d ds_i^1 = |\det(H'(s^1)|^{-1} \wedge_{i=1}^d ds_i^2.$$

Substituting for the integrand in (6.19), one checks the desired result.

To prove the second part, it suffices again to prove it locally for an open subset S included in a unique chart. Let $\big((S, \phi)$ be a chart and let $Y(s)$ again be the process written in this chart. It suffices to prove that

$$\mathrm{E}\left(M_{u,k}^X(S)\right) = \int_{\phi(S)} d\lambda(s) \int_u^{+\infty} dx \mathrm{E}\left(\delta^k(Y''(s)) \middle| Y(s) = x, Y'(s) = 0\right)$$
$$\times\ p_{Y(s),Y'(s)}(x, 0). \tag{6.20}$$

Since $M_{u,k}^X(S)$ is equal to $M_{u,k}^Y\{\phi(S)\}$, the result is a direct consequence of Theorem 6.4. □

Riemannian Manifold. The form in (6.19) is independent of the parameterization, but the terms inside the integrand are not. It is possible to give an expression that consists of three terms independent of the parameterization in the case when W is a Riemannian manifold. When such a Riemannian metric is not given, it is always possible to use the metric g induced by the process itself (see Taylor and Adler, 2003) by setting

$$g(s)(Y, Z) = \mathrm{E}\Big(\big(Y(X)\big)\big(Z(X)\big)\Big),$$

Y and Z being two tangent vectors belonging to the tangent space $T(s)$ at $s \in W$. $Y(X)$ [respectively, $Z(X)$] denotes the action of the tangent vector Y (respectively, Z) on the function X. This metric leads to very simple expressions for centered variance-1 Gaussian processes.

The main point is that at a singular point of X, the second-order derivative D^2X does not depend on the parameterization since it defines locally the Taylor expansion of the function X. Given the Riemannian metric g_s, the second differential can be represented by an endomorphism that will be denoted $\nabla^2 X(s)$:

$$(D^2X)(s)[Y, Z] = g_s(\nabla^2 X(s)Y, Z).$$

This endomorphism is independent of the parameterization and of course its determinant. So in a chart,

$$\det\left(\nabla^2 X(s)\right) = \det(D^2X(s)) \det(g_s)^{-1}, \tag{6.21}$$

and $\nabla^2 X(s)$ is negative definite if and only if $D^2X(s)$ is. Hence,

$$\delta^k\left(\nabla^2 X(s)\right) = \delta^k\left(D^2X(s)\right)\det(g_s)^{-1} \qquad k = 1, 2.$$

We turn now to the density in (6.19). The gradient at some location s is defined as the unique vector $\nabla X(s) \in T(s)$ such that $g_s(\nabla X(s), Y) = DX(s)[Y]$. In a chart the vector of coordinates of the gradient in the basis $\partial x_i, i = 1, d$ is given by $(g_s)^{-1} DX(s)$, where $DX(s)$ is now the vector of coordinates of the derivative in the basis dx^i, $i = 1, d$. The joint density at $(x, 0)$ of $\big(X(s), \nabla X(s)\big)$ is independent of the parameterization only if expressed in an orthonormal basis of the tangent space. In that case, the vector of coordinates is given by

$$\widetilde{\nabla X}(s) = (g_s)^{1/2} \nabla X(s) = (g_s)^{-1/2} DX.$$

By the change-of-variables formula,

$$p_{X(s),\widetilde{\nabla X}(s)}(x, 0) = p_{X(s),DX(s)}(x, 0)\sqrt{\det(g_s)}.$$

Recalling that the Riemannian volume Vol satisfies

$$d\,\mathrm{Vol} = \sqrt{\det(g_s)}\ \wedge_{i=1}^{d} ds_i^2,$$

we can rewrite expression (6.19) as

$$\int_u^{+\infty} dx \mathrm{E}\big(\delta^k(\nabla^2 X(s)\big|X(s) = x, \nabla X(s) = 0\big) p_{X(s),\nabla X(s)}(x, 0) d\,\mathrm{Vol}, \qquad (6.22)$$

where we have omitted the tilde above $\nabla X(s)$ for simplicity. This is the Riemannian expression.

Embedded Manifold. In most practical applications, W is naturally embedded in a Euclidean space: say, $\mathbb{R}^m$. In this case we look for an expression for (6.22) as a function of the natural derivative on $\mathbb{R}^m$. The manifold is equipped with the metric induced by the Euclidean metric in $\mathbb{R}^m$. Considering the form (6.22), clearly the Riemannian volume is just the geometric measure σ on W.

Following Milnor (1965), we assume that the process $X(t)$ is defined on an open neighborhood of W so that the ordinary derivatives $X'(s)$ and $X''(s)$ are well defined for $s \in W$. Denoting the projector onto the tangent and normal spaces by $P_{T(s)}$ and $P_{N(s)}$, we have

$$\nabla X(s) = P_{T(s)}(X'(s)).$$

The next formula is well known and gives the expression of the second differential form at a singular point:

$$Y, Z \in T(s) \rightsquigarrow X''(s)[Y, Z] + <\mathrm{I\!I}[Y, Z], X'(s)>, \qquad (6.23)$$

where $\mathrm{I\!I}$ is the second fundamental form of W embedded in $\mathbb{R}^m$, which can be defined in our simple case by

$$Y, Z \in T(s) \rightsquigarrow P_{N(s)}(D_X Y).$$

The determinant of the bilinear form given by (6.23), expressed on an orthonormal basis, gives the value of $\det\big(\nabla^2 X(s)\big)$. As a conclusion, we get the expression of every term involved in (6.22).

Examples

1. With a given orientation we get

$$\nabla^2 X = X_T'' + \mathrm{II} X_N',$$

where X_T'' is the tangent projection of the second derivative and X_N' is the normal component of the gradient.

2. *Sphere*. When W is a sphere of radius $r > 0$ in $\mathbb{R}^{d+1}$ oriented toward the inside,

$$\nabla^2 X = X_T'' + r I_d X_N'. \tag{6.24}$$

3. *Curve*. When the manifold is a curve parameterized by the arc length,

$$\begin{aligned} \mathrm{E}\big(M_u^k(W)\big) = \int_u^{+\infty} dx \int_0^L dt \\ \cdot \mathrm{E}\left(\delta^k\left(X_T''(t) + C(t)X_N'(t)\middle|X(t) = x, X_T'(t) = 0\right)\right. \\ \times\ p_{X(t),X_T'(t)}(x, 0), \end{aligned} \tag{6.25}$$

where $C(t)$ is the curvature at location t and $X_N'(t)$ is the derivative taken is the direction of the normal to the curve at point t.

6.1.4. Extensions to Certain Non-Gaussian Random Fields

It is easy to adapt the proofs of Rice formulas above to certain classes of Gaussian-related random fields that do not need to be Gaussian. We exemplify this with the statement of Theorem 6.2, but the same holds true, *mutatis mutandis*, for the other theorems.

To be precise, the conclusion of Theorem 6.2 remains valid if we replace hypotheses (i) to (iv) by the following (we keep the same notations as in the statement of the theorem):

(i) $Z(t) = H[Y(t)]$ for $t \in W$, where:

- $\{Y(t) : t \in W\}$ is a Gaussian random field having values in $\mathbb{R}^n$ and C^1-paths such that for each $t \in W$, the distribution of $Y(t)$ does not degenerate.
- $H : \mathbb{R}^n \to \mathbb{R}^d$ is a C^1-function.

(ii) For each $t \in W$, $Z(t)$ has a density $p_{Z(t)}(x)$, which is a continuous function of the pair $(t, x) \in W \times \mathbb{R}^d$.

(iii) $\mathrm{P}\{\exists t \in W, Z(t) = u, \det\big(Z'(t)\big) = 0\} = 0$.

Notice that these hypotheses imply that one must have $n \geq d$. The only change to be introduced in the proof is that instead of performing the regression on $Z(t)$, one should do it on $Y(t)$.

As for the validity of Rice formulas for more general non-Gaussian random fields, a careful analysis of the proof of Theorem 6.2, shows that, in fact, Gaussianity plays a role only in assuring the continuity (as functions of u) of the conditional expectation and the density that appear in the integrands on the right-hand sides of (6.4) and (6.10). The continuity of the density is obvious and that of the conditional expectation is a consequence of the possibility of using regression to get rid of the conditioning, which is a specifically Gaussian property. Otherwise, the proof is independent of the nature of the law of the random field $\{Z(t) : t \in W\}$. From the standpoint of applications, one must also consider that if the random field is non-Gaussian, the actual computation of the conditional expectation can be difficult or impossible to actually perform, and interest in the formula remains limited.

Because of this, we next state as a theorem the Rice formula for the expectation of the number or roots of non-Gaussian random fields. Similar expressions hold true for higher moments and for weighted roots as well as for random fields parameterized on manifolds. The proof strictly follows that of Theorem 6.2, except for the points just mentioned.

Theorem 6.7. *Let $Z : W \to \mathbb{R}^d$ be a random field, W an open subset of $\mathbb{R}^d$, and $u \in \mathbb{R}^d$ a fixed point. Assume that:*

(i) *Almost surely, the function $t \rightsquigarrow Z(t)$ is of class $\mathcal{C}^1$.*
(ii) *For each $t \in W$, $Z(t)$ has a density $p_{Z(t)}(\cdot)$ and the function $(t,x) \rightsquigarrow p_{Z(t)}(x)$ is continuous for $t \in U$ and x is some neighborhood of u.*
(iii) *Let $\alpha : \mathcal{C}^1(U) \to \mathbb{R}$ be a real-valued functional defined on $\mathcal{C}^1(W)$, which is continuous in the sense that if $\{f_n\}_{n=1,2,\dots}$ is a sequence of functions in $\mathcal{C}^1(W)$ such that $f_n \to f$, $f_n' \to f'$ as $n \to \infty$, uniformly on the compact subsets of W, then $\alpha(f_n) \to \alpha(f)$. Our assumption is that for such a functional α there exists a version of the conditional expectation*

$$\mathrm{E}(\alpha(\mathrm{Z}) \big| Z(t) = x),$$

which is continuous as a function of the pair (t,x), for $t \in W$ and x in some neighborhood of u.
(iv) $\mathrm{P}\{\exists t \in W, Z(t) = u, \det\big(Z'(t)\big) = 0\} = 0$. *Then for every Borel set B contained in W, one has*

$$\mathrm{E}\left(N_u(Z,B)\right) = \int_B \mathrm{E}\left(|\det(Z'(t))| \big| Z(t) = u\right) p_{Z(t)}(u)\, dt. \qquad (6.26)$$

If B is compact, both sides are finite.

6.2. RANDOM FIELDS FROM $\mathbb{R}^d$ TO $\mathbb{R}^{d'}$, $d > d'$

We follow a method similar to that of Section 6.1, but a certain number of new problems arise. For f a $\mathcal{C}^1$-function defined on W and u a regular value of f, we denote by $\sigma_u(f, T)$ the $(d - d')$ geometric measure of the intersection of the set T with the level set $C_u(f, U) = \{t \in U : f(t) = u\}$. Note that since u is a regular value at each point $t \in C_u(f, U)$, the jacobian matrix $f'(t)$ is of full rank d'. Thus, we can choose a subset γ of $\{1, \dots, d\}$ of size d' such that the matrix $\{\partial f_i/\partial t_j, i = 1, \dots, d', j \in \gamma\}$ is invertible. For simplicity, and without loss of generality, we will assume that $\gamma^c = \{1, \dots, (d - d')\}$. Using the implicit function theorem we know that there exist a neighborhood V_t of $t_1, \dots, t_{d-d'}$ and a function $g : \mathbb{R}^{d-d'} \to \mathbb{R}^{d'}$ (which depend on t) such that $s_1, \dots, s_{d-d'}, g(s_1, \dots, s_{d-d'})$ is a local parameterization of the level set $C_u(f, U)$. This defines a chart and proves that the level set is a $\mathcal{C}^1$-manifold of dimension $d - d'$.

6.2.1. Theorems for Gaussian Random Fields

We start with three statements for Gaussian random fields that are analogous to those of Theorems 6.2, 6.3, and 6.4.

Theorem 6.8 (Rice Formula for the Expectation of the Geometric Measure of the Level Set). *Let $Z : W \to \mathbb{R}^{d'}$ be a random field, W an open subset of $\mathbb{R}^d$, and $u \in \mathbb{R}^{d'}$ a fixed point. Assume that:*

(i) *Z is Gaussian.*
(ii) *Almost surely, the function $t \rightsquigarrow Z(t)$ is of class $\mathcal{C}^1$.*
(iii) *For each $t \in W$, $Z(t)$ has a nondegenerate distribution [i.e., $\text{Var}(Z(t)) \succ 0$].*
(iv) $P\{\exists t \in W, Z(t) = u, Z'(t)$ *does not have full rank*$\} = 0$.

Then, for every Borel set B contained in W, one has

$$E(\sigma_u(Z, B)) = \int_B E\left(\left[\det\left(Z'(t)(Z'(t))^{T}\right)\right]^{1/2} \big| Z(t) = u\right) p_{Z(t)}(u)\, dt. \quad (6.27)$$

If B is compact, both sides in (6.27) *are finite.*

Theorem 6.9 (Rice Formula for the kth Moment). *Let k, $k \geq 2$ be an integer. Assume the same hypotheses as in Theorem 6.8 except that* (iii) *is replaced by:*

(iii′) *For distinct values $t_1, \dots, t_k \in W$ of the parameter, the distribution of $(Z(t_1), \dots, Z(t_k))$ does not degenerate in $(\mathbb{R}^{d'})^k$.*

Then for every Borel set B contained in W and levels $u_1, \dots, u_k$, one has

$$\mathrm{E}\left(\prod_{j=1}^{k} \sigma_{u_j}(Z, B)\right) = \int_{B^k} \mathrm{E}\left(\prod_{j=1}^{k} \left[\det\left(Z'(t_j)(Z'(t_j))^{\mathrm{T}}\right)\right]^{1/2} \Big| Z(t_1) = u_1, \dots, Z(t_k) = u_k\right) \cdot p_{Z(t_1),\dots,Z(t_k)}(u_1, \dots, u_k)\, dt_1 \cdots dt_k, \tag{6.28}$$

where both members may be infinite.

The same kind of result holds true for integrals over the level set, as stated in the next theorem.

Theorem 6.10 (Expected Integral on the Level Set). *Let Z be a random field that verifies the hypotheses of Theorem 6.8. Assume that for each $t \in W$ one has another random field $Y^t : V \to \mathbb{R}^n$, where V is a topological space, verifying the following conditions:*

(a) *$Y^t(v)$ is a measurable function of (ω, t, v) and a.s. $(t, v) \rightsquigarrow Y^t(v)$ is continuous.*
(b) *For each $t \in W$, the random process $(s, v) \rightsquigarrow \left(Z(s), Y^t(v)\right)$ defined on $W \times V$ is Gaussian.*

Moreover, assume that $g : W \times \mathcal{C}(V, \mathbb{R}^n) \to \mathbb{R}$ is a bounded function, which is continuous when one puts on $\mathcal{C}(V, \mathbb{R}^n)$ the topology of uniform convergence on compact sets. Then for each compact subset I of W, one has

$$\mathrm{E}\left(\int_{I \cap Z^{-1}(u)} g(t, Y^t)\sigma_u(Z, dt)\right) = \int_I \mathrm{E}\left([\det(Z'(t)(Z'(t))^{\mathrm{T}})]^{1/2} g(t, Y^t) \big| Z(t) = u\right) p_{Z(t)}(u)\, dt. \tag{6.29}$$

6.2.2. Remark on Hypothesis (iv) of Theorems 6.8, 6.9, and 6.10

Let us give sufficient conditions to assure that hypothesis (iv) holds true: that is, that with probability 1 the given level u is not a critical value of the random field. They are more restrictive than those for $d = d'$ and based on the following proposition, which is a generalization of Bulinskaya's Lemma 1.20 (see Exercise 6.4 for a proof).

Proposition 6.11. *Let $\mathcal{Y} = \{Y(t) : t \in W\}$ be a random field with values in $\mathbb{R}^{m+k}$ and W an open subset of $\mathbb{R}^d$. m and k are positive integers. Let $u \in \mathbb{R}^{m+k}$ and I a subset of W. We assume that $\mathcal{Y}$ satisfies the following conditions:*

- *The paths $t \rightsquigarrow Y(t)$ are of class $\mathcal{C}^1$.*
- *For each $t \in W$, the random vector $Y(t)$ has a density and there exists a constant C such that*

$$p_{Y(t)}(x) \leq C$$

for $t \in I$ and x in some neighborhood of u.
- *The Hausdorff dimension of I is smaller or equal than m.*

Then, almost surely, there is no point $t \in I$ such that $Y(t) = u$

This implies the following:

Proposition 6.12. *Let $\mathcal{Z} = \{Z(t) : t \in W\}$ be a random field, W an open subset of $\mathbb{R}^d$, with values in $\mathbb{R}^{d'}$. Let $u \in \mathbb{R}^{d'}$. We assume the following:*

- *The paths of $\mathcal{Z}$ are of class $\mathcal{C}^2$.*
- *For each $t \in W$, the pair $(Z(t), Z'(t))$ has a joint density $p_{Z(t),Z'(t)}(x, x')$ in $\mathbb{R}^{d'} \times \mathbb{R}^{d.d'}$, which is bounded for (t, x') varying in a compact subset of $W \times \mathbb{R}^{d.d'}$ and x in some neighborhood of u.*

Then (iv) *holds true.*

Proof. Apply Proposition 6.11 to the random field

$$Y(t, \lambda) = \left(Z(t) : (Z'(t))^{\mathrm{T}}\lambda\right)$$

defined for $(t, \lambda) \in W \times S^{d'-1}$ with values in $\mathbb{R}^{d'} \times \mathbb{R}^d$. □

6.2.3. Scheme of the Proofs of Theorems 6.8, 6.9, and 6.10

We are not going to give full proofs of these theorems, since as we have already mentioned, they follow the same lines as those for $d' = d$. We limit ourselves to pointing out the differences between both situations.

First, we need a proposition replacing the area formula (6.1) for nonrandom functions. We state it for $\mathcal{C}^1$-functions, since this will be sufficient for our purposes.

Proposition 6.13 (Co-area Formula). *Let f be a $\mathcal{C}^1$-function defined on an open subset W of $\mathbb{R}^d$ taking values in $\mathbb{R}^{d'}$. Assume that the set of critical values of f has zero Lebesgue measure. Let g: $\mathbb{R}^{d'} \to \mathbb{R}$ be continuous and bounded. Then*

$$\int_{\mathbb{R}^{d'}} g(u)\sigma_u(f, B)\, du = \int_B \left[\det\left(f'(t)(f'(t))^{\mathrm{T}}\right)\right]^{1/2} g(f(t))\, dt \qquad (6.30)$$

for any Borel subset B of W whenever the integral on the right-hand side is well defined.

Remarks on Formula (6.30)

1. Clearly, this extends the area formula since if $d = d'$, $\sigma_u(f, T)$ is the number of points of the level set on T [i.e., $N_u(f, T)$].

2. For a proof of Proposition 6.13 under more general conditions, we refer the reader to Federer's book (1969).

3. Remark 3 after Proposition 6.1 applies here in the same way. This means that if we replace the function $g(u)$ in (6.30) by any measurable function $h(t, u)$, we obtain the weighted co-area formula

$$\int_{\mathbb{R}^{d'}} du \int_{R^d} h(t,u)\sigma_u(f, dt) = \int_{\mathbb{R}^d} \left[\det\left(f'(t)(f'(t))^{\mathrm{T}}\right)\right]^{1/2} h((t, f(t))\, dt \tag{6.31}$$

whenever the right-hand side is well defined.

Let us now enumerate the changes required in the proof of Theorem 6.2 to obtain Theorem 6.8.

- Replace $\Delta(t)$ by

$$\overline{\Delta}(t) = \left[\det\left(Z'(t)(Z'(t))^{\mathrm{T}}\right)\right]^{1/2}.$$

- Whenever u is not a critical value of $Z(\cdot)$, instead of $C_u^m(B)$ and $Q_u^{n,m}(B)$ that were defined in (6.8) and (6.9), we set, respectively;

$$c_u^m(B) = \int_B F_m(\overline{\Delta}(s))\sigma_u(Z, ds) \tag{6.32}$$

and

$$q_u^{n,m}(B) = c_u^m(B)G_n(c_u^m(B)). \tag{6.33}$$

- Instead of (6.10) we have

$$\mathrm{E}(q_u^{n,m}(B)) = \int_B \mathrm{E}\left[\overline{\Delta}(t)F_m(\overline{\Delta}(t)G_n(c_u^m(B))\middle|Z(t) = u\right]p_{Z(t)}(u)\, dt, \tag{6.34}$$

which holds true for almost every $u \in \mathbb{R}^{d'}$. This follows from the weighted co-area formula (6.31).

To finish, one performs two additional steps: (1) proving that both sides in equality (6.34) are continuous functions of u, so that equality holds for all $u \in \mathbb{R}^{d'}$, and (2) passing to the limit as $n \to \infty$, $m \to \infty$, in that order. On the left-hand side of (6.34), the first step follows from the continuity provided by the implicit function theorem and the obvious inequality $q_u^{n,m}(B) \leq n$. The second step follows by monotone convergence. For the remainder of the proof, we proceed as in the proof of Theorem 6.2.

Now consider formula (6.28) for the higher moments of $\sigma_u(Z, B)$ in Theorem 6.9. Define (as in the case $d' = d$) the random process

$$\widetilde{Z}(t_1, \ldots, t_k) = \big(Z(t_1), \ldots, Z(t_k)\big)$$

with parameter set $W^k \subset (\mathbb{R}^d)^k$ and values in $(\mathbb{R}^{d'})^k$. If $u_1, \ldots, u_k$ are regular values of Z, then $(u_1, \ldots, u_k) \in (\mathbb{R}^{d'})^k$ is a regular value of $\widetilde{Z}$ and

$$\sigma_{(u_1,\ldots,u_k)}(\widetilde{Z}, B_1 \times \cdots \times B_k) = \sigma_{u_1}(B_1) \cdots \sigma_{u_k}(B_k)$$

for any choice of the Borel subsets $B_1, \ldots, B_k$ of W. Following the same reasoning as in the proof of Theorem 6.3, one only needs to prove that the measure of the diagonal set

$$D_k(I) = \left\{(t_1, \ldots, t_k) \in I^k, t_i = t_j \text{ for some pair } i, j, i \neq j\right\},$$

that is, $\sigma_{(u_1,\ldots,u_k)}(\widetilde{Z}, D_k(I))$, vanishes for any rectangle $I \subset W$. To see this, notice that $\widetilde{Z}^{-1}(u_1, \ldots, u_k)$ is a differentiable manifold with dimension $k(d - d')$ which carries the geometric measure $\sigma_{(u_1,\ldots,u_k)}(\widetilde{Z}, \cdot)$ and its intersection with $D_k(I)$ is a finite union of submanifolds having dimension smaller or equal to $(k-1)(d-d')$, so its geometric measure is zero.

One should notice that when $u_1 = \cdots = u_k$, there is a difference between the case $d = d'$ and the case $d > d'$, since in the first, the diagonal charges a positive geometric measure. In fact, in this case, all the manifolds are zero-dimensional and the argument in the preceding paragraph does not work. That is why when $d = d'$, one actually gets the integral formula for the *factorial* moments of the number of roots. The difference between ordinary and factorial moments of order k is the expectation of the measure carried by the diagonal $D_k(B)$.

Finally, the proof of Theorem 6.10 does not require any new ingredients with respect to the proof of Theorem 6.4.

EXERCISES

6.1. (**a**) Assume that Z_1 and Z_2 are $\mathbb{R}^d$-valued random fields defined on compact subsets I_1 and I_2 of $\mathbb{R}^d$ and suppose that $(Z_i, I_i)(i = 1, 2)$ satisfy the hypotheses of Theorem 6.2 and that for every $s \in I_1$ and $t \in I_2$, the distribution of $\big(Z_1(s), Z_2(t)\big)$ does not degenerate. Prove that for each pair $u_1, u_2 \in \mathbb{R}^d$:

$$\begin{aligned}
&\mathrm{E}\left(N_{u_1}^{Z_1}(I_1)N_{u_2}^{Z_2}(I_2)\right) \\
&= \int_{I_1\times I_2} \mathrm{E}\left(|\det(Z_1'(t_1))||\det(Z_2'(t_2))|\big|Z_1(t_1)=u_1, Z_2(t_2)=u_2\right) \\
&\times\ p_{Z_1(t_1),Z_2(t_2)}(u_1,u_2)\,dt_1\,dt_2. \qquad (6.35)
\end{aligned}$$

(b) Extend part (a) to higher moments.

6.2. Let $\mathcal{X} = \{X(t) : t \in \mathbb{R}^d\}$ be a real-valued centered Gaussian random field. We denote

$$r(s,t) = \mathrm{E}\big(X(s)X(t)\big) \qquad s,t \in \mathbb{R}^d$$

as its covariance. We assume that the process is stationary in the sense that $r(s,t) = \Gamma(s-t)$ for all $s,t \in \mathbb{R}^d$.

(a) Let the function Γ be continuous. Prove Bochner's theorem (see Chapter 1), that is, there exists a unique Borel measure on $\mathbb{R}^d$, say μ, such that for all $\tau \in \mathbb{R}^d$,

$$\Gamma(\tau) = \int_{\mathbb{R}^d} \exp[i\langle\tau, x\rangle]\mu(dx).$$

μ is called the spectral measure of the random field $\mathcal{X}$.

(b) Denote $\Lambda_2 = \int_{\mathbb{R}^d} \|x\|^2\mu(dx)$, which can be finite or infinite. Prove that Γ is twice differentiable at the origin if and only if Λ_2 is finite, and in this case Γ is a $\mathcal{C}^2$-function and its partial derivatives can be computed by means of the formula

$$\frac{\partial^2\Gamma}{\partial\tau_j\,\partial\tau_k}(\tau) = -\int_{\mathbb{R}^d} x_j x_k \exp[i\langle\tau, x\rangle]\mu(dx)$$

for $j,k = 1,\ldots,d$, with the notation $\tau = (\tau_1,\ldots,\tau_d)^{\mathrm{T}}, x = (x_1,\ldots, x_d)^{\mathrm{T}}$.

(c) Prove that if the field has $\mathcal{C}^1$ sample paths, $\Lambda_2 < \infty$. Let I be a Borel subset of $\mathbb{R}^d$; then

$$\mathrm{E}\big(\sigma_u(X,I)\big) = \lambda_d(I)\phi(u)\mathrm{E}\big(\|\xi\|\big),$$

where λ_d denotes a Lebesgue measure in $\mathbb{R}^d$, $\phi(u)$ is the standard normal density, and ξ is a $N[0,\Lambda_2]$ Gaussian random variable with values in $\mathbb{R}^d$.

6.3. Let $\mathcal{X} = \{X(t) : t \in \mathbb{R}^d\}$ be a random field defined as

$$X(t) = X_1^2(t) + \cdots + X_m^2(t),$$

where $\{X_k(t) : t \in \mathbb{R}^d\}_{k=1,\ldots,m}$ are m independent random fields, each being centered Gaussian stationary with covariance function Γ (see Exercise 6.2). Prove that for each Borel subset I of $\mathbb{R}^d$, one has

$$\mathrm{E}\left(\sigma_u(X, I)\right) = 2\sqrt{u}\lambda_d(I)\chi_m^2(u)\mathrm{E}\big(\|\xi\|\big),$$

where ξ is as in Exercise 6.2 and $\chi_m^2(u)$ is the χ^2 density with m degrees of freedom, that is, the density of the random variable $\|\eta\|^2$, where η is standard normal in $\mathbb{R}^m$.

6.4. Prove Propositions 6.11 and 6.12.

CHAPTER 7

REGULARITY OF THE DISTRIBUTION OF THE MAXIMUM

In this chapter, except in Theorem 7.4, we consider only Gaussian processes, and our purpose is to give an account of what is known of the regularity of the probability distribution of the supremum. The main classical result is Tsirelson's theorem (1975). We begin with a statement of this theorem as it is given in Lifshits' book (1995), to which we also refer for the proof.

Theorem 7.1 (Tsirelson). *Let $\{X(t), t \in T\}$ be a real-valued bounded Gaussian process defined on a countable parameter set T. Then the distribution F_M of the random variable $M = \sup_{t\in T} X(t)$ has the following properties:*

(1) *It is continuous on $\mathbb{R}$, except at most at one point: the left limit of its support, that is,*

$$u_0 := \inf\{u : F_M(u) > 0\}.$$

(2) *It is absolutely continuous on the half-line $(u_0, +\infty)$.*
(3) *It is differentiable on $(u_0, +\infty)$ except for at most a countable set E.*
(4) *The derivative F' is positive and continuous on $(u_0, +\infty)\backslash E$. At each point of E, the derivative F' has left and right limits and jumps downward.*
(5) *For each $u > u_0$, F' has finite variation on $[u, +\infty)$.*
(6) *F' is the density of M on $(u_0, +\infty)$.*

Further improvements are given by Weber (1985), Lifshits (1995), Diebolt and Posse (1996), and references therein. One should notice that in this statement

Level Sets and Extrema of Random Processes and Fields, By Jean-Marc Azaïs and Mario Wschebor

of Tsirelson's theorem, the parameter set is countable. This says that the same result holds true for separable bounded Gaussian processes, since in this case, the distribution of the supremum coincides a.s. with the one of the supremum on some countable nonrandom set.

Our aim in this chapter is to go beyond these regularity properties of the distribution of M, at the cost of imposing a certain number of conditions on the process. In fact, we require the parameter set to have a certain geometric structure and the paths of the process to have a certain regularity. This will allow us to exploit the analytic properties of the paths to obtain results about the distribution function F_M.

The theorems we present are much stronger in the case of one-parameter processes than in the case of random fields. In the first case we are able to extend Tsirelson-type properties considerably. For example, we prove that if a Gaussian process defined on a compact interval T of the real line has $\mathcal{C}^\infty$-paths and its law satisfies a quite general nondegeneracy condition, the distribution of its maximum is a function of class $\mathcal{C}^\infty$. For multiparameter processes (random fields) much less is known and we will only prove results on the first derivative of F_M.

For one-parameter processes the main results are taken from Azaïs and Wschebor (2001). The proofs here are simpler than the original version, due to some technical improvements that we present in Section 7.1, where we start with an implicit formula for the density of the maximum of a Gaussian random field defined on a subset of $\mathbb{R}^d$. This will be our main tool in this chapter, and we will see later that it is also useful as a tool to study the asymptotic properties of the tails of the distribution of the maximum. Its proof is extracted from Azaïs and Wschebor (2008).

7.1. IMPLICIT FORMULA FOR THE DENSITY OF THE MAXIMUM

Assumptions and Notation. $\mathcal{X} = \{X(t) : t \in S\}$ denotes a real-valued Gaussian field defined on the parameter set S. We assume that S satisfies the hypotheses:

(A1):

- S is a compact subset of $\mathbb{R}^d$
- S is the disjoint union of $S_d, S_{d-1} \dots, S_0$, where S_j is an orientable C^3 manifold of dimension j without boundary. The S_j's will be called *faces*. Let S_{d_0}, $d_0 \le d$ be the nonempty face having largest dimension. σ_j denotes the j-dimensional geometric measure on S_j.
- We will assume that each S_j has an atlas such that the second derivatives of the inverse functions of all charts (viewed as diffeomorphisms from an open set in $\mathbb{R}^j$ to S_j) are bounded by a fixed constant. For $t \in S_j$ we denote L_t the maximum curvature of S_j at the point t. It follows that L_t is bounded for $t \in S$.

Notice that the decomposition $S = S_d \cup \cdots \cup S_0$ is not unique.

Concerning the random field we make assumptions (A2) to (A5):

(A2) $\mathcal{X}$ is defined on an open set containing S and has C^2 paths.

(A3) For every $t \in S$ the distribution of $(X(t), X'(t))$ does not degenerate; for every $s, t \in S$, $s \neq t$, the distribution of $(X(s), X(t))$ does not degenerate.

(A4) Almost surely the maximum of $X(t)$ on S is attained at a single point.

For $t \in S_j$, $X'_j(t)$ and $X'_{j,N}(t)$ denote, respectively, the derivative along S_j and the normal derivative. Both quantities are viewed as vectors in $\mathbb{R}^d$, and the density of their distribution will be expressed, respectively, with respect to an orthonormal basis of the tangent space $T_{t,j}$ of S_j at the point t, or its orthogonal complement $N_{t,j}$. $X''_j(t)$ will denote the second derivative of X along S_j, at the point $t \in S_j$ and will be viewed as a matrix expressed in an orthogonal basis of $T_{t,j}$. Similar notations will be used for any function defined on S_j.

(A5) Almost surely, for every $j = 1, \ldots, d$ there is no point t in S_j such that $X'_j(t) = 0$, $\det(X''_j(t)) = 0$.

The fundamental property that we use is the representation of the density of the maximum given in Theorem 7.2.

Theorem 7.2. *Let $M = \max_{t \in S} X(t)$. Under assumptions (A1) to (A5), the distribution of M has the density*

$$p_M(x) = \sum_{t \in S_0} \mathrm{E}\big(\mathbf{I}_{A_x} sX(t) = x\big) p_{X(t)}(x)$$
$$+ \sum_{j=1}^{d} \int_{S_j} \mathrm{E}\big(|\det(X''_j(t))| \mathbf{I}_{A_x} | X(t) = x, X'_j(t) = 0\big)$$
$$\times\ p_{X(t), X'_j(t)}(x, 0) \sigma_j(dt), \tag{7.1}$$

where $A_x = \{M \le x\}$.

Remarks. This equality is stated in terms of the density, but it is obvious that one also obtains an exact (implicit) formula for the distribution of the maximum on integrating once both sides of (7.1). One can replace $|\det(X''_j(t))|$ in the conditional expectation by $(-1)^j \det(X''_j(t))$, since under the conditioning and whenever the event $\{M \le x\}$ holds true, $X''_j(t)$ is negative semidefinite.

Proof of Theorem 7.2. Let $N_j(u)$, $j = 0, \ldots, d$ be the number of global maxima of $X(\cdot)$ on S that belong to S_j and are larger than u. From the hypotheses it follows that a.s. $\sum_{j=0,\dots,d} N_j(u)$ is equal to 0 or to 1, so that

$$P\{M > u\} = \sum_{j=0,\dots,d} P\{N_j(u) = 1\} = \sum_{j=0,\dots,d} E(N_j(u)). \tag{7.2}$$

The proof will be finished as soon as we show that each term in (7.2) is the integral over $(u, +\infty)$ of the corresponding term in (7.1). This is self-evident for $j = 0$.

Let us consider the term $j = d$. We apply the weighted Rice formula (6.6) as follows:

- Z is the random field X' defined on S_d.
- For each $t \in S_d$, set $W = S$ and $Y^t : S \to \mathbb{R}^2$, defined as

$$Y^t(w) := \big(X(w) - X(t), X(t)\big).$$

  Notice that the second coordinate in the definition of Y^t does not depend on w.
- In the place of the function g, we take for each $n = 1, 2, \dots$ the function g_n, defined as follows:

$$g_n(t, f_1, f_2) = g_n(f_1, f_2) = \Big(1 - \mathcal{F}_n\Big(\sup_{w\in S} f_1(w)\Big)\Big)\big(1 - \mathcal{F}_n(u - f_2(\overline{w}))\big),$$

  where $\overline{w}$ is any point in W and for n a positive integer and $x \geq 0$, we define as in formula (6.7),

$$\mathcal{F}_n(x) := \mathcal{F}(nx) \qquad \text{with } \mathcal{F}(x) = 0 \text{ if } 0 \leq x \leq \tfrac{1}{2}, \mathcal{F}(x) = 1 \text{ if } x \geq 1, \tag{7.3}$$

  and $\mathcal{F}$ monotone nondecreasing and continuous.

It is easy to check that all the requirements in Theorem 6.4 are satisfied, so that for the value 0 instead of u in formula (6.6), we get

$$E\left(\sum_{t\in S_d, X'(t)=0} g_n(Y^t)\right) = \int_{S_d} E(|\det(X''(t)|g_n(Y^t)|X'(t) = 0)p_{X'(t)}(0)\lambda_d(dt). \tag{7.4}$$

Notice that the formula holds true for each compact subset of S_d in the place of S_d, hence for S_d itself by monotone convergence.

Now let $n \uparrow \infty$ in (7.4). Clearly, $g_n(Y^t) \downarrow \mathbb{I}_{X(s)-X(t)\leq 0, \forall s\in S} \cdot \mathbb{I}_{X(t)\geq u}$. The passage to the limit does not present any difficulty since $0 \leq g_n(Y^t) \leq 1$ and the sum in the left-hand side is bounded by the random variable $N_0^{X'}(\overline{S_d})$, which is in L^1 because of the Rice formula. We get

$$\mathrm{E}(N_d(u)) = \int_{S_d} \mathrm{E}(|\det(X''(t)|\mathbb{I}_{X(s)-X(t)\le 0, \forall s\in S}\,\mathbb{I}_{X(t)\ge u}|X'(t)=0)$$
$$\times\, p_{X'(t)}(0)\lambda_d(dt).$$

Conditioning on the value of $X(t)$, we obtain the desired formula for $j=d$.

The proof for $1 \le j \le d-1$ is essentially the same, but one must take care of the parameterization of the manifold S_j. One can first establish locally the formula on a chart of S_j, using local coordinates. It can be proved as in Proposition 6.6 (the only modification is due to the term $\mathbb{I}_{A_x}$) that the quantity written in some chart as

$$\mathrm{E}\big(\det(Y''(s))\mathbb{I}_{A_x}|Y(s)=x,\, Y'(s)=0\big)p_{Y(s),Y_j'(s)}(x,0)\,ds,$$

where the random field $Y(s)$ is $\mathcal{X}$ written in some chart of S_j [i.e., $Y(s) = X(\phi^{-1}(s))$] defines a j-form, that is, a measure on S_j that does not depend on the parameterization and which has a density with respect to the Lebesgue measure ds in every chart. It can be proved that the integral of this j-form on S_j gives the expectation of $N_j(u)$.

To get formula (7.1) it suffices to consider locally around a precise point $t \in S_j$ the chart ϕ given by the projection on the tangent space at t. In this case we obtain that at t, ds is in fact $\sigma_j(dt)$ and $Y'(s)$ is isometric to $X'_j(t)$, where $s = \phi(t)$. This completes the proof. □

7.2. ONE-PARAMETER PROCESSES

In this section we restrict the scope of our study to random processes defined on a compact interval of the line. Without loss of generality, we assume this interval to be [0, 1]. As announced, this will enable us to obtain deeper results on the regularity of the distribution of M. The statement of the main theorem is the following:

Let $\mathcal{X} = \{X(t) : t \in [0,1]\}$ be a stochastic process with real values. It is said to satisfy the hypothesis H_k, k a positive integer, if:

1. *$\mathcal{X}$ is Gaussian.*
2. *a.s. $\mathcal{X}$ has C^k-sample paths.*
3. *For every integer $n \ge 1$ and any set $t_1, \dots, t_n$ of distinct parameter values, the distribution of the random vector*

   $$X(t_1), \dots, X(t_n), X'(t_1), \dots, X'(t_n), \dots, X^{(k)}(t_1), \dots, X^{(k)}(t_n)$$

   is nondegenerate.

We denote by $m(t)$ and $r(s,t)$ the mean and covariance functions of X and use the notation

$$r_{ij} := \frac{\partial^{i+j}}{\partial s^i\,\partial t^j} r \ (i, j = 0, 1, \ldots)$$

for the derivatives, whenever they exist. It is in general a nontrivial task to verify condition 3. However, for stationary Gaussian processes a simple and sufficient condition on the spectral measure which implies condition 3 is given in Exercise 3.5.

Theorem 7.3. *Assume that $\mathcal{X}$ satisfies H_{2k}. Denote by $F(u) = \mathrm{P}(M \le u)$ the distribution function of $M = \max_{t\in[0,1]} X(t)$. Then F is of class $\mathcal{C}^k$ and its successive derivatives can be computed by repeated application of Lemma 7.7.*

Theorem 7.3 for random processes with one parameter appears to be a considerable extension of Theorem 7.1. For example, it implies that if the process is Gaussian with $\mathcal{C}^\infty$-paths and satisfies the nondegeneracy condition for every $k = 1, 2, \ldots$, the distribution of the maximum is $\mathcal{C}^\infty$. The same methods we will be using in the proof also provide bounds for the successive derivatives. The asymptotic behavior as their argument tends to $+\infty$ is considered in Chapter 8, where we study a certain number of asymptotic methods related to distribution of the maximum.

Before proceeding to the proof of Theorem 7.3, which turns out to be quite long and presents a number of technical difficulties, let us digress to state two theorems on the density of the maximum which are easier to prove and provide simple inequalities for the density of the maximum. The first, Theorem 7.4, refers to general, not necessarily Gaussian processes. The second, Theorem 7.5, concerns Gaussian processes. As applications, one gets upper and lower bounds for the density of M under conditions which otherwise have required complicated calculations and unnecessary restrictions.

Theorem 7.4. *Assume that the process $\mathcal{X} = \{X(t) : t \in [0,1]\}$ has $\mathcal{C}^2$-paths, that for each $t \in [0,1]$, the triplet $(X(t), X'(t), X''(t))$ admits a joint density, and $X'(t)$ has a bounded density $p_{X'(t)}(\cdot)$. We also assume that the function*

$$I(x,z) := \int_0^1 \mathrm{E}\big(X''^{-}(t)|X(t) = x, X'(t) = z\big)\, p_{X(t),X'(t)}(x,z)\,dt$$

is uniformly continuous in z for (x,z) in some neighborhood of $(u,0)$. Then the distribution of M admits a density $p_M(\cdot)$ satisfying a.e.

$$p_M(u) \le \mathrm{P}(X'(0) < 0|X(0) = u)\,p_{X(0)}(u) + \mathrm{P}(X'(1) > 0|X(1) = u)\,p_{X(1)}(u)$$
$$+ \int_0^1 \mathrm{E}\big(X''^{-}(t)|X(t) = x, X'(t) = 0\big)\,p_{X(t),X'(t)}(x,0)\,dt. \tag{7.5}$$

Proof. Let $u \in \mathbb{R}$ and $h > 0$. We have

$$\begin{aligned}P(M \le u) - P(M \le u - h) = P(u - h < M \le u) &\le P(u - h < X(0) \le u, X'(0) < 0)\\ &\quad + P(u - h < X(1) \le u, X'(1) > 0)\\ &\quad + P(M^+_{u-h,u} > 0),\end{aligned}$$

where $M^+_{u-h,u} = M^+_{u-h,u}(0,1)$, since if $u - h < M \le u$, either the maximum occurs in the interior of the interval $[0, 1]$ or at 0 or 1, with the derivative taking the sign indicated. Notice that

$$P(M^+_{u-h,u} > 0) \le E(M^+_{u-h,u}).$$

Using Proposition 1.20, with probability 1, $X'(\cdot)$ has no tangencies at level 0; thus, an upper bound for this expectation follows from Kac's formula:

$$M^+_{u-h,u} = \lim_{\delta \to 0} \frac{1}{2\delta} \int_0^1 1_{\{X(t)\in[u-h,u]\}}\, \mathbb{I}_{\{X'(t)\in[-\delta,\delta]\}}\, \mathbb{I}_{\{X''(t)<0\}} |X''(t)|\, dt \quad \text{a.s.,}$$

which together with Fatou's lemma implies that

$$E(M^+_{u-h,u}) \le \liminf_{\delta \longrightarrow 0} \frac{1}{2\delta} \int_{-\delta}^{\delta} dz \int_{u-h}^{u} I(x, z)\, dx = \int_{u-h}^{u} I(x, 0)\, dx.$$

Combining this bound with the preceding one, we get

$$\begin{aligned}P(M \le u) - P(M \le u - h) \le \int_{u-h}^{u} &[P(X'(0) < 0|X(0) = x) p_{X(0)}(x)\\ &+ P(X'(1) > 0|X(1) = x) p_{X(1)}(x) + I(x, 0)]\, dx,\end{aligned}$$

which gives the result. □

Despite the simplicity of the proof, in the case of Gaussian processes, this theorem provides under quite general conditions, an upper bound for the density which is difficult to improve (see, e.g., Diebolt and Posse, 1996). If we consider a Gaussian centered process with unit variance, by means of a deterministic time change, we can also assume that the process has *unit speed* [which means that $\mathrm{Var}(X'(t)) \equiv 1$]. This transforms the interval [0,1] into an interval having length, say, L. Then one can prove (see Exercise 7.2) that (7.5) reduces to

$$p_M(u) \le p^+(u) := \varphi(u)\left[1 + (2\pi)^{-1/2} \int_0^L C(t)\varphi(u/C(t)) + u\Phi(u/C(t))\, dt\right] \tag{7.6}$$

with $C(t) := \sqrt{r_{22}(t,t) - 1}$.

As $u \to +\infty$,

$$p^+(u) = \varphi(u)\left[1 + Lu(2\pi)^{-1/2} + (2\pi)^{-1/2}u^{-2}\int_0^L C^3(t)\varphi(u/C(t))\,dt\right] + O(u^{-4}\varphi\left(u/C^+\right)) \tag{7.7}$$

with $C^+ := \sup_{t\in[0,L]} C(t)$.

The following theorem is a direct consequence of Theorem 7.2. The only point that is new is the continuity of the density, which will be proved later.

Theorem 7.5. *Suppose that X is a Gaussian process with $\mathcal{C}^2$-paths and such for all $s, t, s \neq t \in [0, 1]$, $X(s), X(t), X'(t)$ and $X(t), X'(t), X''(t)$ admit a joint density. Then M has a* continuous *density p_M given for every u by*

$$p_M(u) = \mathrm{P}(M \le u|X(0) = u)p_{X(0)}(u) + \mathrm{P}(M \le u|X(1) = u)p_{X(1)}(u) + \int_0^1 \mathrm{E}(X''^{-}(t)\mathbf{I}_{M\le u}|X(t) = u, X'(t) = 0)p_{X(t),X'(t)}(u, 0)\,dt. \tag{7.8}$$

Using (7.8), one can obtain sharper upper bounds than those produced by (7.5) (see Exercise 7.1).

We turn next to proofs of our main results.

Proof. We begin with an auxiliary technical lemma.

Lemma 7.6. *(a) Let $\mathcal{Z} = \{Z(t) : t \in [0, 1]$ be a* centered *stochastic process satisfying H_k $(k \ge 2)$ and t a point in $(0, 1)$. Define the Gaussian processes $Z^0(s), Z^1(s), Z^t(s)$ by means of the orthogonal decompositions*

$$Z(s) = \begin{cases} a^0(s)Z(0) + sZ^0(s) & s \in (0,1] \quad (7.9)\\ a^1(s)Z(1) + (1-s)Z^1(s) & s \in [0,1) \quad (7.10)\\ b^t(s)Z(t) + c^t(s)\,Z'(t) + \frac{(s-t)^2}{2}Z^t(s) & s \in [0,1], s \neq t. \quad (7.11)\end{cases}$$

Then the processes Z^0, Z^1, and Z^t can be extended continuously at $s = 0$, $s = 1$, and $s = t$, respectively, so that they satisfy H_{k-1}, H_{k-1}, and H_{k-2}, respectively. Notice that, in fact, the functions a^0, a^1, b^t, and c^t are the ordinary regression coefficients.

(b) Let f be any function of class $\mathcal{C}^k$. When there is no ambiguity on the process $\mathcal{Z}$, we will define f^0, f^1, and f^t in the same manner, putting f instead of Z in (7.9), (7.10), and (7.11), but still keeping the regression coefficients corresponding to $\mathcal{Z}$. Then f^0, f^1, and f^t can be extended by continuity in the same way to functions in $\mathcal{C}^{k-1}$, $\mathcal{C}^{k-1}$, and $\mathcal{C}^{k-2}$, respectively.

(b′) As a consequence, if $\mathcal{Z}$ is a process satisfying H_k which is not centered, we can define Z^0, Z^1, and Z^t using (7.9), (7.10), and (7.11) applied separately to

the centered process $t \to \mathrm{Z}(t) - \mathrm{E}(\mathrm{Z}(t))$ *and to the mean* $\mathrm{E}(\mathrm{Z}(t))$ *and summing up the two components. In fact, we again obtain* (7.9), (7.10), *and* (7.11)

(c) *Let* m *be a positive integer; suppose that* $\mathrm{Z}(t)$ *satisfies* H_{2m+1} *and* $t_1, \dots, t_m$ *belong to* $[0, 1]$. *Denote by* $\mathrm{Z}^{t_1,\dots,t_m}(s)$ *the process obtained by repeated application of the operation of part* (a); *that is,*

$$\mathrm{Z}^{t_1,\dots,t_m}(s) = (\mathrm{Z}^{t_1,\dots,t_{m-1}})^{t_m}(s).$$

Denote by $s_1, \dots, s_p$ $(p \le m)$ *the ordered p-tuple of the elements of* $t_1, \dots, t_m$ *that belong to* $(0, 1)$ (*i.e., they are neither* 0 *nor* 1). *Then, a.s. the application*

$$(s_1, \dots, s_p, s) \rightsquigarrow \left(\mathrm{Z}^{t_1,\dots,t_m}(s), (\mathrm{Z}^{t_1,\dots,t_m})'(s)\right)$$

is continuous.

Proof. Parts (a) and (b) follow directly by computing the regression coefficients $a^0(s), a^1(s)$, $b^t(s)$, and $c^t(s)$, substituting into formulas (7.9), (7.10), and (7.11) and using the arguments above. We now prove part (c), which is a consequence of the following. Suppose that $\mathrm{Z}(t_1, \dots, t_k)$ is a Gaussian field with $\mathcal{C}^p$-sample paths ($p \ge 2$) defined on $[0, 1]^k$ with no degeneracy in the same sense that in the definition of hypothesis $H_k(3)$ for one-parameter processes. Then the Gaussian fields defined by means of

$$\begin{aligned}
\mathrm{Z}^0(t_1, \dots, t_k) &= (t_k)^{-1}(\mathrm{Z}(t_1, \dots, t_{k-1}, t_k) \\
&\quad - a^0(t_1, \dots, t_k)\mathrm{Z}(t_1, \dots, t_{k-1}, 0)) \qquad \text{for } t_k \ne 0, \\
\mathrm{Z}^1(t_1, \dots, t_k) &= (1 - t_k)^{-1}(\mathrm{Z}(t_1, \dots, t_{k-1}, t_k) \\
&\quad - a^1(t_1, \dots, t_k)\mathrm{Z}(t_1, \dots, t_{k-1}, 1)) \qquad \text{for } t_k \ne 1, \\
\widetilde{\mathrm{Z}}(t_1, \dots, t_k, t_{k+1}) &= 2(t_{k+1} - t_k)^{-2}(\mathrm{Z}(t_1, \dots, t_{k-1}, t_{k+1}) \\
&\quad - b(t_1, \dots, t_k, t_{k+1})\mathrm{Z}(t_1, \dots, t_k) \\
&\quad - c(t_1, \dots, t_k, t_{k+1})\frac{\partial \mathrm{Z}}{\partial t_k}(t_1, \dots, t_k)) \qquad \text{for } t_{k+1} \ne t_k
\end{aligned}$$

can be extended to $[0, 1]^k$ (respectively, $[0, 1]^k$ and $[0, 1]^{k+1}$) into fields with paths in $\mathcal{C}^{p-1}$ (respectively, $\mathcal{C}^{p-1}$ and $\mathcal{C}^{p-2}$). In the formulas above:

- $a^0(t_1, \dots, t_k)$ is the regression coefficient of $\mathrm{Z}(t_1, \dots, t_k)$ on $\mathrm{Z}(t_1, \dots, t_{k-1}, 0)$.
- $a^1(t_1, \dots, t_k)$ is the regression coefficient of $\mathrm{Z}(t_1, \dots, t_k)$ on $\mathrm{Z}(t_1, \dots, t_{k-1}, 1)$.
- $b(t_1, \dots, t_k, t_{k+1})$ and $c(t_1, \dots, t_k, t_{k+1})$ are the regression coefficients of $\mathrm{Z}(t_1, \dots, t_{k-1}, t_{k+1})$ on the pair $(\mathrm{Z}(t_1, \dots, t_k), \partial \mathrm{Z}(t_1, \dots, t_k)/\partial t_k)$.

Let us prove the statement about $\widetilde{Z}(t_1, \ldots, t_k, t_{k+1})$. The other two are simpler. Suppose for the moment that $\widetilde{Z}(t_1, \ldots, t_k, t_{k+1})$ is centered. Denote by V the subspace of $L^2(\Omega, \Im, P)$ generated by the pair $(Z(t_1, \ldots, t_k), \partial Z(t_1, \ldots, t_k)/\partial t_k)$. Denote by $\Pi_{V\perp}$ the version of the orthogonal projection of $L^2(\Omega, \Im, P)$ on the orthogonal complement of V, which is

$$\Pi_{V\perp}(Y) := Y - \left[bZ(t_1, \ldots, t_k) + c\frac{\partial Z}{\partial t_k}(t_1, \ldots, t_k)\right],$$

where b and c are the regression coefficients of Y on the pair

$$Z(t_1, \ldots, t_k), \ \frac{\partial Z}{\partial t_k}(t_1, \ldots, t_k).$$

If $\{Y(\theta) : \theta \in \Theta\}$ is a random field with continuous paths and such that $\theta \to Y(\theta)$ is continuous in $L^2(\Omega, \Im, P)$, then a.s.

$$(\theta, t_1, \ldots, t_k) \to \Pi_{V\perp}(Y_\theta)$$

is continuous. From the definition,

$$\widetilde{Z}(t_1, \ldots, t_k, t_{k+1}) = 2\,(t_{k+1} - t_k)^{-2}\, \Pi_{V\perp}\left(Z(t_1, \ldots, t_{k-1}, t_{k+1})\right).$$

On the other hand, by Taylor's formula,

$$Z(t_1, \ldots, t_{k-1}, t_{k+1}) = Z(t_1, \ldots, t_k) + (t_{k+1} - t_k)\frac{\partial Z}{\partial t_k}(t_1, \ldots, t_k) + R_2(t_1, \ldots, t_k, t_{k+1})$$

with

$$R_2(t_1, \ldots, t_k, t_{k+1}) = \int_{t_k}^{t_{k+1}} \frac{\partial^2 Z}{\partial t_k^2}(t_1, \ldots, t_{k-1}, \tau)\,(t_{k+1} - \tau)\,d\tau,$$

so that

$$\widetilde{Z}(t_1, \ldots, t_k, t_{k+1}) = \Pi_{V\perp}\left[2\,(t_{k+1} - t_k)^{-2} R_2(t_1, \ldots, t_k, t_{k+1})\right]. \tag{7.12}$$

It is clear that the paths of the random field $\widetilde{Z}$ are $p-1$ times continuously differentiable for $t_{k+1} \neq t_k$. Relation (7.12) shows that they have a continuous extension to $[0, 1]^{k+1}$ with $\widetilde{Z}(t_1, \ldots, t_k, t_k) = \Pi_{V\perp}\left(\partial^2 Z/\partial t_k^{2(t_1, \ldots, t_k)}\right)$. In fact,

$$\Pi_{V\perp}\left(2\,(s_{k+1} - s_k)^{-2} R_2(s_1, \ldots, s_k, s_{k+1})\right)$$
$$= 2\,(s_{k+1} - s_k)^{-2} \int_{s_k}^{s_{k+1}} \Pi_{V\perp}\left(\frac{\partial^2 Z}{\partial t_k^2}(s_1, \ldots, s_{k-1}, \tau)\right)(s_{k+1} - \tau)\,d\tau.$$

The integrand is a continuous function of the parameters therein, so that, a.s.,

$$\widetilde{Z}(s_1, \ldots, s_k, s_{k+1}) \to \Pi_{V^\perp}\left(\frac{\partial^2 Z}{\partial t_k^2}(t_1, \ldots, t_k)\right)$$

when $(s_1, \ldots, s_k, s_{k+1}) \to (t_1, \ldots, t_k, t_k)$. This proves (c) in case $\widetilde{Z}$ is centered.

It remains to consider the case when $\widetilde{Z}$ is purely deterministic, say $\widetilde{Z}(t_1, \ldots, t_k) = f(t_1, \ldots, t_k)$. Making t_{k+1} tend to t_k in the regression equation, we see that

$$b(t_1, \ldots, t_k, t_k) = 1 \qquad c(t_1, \ldots, t_k, t_k) = 0 \tag{7.13}$$

$$\frac{\partial b}{\partial t_k}(t_1, \ldots, t_k, t_k) = 0 \qquad \frac{\partial c}{\partial t_k}(t_1, \ldots, t_k, t_k) = 1, \tag{7.14}$$

so $b(t_1, \ldots, t_k, t_{k+1}) = 1 + O\big((t_k - t_{k+1})^2\big)$ and $c(t_1, \ldots, t_k, t_{k+1}) = t_{k+1} - t_k + O\big((t_k - t_{k+1})^2\big)$ and

$$\begin{aligned}\widetilde{f}(t_1, \ldots, t_k, t_{k+1}) = 2(t_{k+1} - t_k)^{-2}\big[f(t_1, \ldots, t_{k-1}, t_{k+1}) - Z(t_1, \ldots, t_k)\\ - (t_{k+1} - t_k)\frac{\partial Z}{\partial t_k}(t_1, \ldots, t_k)\big] + O(1).\end{aligned}$$

The result is now a simple consequence of the Taylor formula. In the same way, when $p \geq 3$, we obtain the continuity of the partial derivatives of $\widetilde{Z}$ up to the order $p - 2$. □

Proof of Theorem 7.5. We apply Theorem 7.2. It is easy to check that whenever the parameter set is a compact interval in the line, it is not necessary that the process be defined in a neighborhood of S, since this assumption is in fact used to define the derivative at the boundary of S. In the case of an interval, we simply use one-sided derivatives at the extremes. The conditions required for (7.8) to hold true are fulfilled.

Set $\beta(t) \equiv 1$. Then:

- For $t \neq 0, 1$, under the condition $X(t) = u$, $X'(t) = 0$, the event $\{M \leq u\}$ can be written $\{\forall s \in [0, 1], X^t(s) \leq \beta^t(s)u\}$ with the notation of Lemma 7.6. This event will be denoted $A_u(X^t, \beta^t)$.
- For $t \neq 0, 1$, under the condition $X(t) = u$, $X'(t) = 0$, $X''(t)$ is equal to $X^t(t)$.
- Under the condition $X(0) = u$, the event $\{M \leq u\}$ is equal to $A_u(X^0, \beta^0)$.
- Under the condition $X(1) = u$, the event $\{M \leq u\}$ is equal to $A_u(X^1, \beta^1)$.

We prove, for example, that $\mathrm{P}\{A_u(X^t, \beta^t)\}$ is a continuous function of u. Let $h > 0$. We have the inequalities:

$$\begin{aligned} &|\mathrm{P}\{A_u(X^t, \beta^t)\} - \mathrm{P}\{A_{u-h}(X^t, \beta^t)\}| \qquad (7.15)\\ &\leq \mathrm{P}(\{A_u(X^t, \beta^t)\}\backslash\{A_{u-h}(X^t, \beta^t)\}) + \mathrm{P}(\{A_{u-h}(X^t, \beta^t)\}\backslash\{A_u(X^t, \beta^t)\})\\ &\leq \mathrm{P}\left\{\sup_{s\in[0,1]} (X^t(s) - \beta^t(s)u) \in [-h\|\beta^t\|_\infty, 0]\right\}\\ &\quad + \mathrm{P}\left\{\sup_{s\in[0,1]} (X^t(s) - \beta^t(s)u) \in [0, h\|\beta^t\|_\infty]\right\}. \end{aligned}$$

Now, apply Ylvisaker's Theorem 1.21 to prove that the expression above tends to zero as $h \to 0$, which proves the continuity of $\mathrm{P}\{A_u(X^t, \beta^t)\}$. Similar arguments can be applied to prove that each of the three terms on the right-hand side of (7.8) is a continuous function of u. □

Our next lemma is the basic technical tool to prove the fundamental Theorem 7.3.

Lemma 7.7. *Suppose that $\mathcal{Z} = \{Z(t) : t \in [0, 1]\}$ is a stochastic process that verifies H_2. Define*

$$F_v(u) = \mathrm{E}\left\{\xi_v.\mathbf{I}_{A_u}\right\},$$

where

$$A_u = A_u(Z, \beta) = \{Z(t) \leqq \beta(t)u \text{ for all } t \in [0, 1]\},$$

$\beta(\cdot)$ is a real-valued C^2-function defined on $[0, 1]$, and $\xi_v = G(Z(t_1)-\beta(t_1)v, \ldots, Z(t_m) - \beta(t_m)v)$ for some positive integer m, $t_1, \ldots, t_m \in [0, 1]$, $v \in \mathbb{R}$ and some C^∞-function $G : \mathbb{R}^m \to \mathbb{R}$ having at most polynomial growth at ∞, that is, $| G(x) |\leq C(1+ \| x \|^p)$ for some positive constants C and p and all $x \in \mathbb{R}^m$. Then, for each $v \in \mathbb{R}$, F_v is of class C^1 and its derivative is a continuous function of the pair (u, v), which can be written in the form

$$\begin{aligned} F_v'(u) &= \beta(0)\mathrm{E}\{\xi_{v,u}^0\ \mathbf{I}_{A_u(Z^0,\beta^0)}\}p_{Z(0)}(\beta(0)u) \qquad (7.16)\\ &+ \beta(1)\mathrm{E}\{\xi_{v,u}^1\ \mathbf{I}_{A_u(Z^1,\beta^1)}\}p_{Z(1)}(\beta(1)u)\\ &- \int_0^1 \beta(t)\mathrm{E}\{\xi_{v,u}^t(Z_t^t - \beta^t(t)u)\ \mathbf{I}_{A_u(Z^t,\beta^t)}\}\\ &\times p_{Z(t),Z'(t)}(\beta(t)u, \beta'(t)u)\, dt. \end{aligned}$$

Proof.

$$A_u \backslash A_{u-h} = A_u \cap \left[u - h < \sup_{t:\beta(t)>0} \frac{Z(t)}{\beta(t)} \leq u \right].$$

The set $B^+ := \{t \in [0, 1] : \beta(t) > 0\}$ is open in $[0, 1]$, so it is a countable union of disjoint open intervals $B^+ = \bigcup_n (a_n, b_n)$. (The reader may notice that intervals having the form $[0, b_n)$ or $(a_n, 1]$ may be present.) By a monotone convergence argument,

$$\mathrm{E}\left(\xi_v \mathbb{I}_{A_u \backslash A_{u-h}}\right) = \lim_{N \to +\infty} \lim_{\varepsilon \to 0} \mathrm{E}\left[\xi_v \mathbb{I}_{\{u-h < \sup_{t \in B_\varepsilon^N} Y(t) \leq u\}}\right],$$

where

$$Y(t) := \frac{Z(t)}{\beta(t)}, \qquad B_\varepsilon^N := \cup_{n=1}^N (a_n + \varepsilon; b_n - \varepsilon)$$

and ε is small enough that B_ε^N is well defined. Since $\beta(t)$ is bounded away from zero on B_ε^N, the conditions of Theorem 7.2 are fulfilled by the process $Y(t)$. Putting the weights ξ_v at every global maximum, we get

$$\begin{aligned} \mathrm{E}(\xi_v \mathbb{I}_{A_u \backslash A_{u-h}}) = &\lim_{N \to +\infty} \lim_{\varepsilon \to 0} \sum_{n=1}^N \int_{a_n - \varepsilon}^{b_n + \varepsilon} dt \int_{u-h}^{u} dx \\ &\times \mathrm{E}[\xi_v |Y''(t)| \mathbb{I}_{A_x} | Y(t) = x, Y'(t) = 0] p_{Y(t), Y'(t)}(x, 0) \\ &+ \sum_{n=1}^N \int_{u-h}^{u} \left[\mathrm{E}[\xi_v \mathbb{I}_{A_x} | Y(a_n + \varepsilon) = x] p_{Y(a_n+\varepsilon)}(x)\right. \\ &\left.+ \mathrm{E}[\xi_v \mathbb{I}_{A_x} | Y(b_n - \varepsilon) = x] p_{Y(b_n-\varepsilon)}(x)\right] dx. \end{aligned} \tag{7.17}$$

Changing variables in each integral, we get

$$\begin{aligned} \mathrm{E}[\xi_v \mathbb{I}_{A_u \backslash A_{u-h}}] = &\lim_{N \to +\infty} \lim_{\varepsilon \to 0} \sum_{n=1}^N \int_{a_n - \varepsilon}^{b_n + \varepsilon} dt \int_{u-h}^{u} dx \\ &\cdot \beta(t) \mathrm{E}[\xi_v |Z''(t) - \beta''(t)x| \mathbb{I}_{A_x} | Z(t) \\ &= \beta(t)x, Z'(t) = \beta'(t)x] p_{Z(t), Z'(t)}(\beta(t)x, \beta'(t)x) \\ &+ \sum_{n=1}^N \int_{u-h}^{u} [\beta(a_n + \varepsilon) \mathrm{E}[\xi_v \mathbb{I}_{A_x} | Z(a_n + \varepsilon) \\ &= \beta(a_n + \varepsilon)x] p_{Z(a_n+\varepsilon)}(\beta(a_n + \varepsilon)x) \\ &+ \beta(b_n - \varepsilon) \mathrm{E}[\xi_v \mathbb{I}_{A_x} | Z(b_n - \varepsilon) \\ &= \beta(b_n - \varepsilon)x] p_{Z(b_n-\varepsilon)}(\beta(b_n - \varepsilon)x)] \, dx. \end{aligned} \tag{7.18}$$

We get, as in the proof of Theorem 7.8,

$$\mathrm{E}[\xi_v \mathbb{I}_{A_u \setminus A_{u-h}}] = \lim_{N\to+\infty} \lim_{\varepsilon\to 0} \sum_{n=1}^{N} \int_{a_n-\varepsilon}^{b_n+\varepsilon} dt \int_{u-h}^{u} dx\, \beta(t) \tag{7.19}$$
$$\times\ \mathrm{E}\left[\xi_{v,u}^t \left|Z^t(t) - \beta^t(t)x\right| \mathbb{I}_{A_x(Z^t,\beta^t)}\right] p_{Z(t),Z'(t)}(\beta(t)x, \beta'(t)x)$$
$$+ \sum_{n=1}^{N} \int_{u-h}^{u} [\beta(a_n+\varepsilon)\mathrm{E}[\xi_v \mathbb{I}_{A_x} | Z(a_n+\varepsilon) = \beta(a_n+\varepsilon)x]$$
$$\times p_{Z(a_n+\varepsilon)}(\beta(a_n+\varepsilon)x)$$
$$+\beta(b_n-\varepsilon)\mathrm{E}\left[\xi_v \mathbb{I}_{A_x} | Z(b_n-\varepsilon) = \beta(b_n-\varepsilon)x\right]$$
$$\times\ p_{Z(b_n-\varepsilon)}(\beta(b_n-\varepsilon)x)]\, dx.$$

In the present form we can see that the conditional expectations and the densities appearing on the right-hand side of this formula are bounded, so excepting the case where $a_n = 0$ or $b_n = 1$, the contribution of the points $a_n + \varepsilon$ and $b_n - \varepsilon$ tends to zero as $\varepsilon \to 0$ since $\beta(a_n) = \beta(b_n) = 0$. Letting $\varepsilon \to 0$ and N tend to $+\infty$ in that order, we obtain

$$\mathrm{E}(\xi_v \mathbb{I}_{A_u \setminus A_{u-h}}) = \int_{B+} \beta(t)\, dt \int_{u-h}^{u} dx\ \mathrm{E}[\xi_{v,x}^t |Z^t(t) - \beta^t(t)x| \tag{7.20}$$
$$\times\ \mathbb{I}_{A_x(Z^t,\beta^t)}] p_{Z(t),Z'(t)}(\beta(t)x, \beta'(t)x)$$
$$+ \int_{u-h}^{u} dx\ (\beta(0))^+\ \mathrm{E}[\xi_{v,x}^0\ \mathbb{I}_{A_x(Z^0,\beta^0)}] p_{Z(0)}(\beta(0)x)$$
$$+ \int_{u-h}^{u} dx\ (\beta(1))^+ \mathrm{E}[\xi_{v,x}^1\ \mathbb{I}_{A_x(Z^1,\beta^1)}] p_{Z(1)}(\beta(1)x).$$

Lemma 7.7 shows that the integrand is a continuous function of x, so that also taking into account the sign of $Z^t(t) - \beta^t(t)u$ inside the expectation, we obtain

$$\lim_{h\downarrow 0} \frac{1}{h}\mathrm{E}(\xi_v \mathbb{I}_{A_u \setminus A_{u-h}}) = -\int_{B+} \beta(t)\mathrm{E}[\xi_{v,u}^t (Z^t(t) - \beta^t(t)u)\ \mathbb{I}_{A_u(Z^t,\beta^t)}] \tag{7.21}$$
$$\times p_{Z(t),Z'(t)}(\beta(t)u, \beta'(t)x)\, dt$$
$$+\ (\beta(0))^+\ \mathrm{E}(\xi_{v,u}^0\ \mathbb{I}_{A_u(Z^0,\beta^0)}) p_{Z(0)}(\beta(0)u)$$
$$+\ (\beta(1))^+\ \mathrm{E}(\xi_{v,u}^1\ \mathbb{I}_{A_u(Z^1,\beta^1)}) p_{Z(1)}(\beta(1)u).$$

Similar computations (which we do not perform here) show an analogous result for

$$\lim_{h\downarrow 0} \frac{1}{h}\mathrm{E}\left\{\xi_v\ \mathbb{I}_{A_{u-h} \setminus A_u}\right\}.$$

This completes the proof of the lemma. □

Proof of Theorem 7.3. We proceed by induction on k. We give some details for the first two derivatives, including some implicit formulas that will illustrate the procedure for general k. We introduce the following additional notation. Put $Y(t) := X(t) - \beta(t)u$ and define, on the interval $[0, 1]$, the processes $X^0, X^1, X^t, Y^0, Y^1, Y^t$ and the functions $\beta^0, \beta^1, \beta^t$, as in Lemma 7.6. Notice that the regression coefficients corresponding to the processes X and Y are the same, so that anyone of them may be used to define the functions β^0, β^1, and β^t. One can easily check that

$$Y^0(s) = X^0(s) - \beta^0(s)u$$
$$Y^1(s) = X^1(s) - \beta^1(s)u$$
$$Y^t(s) = X^t(s) - \beta^t(s)u.$$

For $t_1, \ldots, t_m \in [0, 1]$, $m \geq 2$, we define by induction the stochastic processes $X^{t_1,\ldots,t_m} = (X^{t_1,\ldots,t_{m-1}})^{t_m}$, $Y^{t_1,\ldots,t_m} = (Y^{t_1,\ldots,t_{m-1}})^{t_m}$, and the function $\beta^{t_1,\ldots,t_m} = (\beta^{t_1,\ldots,t_{m-1}})^{t_m}$, using Lemma 7.6 for the computations at each stage.

With the aim of somewhat reducing the size of the formulas, we express the successive derivatives in terms of the processes $Y^{t_1,\ldots,t_m}$ instead of $X^{t_1,\ldots,t_m}$. The reader must keep in mind that for each m-tuple $t_1, \ldots, t_m$, the results depend on u through the expectation of the stochastic process $Y^{t_1,\ldots,t_m}$. Also, for a stochastic process Z, we use the notation

$$A(Z) = A_0(Z, \beta) = \{Z(t) \leq 0 : \text{ for all } t \in [0, 1]\}.$$

First Derivative. Suppose that X satisfies H_2. We use formula (7.16) in Lemma 7.7 for $\xi \equiv 1$, $Z = X$, and $\beta(\cdot) \equiv 1$, obtaining for the first derivative:

$$F'(u) = \mathrm{E}[\,\mathbf{1}_{A(Y^0)}]p_{Y(0)}(0) + \mathrm{E}\left[\,\mathbf{1}_{A(Y^1)}\right] p_{Y(1)}(0) - \int_0^1 \mathrm{E}\left[Y^{t_1}(t_1)\,\mathbf{1}_{A(Y^{t_1})}\right] p_{Y(t_1),Y'(t_1)}(0,0)\,dt_1. \tag{7.22}$$

This expression is exactly the expression in (7.8) with the notational changes just mentioned and after taking note of the fact that the process is Gaussian, via regression on the condition in each term. Notice that according to the definition of the Y-processes,

$$\mathrm{E}[\mathbf{1}_{A(Y^0)}] = \mathrm{E}[\mathbf{1}_{A_u(X^0,\beta^0)}]$$
$$\mathrm{E}[\mathbf{1}_{A(Y^1)}] = \mathrm{E}[\,\mathbf{1}_{A_u(X^1,\beta^1)}]$$
$$\mathrm{E}[Y^{t_1}(t_1)\,\mathbf{1}_{A(Y^{t_1})}] = \mathrm{E}[Y^{t_1}(t_1)\,\mathbf{1}_{A_u(X^{t_1},\beta^{t_1})}].$$

Second Derivative. Suppose that X satisfies H_4. Then, X^0, X^1, X^{t_1} satisfy H_3, H_3, H_2, respectively. Therefore, Lemma 7.7 applied to these processes can be used to show the existence of $F''(u)$ and to compute a similar formula, except for the necessity of justifying differentiation under the integral sign in the third term. We get the expression

$$\begin{aligned}
F''(u) = & -\mathrm{E}[\,\mathbb{I}_{A(Y^0)}]p^{(1)}_{Y(0)}(0) - \mathrm{E}[\,\mathbb{I}_{A(Y^1)}]p^{(1)}_{Y(1)}(0) \qquad (7.23)\\
& + \int_0^1 \mathrm{E}[Y^{t_1}(t_1)\,\mathbb{I}_{A(Y^{t_1})}]p^{(1,0)}_{Y(t_1),Y'(t_1)}(0,0)\,dt_1\\
& + p_{Y(0)}(0)\{\beta^0(0)\mathrm{E}[\,\mathbb{I}_{A(Y^{0,0})}]p_{Y^0(0)}(0) + \beta^0(1)\mathrm{E}[\,\mathbb{I}_{A(Y^{0,1})}]p_{Y^0(1)}(0)\}\\
& - \int_0^1 \beta^0(t_2)\mathrm{E}[Y^{0,t_2}(t_2)\,\mathbb{I}_{A(Y^{0,t_2})}]p_{Y^0(t_2),(Y^0)'(t_2)}(0,0)\,dt_2\\
& + p_{Y(1)}(0)\{\beta^1(0)\mathrm{E}[\,\mathbb{I}_{A(Y^{1,0})}]p_{Y^1(0)}(0) + \beta^1(1)\mathrm{E}[\,\mathbb{I}_{A(Y^{1,1})}]p_{Y^1(1)}(0)\}\\
& - \int_0^1 \beta^1(t_2)\mathrm{E}[Y^{1,t_2}(t_2)\,\mathbb{I}_{A(Y^{1,t_2})}]p_{Y^1(t_2),(Y^1)'(t_2)}(0,0)\,dt_2\\
& - \int_0^1 p_{Y(t_1),Y'(t_1)}(0,0)\left\{\begin{array}{l}
-\beta^{t_1}(t_1)\mathrm{E}[\,\mathbb{I}_{A(Y^{t_1})}]\\
\;+\beta^{t_1}(0)\,\mathrm{E}[Y^{t_1,0}(t_1)\mathbb{I}_{A(Y^{t_1,0})}]p_{Y^{t_1}(0)}(0)\\
\;+\beta^{t_1}(1)\mathrm{E}[Y^{t_1,1}(t_1)\,\mathbb{I}_{A(Y^{t_1,1})}]p_{Y^{t_1}(1)}(0)\\
\;-\displaystyle\int_0^1 \beta^{t_1}(t_2)\mathrm{E}[Y^{t_1,t_2}(t_1)Y^{t_1,t_2}(t_2)\\
\;\times\,\mathbb{I}_{A(Y^{t_1,t_2})}]p_{Y^{t_1}(t_2),(Y^{t_1})'(t_2)}(0,0)\,dt_2
\end{array}\right\}dt_1.
\end{aligned}$$

In this formula, $p^{(1)}_{Y(t_0)}$, $p^{(1)}_{Y(t_1)}$, and $p^{(1,0)}_{Y(t_1),Y'(t_1)}(0,0)$ stand, respectively, for the derivative of $p_{Y(t_0)}(\cdot)$, the derivative of $p_{Y(t_1)}(\cdot)$, and the derivative with respect to the first variable of $p_{Y(t_1),Y'(t_1)}(\cdot,\cdot)$. To validate the formula above, notice the following points:

- The first two lines are obtained by differentiating with respect to u, the densities $p_{Y(0)}(0) = p_{X(0)}(-u)$, $p_{Y(1)}(0) = p_{X(1)}(-u)$, and $p_{Y(t_1),Y'(t_1)}(0,0) = p_{X(t_1),X'(t_1)}(-u,0)$.
- Lines 3 and 4 come from the application of Lemma 7.7 to differentiate $\mathrm{E}[1_{A(Y^0)}]$. The lemma is applied with $Z = X^0$, $\beta = \beta^0$, and $\xi = 1$.
- Similarly, lines 5 and 6 contain the derivative of $\mathrm{E}[1_{A(Y^1)}]$.
- The remainder corresponds to differentiating the function

$$\mathrm{E}[Y^{t_1}(t_1)\,\mathbb{I}_{A(Y^{t_1})}] = \mathrm{E}[(X^{t_1}(t_1) - \beta^{t_1}(t_1)u)\,\mathbb{I}_{A_u(X^{t_1},\beta^{t_1})}]$$

in the integrand of the third term in (7.22). The first term in line 7 comes from the simple derivative

$$\frac{\partial}{\partial v}\mathrm{E}[(X^{t_1}(t_1) - \beta^{t_1}(t_1)v)\,\mathbb{I}_{A_u(X^{t_1},\beta^{t_1})}] = -\beta^{t_1}(t_1)\mathrm{E}(\,\mathbb{I}_{A(Y^{t_1})}).$$

The other terms are obtained by applying Lemma 7.7 to compute

$$\frac{\partial}{\partial u}\mathrm{E}[(X^{t_1}(t_1) - \beta^{t_1}(t_1)v)\,\mathbb{I}_{A_u(X^{t_1},\beta^{t_1})}],$$

setting $Z = X^{t_1}$, $\beta = \beta^{t_1}$, and $\xi = X^{t_1}(t_1) - \beta^{t_1}(t_1)v$.

- Finally, differentiation under the integral sign is valid since because of Lemma 7.6, the derivative of the integrand is a continuous function of (t_1, t_2, u), due to regularity and nondegeneracy of the Gaussian distributions involved, and the use of Ylvisaker's theorem.

General Case. With the notation above, given the m-tuple $t_1, \dots, t_m$ of elements of $[0, 1]$, we call the processes $Y, Y^{t_1}, Y^{t_1,t_2}, \dots, Y^{t_1,\dots,t_{m-1}}$ the *ancestors* of $Y^{t_1,\dots,t_m}$. In the same way, we define the *ancestors* of the function $\beta^{t_1,\dots,t_m}$.

Assume the following induction hypothesis: If X satisfies H_{2k}, F is k times continuously differentiable and $F^{(k)}$ is the sum of a finite number of terms belonging to the class D_k, which consists of all expressions of the form

$$\int_0^1 \cdots \int_0^1 ds_1 \cdots ds_p Q(s_1, \dots, s_p)\mathrm{E}[\xi\,\mathbb{I}_{A(Y^{t_1,\dots,t_m})}]K_1(s_1, \dots, s_p)K_2(s_1, \dots, s_p), \tag{7.24}$$

where:

- $1 \le m \le k$.
- $t_1, \dots, t_m \in [0, 1]$, $m \ge 1$.
- $s_1, \dots, s_p$, $0 \le p \le m$, are the elements in $\{t_1, \dots, t_m\}$ that belong to $(0, 1)$ (i.e., which are neither "0" nor "1"). When $p = 0$, no integral sign is present.
- $Q(s_1, \dots, s_p)$ is a polynomial in the variables $s_1, \dots, s_p$.
- ξ is a product of values of $Y^{t_1,\dots,t_m}$ at some locations belonging to $\{s_1, \dots, s_p\}$.
- $K_1(s_1, \dots, s_p)$ is a product of values of some ancestors of $\beta^{t_1,\dots,t_m}$ at some locations belonging to the set $\{s_1, \dots, s_p\} \cup \{0, 1\}$.
- $K_2(s_1, \dots, s_p)$ is a sum of products of densities and derivatives of densities of the random variables $Z(\tau)$ at the point 0, or the pairs $(Z(\tau), Z'(\tau))$ at the point $(0, 0)$, where $\tau \in \{s_1, \dots, s_p\} \cup \{0, 1\}$ and the process Z is an ancestor of $Y^{t_1,\dots,t_m}$.

Notice that K_1 does not depend on u, but K_2 is a function of u. It is clear that the induction hypothesis is verified for $k = 1$. Assume that it is true up to the integer k and that X satisfies H_{2k+2}. Then $F^{(k)}$ can be written as a sum of terms of the form (7.24). Consider a term of this form and observe that the variable u may appear in three locations:

1. In ξ, where differentiation is simple given its product form, the fact that $\partial(Y^{t_1,\ldots,t_q}(s))/\partial u = -\beta^{t_1,\ldots,t_q}(s)$, $q \le m$, $s \in \{s_1, \ldots, s_p\}$, and the boundedness of moments, allowing us to differentiate under the integral and expectation signs.
2. In $K_2(s_1, \ldots, s_p)$, which is clearly C^∞ as a function of u. Its derivative with respect to u takes the form of a product of functions of the types $K_1(s_1, \ldots, s_p)$ and $K_2(s_1, \ldots, s_p)$ defined above.
3. In $1_{A(Y^{t_1,\ldots,t_m})}$. Lemma 7.7 shows that differentiation produces three terms, depending on the processes $Y^{t_1,\ldots,t_m,t_{m+1}}$, with t_{m+1} belonging to $(0, 1) \cup \{0, 1\}$. Each term obtained in this way belongs to D_{k+1}.

The proof is achieved by taking into account that as in the computation of the second derivative, Lemma 7.6 implies that the derivatives of the integrands are continuous functions of u that are bounded as functions of $(s_1, \ldots, s_p, t_{m+1}, u)$ if u varies in a bounded set. □

The statement and proof of Theorem 7.3 cannot, of course, be used to obtain explicit expressions for the derivatives of the distribution function F. However, the implicit formula for $F^{(k)}(u)$ as a sum of elements of D_k can be transformed into explicit upper bounds if one replaces the indicator functions $\mathbf{I}_{A(Y^{t_1,\ldots,t_m})}$ everywhere by 1 and the functions $\beta^{t_1,\ldots,t_m}(\cdot)$ by their absolute value.

On the other hand, Theorem 7.3 permits us to have the exact asymptotic behavior of $F^{(k)}(u)$ as $u \to +\infty$ in case $\mathrm{Var}(X_t)$ is constant. Even though the number of terms in the formula increases rapidly with k, there is exactly one term that is dominant. It turns out that as $u \to +\infty$, $F^{(k)}(u)$ is equivalent to the kth derivative of the equivalent of $F(u)$. We come back to this point in Chapter 8.

7.3. CONTINUITY OF THE DENSITY OF THE MAXIMUM OF RANDOM FIELDS

Let $S \subset \mathbb{R}^d$, $d > 1$ and $\mathcal{X} = \{X(t) : t \in U\}$, a real-valued random field defined on some open neighborhood U of S. We assume that S and $\mathcal{X}$ satisfy assumptions (A1) to (A5) of Section 7.1. We know from Theorem 7.2 that the probability distribution of the maximum $M = \max_{t \in S} X(t)$ has a density p_M that verifies equality (7.1).

As we did in Theorem 7.5, corresponding to the one-dimensional case, we will now prove that p_M is continuous. The problem is more difficult here, since the equivalent of Lemma 7.6 is more difficult. The result is the following:

Theorem 7.8. *In addition to the foregoing assumptions, for every $s \in S$, $t \in S_j$, $j = 0, 1, \ldots, d_0$, $s \neq t$, the joint distributions of the triplets $(X(t), X'(t), X''(t))$, $(X(s), X(t), X'_j(t))$ do not degenerate. Then the density p_M given by (7.1) is continuous.*

Proof. For $t \in S_j$ and $s \in S$, we define the normalization

$$n(t,s) := \|(s-t)_{j,N}\| + \frac{1}{2}\|s-t\|^2,$$

where $(s-t)_{j,N}$ is the normal component of $s-t$ (i.e., its orthogonal projection onto the subspace $N_{t,j}$; see Section 7.1 for the notation). For $s \neq t, t \in S_j$, define $X^t(s)$ by means of the Gaussian regression

$$X(s) = a^t(s)X(t) + < b^t(s), X'(t) > + n(t,s)X^t(s). \tag{7.25}$$

The reader should notice that $X^t(s)$ is not the same as before, since the present normalization is different. Let us consider the expression on the right-hand side of (7.1). A dominated convergence argument shows that it is enough to prove the continuity of the integrands appearing in that formula. Such an integrand has the form

$$\mathrm{E}(|\det(X_j''(t))|\, \mathbf{I}_{A_x} | X(t) = x, X_j'(t) = 0)\, p_{X(t),X_j'(t)}(x, 0).$$

The density $p_{X(t),X_j'(t)}(x,0)$ is clearly a continuous function of x. The regression of $X_j''(t)$ on the condition does not present any problem, and the only remaining point is to check that

$$\mathrm{P}(A_x | X(t) = x, X_j'(t) = 0)$$

is a continuous function of x. We write this as an unconditional probability, using the regression formula (7.25), so that it becomes

$$\mathrm{P}\{Y^t(s) \leq \gamma^t(s)x \text{ for all } s \in S, s \neq t\},$$

where

$$Y^t(s) = X^t(s) + \frac{\langle b^t(s), X'_{j,N}(t)\rangle}{n(t,s)}$$

and

$$\gamma^t(s) = \frac{1 - a^t(s)}{n(t,s)}.$$

On the other hand, from the regression formula it follows easily that the $\mathcal{C}^2$-functions a^t and b^t verify

$$a^t(t) = 1, \quad (a^t)'(t) = 0, \quad b^t(t) = 0, \quad (b^t)'(t) = \mathrm{Id},$$

where Id is the identity in $\mathbb{R}^d$.

The reader can check that there exist positive constants K, c, and $\bar{\gamma}$ such that for all $s \in S, s \neq t$, one has

$$\mathrm{E}(Y^t(s)) \leq K$$

$$\operatorname{Var}(Y^t(s)) \geq c \tag{7.26}$$
$$|\gamma(s,t)| \leq \bar{\gamma}.$$

Let us denote $C_x := \{Y^t(s) \leq \gamma^t(s)x \text{for all} s \in S, s \neq t\}$ and let $h > 0$. Clearly,

$$|\mathrm{P}\{C_x\} - \mathrm{P}\{C_{x-h}\}| \leq \mathrm{P}\{C_x \setminus C_{x-h}\} + \mathrm{P}\{C_{x-h} \setminus C_x\},$$

and it suffices to prove that both terms on the right-hand side are small if h is small. Let us do it for the first one; the second is similar. We have

$$\mathrm{P}\{C_x \backslash C_{x-h}\} \leq \mathrm{P}\{-h\bar{\gamma} < \sup_{s \in S, s \neq t} [Y^t(s) - \gamma(s,t)x] \leq 0\}. \tag{7.27}$$

To finish, observe that the random field $\{Y^t(s) - \gamma(s,t)x : s, t \in S, s \neq t\}$ verifies the hypothesis of Theorem 1.22. So the distribution of its supremum has no atom in $\mathbb{R}$. This shows that the right-hand side of (7.27) tends to zero as $h \to 0$, and we are done.

Remark. The last proof exhibits the main technical difference between one-parameter processes and multidimensional-parameter processes in what concerns the study of the regularity of the distribution of the maximum. The random field $\{X^t(s) : s \in S, s \neq t\}$ is constructed after Gaussian regression and renormalization. In the multidimensional-parameter case, it does not have a limit as $s \to t$; the paths present a "helix behavior" as one approaches point t. In the one-dimensional parameter case, this new process can be extended continuously to $s = t$, and the extension also preserves a part of the regularity of the original process. So we are able to iterate the procedure, renormalize, and continue in this way. This is the basis of the proof of Theorem 7.3 that we are unable to reproduce for general Gaussian random fields.

EXERCISES

7.1. (**a**) Check the following inequalities:

$$\begin{aligned}
\mathrm{P}(M \leq u | X_0 = u) &= \mathrm{P}(M \leq u, X'(0) < 0 | X(0) = u) \\
&\geq \mathrm{P}\big(X'(0) < 0 | X(0) = u\big) \\
&\quad - \mathrm{E}(U_u[0,1]\, \mathbb{1}_{\{X'(0)<0\}} | X(0) = u). \\
\mathrm{P}(M \leq u | X(1) = u) &= \mathrm{P}(M \leq u, X'(1) > 0 | X(1) = u) \\
&\geq \mathrm{P}\big(X'(1) > 0 | X(1) = u\big) \\
&\quad - \mathrm{E}(D_u[0,1]\, \mathbb{1}_{\{X'(1)>0\}} | X(1) = u).
\end{aligned}$$

If $x'' < 0$:

$$\begin{aligned} &\mathrm{P}(M \le u | X(t) = u, X'(t) = 0, X''(t) = x'') \\ &\quad \ge 1 - \mathrm{E}([D_u([0,t]) + U_u([t,1])] | X(t) = u, X'(t) = 0, X''(t) = x''). \end{aligned}$$

(b) Using the inequalities in part (a), prove the following lower bound for the density of the maximum:

$$\begin{aligned} p_M(u) \ge\ &\mathrm{P}(M \le u | X(0) = u) p_{X(0)}(u) \\ &+ \mathrm{P}(M \le u | X(1) = u) p_{X(1)}(u) \\ &+ \int_0^1 \mathrm{E}(X''^-(t)\, \mathbf{1}_{M \le u} | X(t) = x, X'(t) = 0) p_{X(t),X'(t)}(x, 0)\, dt \\ &- \int_0^1 ds \int_{-\infty}^0 dx' \int_0^{+\infty} x'_s p_{X(s),X'(s),X(0),X'(0)}(u, x'_s, u, x')\, dx'_s \\ &- \int_0^1 dt \int_{-\infty}^0 |x''| dx'' \\ &\quad \left[\int_0^t ds \int_{-\infty}^0 |x'| p_{X(s),X'(s),X(t),X'(t),X''(t)}(u, x', u, 0, x'')\, dx' \right. \\ &\quad \left. + \int_t^1 ds \int_0^{+\infty} x' p_{X(s),X'(s),X(t),X'(t),X''(t)}(u, x', u, 0, x'')\, dx' \right]. \end{aligned} \tag{7.28}$$

7.2. **(a)** Prove inequality (7.6).
(b) Prove (7.7), which gives an asymptotic bound for $p_M(u)$ as $u \to +\infty$.

7.3. Let $\{X(s,t) : s, t \in \mathbb{R}\}$ be a real-valued two-parameter Gaussian centered stationary isotropic random field with covariance Γ, $\Gamma(0) = 1$. Assume that its spectral measure μ is absolutely continuous with respect to Lebesgue measure in $\mathbb{R}^2$ with density

$$\mu(dx, dy) = f(\rho)\, ds\, dt, \qquad \rho = (x^2 + y^2)^{1/2},$$

so that

$$2\pi \int_0^{+\infty} \rho f(\rho)\, d\rho = 1.$$

Assume further that $J_k = \int_0^{+\infty} \rho^k f(\rho)\, d\rho < \infty$, for $1 \le k \le 5$. Denote by $X, X_s, X_t, X_{ss}, X_{st}, X_{tt}$ the values of X and the first and second partial derivatives at the point (s,t) and $X' = (X_s, X_t)^{\mathrm{T}}$ and X'' the matrix of second-order partial derivatives. Let $S = \{(s,t) : s^2 + t^2 \leq 1\}$ be the closed disk of radius 1 centered at the origin and $M = \max_{(s,t)\in S} X(s,t)$.

(a) Prove that:

- X' is independent of X and X'' and has variance $\pi J_3 I_d$.
- X''_{st} is independent of X, X', X''_{ss}, and X''_{tt} and has variance $(\pi/4)J_5$.
- Conditionally on $X = u$, the random variables X''_{ss} and X''_{tt} have
  (1) Expectation: $-\pi J_3$.
  (2) Variance: $(3\pi/4)J_5 - (\pi J_3)^2$.
  (3) Covariance: $(\pi/4)J_5 - (\pi J_3)^2$.

(b) Prove that

$$p_M(u) \leq (I_1 + I_2)\varphi(u),$$

where I_1 and I_2 are computed by the formulas

$$I_1 = \frac{1}{4\sqrt{2\pi}J_3}\int_0^\infty [(\alpha^2 + a^2 - c^2x^2)\Phi(a - cx) + [2a\alpha - \alpha^2(a - cx)]\varphi(a - cx)]x\varphi(x)\,dx,$$

with

$$a = 2\pi J_3 u, \qquad c = \sqrt{\frac{\pi J_5}{4}}$$

$$I_2 = \sqrt{\frac{2}{J_3}}\left[\left(\frac{3\pi}{4}J_5 - (\pi J_3)^2\right)^{(1/2)}\varphi(bu) + \pi J_3 u\Phi(bu)\right]$$

$$b = \frac{\pi J_3}{((3\pi/4)J_5 - (\pi J_3)^2)^{1/2}}.$$

CHAPTER 8

THE TAIL OF THE DISTRIBUTION OF THE MAXIMUM

Let $\mathcal{X} = \{X(t) : t \in S\}$ be a real-valued random field defined on some parameter set S and $M := \sup_{t \in S} X(t)$ its supremum. In this chapter we present recent results which allow us, in certain cases, to give more precise approximations of the tails of the distribution of the random variable M. We will be especially interested in the approximation of $P(M > u)$ for large u, but we also give results that can be used for all u.

For Gaussian processes, a number of fundamental results have been presented in Chapter 2 that we have called the basic inequalities. These are essential for the development of most of the mathematical theory. However, in a wide number of applications, the general situation is that these inequalities are not good enough, one reason being that they depend on certain constants that one is unable to estimate or for which estimations differ substantially from the true values. Some refinements of the same subject, either for certain classes of processes or touching special topics concerning the computation of the distribution of the maximum, are considered in Chapters 3, 4, 5, and 9.

Since the 1990s several methods have been introduced with the aim of obtaining more precise results than those arising from the classical theory, at least under certain restrictions on the process $\mathcal{X}$. These results are interesting both from the standpoint of the mathematical theory and of their use in significant applications. The restrictions on $\mathcal{X}$ include the requirement that the domain S have some finite-dimensional geometrical structure and the paths of the random field, a certain regularity.

Level Sets and Extrema of Random Processes and Fields, By Jean-Marc Azaïs and Mario Wschebor

Examples of these contributions are the double-sum method of Piterbarg (1996a), the Euler–Poincaré characteristic (EPC) approximation (Taylor et al., 2005; Adler and Taylor, (2007); the tube method (Sun, 1993), and the methods contained in the chapters mentioned above. We refer to these books and papers for an account of these results.

This chapter is divided into two parts. In the first part we consider two special topics which concern only the tails of the distribution of one-parameter Gaussian processes: In Section 8.1 we look at the asymptotic behavior of the successive derivatives of the distribution of the maximum and related questions, using the methods of Chapter 7; in Section 8.2 we again use similar tools to study the tails of the distribution of the maximum of certain unbounded Gaussian processes, that is, processes for which the probability q that the supremum is finite is strictly smaller than 1, and we are willing to understand the speed at which $\mathrm{P}(M \leq u)$ approaches the limiting value q as $u \to +\infty$. This section opens what seems to be a quite unexplored subject.

The second part, which begins in Section 8.3, is the main body of the chapter. It is based on Theorem 7.2, allowing us to express the density p_M of F_M by means of a general formula. Even though this is an exact formula, it is only implicit as an expression for the density, since the relevant random variable M appears on the right-hand side. However, it can be usefully employed for our purposes.

First, one can use Theorem 7.2 to obtain bounds for $p_M(u)$ and thus for $\mathrm{P}\{M > u\}$ for *every* u by means of replacing some indicator function in (7.1) by the condition that the normal derivative is "extended outward" (see below for the precise meaning). This will be called the *direct method*. Of course, this may be interesting whenever the expression one obtains can be handled, which is the actual situation when the random field has a law that is stationary and isotropic. For this family of random fields, our method relies on the application of some known results on the spectrum of random matrices, which we will use without proof and which one can find in Mehta's book (2004).

Second, one can use Theorem 7.2 to study the asymptotics of $\mathrm{P}\{M > u\}$ as $u \to +\infty$. More precisely, whenever possible one wants to write

$$\mathrm{P}\{M > u\} = A(u) \exp\left(-\frac{1}{2}\frac{u^2}{\sigma^2}\right) + B(u), \tag{8.1}$$

where $A(u)$ is a known function having polynomially bounded growth as $u \to +\infty$, $\sigma^2 = \sup_{t \in S} \mathrm{Var}(X(t))$, and $B(u)$ is an error bounded by a centered Gaussian density with variance σ_1^2, $\sigma_1^2 < \sigma^2$. We will call the first (respectively, the second) term on the right-hand side of (8.1) the *first* (respectively, *second*)*-order approximation of* $\mathrm{P}\{M > u\}$.

First-order approximation has been considered by Taylor et al. (2005) and by Adler and Taylor (2007) by means of the expectation of the EPC of the excursion set $E_u := \{t \in S : X(t) > u\}$. This works for large values of u. The same authors have considered the second-order approximation; that is, how fast does the difference between $\mathrm{P}\{M > u\}$ and the expected EPC tend to zero when $u \to +\infty$?

We address the same question for both the direct method and the EPC approximation method. Our results on the second-order approximation only speak about the size of the variance of the Gaussian bound. More precise results are only known in the special case where S is a compact interval of the real line and where the Gaussian process $\mathcal{X}$ is stationary and satisfies a certain number of additional requirements. We have stated this special result without proof in Chapter 4. It is due to Piterbarg (1981) when the domain interval is small enough and to Azaïs et al. (2002) in its general form. See the remarks after Proposition 4.1.

The first-order approximation is computed for the direct method in Theorem 8.8 in the case of stationary isotropic random fields defined on a polyhedron, from which a new upper bound for $\mathrm{P}\{M > u\}$ for all real u follows. As for second-order approximation, Theorem 8.10 is the first result here in this direction. It gives a rough bound for the error $B(u)$ as $u \to +\infty$ in case the maximum variance is attained at some strict subset of the face in S having the largest dimension. In Theorem 8.12 we consider random fields with constant variance. This is close to Theorem 4.3 of Taylor et al. (2005). In Theorem 8.15, S is convex, the random field is stationary and isotropic, and we are able to compute the exact asymptotic rate for the second-order approximation as $u \to +\infty$ corresponding to the direct method. In all cases, the second-order approximation for the direct method provides an upper bound for the one arising from the EPC method.

From a technical point of view, the proofs of the results about the supremum of random fields contained in Sections 8.3 to 8.5 require a minimum of elementary differential geometry, which does not go beyond the definitions of embedded differentiable manifold, differentiation of functions defined on it (as in Chapter 6), and curvature. The reader can consult any introductory book on the subject.

8.1. ONE-DIMENSIONAL PARAMETER: ASYMPTOTIC BEHAVIOR OF THE DERIVATIVES OF F_M

Theorem 8.1. *Let $\mathcal{X}$ be a stochastic process with parameter set $[0,1]$ verifying the hypotheses H_{2k} of Section 7.2. We also assume that $E(X(t)) = 0$ and $\mathrm{Var}(X(t)) = 1$. Then, as $u \to +\infty$,*

$$F^{(k)}(u) \approx (-1)^{k-1} \frac{u^k}{2\pi} e^{-u^2/2} \int_0^1 \sqrt{r_{11}(t,t)}\, dt. \qquad (8.2)$$

Proof. We use the notation and results of Chapter 7. To prove the result for $k = 1$, notice first that under the hypothesis of the theorem, one has $r(t,t) = 1$, $r_{01}(t,t) = 0$, and $r_{02}(t,t) = -r_{11}(t,t)$. An elementary computation of the regression (7.11) replacing Z by X shows that:

$$b^t(s) = r(s,t), \qquad c^t(s) = \frac{r_{01}(s,t)}{r_{11}(t,t)}$$

and

$$\beta^t(s) = 2\frac{1 - r(s,t)}{(t-s)^2}$$

since we start with $\beta(t) = 1$ for every t.

This shows that for every $t \in [0,1]$, one has $\inf_{s\in[0,1]}(\beta^t(s)) > 0$ because of the nondegeneracy condition and $\beta^t(t) = -r_{02}(t,t) = r_{11}(t,t) > 0$. The expression for F' becomes

$$F'(u) = \varphi(u)L(u), \tag{8.3}$$

where

$$\begin{aligned} L(u) &= L_1(u) + L_2(u) + L_3(u) \\ L_1(u) &= \mathrm{P}(A_u(X^0, \beta^0)) \\ L_2(u) &= \mathrm{P}(A_u(X^1, \beta^1)) \\ L_3(u) &= -\int_0^1 \mathrm{E}\left\{(X_t^t - \beta^t(t)u)\,\mathbb{I}_{A_u(X^t,\beta^t)}\right\} \frac{dt}{(2\pi r_{11}(t,t))^{1/2}}. \end{aligned}$$

Since for each $t \in [0,1]$ the process X^t is bounded, it follows that

$$\text{a.s.} \quad \mathbb{I}_{A_u(X^t,\beta^t)} \to 1 \text{ as } u \to +\infty.$$

A dominated convergence argument shows now that $L_3(u)$ is equivalent to

$$-\frac{u}{(2\pi)^{1/2}} \int_0^1 \frac{r_{02}(t,t)}{(r_{11}(t,t))^{1/2}}\,dt = \frac{u}{(2\pi)^{1/2}} \int_0^1 \sqrt{r_{11}(t,t)}\,dt.$$

Since $L_1(u)$ and $L_2(u)$ are bounded by 1, (8.2) follows for $k = 1$.

For $k \geq 2$, write

$$F^{(k)}(u) = \varphi^{(k-1)}(u)L(u) + \sum_{h=2}^{h=k} \binom{k-1}{h-1} \varphi^{(k-h)}(u)L^{(h-1)}(u). \tag{8.4}$$

As $u \to +\infty$, for each $j = 0, 1, \dots, k-1$, $\varphi^{(j)}(u) \simeq (-1)^j u^j \varphi(u)$, so that the first term in (8.4) is equivalent to the expression in (8.2). Hence, to prove the theorem, it suffices to show that the successive derivatives of the function L are bounded. In fact, we prove the stronger inequality

$$|L^{(j)}(u)| \leq l_j \varphi\left(\frac{u}{a_j}\right) \qquad j = 1, \dots, k-1 \tag{8.5}$$

for some positive constants l_j, a_j, $j = 1, \dots, k-1$.

We first consider the function L_1. One has

$$\beta^0(s) = \frac{1 - r(s,0)}{s} \qquad \text{for } 0 < s \le 1,\ \beta^0(0) = 0,$$

$$(\beta^0)'(s) = \frac{-1 + r(s,0) - s.r_{10}(s,0)}{s^2} \qquad \text{for } 0 < s \le 1,\ (\beta^0)'(0) = \frac{1}{2} r_{11}(0,0).$$

The derivative $L_1'(u)$ becomes

$$L_1'(u) = \beta^0(1)\mathrm{E}\left\{\mathbb{I}_{A_u(X^{0,1},\beta^{0,1})}\right\} p^0_{X(1)}(\beta^0(1)u)$$
$$- \int_0^1 \beta^0(t)\mathrm{E}\left\{(X(t)^{0,t} - \beta^{0,t}(t)u)\mathbb{I}_{A_u(X^{0,t},\beta^{0,t})}\right\}$$
$$\times\, p_{X^0(t),(X^0)'(t)}(\beta^0(t)u, (\beta^0)'(t)u)\, dt.$$

Notice that $\beta^0(1)$ is nonzero, so that the first term is bounded by a constant times a nondegenerate Gaussian density. Even though $\beta^0(0) = 0$, the second term is also bounded by a constant times a nondegenerate Gaussian density because the joint distribution of the pair $(X^0(t), (X^0)'(t))$ is nondegenerate, and the pair $(\beta^0(t), (\beta^0)'(t)) \ne (0,0)$ for every $t \in [0,1]$.

Applying a similar argument to the succesive derivatives, we obtain (8.5) with L_1 instead of L. The same follows with no changes for

$$L_2(u) = \mathrm{P}(A_u(X^1, \beta^1)).$$

For the third term

$$L_3(u) = -\int_0^1 \mathrm{E}\left\{(X^t(t) - \beta^t(t)u)\mathbb{I}_{A_u(X^t,\beta^t)}\right\} \frac{dt}{(2\pi r_{11}(t,t))^{1/2}}$$

we proceed similarly, taking into account that $\beta^t(s) \ne 0$ for every $s \in [0,1]$. So (8.5) follows and we are done. □

A Refinement. It is possible to refine the result, obtaining the exact first-order approximation and a bound for the second-order approximation, not only for the tail of the distribution, but also for its derivatives. That is, the following refinement allows us to write the successive derivatives of F_M in a form analogous to (8.1) in which the second term on the right-hand side is bounded by a Gaussian density having a variance that is smaller than the supremum of the variance of the process. Let $n = 0, 1, 2, \ldots$. We will repeatedly use the Hermite polynomials, defined as

$$H_n(x) := e^{x^2}\left(-\frac{\partial}{\partial x}\right)^n e^{-x^2} \tag{8.6}$$

and the modified Hermite polynomials,

$$\bar{H}_n(x) := e^{x^2/2}\left(-\frac{\partial}{\partial x}\right)^n e^{-x^2/2}. \tag{8.7}$$

For the properties of the Hermite polynomials, we refer to Mehta's book (2004).

Theorem 8.2. *Suppose that $\mathcal{X}$ satisfies the hypotheses of the theorem with $k \geq 2$. For $j = 1, \ldots, k$, one has*

$$\begin{aligned} F^{(j)}(u) &= (-1)^{j-1}\overline{H}_{j-1}(u)\left[1 + (2\pi)^{-1/2}u\int_0^1 (r_{11}(t,t))^{1/2}\,dt\right]\varphi(u) \\ &\quad + \rho_j(u)\varphi(u), \end{aligned} \tag{8.8}$$

where

$$|\, \rho_j(u) \,| \leq C_j \exp(-\delta u^2)$$

with $C_1, C_2, \ldots$ positive constants, and $\delta > 0$ does not depend on j.

Proof. The proof of (8.8) is a slight modification of the one in Theorem 8.1. Notice first that from the computation of $\beta^0(s)$ above it follows that (1) if $X^0(0) < 0$, then if u is large enough, $X^0(s) - \beta^0(s)u \leq 0$ for all $s \in [0, 1]$, and (2) if $X^0(0) > 0$, then $X^0(0) - \beta^0(0)u > 0$ so that

$$L_1(u) = \mathrm{P}(X^0(s) - \beta^0(s).u \leq 0 \text{ for all } s \in [0,1]) \uparrow \tfrac{1}{2} \quad \text{as } u \uparrow +\infty.$$

Based on (8.5), this implies that if $u \geq 0$,

$$0 \leq \tfrac{1}{2} - L_1(u) = \int_u^{+\infty} L_1'(v)\,dv \leq D_1 \exp(-\delta_1 u^2)$$

with D_1 and δ_1 positive constants.

$L_2(u)$ is similar. Finally,

$$\begin{aligned} L_3(u) &= -\int_0^1 \mathrm{E}\left\{X^t(t) - \beta^t(t)u\right\} \frac{dt}{(2\pi r_{11}(t,t))^{1/2}} \\ &\quad - \int_0^1 \mathrm{E}\left\{(X^t(t) - \beta^t(t)u)\mathbf{I}_{(A_u(X^t,\beta^t))^C}\right\} \frac{dt}{(2\pi r_{11}(t,t))^{1/2}}. \end{aligned} \tag{8.9}$$

The first term in (8.9) is equal to

$$(2\pi)^{-1/2}u\int_0^1 (r_{11}(t,t))^{1/2}\,dt.$$

As for the second term in (8.9), denote $\beta_{\#} = \inf_{s,t\in[0,1]} \beta^t(s) > 0$ and let $u > 0$. Then

$$\mathrm{P}\left(\left(A_u(X^t,\beta^t)\right)^C\right) \le \mathrm{P}(\exists\ s \in [0,1] \text{ such that } X^t(s) > \beta_{\#}.u) \le D_3\exp(-\delta_3 u^2)$$

with D_3 and δ_3 positive constants, the last inequality being a consequence of the basic inequality (2.33).

The remainder follows in the same way as the proof of Theorem 8.1. □

8.2. AN APPLICATION TO UNBOUNDED PROCESSES

As we have seen in Chapter 2, the tails of the probability distribution of the supremum of a Gaussian processes which a.s. has bounded paths have a nice behavior for large values of the argument, since they are bounded by the tails of the distribution of one Gaussian variable. As we have seen, this has important consequences to handle the supremum of a Gaussian process, via the exponential bound for the tails of its probability distribution.

In this section we consider a less explored subject, that is, the behavior of large values of the supremum of Gaussian processes in case they are unbounded. One should observe that if a separable centered Gaussian process $\{Z(t) : t \in T\}$ is unbounded, then $\mathrm{P}(\{\sup|Z(t)| < \infty, t \in T\}) = 0$, but this does not imply that $q = \mathrm{P}(\{\sup Z(t) < \infty, t \in T\}) = 0$. [Notice that $\{\sup Z(t) < +\infty, t \in T\}$ is *not* a subspace of the space of possible paths, so that the Gaussian 0 or 1 law can not be applied to it.]

We will see that the methods of Chapter 7, which a priori had the purpose of studying the regularity of the distribution of M_T, are useful to understand the asymptotic behavior of an interesting family of one-parameter unbounded processes. In fact, the tail of the probability distribution of M_T, having total mass q strictly smaller than 1, that is,

$$q - \mathrm{P}(M_T \le u),$$

which tends to zero as $u \to +\infty$, is not necessarily bounded by a Gaussian tail and can be estimated using the differentiation theorems above.

Let $\mathcal{X} = \{X(t) : t \in [0,1]\}$ satisfy hypothesis (H2) of Chapter 7. We also assume that the process is centered. Let $\beta : [0,1] \to \mathbb{R}^+$ be a continuous function that vanishes only at $t = 0$, twice continuously differentiable for $t \in (0,1]$. We are going to study the behavior as $u \to +\infty$ of the function

$$F(u) = \mathrm{P}(X(t) \le \beta(t)u \text{ for all } t \in (0,1]),$$

which is obviously the (defective) distribution function of the supremum over $[0, 1]$ of the unbounded Gaussian process

$$Z(t) = \frac{X(t)}{\beta(t)}$$

that has exploding paths at the only point $t = 0$.

Clearly,

$$\text{a.s.} \quad \lim_{u \to +\infty} \mathbb{1}_{\{X(t) \le \beta(t)u \text{ for all } t \in [0,1]\}} = \mathbb{1}_{\{X(0)<0\}},$$

so that

$$q = F(+\infty) = \tfrac{1}{2}.$$

Theorem 8.3. *Assume that if $t \downarrow 0$, one has $\beta(t) \approx Ct^{\alpha}$, $\beta'(t) \approx C\alpha t^{\alpha-1}$, and $\beta''(t) \approx C\alpha(\alpha - 1)t^{\alpha-2}$, where C and α are positive constants. Then we have:*

(i) *If $0 < \alpha \le 1$,*

$$\tfrac{1}{2} - F(u) \le a \exp(-bu^2) \qquad \textit{for } u > 0,$$

where a and b are positive constants.

(ii) *If $\alpha > 1$,*

$$\frac{1}{2} - F(u) \approx \frac{K}{u^{1/(\alpha-1)}} \quad \textit{when } u \to +\infty, \tag{8.10}$$

where K is a positive constant depending on C, α, $\operatorname{Var}\{X(0)\}$, and $\operatorname{Var}\{X'(0)\}$ that we will obtain explicitly in the proof.

Proof. With no loss of generality, we may assume that $\operatorname{Var}\{X(t)\} = 1$ for all $t \in [0, 1]$. If this is not the case, it suffices to replace the process by $X(t)/[\operatorname{Var}\{X(t)\}]^{1/2}$ and the function $\beta(t)$ by $\beta(t)/[\operatorname{Var}\{X(t)\}]^{1/2}$.

To study the behavior of

$$\tfrac{1}{2} - F(u) = \int_u^{+\infty} F'(x)\,dx$$

when $u \to +\infty$, we write $F'(x)$ using Lemma 7.7, that is, because $\beta(0) = 0$,

$$\begin{aligned} F'(x) &= \beta(1)\mathrm{P}(X^1(t) \le \beta^1(t)x \;\forall t \in [0, 1])\,p_{X(1)}(\beta(1)x) \\ &\quad - \int_0^1 \beta(t)\mathrm{E}\left\{\left[X^t(t) - \beta^t(t)x\right] \mathbb{1}_{\{X^t(s) \le \beta^t(s)x \forall s \in [0,1]\}}\right\} \\ &\quad \times p_{X(t),X'(t)}(\beta(t)x, \beta'(t)x)\,dt. \end{aligned} \tag{8.11}$$

A simple computation shows that in the present case the process X^1 and the function β^1 are the continuous extensions to $[0, 1]$ of

$$X^1(t) = \frac{X(t) - r(t,1)X(1)}{1-t} \quad \text{and} \quad \beta^1(t) = \frac{\beta(t) - r(t,1)\beta(1)}{1-t} \tag{8.12}$$

defined for $t \in [0, 1)$.

In the same way, for $t \in [0, 1]$, X^t and β^t are the continuous extensions to $[0, 1]$ of

$$X^t(s) = \frac{2}{(s-t)^2}\left[X(s) - r(s,t)X(t) - \frac{r_{10}(s,t)}{r_{11}(t,t)}X'(t)\right]$$

$$\beta^t(s) = \frac{2}{(s-t)^2}\left[\beta(s) - r(s,t)\beta(t) - \frac{r_{10}(s,t)}{r_{11}(t,t)}\beta'(t)\right]$$

defined for $s \in [0, 1]$, $s \neq t$.

In fact, the proof of formula (8.11) requires the function $\beta(\cdot)$ to be of class C^2 on the entire interval $[0, 1]$, which need not be our case. The proof that (8.11) holds true for $x > 0$ is left to the reader and is a consequence of the hypotheses on $\beta(\cdot)$ near $t = 0$.

It is easily seen that the first term on the right-hand side of (8.11) is bounded above by $a_1 \exp(-b_1 x^2)$, where a_1 and b_1 are positive constants. As for the second term, the density in the integrand is

$$p_{X(t),X'(t)}(\beta(t)x, \beta'(t)x) = \frac{1}{2\pi\,[\text{Var}\{X'(t)\}]^{1/2}} \times \exp\left[-\frac{x^2}{2}\left(\beta^2(t) + \frac{(\beta'(t))^2}{\text{Var}\{X'(t)\}}\right)\right] \tag{8.13}$$

and

$$\beta^t(s) = \frac{2}{(s-t)^2}\left[\beta(s) - \beta(t) - (s-t)\beta'(t)\right] - r_{20}(t + \theta_1(s-t), t)\beta(t) - \frac{r_{12}(t, t+\theta_2(s-t))}{r_{11}(t,t)}\beta'(t) \qquad 0 < \theta_1, \theta_2 < 1 \tag{8.14}$$

so that

$$\beta^t(t) = \beta''(t) - r_{20}(t,t)\beta(t) - \frac{r_{12}(t,t)}{r_{11}(t,t)}\beta'(t),$$

so $0 < t \leq 1$.

Now assume hypothesis (i). Using the conditions on the function $\beta(\cdot)$, one can verify that if $x > 1$, the absolute value of the integrand on the right-hand side of (8.11) is bounded by $a_2 \exp(-b_2 x^2)$, a_2 and b_2 positive constants. This proves the first part of the theorem.

If condition (ii) holdxs, take $\varepsilon, 0 < \varepsilon < 1$ and use the splitting

$$\tfrac{1}{2} - F(u) = \tfrac{1}{2} - F_\varepsilon(u) + G_\varepsilon(u),$$

where

$$F_\varepsilon(u) = \mathrm{P}(X(t) \le \beta(t)u \text{ for all } t \in [0, \varepsilon])$$

$$0 \le G_\varepsilon(u) \le \mathrm{P}\left(\sup_{t\in[\varepsilon,1]} \frac{X(t)}{\beta(t)} > u\right) \le g_1 \exp(-g_2 u^2),$$

with g_1 and g_2 positive constants depending on ε, the last inequality resulting from an application of (2.33) to the bounded process $X(t)/\beta(t)$ for $t \in [\varepsilon, 1]$.

Hence, to finish the proof of (ii) it suffices to show (ii) when one replaces $F(u)$ by $F_\varepsilon(u)$, where $\varepsilon > 0$ is fixed and small enough. With that purpose, we apply the same formula (8.11), *mutatis mutandis*, that is changing the interval $[0, 1]$ by $[0, \varepsilon]$. We obtain

$$\begin{aligned} F'_\varepsilon(x) &= \beta(\varepsilon)\mathrm{P}(X^1(t) \le \beta^1(t)x \text{ for all } t \in [0, \varepsilon])p_{X(\varepsilon)}(\beta(\varepsilon)x) \\ &\quad - \int_0^\varepsilon \beta(t)\mathrm{E}\left\{\left[X^t(t) - \beta^t(t)x\right] \mathbf{I}_{\{X^t(s)\le\beta^t(s)x \text{ for all } s\in[0,\varepsilon]\}}\right\} \\ &\quad \cdot p_{X(t),X'(t)}(\beta(t)x, \beta'(t)x)\, dt. \end{aligned} \tag{8.15}$$

In the integral on the right-hand side of (8.15), make the change of variable $z = t^{\alpha-1}x$. Then $t(x) = (z/x)^{(1/\alpha-1)} \to 0$ when $x \to +\infty$ for each $z > 0$ fixed. For an adequate choice of ε, we have, for each $z > 0$,

$$\text{a.s.} \quad \mathbf{I}_{\{X^{t(x)}(s)\le\beta^{t(x)}(s)x \text{ for all } s\in[0,\varepsilon]\}} \to 1 \qquad \text{when } x \to +\infty.$$

This follows from the statement

$$\text{for } z > 0 \text{ fixed, } \beta^{t(x)}(s)x \to +\infty \text{ when } x \to +\infty, \text{ uniformly in } s \in [0, \varepsilon]. \tag{8.16}$$

To prove (8.16), use (8.14) and the form of $\beta(t)$ near $t = 0$. We have (elementary checking):

- If $1 < \alpha < 2$ and L is any positive real number, one can choose $\varepsilon > 0$ small enough so that

$$\beta^t(s)x \ge Lx - C_1tz - C_2z \qquad \text{for all } s, t \in [0, \varepsilon], \tag{8.17}$$

  and C_1 and C_2 are some positive constants.

- If $\alpha \geq 2$, one can choose small enough $\varepsilon > 0$ in such a way that

$$\beta^t(s)x \geq C_1' \frac{z}{t} - C_2' tz - C_3' z \quad \text{for all } s, t \in [0, \varepsilon], \tag{8.18}$$

and C_1', C_2', and C_3' are some positive constants. Either (8.17) or (8.18) implies (8.16).

To find the equivalent of the right-hand side of (8.15) is now an exercise (apply dominated convergence). We get

$$F_\varepsilon'(x) \approx \left(\frac{\sigma}{\alpha}\right)^{\alpha/(\alpha-1)} \frac{1}{\sqrt{2\pi}} C^{1/(1-\alpha)} I_{\alpha/(\alpha-1)} x^{\alpha/(1-\alpha)} \qquad \text{when } x \to +\infty,$$

where $\sigma = \left[\operatorname{Var}\left\{X'(0)\right\}\right]^{1/2}$ and I_a is defined for $a \geq 0$ by means of

$$I_a = \int_0^{+\infty} y^a \frac{\exp(-y^2/2)}{\sqrt{2\pi}} dy.$$

Finally, (8.10) follows integrating once. One also obtains the value of the constant

$$K = (\alpha - 1) \left(\frac{\sigma}{\alpha}\right)^{\alpha/(\alpha-1)} \frac{1}{\sqrt{2\pi}} C^{1/(1-\alpha)} I_{\alpha/(\alpha-1)}.$$

8.3. A GENERAL BOUND FOR p_M

We need to introduce some further notations. For t in S_j, $j \leq d_0$, we define $C_{t,j}$ as the closed convex cone generated by the set of directions

$$\left\{\lambda \in \mathbb{R}^d : \|\lambda\| = 1; \exists\, s_n \in S (n = 1, 2, \dots) \text{ such that } s_n \to t, \right.$$
$$\left. \frac{t - s_n}{\|t - s_n\|} \to \lambda \text{ as } n \to +\infty \right\},$$

whenever this set is nonempty and $C_{t,j} = \{0\}$ if it is empty. We will denote by $\widehat{C}_{t,j}$ the dual cone of $C_{t,j}$, that is,

$$\widehat{C}_{t,j} := \{z \in \mathbb{R}^d : \langle z, \lambda \rangle \geq 0 \text{ for all } \lambda \in C\}.$$

Notice that these definitions easily imply that $T_{t,j} \subset C_{t,j}$ and $\widehat{C}_{t,j} \subset N_{t,j}$. Remark also that for $j = d_0$, $\widehat{C}_{t,j} = N_{t,j}$.

We will say that the function $X(\cdot)$ has an "extended outward" derivative at point t in S_j, $j \leq d_0$ if $X'_{j,N}(t) \in \widehat{C}_{t,j}$.

Theorem 8.4. *Under assumptions (A1) to (A5) of Section 7.1, one has:*
(a) $p_M(x) \le \overline{p}(x)$, *where*

$$\overline{p}(x) := \sum_{t \in S_0} \mathrm{E}\big(\mathbf{I}_{X'(t) \in \widehat{C}_{t,0}} \big| X(t) = x\big) p_{X(t)}(x)$$
$$+ \sum_{j=1}^{d_0} \int_{S_j} \mathrm{E}\big(|\det(X_j''(t))| \mathbf{I}_{X'_{j,N}(t) \in \widehat{C}_{t,j}} \big| X(t) = x, X_j'(t) = 0\big)$$
$$\times p_{X(t),X_j'(t)}(x, 0) \sigma_j(dt). \tag{8.19}$$

(b) $\mathrm{P}\{M > u\} \le \displaystyle\int_u^{+\infty} \overline{p}(x)\, dx.$

Proof. Part (a) follows from Theorem 7.2 and the observation that if $t \in S_j$, one has $\{M \le X(t)\} \subset \{X'_{j,N}(t) \in \widehat{C}_{t,j}\}$. Part (b) is an obvious consequence of Part (a). □

The actual interest in this theorem depends on the feasibility of computing $\overline{p}(x)$. It turns out that this can be done in some relevant cases, as we will see in the next section. The results can be compared with the approximation of $\mathrm{P}\{M > u\}$ by means of $\int_u^{+\infty} p^E(x)\, dx$ given by Taylor et al. (2005) and Adler and Taylor (2007), where

$$p^E(x) := \sum_{t \in S_0} \mathrm{E}\big(\mathbf{I}_{X'(t) \in \widehat{C}_{t,0}} \big| X(t) = x\big) p_{X(t)}(x)$$
$$+ \sum_{j=1}^{d_0} (-1)^j \int_{S_j} \mathrm{E}\big(\det(X_j''(t)) \mathbf{I}_{X'_{j,N}(t) \in \widehat{C}_{t,j}} \big| X(t) = x, X_j'(t) = 0\big)$$
$$\times p_{X(t),X_j'(t)}(x, 0) \sigma_j(dt). \tag{8.20}$$

Under certain conditions, $\int_u^{+\infty} p^E(x)\, dx$ is the expected value of the EPC of the excursion set E_u. The advantage of $p^E(x)$ over $\overline{p}(x)$ is that one can have nice expressions for it (see Exercise 8.1). Conversely, $\overline{p}(x)$ has the obvious advantage that it is an upper bound of the true density $p_M(x)$. Hence, upon integrating once, it provides an upper bound for the tail probability, for every u value.

We can go a bit farther, and show that $[\overline{p}(x) + p^E(x)]/2$ is also an upper bound for $p^M(x)$. This follows easily upon the following observation: Denote by E_j the event that the random linear operator $X_j''(t)$ is nonnegative definite. In the formula giving $p^M(x)$ (see Theorem 7.2), one can replace inside the conditional expectation $|\det(X_j''(t))| \mathbf{I}_{A_x}$ by $(-1)^j \det(X_j''(t)) \mathbf{I}_{A_x} \mathbf{I}_{E_j}$, since under the conditioning E_j occurs, and obviously, by

$$\frac{|\det(X''_j(t))| + (-1)^j \det(X''_j(t))}{2} \mathbb{I}_{A_x} \mathbb{I}_{E_j}.$$

Replacing $\mathbb{I}_{A_x} \mathbb{I}_{E_j}$ by $\mathbb{I}_{X'_{j,N}(t) \in \widehat{C}_{t,j}}$, we get for $p^M(x)$ the more precise upper bound $[\bar{p}(x) + p^E(x)]/2$.

Under additional conditions, these upper bounds both provide good first-order approximations for $p_M(x)$ as $x \to \infty$, as we will see in the remainder of the chapter. For the special case in which the random field $\mathcal{X}$ is centered and has a law that is invariant under orthogonal linear transformations and translations, in the next section we present a procedure to compute $\overline{p}(x)$.

8.4. COMPUTING $\overline{p}(x)$ FOR STATIONARY ISOTROPIC GAUSSIAN FIELDS

For one-parameter centered Gaussian process having constant variance and satisfying certain regularity conditions, a bound for $p_M(x)$ has been given by inequality (7.6). In the two-dimensional parameter case, Mercadier (2005) has given a bound for $\mathrm{P}\{M > u\}$, using a method especially suited to dimension 2 that will be presented in Chapter 9. When the parameter is one- or two-dimensional, these bounds are sharper than those below, which apply to any dimension.

We will assume that the process $\mathcal{X}$ is centered Gaussian, with a covariance function that can be written as

$$\mathrm{E}\big(X(s)X(t)\big) = \rho\big(\|s - t\|^2\big), \tag{8.21}$$

where $\rho : \mathbb{R}^+ \to \mathbb{R}$ is of class C^4. Without loss of generality, we assume that $\rho(0) = 1$. Assumption (8.21) is equivalent to saying that the law of $\mathcal{X}$ is invariant under orthogonal linear transformations and translations of the underlying parameter space $\mathbb{R}^d$. We also assume that the set S has a polyhedral shape. More precisely, we assume that each $S_j (j = 1, \ldots, d)$ is a union of subsets of affine manifolds of dimension j in $\mathbb{R}^d$.

For the proof of Theorem 8.8, which gives an expression for the bound $\overline{p}(x)$ of the density, we need some auxiliary computational lemmas. The first lemma is elementary and the proof is left to the reader. Here, and in the remainder of this chapter, we use the abridged notation: $\rho' := \rho'(0)$, $\rho'' := \rho''(0)$.

Lemma 8.5. *Under the conditions above, for each $t \in U$, $i, i', k, k', j = 1, \ldots, d$:*

(1) $\mathrm{E}\left(\dfrac{\partial X}{\partial t_i}(t) X(t)\right) = 0.$

(2) $\mathrm{E}\left(\dfrac{\partial X}{\partial t_i}(t) \dfrac{\partial X}{\partial t_k}(t)\right) = -2\rho' \delta_{ik}$ *and* $\rho' < 0$.

(3) $\mathrm{E}\left(\frac{\partial^2 X}{\partial t_i \partial t_k}(t) X(t)\right) = 2\rho'\delta_{ik}, \mathrm{E}\left(\frac{\partial^2 X}{\partial t_i \partial t_k}(t)\frac{\partial X}{\partial t_j}(t)\right) = 0.$

(4) $\mathrm{E}\left(\frac{\partial^2 X}{\partial t_i \partial t_k}(t)\frac{\partial^2 X}{\partial t_{i'} \partial t_{k'}}(t)\right) = 24\rho''\left[\delta_{ii'}\delta_{kk'} + \delta_{i'k}\delta_{ik'} + \delta_{ik}\delta_{i'k'}\right].$

(5) $\rho'' - \rho'^2 \geq 0.$

(6) *If $t \in S_j$, the conditional distribution of $X_j''(t)$ given $X(t) = x$ and $X_j'(t) = 0$ is the same as the unconditional distribution of the random matrix*

$$Z + 2\rho' x I_j,$$

where $Z = (Z_{ik} : i, k = 1, \ldots, j)$ is a symmetric $j \times j$ matrix with centered Gaussian entries, independent of the pair $\left(X(t), X'(t)\right)$, such that for $i \leq k$, $i' \leq k'$ one has

$$\mathrm{E}(Z_{ik} Z_{i'k'}) = 4\left[2\rho''\delta_{ii'} + (\rho'' - \rho'^2)\right]\delta_{ik}\delta_{i'k'} + 4\rho''\delta_{ii'}\delta_{kk'}(1 - \delta_{ik}).$$

Our second lemma is the following:

Lemma 8.6. *Let*

$$J_n(x) := \int_{-\infty}^{+\infty} e^{-y^2/2} H_n(z)\, dy \qquad n = 0, 1, 2, \ldots, \tag{8.22}$$

where z stands for the linear form $z = ay + bx$ and a, b are real parameters that satisfy $a^2 + b^2 = \frac{1}{2}$. Then

$$J_n(x) = (2b)^n \sqrt{2\pi}\, \overline{H}_n(x).$$

Proof. From the definitions of H_n and $\overline{H}_n$, we get

$$\sum_{n=0}^{\infty} \frac{(w)^n}{n!} H_n(z) = e^{-w^2 + 2wz}, \qquad \sum_{n=0}^{\infty} \frac{(w)^n}{n!} \overline{H}_n(z) = e^{-w^2/2 + wz},$$

using the Taylor expansion of $e^{(z-w)^2}$ and $e^{(z-w)^2/2}$ in w around 0. Therefore,

$$\begin{aligned}
\sum_{n=0}^{\infty} \frac{(w)^n}{n!} J_n(x) &= \int_{\mathbb{R}} e^{-y^2/2 - w^2 + 2w(ay+bx)} dy \\
&= e^{2wbx - 2(bw)^2} \int_{\mathbb{R}} e^{(y-2wa)^2/2} dy \\
&= \sqrt{2\pi} \sum_{n=0}^{\infty} \frac{(2bw)^n}{n!} \overline{H}_n(x).
\end{aligned}$$

This implies that $J_n(x) = (2b)^n\sqrt{2\pi}\,H_n(x)$. □

We also need the integrals

$$I_n(v) = \int_v^{+\infty} e^{-t^2/2} H_n(t)\,dt.$$

They are computed in the next lemma, which can be proved easily, using standard properties of Hermite polynomials. This is also left to the reader.

Lemma 8.7

(a)

$$I_n(v) = 2e^{-v^2/2} \sum_{k=0}^{[(n-1)/2]} 2^k \frac{(n-1)!!}{(n-1-2k)!!} H_{n-1-2k}(v)$$
$$+ \mathbb{I}_{\{n \text{ even}\}} 2^{n/2}(n-1)!!\sqrt{2\pi}(1-\Phi(x)). \tag{8.23}$$

(b)

$$I_n(-\infty) = \mathbb{I}_{\{n \text{ even}\}} 2^{n/2}\,(n-1)!!\,\sqrt{2\pi}. \tag{8.24}$$

We are now ready to state and prove the announced expression for $\overline{p}(x)$.

Theorem 8.8. *Assume that the random field $\mathcal{X}$ is centered Gaussian, satisfies conditions (A1) to (A5) of Section 7.1, and has a covariance having the form (8.21), which verifies the regularity conditions of the beginning of this section. Moreover, assume S has polyhedral shape. Then $\overline{p}(x)$ can be expressed by means of the formula*

$$\overline{p}(x) = \varphi(x)\left\{\sum_{t\in S_0}\widehat{\sigma}_0(t) + \sum_{j=1}^{d_0}\left[\left(\frac{|\rho'|}{\pi}\right)^{j/2}\overline{H}_j(x) + R_j(x)\right]g_j\right\}, \tag{8.25}$$

where:

- *g_j is a geometric parameter of the face S_j defined by*

$$g_j = \int_{S_j}\widehat{\sigma}_j(t)\sigma_j(dt), \tag{8.26}$$

where $\widehat{\sigma}_j(t)$ is the normalized solid angle of the cone $\widehat{C}_{t,j}$ in $N_{t,j}$; that is,

$$\widehat{\sigma}_j(t) = \frac{\sigma_{d-j-1}(\widehat{C}_{t,j}\cap \mathcal{S}^{d-j-1})}{\sigma_{d-j-1}(\mathcal{S}^{d-j-1})} \qquad \textit{for } j = 0,\dots,d-1 \tag{8.27}$$

$$\widehat{\sigma}_d(t) = 1. \tag{8.28}$$

Notice that for convex or other usual polyhedra, $\widehat{\sigma}_j(t)$ is constant for $t \in S_j$, so that g_j is equal to this constant multiplied by the j-dimensional geometric measure of S_j.

- *For $j = 1, \dots d$,*

$$R_j(x) = \left(\frac{2\rho''}{\pi|\rho'|}\right)^{j/2} \frac{\Gamma((j+1)/2)}{\pi} \int_{-\infty}^{+\infty} T_j(v) \exp\left(-\frac{y^2}{2}\right) dy, \quad (8.29)$$

where

$$v := -2^{-1/2}\big((1-\gamma^2)^{1/2} y - \gamma x\big) \text{ with } \gamma := |\rho'|(\rho'')^{-1/2} \quad (8.30)$$

$$T_j(v) := \left[\sum_{k=0}^{j-1} \frac{H_k^2(v)}{2^k k!}\right] e^{-v^2/2} - \frac{H_j(v)}{2^j (j-1)!} I_{j-1}(v), \quad (8.31)$$

where I_n is given in Lemma 8.7.

For the proof of the theorem, we still need some additional ingredients from random matrix theory. The random $n \times n$ real random matrix G_n is said to have *the GOE* (Gaussian orthogonal ensemble) *distribution* if it is symmetric, has centered Gaussian entries $g_{ik}, i, k = 1, \dots, n$ satisfying $\mathrm{E}(g_{ii}^2) = 1$, $\mathrm{E}(g_{ik}^2) = \frac{1}{2}$ if $i < k$ and the random variables $\{g_{ik}, 1 \le i \le k \le n\}$ are independent.

Following Mehta (2004), we denote as $q_n(\nu)$ the density of eigenvalues of $n \times n$ GOE matrices at the point ν; that is, $q_n(\nu)\,d\nu$ is the probability of G_n having an eigenvalue in the interval $(\nu, \nu + d\nu)$. One has the formula

$$\begin{aligned} e^{\nu^2/2} q_n(\nu) = {} & e^{-\nu^2/2} \sum_{k=0}^{n-1} c_k^2 H_k^2(\nu) + \tfrac{1}{2}(n/2)^{1/2} c_{n-1} c_n H_{n-1}(\nu) \\ & \times \left[\int_{-\infty}^{+\infty} e^{-y^2/2} H_n(y)\,dy - 2\int_{\nu}^{+\infty} e^{-y^2/2} H_n(y)\,dy\right] \\ & + \mathbb{1}_{\{n \text{ odd}\}} \frac{H_{n-1}(\nu)}{\int_{-\infty}^{+\infty} e^{-y^2/2} H_{n-1}(y)\,dy}, \end{aligned} \quad (8.32)$$

where $c_k := (2^k k! \sqrt{\pi})^{-1/2}, k = 0, 1, \dots$. The proof can be found in Chapter 7 of Mehta's book.

We will use the following remark due to Fyodorov (2006), which we state as a lemma.

Lemma 8.9. *Let G_n be a GOE $n \times n$ matrix. Then for $\nu \in \mathbb{R}$ one has*

$$\mathrm{E}\big(|\det(G_n - \nu I_n)|\big) = 2^{3/2}\Gamma\big((n+3)/2\big) \exp(\nu^2/2) \frac{q_{n+1}(\nu)}{n+1}. \quad (8.33)$$

Proof. Denote by $\nu_1, \ldots, \nu_n$ the eigenvalues of G_n. It is well known (Kendall et al., 1983; Mehta, 2004) that the joint density f_n of the n-tuple of random variables $(\nu_1, \ldots, \nu_n)$ is given by the formula

$$f_n(\nu_1, \ldots, \nu_n) = k_n \exp\left(-\frac{\sum_{i=1}^n \nu_i^2}{2}\right) \prod_{1 \le i < k \le n} |\nu_k - \nu_i|$$

$$\text{with } k_n := (2\pi)^{-n/2}(\Gamma(3/2))^n \left(\prod_{i=1}^n \Gamma(1 + i/2)\right)^{-1}.$$

Then

$$\begin{aligned}
\mathrm{E}\big(|\det(G_n - \nu I_n)|\big) &= \mathrm{E}\left(\prod_{i=1}^n |\nu_i - \nu|\right) = \int_{\mathbb{R}^n} \prod_{i=1}^n |\nu_i - \nu| k_n \exp\left(-\frac{\sum_{i=1}^n \nu_i^2}{2}\right) \\
&\quad \times \prod_{1 \le i < k \le n} |\nu_k - \nu_i| d\nu_1, \ldots, d\nu_n \\
&= e^{\nu^2/2} \frac{k_n}{k_{n+1}} \int_{\mathbb{R}^n} f_{n+1}(\nu_1, \ldots, \nu_n, \nu)\, d\nu_1, \ldots, d\nu_n \\
&= e^{\nu^2/2} \frac{k_n}{k_{n+1}} \frac{q_{n+1}(\nu)}{n+1}.
\end{aligned}$$

The remainder is obvious. □

Proof of Theorem 8.8. We use the definition (8.19) given in Theorem 8.4 and the moment computations of Lemma 5, which imply that:

$$p_{X(t)}(x) = \varphi(x) \tag{8.34}$$

$$p_{X(t),X'_j(t)}(x, 0) = \varphi(x)(2\pi)^{-j/2}(-2\rho')^{-j/2} \tag{8.35}$$

$$X'(t) \text{ is independent of } X(t) \tag{8.36}$$

$$X'_{j,N}(t) \text{ is independent of } (X''_j(t), X(t), X'_j(t)). \tag{8.37}$$

Since the distribution of $X'(t)$ is centered Gaussian with variance $-2\rho' I_d$, it follows that

$$\mathrm{E}(\mathbb{1}_{X'(t) \in \widehat{C}_{t,0}} | X(t) = x) = \widehat{\sigma}_0(t) \qquad \text{if } t \in S_0,$$

and if $t \in S_j$, $j \ge 1$,

$$\begin{aligned}
&\mathrm{E}(|\det(X''_j(t))| \mathbb{1}_{X'_{j,N}(t) \in \widehat{C}_{t,j}} | X(t) = x, X'_j(t) = 0) \\
&= \widehat{\sigma}_j(t) \mathrm{E}(|\det(X''_j(t))| | X(t) = x, X'_j(t) = 0) \\
&= \widehat{\sigma}_j(t) \mathrm{E}(|\det(Z + 2\rho' x I_j)|).
\end{aligned} \tag{8.38}$$

where $\widehat{\sigma}_j(t)$ is the normalized solid angle defined in the statement of the theorem and the random $j \times j$ real matrix Z has the distribution of Lemma 8.5.

Standard moment computations show that Z has the same distribution as the random matrix $\sqrt{8\rho''}\, G_j + 2\sqrt{\rho'' - \rho'^2}\, \xi I_j$, where G_j is a $j \times j$ GOE random matrix and ξ is standard normal in $\mathbb{R}$ and independent of G_j. So for $j \geq 1$, one has

$$\mathrm{E}\big(|\det(Z + 2\rho' x I_j)|\big) = (8\rho'')^{j/2} \int_{-\infty}^{+\infty} \mathrm{E}\big(|\det(G_j - \nu I_j)|\big)\varphi(y)\, dy,$$

where ν is given by (8.30).

For the conditional expectation in (8.19), plug the last expression into (8.38) and use (8.33), (8.32), and Lemma 8.7. For the density in (8.19), use (8.35). Then, after some algebra, Lemma 8.6 gives (8.25). □

Remarks on Theorem 8.8

1. The *principal term* is

$$\varphi(x)\left\{\sum_{t\in S_0} \widehat{\sigma}_0(t) + \sum_{j=1}^{d_0}\left[\left(\frac{|\rho'|}{\pi}\right)^{j/2} \overline{H}_j(x)\right] g_j\right\}, \tag{8.39}$$

which is the product of a standard normal density times a polynomial with degree d_0. Integrating once, we get, in our special case, the formula for the expectation of the EPC of the excursion set given by Adler and Taylor (2007).

2. The *complementary term*

$$\varphi(x)\sum_{j=1}^{d_0} R_j(x) g_j, \tag{8.40}$$

can be computed by means of a formula, as follows from the statement of the theorem. These formulas will, in general, be quite unpleasant, due to the complicated form of $T_j(v)$. However, for low dimensions they are simple. For example,

$$T_1(v) = \sqrt{2\pi}\big[\varphi(v) - v(1 - \Phi(v))\big], \tag{8.41}$$

$$T_2(v) = 2\sqrt{2\pi}\,\varphi(v), \tag{8.42}$$

$$T_3(v) = \sqrt{\frac{\pi}{2}}\big[3(2v^2 + 1)\varphi(v) - (2v^2 - 3)v(1 - \Phi(v))\big]. \tag{8.43}$$

3. Second-order asymptotics for $p_M(x)$ as $x \to +\infty$ will be the focus of the following sections. However, we can state already that the complementary term

(8.40) is equivalent, as $x \to +\infty$, to

$$\varphi(x) g_{d_0} K_{d_0} x^{2d_0-4} \exp\left(-\frac{1}{2}\frac{\gamma^2}{3-\gamma^2}x^2\right), \tag{8.44}$$

where the constant K_j, $j = 1, 2, \ldots$ is given by

$$K_j = 2^{3j-2} \frac{\Gamma((j+1)/2)}{\sqrt{\pi}(2\pi\gamma)^{j/2}(j-1)!} \rho''^{j/4} \left(\frac{\gamma}{3-\gamma^2}\right)^{2j-4}. \tag{8.45}$$

We are not going to go through this calculation, which is completely elementary but requires some extra work, which is left to the reader. An outline is as follows: Replace the Hermite polynomials in the expression for $T_j(v)$ given by (8.31) by the well-known expansion (again, see Mehta, 2004)

$$H_j(v) = j! \sum_{i=0}^{[j/2]} (-1)^i \frac{(2v)^{j-2i}}{i!(j-2i)!} \tag{8.46}$$

and $I_{j-1}(v)$ by means of the formula in Lemma 8.7.

Evaluating the term of highest degree in the polynomial part, one proves that as $v \to +\infty$, $T_j(v)$ is equivalent to

$$\frac{2^{j-1}}{\sqrt{\pi}(j-1)!} v^{2j-4} e^{-v^2/2}. \tag{8.47}$$

Using the definition of $R_j(x)$ and changing variables in the integral in (8.29), one gets for $R_j(x)$ the equivalent:

$$K_j x^{2j-4} \exp\left(-\frac{1}{2}\frac{\gamma^2}{3-\gamma^2}x^2\right). \tag{8.48}$$

In particular, the equivalent of (8.40) is given by the highest-order nonvanishing term in the sum.

4. Now consider the case in which S is the sphere S^{d-1} and the process satisfies the same conditions as in the theorem. Even though the theorem cannot be applied directly, it is possible to deal with this example to compute $\overline{p}(x)$, performing only some minor changes. In this case, only the term that corresponds to $j = d-1$ in (8.19) does not vanish, $\widehat{C}_{t,d-1} = N_{t,d-1}$, so that $\mathbf{I}_{X'_{d-1,N}(t) \in \widehat{C}_{t,d-1}} = 1$ for each $t \in S^{d-1}$, and one can use invariance under the orthogonal group to obtain

$$\overline{p}(x) = \varphi(x) \frac{\sigma_{d-1}(S^{d-1})}{(2\pi)^{(d-1)/2}} \mathrm{E}\left(|\det(Z + 2\rho' x I_{d-1}) + (2|\rho'|)^{1/2} \eta I_{d-1}|\right), \tag{8.49}$$

where Z is a $(d-1) \times (d-1)$ centered Gaussian matrix with the covariance structure of Lemma 8.5 and η is a standard normal real random variable, independent of Z. Equality (8.49) follows from the fact that the normal derivative at each point is centered Gaussian with variance $2|\rho'|$ and independent of the tangential derivative. So we apply the previous computation, replacing x by $x + (2|\rho'|)^{-1/2}\eta$ and obtain the expression

$$\overline{p}(x) = \varphi(x)\frac{2\pi^{d/2}}{\Gamma(d/2)} \cdot \int_{-\infty}^{+\infty} \left[\left(\frac{|\rho'|}{\pi}\right)^{(d-1)/2} \overline{H}_{d-1}(x + (2|\rho'|)^{-1/2}y) + R_{d-1}(x + (2|\rho'|)^{-1/2}y) \right] \varphi(y)\,dy. \tag{8.50}$$

8.5. ASYMPTOTICS AS $x \to +\infty$

In this section we consider the errors in the direct and EPC methods (see Section 8.3) for large values of the argument x. Theses errors are

$$\overline{p}(x) - p_M(x) = \sum_{t \in S_0} \mathrm{E}\big(\mathbf{1}_{X'(t) \in \widehat{C}_{t,0}} \cdot \mathbf{1}_{M>x} \big| X(t) = x\big) p_{X(t)}(x) + \sum_{j=1}^{d_0} \int_{S_j} \mathrm{E}\big(|\det(X_j''(t)| \mathbf{1}_{X'_{j,N}(t) \in \widehat{C}_{t,j}} \cdot \mathbf{1}_{M>x}\big) \big| X(t) = x, X_j'(t) = 0\big) p_{X(t),X_j'(t)}(x, 0)\sigma_j(dt). \tag{8.51}$$

$$p^E(x) - p_M(x) = \sum_{t \in S_0} \mathrm{E}\big(\mathbf{1}_{X'(t) \in \widehat{C}_{t,0}} \cdot \mathbf{1}_{M>x} \big| X(t) = x\big) p_{X(t)}(x) + \sum_{j=1}^{d_0} (-1)^j \int_{S_j} \mathrm{E}\big(\det(X_j''(t) \mathbf{1}_{X'_{j,N}(t) \in \widehat{C}_{t,j}} \cdot \mathbf{1}_{M>x}\big) \big| X(t) = x, X_j'(t) = 0\big) p_{X(t),X_j'(t)}(x, 0)\sigma_j(dt). \tag{8.52}$$

It is clear that for every real x,

$$|p^E(x) - p_M(x)| \le \overline{p}(x) - p_M(x),$$

so that the upper bounds for $\overline{p}(x) - p_M(x)$ will automatically be upper bounds for $|p^E(x) - p_M(x)|$.

Our next theorem gives sufficient conditions, allowing us to ensure that the error $\overline{p}(x) - p_M(x)$ is bounded by a constant times a Gaussian density having strictly smaller variance than the maximum variance of the given process $\mathcal{X}$.

In this theorem we assume that the maximum of the variance is not attained in $S \backslash S_{d_0}$. This excludes constant variance or some other stationary-like condition. This type of process is considered later in Theorem 8.12. For parameter dimension $d_0 > 1$, a result of this type for nonconstant variance processes is Theorem 3.3 of Taylor et al. (2005).

Theorem 8.10. *Assume that the process $\mathcal{X}$ satisfies conditions (A1) to (A5) of Section 7.1. With no loss of generality, we assume that* $\max_{t \in S} \mathrm{Var}(X(t)) = 1$. *In addition, we assume that the set S_v of points $t \in S$ where the variance of $X(t)$ attains its maximal value is contained in S_{d_0} $(d_0 > 0)$, the nonempty face having the largest dimension, and that no point in S_v is a boundary point of $S \backslash S_{d_0}$. Then there exist some positive constants C and δ such that for every $x > 0$,*

$$|p^E(x) - p_M(x)| \leq \overline{p}(x) - p_M(x) \leq C\varphi(x(1+\delta)), \tag{8.53}$$

where $\varphi(\cdot)$ is the standard normal density.

Proof. Let W be an open neighborhood of the compact subset S_v of S such that $\mathrm{dist}(W, (S \backslash S_{d_0})) > 0$, where dist denotes the Euclidean distance in $\mathbb{R}^d$. For $t \in S_j \cap W^c$, the density

$$p_{X(t), X'_j(t)}(x, 0)$$

can be written as the product of the density of $X'_j(t)$ at the point 0, times the conditional density of $X(t)$ at the point x given that $X'_j(t) = 0$, which is Gaussian with some bounded expectation and a conditional variance which is smaller than the unconditional variance, hence bounded by some constant smaller than 1. Since the conditional expectations in (8.51) are uniformly bounded by some constant, due to the bounds on the moments of the Gaussian law (see Chapter 2), one can deduce that

$$\overline{p}(x) - p_M(x) = \int_{W \cap S_{d_0}} \mathrm{E}\Big(|\det(X''_{d_0}(t))| \mathbf{I}_{X'_{d_0,N}(t) \in \widehat{C}_{t,d_0}} \mathbf{I}_{M>x} \Big| \\ X(t) = x, X'_{d_0}(t) = 0\Big) \cdot p_{X(t), X'_{d_0}(t)}(x, 0) \sigma_{d_0}(dt) + O(\varphi((1+\delta_1)x)), \tag{8.54}$$

as $x \to +\infty$, for some $\delta_1 > 0$. Our following task is to choose W such that one can assure that the first term on the right-hand side of (8.54) has the same form as the second, with a possibly different positive constant δ_1.

To do this, for $s \in S$ and $t \in S_{d_0}$, let us write the Gaussian regression formula of $X(s)$ on the pair $(X(t), X'_{d_0}(t))$:

$$X(s) = a^t(s)X(t) + \langle b^t(s), X'_{d_0}(t)\rangle + \frac{\|t-s\|^2}{2} X^t(s), \tag{8.55}$$

where the regression coefficients $a^t(s)$ and $b^t(s)$ are, respectively, real-valued and $\mathbb{R}^{d_0}$-valued.

From now on we will only be interested in those $t \in W$. In this case, since W does not contain boundary points of $S \backslash S_{d_0}$, it follows that

$$\widehat{C}_{t,d_0} = N_{t,d_0} \quad \text{and} \quad \mathbb{I}_{X'_{d_0,N}(t) \in \widehat{C}_{t,d_0}} = 1.$$

Moreover, whenever $s \in S$ is close enough to t, it has to belong to S_{d_0}. For each t, $\{X^t(s) : s \in S\}$ is a *helix process* (compare with Section 7.3, where a similar normalization has been introduced).

Let us prove that, almost surely, the paths of the real-valued random field $\{X^t(s) : t \in W \cap S_{d_0}, s \in S\}$ are bounded. With no loss of generality (take a chart) we may assume that s varies in a closed ball in $\mathbb{R}^{d_0}$, containing t as an interior point, and remove the subscript d_0 for the derivative. Let us write the Taylor expansion of $X(\cdot)$ around the point t:

$$X(s) = X(t) + \langle s - t, X'(t) \rangle + \|s - t\|^2 \int_0^1 \langle v, X''((1-\alpha)t + \alpha s)(1-\alpha)v \rangle d\alpha, \tag{8.56}$$

where $v = (s - t)/\|s - t\|$ and $\langle \cdot, \cdot \rangle$ and $\| \cdot \|$ denote, respectively, Euclidean scalar product and norm in $\mathbb{R}^{d_0}$.

For each α, perform the Gaussian regression of $\langle v, X''((1-\alpha)t + \alpha s)(1-\alpha)v \rangle$ on the pair $(X(t), X'(t))$; that is,

$$\begin{aligned} \langle v, X''((1-\alpha)t + \alpha s)(1-\alpha)v \rangle &= \tilde{a}^t(s, \alpha, v) X(t) \\ &\quad + \langle \tilde{b}^t(s, \alpha, v), X'(t) \rangle + \tilde{X}^t(s, \alpha, v), \end{aligned} \tag{8.57}$$

where the notation is, as above, *mutatis mutandis*. From the regression formulas it follows that $\tilde{a}^t(s, \alpha)$ and $\tilde{b}^t(s, \alpha)$ are uniformly bounded, independent of s, t, α and v. Comparing (8.55) with (8.57), it follows that

$$X^t(s) = 2 \int_0^1 \tilde{X}^t(s, \alpha, v)(1 - \alpha) d\alpha$$

and the a.s. boundedness of $X^t(s), t \in W \cap S_{d_0}, s \in S$ follows. [The reader may check that for $d_0 > 1$, even though $\lim_{r \downarrow 0} X^t(t + rv)$ exists for each $v \in S^{d_0 - 1}$, it may depend on v, so that the function $X^t(\cdot)$ can be discontinuous at $s = t$.]

Now let us go back to formula (8.54). Conditionally on $X(t) = x$, $X'_{d_0}(t) = 0$, the event $\{M > x\}$ can be written as

$$\{X^t(s) > \beta^t(s)x \text{ for some s} \in S\},$$

where

$$\beta^t(s) = \frac{2(1 - a^t(s))}{\|t - s\|^2}. \tag{8.58}$$

Our next goal is to prove that if we can choose W in such a way that

$$\inf\{\beta^t(s) : t \in W \cap S_{d_0}, s \in S, s \neq t\} > 0, \tag{8.59}$$

we are done. In fact, apply the Cauchy–Schwarz inequality to the conditional expectation in (8.54). Under the conditioning, the elements of $X''_{d_0}(t)$ are the sums of affine functions of x with bounded coefficients plus centered Gaussian variables with bounded variances; hence, the absolute value of the conditional expectation is bounded by an expression of the form

$$\left(Q(t,x)\right)^{1/2}\left(\mathrm{P}\left(\sup_{s\in S\setminus\{t\}} \frac{X^t(s)}{\beta^t(s)} > x\right)\right)^{1/2}, \tag{8.60}$$

where $Q(t,x)$ is a polynomial in x of degree $2d_0$ with bounded coefficients. For each $t \in W \cap S_{d_0}$, the second factor in (8.60) is bounded by

$$\left(\mathrm{P}\left(\sup\left\{\frac{X^t(s)}{\beta^t(s)} : t \in W \cap S_{d_0}, s \in S, s \neq t\right\} > x\right)\right)^{1/2}.$$

Now, we apply to the bounded separable Gaussian process

$$\left\{\frac{X^t(s)}{\beta^t(s)} : t \in W \cap S_{d_0}, s \in S, s \neq t\right\}$$

the basic inequality (2.33), which gives the bound

$$\mathrm{P}\left(\sup\left\{\frac{X^t(s)}{\beta^t(s)} : t \in W \cap S_{d_0}, s \in S, s \neq t\right\} > x\right) \le C_2 \exp(-\delta_2 x^2)$$

for positive constants C_2 and δ_2 and any $x > 0$. Also, the same argument above for the density $p_{X(t),X'_{d_0}(t)}(x,0)$ shows that it is bounded by a constant times the standard normal density. To finish, it suffices to replace these bounds in the first term on the right-hand side of (8.54).

It remains to choose W for (8.59) to hold true. Consider the auxiliary process

$$Y(s) := \frac{X(s)}{\sqrt{r(s,s)}} \qquad s \in S. \tag{8.61}$$

Clearly, $\mathrm{Var}(Y(s)) = 1$ for all $s \in S$. We set

$$r^Y(s, s') := \mathrm{Cov}(Y(s), Y(s')) \qquad s, s' \in S.$$

Let us assume that $t \in S_v$. Since the function $s \rightsquigarrow \mathrm{Var}(X(s))$ attains its maximum value at $s = t$, it follows that $X(t)$ and $X'_{d_0}(t)$ are independent. This implies that in the regression formula (8.55) the coefficients are easily computed and $a^t(s) = r(s, t)$, which is strictly smaller than 1 if $s \neq t$, because of the nondegeneracy condition. Then

$$\beta^t(s) = \frac{2(1 - r(s,t))}{\|t - s\|^2} \geq \frac{2(1 - r^Y(s,t))}{\|t - s\|^2}. \tag{8.62}$$

Since $r^Y(s, s) = 1$ for every $s \in S$, the Taylor expansion of $r^Y(s, t)$ as a function of s, around $s = t$, takes the form

$$r^Y(s,t) = 1 + \langle s - t, r^Y_{20,d_0}(t,t)(s - t)\rangle + o(\|s - t\|^2), \tag{8.63}$$

where the notation is self-explanatory.

Also, using the fact that $\mathrm{Var}(Y(s)) = 1$ for $s \in S$, we easily obtain

$$-r^Y_{20,d_0}(t,t) = \mathrm{Var}(Y'_{d_0}(t)) = \mathrm{Var}(X'_{d_0}(t)), \tag{8.64}$$

where the last equality follows by differentiation in (8.61) and by setting $s = t$. Equation (8.64) implies that $-r^Y_{20,d_0}(t, t)$ is uniformly positive definite on $t \in S_v$, meaning that its minimum eigenvalue has a strictly positive lower bound. Based on (8.62) and (8.63), this already shows that

$$\inf\{\beta^t(s) : t \in S_v, s \in S, s \neq t\} > 0. \tag{8.65}$$

The foregoing argument also shows that

$$\inf\{-\tau(a^t)''_{d_0}(t)\tau : t \in S_v, \tau \in S^{d_0-1}, s \neq t\} > 0, \tag{8.66}$$

since whenever $t \in S_v$, one has $a^t(s) = r(s, t)$, so that

$$(a^t)''_{d_0}(t) = r_{20,d_0}.(t,t).$$

To end up, assume that there is no neighborhood W of S_v satisfying (8.59). In that case, using a compactness argument, one can find two convergent sequences $\{s_n\} \subset S$, $\{t_n\} \subset S_{d_0}$, $s_n \to s_0$, $t_n \to t_0 \in S_v$ such that

$$\beta^{t_n}(s_n) \to \ell \leq 0.$$

ℓ may be $-\infty$.

$t_0 \neq s_0$ is not possible, since it would imply that

$$\ell = 2\left(\frac{1 - a^{t_0}(s_0)}{\|t_0 - s_0\|^2}\right) = \beta^{t_0}(s_0),$$

which is strictly positive.
If $t_0 = s_0$, on differentiating in (8.55) with respect to s along S_{d_0}, we get

$$X'_{d_0}(s) = (a^t)'_{d_0}(s)X(t) + \langle (b^t)'_{d_0}(s), X'_{d_0}(t)\rangle + \frac{\partial_{d_0}}{\partial s}\frac{\|t - s\|^2}{2}X^t(s),$$

where $(a^t)'_{d_0}(s)$ is a column vector of size d_0 and $(b^t)'_{d_0}(s)$ is a $d_0 \times d_0$ matrix. Then one must have $a^t(t) = 1$, $(a^t)'_{d_0}(t) = 0$. Thus,

$$\beta^{t_n}(s_n) = -u_n^T (a^{t_0})''_{d_0}(t_0)u_n + o(1),$$

where $u_n := (s_n - t_n)/\|s_n - t_n\|$. Since $t_0 \in S_v$, we may apply (8.66), and the limit ℓ of $\beta^{t_n}(s_n)$ must be positive. □

A straightforward application of Theorem 8.10 is the following:

Corollary 8.11. *Under the hypotheses of Theorem 8.10, there exist positive constants C and δ such that for every $u > 0$,*

$$0 \leq \left|\int_u^{+\infty} p^E(x)\,dx - \mathrm{P}(M > u)\right| \leq \int_u^{+\infty} \overline{p}(x)\,dx - \mathrm{P}(M > u) \leq C\mathrm{P}(\xi > u),$$

where ξ is a centered Gaussian variable with variance $1 - \delta$.

The precise order of approximation of $\overline{p}(x) - p_M(x)$ or $p^E(x) - p_M(x)$ as $x \to +\infty$ remains in general an open problem, even if one only asks for the constants σ_d^2 and σ_E^2, respectively, which govern the second-order asymptotic approximation and which are defined by means of

$$\frac{1}{\sigma_d^2} := \lim_{x\to+\infty} -2x^{-2}\log\left[\overline{p}(x) - p_M(x)\right] \tag{8.67}$$

and

$$\frac{1}{\sigma_E^2} := \lim_{x\to+\infty} -2x^{-2}\log\left|p^E(x) - p_M(x)\right| \tag{8.68}$$

whenever these limits exist. In general, we are unable to compute the limits (8.67) or (8.68) or even to prove that they actually exist or differ. In the remainder of this chapter we give some lower bounds for the lim inf as $x \to +\infty$. This is already interesting since it gives some upper bounds for the speed of approximation for

$p_M(x)$ by means of either $\overline{p}(x)$ or $p^E(x)$. A more precise result is Theorem 8.15, where we are able to prove the existence of the limit and compute σ_d^2 when $\mathcal{X}$ is centered Gaussian, defined on a convex parameter set, and has a law that is invariant under isometries and translations of $\mathbb{R}^d$.

For the next theorem we need an additional condition on the parameter set S. For S verifying hypothesis (A1), we define

$$\kappa(S) = \sup_{0\le j\le d_0} \sup_{t\in S_j} \sup_{s\in S, s\neq t} \frac{\operatorname{dist}\big((t-s), \mathcal{C}_{t,j}\big)}{\|s-t\|^2}, \tag{8.69}$$

where dist is the Euclidean distance in $\mathbb{R}^d$.

In Exercise 8.2 it is proved that $\kappa(S) < \infty$ for various relevant classes of parameter sets:

- S is convex [in which case, in fact, $\kappa(S) = 0$].
- S is a C^3 manifold, with or without boundary.
- S verifies a certain kind of local convexity condition, which is described precisely in the exercise.

However, $\kappa(S) < \infty$ can fail in general. A simple example showing what is going on is the following: Take an orthonormal basis of $\mathbb{R}^2$ and set

$$S = \{(\lambda, 0) : 0 \le \lambda \le 1\} \cup \{(\mu\cos\theta, \mu\sin\theta) : 0 \le \mu \le 1\},$$

where $0 < \theta < \pi$; that is, S is the boundary of an angle of size θ. One easily checks that $\kappa(S) = +\infty$.

Theorem 8.12. *Let $\mathcal{X}$ be a stochastic process on S satisfying conditions (A1) to (A5) in Section 7.1. Suppose, in addition, that* $\operatorname{Var}(X(t)) = 1$ *for all $t \in S$ and that $\kappa(S) < +\infty$. Then*

$$\liminf_{x\to+\infty} -2x^{-2}\log\left[\overline{p}(x) - p_M(x)\right] \ge 1 + \inf_{t\in S} \frac{1}{\sigma_t^2 + \overline{\lambda}(t)\kappa_t^2} \tag{8.70}$$

with

$$\sigma_t^2 := \sup_{s\in S\setminus\{t\}} \frac{\operatorname{Var}\big(X(s)\big|X(t), X'(t)\big)}{(1-r(s,t))^2}$$

and

$$\kappa_t := \sup_{s\in S\setminus\{t\}} \frac{\operatorname{dist}\Big(-\Lambda_t^{-1} r_{01}(s,t), \mathcal{C}_{t,j}\Big)}{1-r(s,t)}, \tag{8.71}$$

where $\Lambda_t := \mathrm{Var}(X'(t))$, $\overline{\lambda}(t)$ *is the maximum eigenvalue of* Λ_t, *and in* (8.71), j *is such that* $t \in S_j$ $(j = 0, 1, \ldots, d_0)$. *The quantity on the right-hand side of* (8.70) *is strictly bigger than* 1.

Remark. In formula (8.70) it may happen that the denominator on the right-hand side is identically zero, in which case we put $+\infty$ for the infimum. This is the case of the one-parameter process $X(t) = \xi \cos t + \eta \sin t$, where ξ and η are independent standard normal random variables (the sine–cosine process), and S is an interval having length strictly smaller than π.

Proof. First, let us prove that $\sup_{t \in S} \kappa_t < \infty$. For each $t \in S$, let us write the Taylor expansions

$$\begin{aligned} r_{01}(s,t) &= r_{01}(t,t) + r_{11}(t,t)(s-t) + O(\|s-t\|^2) \\ &= \Lambda_t(s-t) + O(\|s-t\|^2), \end{aligned}$$

where O is uniform on $s, t \in S$, and

$$1 - r(s,t) = (s-t)^{\mathrm{T}} \Lambda_t (s-t) + O(\|s-t\|^2) \geq L_2 \|s-t\|^2,$$

where L_2 is a positive constant. It follows that for $s \in S, t \in S_j, s \neq t$, one has

$$\frac{\operatorname{dist}\left(-\Lambda_t^{-1} r_{01}(s,t), \mathcal{C}_{t,j}\right)}{1 - r(s,t)} \leq L_3 \frac{\operatorname{dist}\left((t-s), \mathcal{C}_{t,j}\right)}{\|s-t\|^2} + L_4, \tag{8.72}$$

where L_3 and L_4 are positive constants. So

$$\frac{\operatorname{dist}\left(-\Lambda_t^{-1} r_{01}(s,t), \mathcal{C}_{t,j}\right)}{1 - r(s,t)} \leq L_3 \kappa(S) + L_4,$$

which implies that $\sup_{t \in S} \kappa_t < \infty$.

With the same notations as in the proof of Theorem 8.10, using (7.1) and (8.19), one has

$$\overline{p}(x) - p_M(x) = \varphi(x) B(x), \tag{8.73}$$

where

$$\begin{aligned} B(x) := \sum_{t \in S_0} \mathrm{E}\left(\mathbb{I}_{X_t'(t) \in \widehat{\mathcal{C}}_{t,0}} \mathbb{I}_{M > x} \middle| X(t) = x\right) + \sum_{j=1}^{d_0} (2\pi)^{-j/2} \\ \cdot \int_{S_j} \mathrm{E}\left(|\det(X_j''(t))| \mathbb{I}_{X_{j,N}'(t) \in \widehat{\mathcal{C}}_{t,j}} \mathbb{I}_{M > x} \middle| X(t) = x, X_j'(t) = 0\right) \\ \times [\det(\mathrm{Var}(X_j'(t)))]^{-1/2} \sigma_j(dt). \end{aligned}$$

Proceeding in a similar way to that of the proof of Theorem 8.10, an application of the Hölder inequality to the conditional expectation in each term on the right-hand side of (8.73) shows that the desired result will follow as soon as we prove that

$$\liminf_{x\to+\infty} -2x^{-2}\log \mathrm{P}\big(\{X'_{j,N} \in \widehat{C}_{t,j}\} \cap \{M > x\} \big| X(t) = x, X'_j(t) = 0\big) \geq \frac{1}{\sigma_t^2 + \overline{\lambda}(t)\kappa_t^2} \tag{8.74}$$

for each $j = 0, 1, \ldots, d_0$, where the lim inf has some uniformity in t.

Let us write the Gaussian regression of $X(s)$ on the pair $(X(t), X'(t))$:

$$X(s) = a^t(s)X(t) + \langle b^t(s), X'(t)\rangle + R^t(s).$$

Since $X(t)$ and $X'(t)$ are independent, one easily computes

$$a^t(s) = r(s,t)$$
$$b^t(s) = \Lambda_t^{-1} r_{01}(s,t).$$

Hence, conditionally on $X(t) = x$, $X'_j(t) = 0$, the events

$$\{M > x\} \quad \text{and} \quad \{R^t(s) > (1 - r(s,t))x - r_{01}^{\mathrm{T}}(s,t)\Lambda_t^{-1}X'_{j,N}(t) \text{ for some } s \in S\}$$

coincide. Denote by $(X'_{j,N}(t)|X'_j(t) = 0)$ the regression of $X'_{j,N}(t)$ on $X'_j(t) = 0$. So the probability in (8.74) can be written as

$$\int_{\widehat{C}_{t,j}} \mathrm{P}\left\{\zeta^t(s) > x - \frac{r_{01}^{\mathrm{T}}(s,t)\Lambda_t^{-1}x'}{1 - r(s,t)} \text{ for some } s \in S\right\} p_{X'_{j,N}(t)|X'_j(t)=0}(x')\,dx', \tag{8.75}$$

where

$$\zeta^t(s) := \frac{R^t(s)}{1 - r(s,t)}$$

and dx' is the Lebesgue measure on $N_{t,j}$. Recall that $\widehat{C}_{t,j} \subset N_{t,j}$.

If $-\Lambda_t^{-1} r_{01}(s,t) \in C_{t,j}$, one has

$$-r_{01}^{\mathrm{T}}(s,t)\Lambda_t^{-1}x' \geq 0$$

for every $x' \in \widehat{C}_{t,j}$, because of the definition of $\widehat{C}_{t,j}$. If $-\Lambda_t^{-1} r_{01}(s,t) \notin C_{t,j}$, since $C_{t,j}$ is a closed convex cone, we can write

$$-\Lambda_t^{-1} r_{01}(s,t) = z' + z''$$

with $z' \in \mathcal{C}_{t,j}$, $z' \perp z''$, and $\|z''\| = \mathrm{dist}(-\Lambda_t^{-1} r_{01}(s,t), \mathcal{C}_{t,j})$. So, if $x' \in \widehat{\mathcal{C}}_{t,j}$,

$$\frac{-r_{01}^{\mathrm{T}}(s,t)\Lambda_t^{-1}x'}{1-r(s,t)} = \frac{z'^{\mathrm{T}}x' + z''^{\mathrm{T}}x'}{1-r(s,t)} \geq -\kappa_t \|x'\|,$$

using the fact that $z'^{\mathrm{T}}x' \geq 0$ and the Cauchy–Schwarz inequality. It follows that in any case, if $x' \in \widehat{\mathrm{C}}_{t,j}$, the expression in (8.75) is bounded by

$$\int_{\widehat{\mathcal{C}}_{t,j}} \mathrm{P}\Big(\zeta^t(s) > x - \kappa_t\|x'\| \text{ for some } s \in S\Big) p_{X'_{j,N}(t)|X'_j(t)=0}(x')\,dx'. \tag{8.76}$$

To obtain a bound for the probability in the integrand of (8.76), we will use the classical inequality for the tail of the distribution of the supremum of a Gaussian process with bounded paths.

The Gaussian process $(s,t)) \rightsquigarrow \zeta^t(s)$, defined on $(S \times S)\backslash\{s = t\}$, has continuous paths. As the pair (s,t) approaches the diagonal of $S \times S$, $\zeta^t(s)$ may not have a limit, but a.s. one can prove that it is bounded using an argument similar to the one used in the proof of Theorem 8.10 for helix processes, that is, Taylor expansion followed by Gaussian regression.

We set

$$m^t(s) := \mathrm{E}(\zeta^t(s))(s \neq t)$$

$$m := \sup_{s,t \in S, s \neq t} |m^t(s)|$$

$$\mu := \mathrm{E}\left(|\sup_{s,t \in S, s \neq t} [\zeta^t(s) - m^t(s)]|\right).$$

The a.s. boundedness of the paths of $\zeta^t(s)$ implies that $m < \infty$ and $\mu < \infty$. Applying the basic inequality (2.25) of Theorem 2.9 to the centered process $s \rightsquigarrow \zeta^t(s) - m^t(s)$ defined on $S\backslash\{t\}$, we get whenever $x - \kappa_t\|x'\| - m - \mu > 0$:

$$\begin{aligned}
&\mathrm{P}\{\zeta^t(s) > x - \kappa_t\|x'\| \text{ for some } s \in S\} \\
&\leq \mathrm{P}\{\zeta^t(s) - m^t(s) > x - \kappa_t\|x'\| - m \text{ for some } s \in S\} \\
&\leq 2\exp\left(-\frac{(x - \kappa_t\|x'\| - m - \mu)^2}{2\sigma_t^2}\right).
\end{aligned}$$

The Gaussian density in the integrand of (8.76) is bounded by

$$(2\pi\underline{\lambda}_j(t))^{(j-d)/2} \exp\frac{\|x' - m'_{j,N}(t)\|^2}{2\overline{\lambda}_j(t)},$$

where $\underline{\lambda}_j(t)$ and $\overline{\lambda}_j(t)$ are, respectively, the minimum and maximum eigenvalues of $\mathrm{Var}(X'_{j,N}(t)|X'_j(t))$ and $m'_{j,N}(t)$ is the conditional expectation

$E(X'_{j,N}(t)|X'_j(t)=0)$. Notice that $\underline{\lambda}_j(t)$, $\overline{\lambda}_j(t)$ and $m'_{j,N}(t)$ are bounded, $\underline{\lambda}_j(t)$ is bounded below by a positive constant, and $\overline{\lambda}_j(t) \le \overline{\lambda}(t)$. Substituting in (8.76), we have the bound

$$\begin{aligned}
&P\big(\{X'_{j,N} \in \widehat{C}_{t,j}\} \cap \{M > x\} \big| X(t) = x, X'_j(t) = 0\big) \\
&\le (2\pi \underline{\lambda}_j(t))^{(j-d)/2} 2 \int_{\widehat{C}_{t,j} \cap \{x - \kappa_t \|x'\| - m - \mu > 0\}} \\
&\quad \times \exp\left(-\frac{(x - \kappa_t \|x'\| - m - \mu)^2}{2\sigma_t^2} + \frac{\|x' - m'_{j,N}(t)\|^2}{2\overline{\lambda}(t)} \right) dx' \\
&\quad + P\left(\|X'_{j,N}(t)|X'_j(t) = 0\| \ge \frac{x - m - \mu}{\kappa_t} \right),
\end{aligned} \tag{8.77}$$

where it is understood that the second term on the right-hand side vanishes if $\kappa_t = 0$. Let us consider the first term on the right-hand side of (8.77). We have

$$\begin{aligned}
&\frac{(x - \kappa_t \|x'\| - m - \mu)^2}{2\sigma_t^2} + \frac{\|x' - m'_{j,N}(t)\|^2}{2\overline{\lambda}(t)} \\
&\ge \frac{(x - \kappa_t \|x'\| - m - \mu)^2}{2\sigma_t^2} + \frac{(\|x'\| - \|m'_{j,N}(t)\|)^2}{2\overline{\lambda}(t)} \\
&= \left[A(t)\|x'\| + B(t)(x - m - \mu) + C(t)\right]^2 + \frac{(x - m - \mu - \kappa_t \|m'_{j,N}(t)\|)^2}{2\sigma_t^2 + 2\overline{\lambda}(t)\kappa_t^2},
\end{aligned}$$

where the last inequality is obtained after some algebra, $A(t)$, $B(t)$, and $C(t)$ are bounded functions and $A(t)$ is bounded below by a positive constant.

So the first term on the right-hand side of (8.77) is bounded by

$$\begin{aligned}
&2(2\pi \underline{\lambda}_j)^{(j-d)/2} \exp\left(-\frac{(x - m - \mu - \kappa_t \|m'_{j,N}(t)\|)^2}{2\sigma_t^2 + 2\overline{\lambda}(t)\kappa_t^2} \right) \\
&\cdot \int_{R^{d-j}} \exp\left[-\left(A(t)\|x'\| + B(t)(x - m - \mu) + C(t)\right)^2 \right] dx' \\
&\le L|x|^{d-j-1} \exp\left(-\frac{(x - m - \mu - \kappa_t \|m'_{j,N}(t)\|)^2}{2\sigma_t^2 + 2\overline{\lambda}(t)\kappa_t^2} \right),
\end{aligned} \tag{8.78}$$

where L is a constant. The last inequality follows easily using polar coordinates.

Consider next the second term on the right-hand side of (8.77). Using the form of the conditional density $p_{X'_{j,N}(t)|X'_j(t)=0}(x')$, it follows that it is bounded by

$$\mathrm{P}\left\{\|(X'_{j,N}(t)|X'_j(t)=0)-m'_{j,N}(t)\|\geq\frac{x-m-\mu-\kappa_t\|m'_{j,N}(t)\|}{\kappa_t}\right\}$$

$$\leq L_1|x|^{d-j-2}\exp\left(-\frac{(x-m-\mu-\kappa_t\|m'_{j,N}(t)\|)^2}{2\overline{\lambda}(t)\kappa_t^2}\right), \tag{8.79}$$

where L_1 is a constant. Putting (8.78) and (8.79) together with (8.77), we obtain (8.74). □

The following two corollaries are straightforward consequences of Theorem 8.12:

Corollary 8.13. *Under the hypotheses of Theorem 8.12, one has*

$$\liminf_{x\to+\infty}-2x^{-2}\log|p^E(x)-p_M(x)|\geq 1+\inf_{t\in S}\frac{1}{\sigma_t^2+\overline{\lambda}(t)\kappa_t^2}.$$

Corollary 8.14. *Let $\mathcal{X}$ be a stochastic process on S satisfying conditions (A1) to (A5). Suppose, in addition, that* $\mathrm{E}(X(t))=0$, $\mathrm{E}(X^2(t))=1$, $\mathrm{Var}(X'(t)=I_d$ *for all t. Then*

$$\liminf_{u\to+\infty}-2u^{-2}\log\left|\mathrm{P}(M>u)-\int_u^{+\infty}p^E(x)\,dx\right|\geq 1+\inf_{t\in S}\frac{1}{\sigma_t^2+\kappa_t^2}$$

and

$$p^E(x)=\left[\sum_{j=0}^{d_0}(-1)^j(2\pi)^{-j/2}g_j\overline{H}_j(x)\right]\varphi(x),$$

where g_j is given by (8.26) and $\overline{H}_j(x)$ has been defined in Section 8.4.

The proof follows directly from Theorem 8.12, the definition of $p^E(x)$, and the calculation of

$$\mathrm{E}\big(|\det(X''_j(t))|\,|X(t)=x,X'_j(t)=0\big), \tag{8.80}$$

which is detailed in Exercise 8.6.

8.6. EXAMPLES

1. A simple application of Theorem 8.10 is the following. Let $\mathcal{X}$ be a one-parameter real-valued centered Gaussian process with regular paths, defined on the interval $[0, T]$ and satisfying an adequate nondegeneracy condition. Assume that the variance $v(t)$ has a unique maximum, say 1 at the interior point t_0, and $k = \min\{j : v^{(2j)}(t_0) \neq 0\} < \infty$. Notice that $v^{(2k)}(t_0) < 0$. Then one can obtain the equivalent of $p_M(x)$ as $x \to \infty$, which is given by

$$p_M(x) \simeq \frac{1 - v''(t_0)/2}{kC_k^{1/k}} \mathrm{E}\left(|\xi|^{(1/2k)-1}\right) x^{1-1/k}\varphi(x), \tag{8.81}$$

where ξ is a standard normal random variable and

$$C_k = -\frac{1}{(2k)!}v^{(2k)}(t_0) + \frac{1}{4}[v''(t_0)]^2 \mathbb{1}_{k=2}.$$

The proof is a direct application of the Laplace method, and the reader can prove it in Exercise 8.3. Integrating the density from u to $+\infty$, one gets the corresponding bound for $\mathrm{P}\{M > u\}$. In Piterbarg (1996a, Section 8) one can find more general results concerning the distribution of the supremum of Gaussian processes having a variance that takes its maximum value in exactly one point.

2. Let the process $\mathcal{X}$ be centered and satisfy conditions (A1) to (A5). Assume that the law of the process is isotropic and stationary, so that the covariance has the form (8.21) and verifies the regularity condition of Section 8.4. To simplify the computations somewhat, with no loss of generality, we add the normalization $\rho' = \rho'(0) = -\frac{1}{2}$. One can easily check that

$$\sigma_t^2 = \sup_{s \in S\setminus\{t\}} \frac{1 - \rho^2(\|s-t\|^2) - 4\rho'^2(\|s-t\|^2)\|s-t\|^2}{[1 - \rho(\|s-t\|^2)]^2}. \tag{8.82}$$

Furthermore, if

$$\rho'(x) \leq 0 \qquad \text{for } x \geq 0, \tag{8.83}$$

one can show that the sup in (8.82) is attained as $\|s - t\| \to 0$ and is independent of t. Its value is

$$\sigma_t^2 = 12\rho'' - 1. \tag{8.84}$$

The proof is elementary (see Exercise 8.4).

Let S be a convex set. For $t \in S_j$, $s \in S$,

$$\mathrm{dist}\big(-r_{01}(s,t), \mathcal{C}_{t,j}\big) = \mathrm{dist}\big(-2\rho'(\|s-t\|^2)(t-s), \mathcal{C}_{t,j}\big). \tag{8.85}$$

The convexity of S implies that $(t-s) \in \mathcal{C}_{t,j}$. Since $\mathcal{C}_{t,j}$ is a convex cone and $-2\rho'(\|s-t\|^2) \geq 0$, one can conclude that $-r_{01}(s,t) \in \mathcal{C}_{t,j}$, so that the distance in (8.85) is equal to zero. Hence,

$$\kappa_t = 0 \qquad \text{for every } t \in S$$

and an application of Theorem 8.12 gives the inequality

$$\liminf_{x\to+\infty} -\frac{2}{x^2}\log\left[\overline{p}(x) - p_M(x)\right] \geq 1 + \frac{1}{12\rho'' - 1}. \tag{8.86}$$

A direct consequence is that the same inequality holds true when replacing $\overline{p}(x) - p_M(x)$ by $|p^E(x) - p_M(x)|$ in (8.86). The bound for the EPC method has been obtained by Taylor and Adler (2003) using other methods.

Our next theorem improves (8.86). In fact, under the same hypotheses, it says that lim inf is an ordinary limit and the sign $\geq$ is an equality sign.

Theorem 8.15. *Assume that $\mathcal{X}$ is centered, satisfies hypotheses (A1) to (A5), and that the covariance has the form (8.21) with $\rho'(0) = -\frac{1}{2}$, $\rho'(x) \leq 0$ for $x \geq 0$. Let S be a convex set and set $d_0 = d \geq 1$. Then*

$$\lim_{x\to+\infty} -\frac{2}{x^2}\log\left[\overline{p}(x) - p_M(x)\right] = 1 + \frac{1}{12\rho'' - 1}. \tag{8.87}$$

Remark. Notice that since S is convex, the added hypothesis that the maximum dimension d_0 such that S_j is not empty is equal to d is not an actual restriction. In fact, in this case, S is a subset of some affine manifold having dimension d_0, that is, the smallest one containing S, and a simple change of parameter allows us to consider S as a subset of $\mathbb{R}^{d_0}$.

Proof. In view of (8.86), it suffices to prove that

$$\limsup_{x\to+\infty} -\frac{2}{x^2}\log\left[\overline{p}(x) - p_M(x)\right] \leq 1 + \frac{1}{12\rho'' - 1}. \tag{8.88}$$

Using (7.1) and the definition of $\overline{p}(x)$ given by (8.19), one has the inequality

$$\overline{p}(x) - p_M(x) \geq (2\pi)^{-d/2}\varphi(x)\int_{S_d} \mathrm{E}\big(|\det(X''(t))|\mathbb{I}_{M>x}\big| \\ X(t) = x, X'(t) = 0\big)\sigma_d(dt), \tag{8.89}$$

where the right-hand side is a lower bound since it contains only the term corresponding to the largest dimension. We have already replaced the density $p_{X(t),X'(t)}(x,0)$ by its explicit expression using the law of the process. Under the condition $\{X(t) = x, X'(t) = 0\}$, if $v_0^{\mathrm{T}} X''(t) v_0 > 0$ for some $v_0 \in \mathcal{S}^{d-1}$, a Taylor expansion around the point t implies that $M > x$. It follows that

$$\mathrm{E}\big(|\det(X''(t))|\,\mathbb{I}_{M>x}\big|X(t) = x, X'(t) = 0\big)$$
$$\geq \mathrm{E}\Big(|\det(X''(t))|\,\mathbb{I}_{\sup_{v\in\mathcal{S}^{d-1}} v^{\mathrm{T}} X''(t) v > 0}\big|X(t) = x, X'(t) = 0\Big). \quad (8.90)$$

We now apply Lemma 8.5, which describes the conditional distribution of $X''(t)$ given $X(t) = x$, $X'(t) = 0$. Using the notation of this lemma, we may write the right-hand side of (8.90) as

$$\mathrm{E}\left(|\det(Z - x\cdot \mathrm{Id})|\,\mathbb{I}_{\sup_{v\in\mathcal{S}^{d-1}} v^{\mathrm{T}} Z v > x}\right),$$

which is obviously bounded below by

$$\mathrm{E}\big(|\det(Z - x\cdot \mathrm{Id})|\,\mathbb{I}_{Z_{11}>x}\big) = \int_x^{+\infty} \mathrm{E}\big(|\det(Z - x\cdot \mathrm{Id})|\big|Z_{11} = y\big)$$
$$(2\pi)^{-1/2}\sigma^{-1}\exp\left(-\frac{y^2}{2\sigma^2}\right)dy, \quad (8.91)$$

where $\sigma^2 := \mathrm{Var}(Z_{11}) = 12\rho'' - 1$. The conditional distribution of Z given $Z_{11} = y$ is easily deduced from Lemma 5. It can be represented by the random $d \times d$ real symmetric matrix

$$\widetilde{Z} := \begin{pmatrix} y & Z_{12} & \cdots & \cdots & Z_{1d} \\ & \xi_2 + \alpha y & Z_{23} & \cdots & Z_{2d} \\ & & \ddots & & \\ & & & & \xi_d + \alpha y \end{pmatrix},$$

where the random variables $\{\xi_2, \ldots, \xi_d, Z_{ik}, 1 \leq i < k \leq d\}$ are independent centered Gaussian with

$$\mathrm{Var}(Z_{ik}) = 4\rho''(1 \leq i < k \leq d), \qquad \mathrm{Var}(\xi_i) = \frac{16\rho''(8\rho'' - 1)}{12\rho'' - 1} \quad (i = 2, \ldots, d),$$
$$\alpha = \frac{4\rho'' - 1}{12\rho'' - 1}.$$

Notice that $0 < \alpha < 1$.

Now choose $\alpha_0 > 0$ such that $(1+\alpha_0)\alpha < 1$. The expansion of $\det(\widetilde{Z} - xId)$ shows that if $x(1+\alpha_0) \le y \le x(1+\alpha_0)+1$ and x is large enough, then

$$\mathrm{E}\big(|\det(\widetilde{Z} - xId)|\big) \ge L\alpha_0(1-\alpha(1+\alpha_0))^{d-1}x^d,$$

where L is a positive constant. This implies that

$$\frac{1}{\sqrt{2\pi}\sigma}\int_x^{+\infty} \exp\left(-\frac{y^2}{2\sigma^2}\right) \mathrm{E}\big(|\det(\widetilde{Z} - x\cdot \mathrm{Id})|\big)\,dy$$
$$\ge \frac{L}{\sqrt{2\pi}\sigma}\int_{x(1+\alpha_0)}^{x(1+\alpha_0)+1} \exp\left(-\frac{y^2}{2\sigma^2}\right)\alpha_0(1-\alpha(1+\alpha_0))^{d-1}x^d\,dy$$

for x large enough. Based on (8.89), (8.90), and (8.91), we conclude that for x large enough,

$$\overline{p}(x) - p_M(x) \ge L_1 x^d \exp\left[-\frac{x^2}{2} - \frac{(x(1+\alpha_0)+1)^2}{2\sigma^2}\right]$$

for some new positive constant L_1. Since α_0 can be chosen arbitrarily small, this implies (8.88). □

3. Consider the same processes as those of Example 2, but now defined on the nonconvex set $\{a \le \|t\| \le b\}$, $0 < a < b$. The same calculations as above show that $\kappa_t = 0$ if $a < \|t\| \le b$ and

$$\kappa_t = \max\left\{\sup_{z\in[2a,a+b]} \frac{-2\rho'(z^2)z}{1-\rho(z^2)}, \sup_{\theta\in[0,\pi]} \frac{-2a\rho'(2a^2(1-\cos\theta))(1-\cos\theta)}{1-\rho(2a^2(1-\cos\theta))}\right\}$$

for $\|t\| = a$.

4. Let us keep the same hypotheses as in Example 2 but without assuming that the covariance is decreasing as in (8.83). The variance is still given by (8.82), but κ_t is not necessarily equal to zero. More precisely, relation (8.85) shows that

$$\kappa_t \le \sup_{s\in S\setminus\{t\}} 2\frac{\rho'(\|s-t\|^2)^+\|s-t\|}{1-\rho(\|s-t\|^2)}.$$

The normalization $\rho' = -\frac{1}{2}$ implies that the process $\mathcal{X}$ is *identity speed*: that is, $\mathrm{Var}(X'(t)) = I_d$, so that $\overline{\lambda}(t) = 1$. An application of Theorem 8.12 gives

$$\liminf_{x\to+\infty} -\frac{2}{x^2}\log\left[\overline{p}(x) - p_M(x)\right] \ge 1 + 1/Z_\Delta, \tag{8.92}$$

where

$$Z_\Delta := \sup_{z\in(0,\Delta]} \frac{1-\rho^2(z^2)-4\rho'^2(z^2)z^2}{[1-\rho(z^2)]^2} + \max_{z\in(0,\Delta]} \frac{4\left[\rho'(z^2)^+z\right]^2}{[1-\rho(z^2)]^2},$$

and Δ is the diameter of S.

Let us show by a numerical example that all these quantities can actually be computed. Suppose that $d = 2$ and let us consider the covariance $r(s,t)$ defined as follows: τ is the Fourier transform of the probability measure on $\mathbb{R}^2$ having the density

$$\frac{1}{\pi}\exp\left(-\frac{\|z\|^2}{2}\right)\left(1-\exp-\frac{\|z\|^2}{2}\right).$$

We set

$$r(s,t) := \tau\left(\sqrt{\tfrac{2}{3}}(s-t)\right).$$

One easily verifies that with our previous notation,

$$r(s,t) = \rho(\|s-t\|^2) = \text{ with } \rho(z) = 2\left(e^{-z/3} - \frac{e^{-z/6}}{2}\right) \qquad z \geq 0.$$

Check that our conditions are satisfied [the change of scale has been chosen so that $\rho'(0) = -\frac{1}{2}$].

Numerically, we find that

$$\frac{2\rho'(z^2)^+z}{1-\rho(z^2)}$$

vanishes for z in the interval $[0; 2.884\ldots]$ and attains its maximum value: $0.0689\ldots$ for $z = 3.7\ldots$ (Figure 8.1). On the other hand,

$$\sup_{z\in(0,\Delta]} \frac{1-\rho^2(z^2)-4\rho'^2(z^2)z^2}{[1-\rho(z^2)]^2}$$

is always attained at $Z = 0^+$ and takes the constant value $\frac{4}{3}$.

As a consequence, for a diameter of S smaller than 2.884, the bound for the exponent $1 + 1/Z_\Delta$ takes the value $7/4 = 1.75$, and takes the minimum value of 1.7473 for a diameter greater or equal to $3.7\ldots$.

5. Suppose that:

- The process $\mathcal{X} = \{X(t) : t \in \mathbb{R}^d\}$ is stationary with covariance having the form

$$\Gamma(t_1,\ldots,t_d) = \prod_{i=1,\ldots,d} \Gamma_i(t_i),$$

where $\Gamma_1,\ldots,\Gamma_d$ are d covariance functions on $\mathbb{R}$ which are monotone, positive on $[0,+\infty)$, and of class $\mathcal{C}^4$.

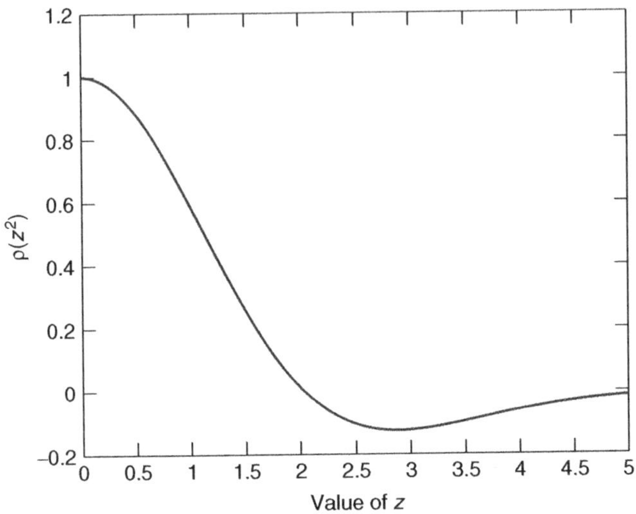

Figure 8.1. Representation of the function $\rho(z^2)$.

- S is a rectangle:

$$S = \prod_{i=1,\dots,d} [a_i, b_i] \qquad a_i < b_i.$$

Then, adding an appropriate nondegeneracy condition, conditions (A2) to (A5) are fulfilled and Theorem 8.12 applies.

Clearly,

$$-r_{0,1}(s,t) = \begin{bmatrix} \Gamma_1'(s_1 - t_1)\Gamma_2(s_2 - t_2)\cdots\Gamma_d(s_d - t_d) \\ \vdots \\ \Gamma_1(s_1 - t_1)\cdots\Gamma_{d-1}(s_{d-1} - t_{d-1})\Gamma_d'(s_d - t_d) \end{bmatrix}$$

belongs to $\mathcal{C}_{t,j}$ for every $s \in S$. As a consequence, $\kappa_t = 0$ for all $t \in S$. On the other hand, standard regressions formulas show that

$$\frac{\operatorname{Var}\big(X(s)\big|X(t), X'(t)\big)}{(1 - r(s,t))^2} = \frac{1 - \Gamma_1^2\cdots\Gamma_d^2 - \Gamma_1'^2\Gamma_2^2\cdots\Gamma_d^2 - \cdots - \Gamma_1^2\cdots\Gamma_{d-1}^2\Gamma_d'^2}{(1 - \Gamma_1\cdots\Gamma_d)^2},$$

where Γ_i stands for $\Gamma_i(s_i - t_i)$. Computation and maximization of σ_t^2 should be performed numerically in each particular case.

EXERCISES

8.1. For $n = 1, 2, \dots$, let G_n be an $n \times n$ GOE random matrix (see Section 8.4). Define

$$D_n(\lambda) = (-2)^n \mathrm{E}\big(\det(G_n - \lambda I_n)\big) \qquad \lambda \in \mathbb{R}$$

(**a**) Prove that (1), $D_1(\lambda) = 2\lambda$, and (2), $D_n'(\lambda) = 2n D_{n-1}(\lambda)(n = 2, 3, \dots)$.

(**b**) Prove that

$$D_n(0) = (-1)^{n/2} \frac{n!}{(n/2)!}$$

if n is even and $D_n(0) = 0$ if n is odd.

(**c**) Using the results of parts (a) and (b) and the fact that the Hermite polynomials satisfy the same relations, conclude that

$$D_n(\lambda) = H_n(\lambda) \ \ \forall \lambda \in \mathrm{R}, \qquad n = 1, 2, \dots.$$

(**d**) Prove that under the conditions of Theorem 8.8,

$$p^E(x) = \varphi(x) \left\{ \sum_{t \in S_0} \widehat{\sigma}_0(t) + \sum_{j=1}^{d_0} \left(\frac{|\rho'|}{\pi} \right)^{j/2} \overline{H}_j(x) g_j \right\}.$$

Hint: Mimic the proof of Theorem 8.8 and take into account part (c).

8.2. Let $\kappa(S)$ be defined in (8.69). Prove that:

(**a**) If S is convex, then $\kappa(S) = 0$.

(**b**) If S is a C^3-manifold, with or without boundary, then $\kappa(S) < \infty$.

(**c**) Assume that S verifies the following condition: For every $t \in S$ there exists an open neighborhood V of t in $\mathbb{R}^d$ and a C^3-diffeomorphism $\psi : V \to B(0, r)$ [where $B(0, r)$ denotes the open ball in $\mathbb{R}^d$ centered at 0 and having radius r, $r > 0$] such that $\psi(V \cap S) = C \cap B(0, r)$, where C is a convex cone. Then $\kappa(S) < \infty$.

8.3. Prove the equivalence (8.81) for the one-parameter processes given in Example 1 of Section 8.5.

8.4. Prove formula (8.84) under the conditions or Example 2 of Section 8.5.

8.5. Perform the details of the computations appearing in Examples 3, 4, and 5 of Section 8.5.

8.6. [*Computation of* $\mathrm{E}\big(\det(X''(t)) \big| X(t) = x, X'(t) = 0\big)$] Let $X(t)$, $t \in \mathbb{R}^d$ be a Gaussian field with real values such that:

- It has $\mathcal{C}^2$-paths.
- $\mathrm{E}(X(t)) = 0$.
- $\mathrm{Var}(X(t))$ is constant and nonsingular. With no loss of generality, we suppose that $\mathrm{Var}(X(t)) = I_d$.

(a) Prove that $X(t)$ and $X'(t)$ are independent.

(b) Prove that $\mathrm{E}(X_{ij}X_{kl})$ is a symmetric function of (i, j, k, l) (it is invariant by permutation).

(c) Admit or prove the following classical result: Let $Y_1, \ldots, Y_n$ be n centered jointly Gaussian variables; then:

- If $n = 2m + 1$; $\mathrm{E}\big(Y_1 \ldots Y_n\big) = 0$.
- If $n = 2m$,

$$\mathrm{E}\big(Y_1 \cdots Y_n\big) = \sum \mathrm{E}\big(Y_{i_1}Y_{i_2}\big) \cdots \mathrm{E}\big(Y_{i_{2m-1}}Y_{i_{2m}}\big),$$

where the sum is over the $(2m)!/m!2^m$ ways of grouping the $2m$ variables pairwise.

Hint: Compute moments using the characteristic function, or see Adler (1981, pp. 108–109).

(d) Let Δ be a $n \times n$ centered Gaussian matrix with entries Δ_{ij}. Suppose that

$$\mathrm{E}\big(\Delta_{ij}\Delta_{kl}\big) = \mathcal{E}(i, j, k, l) - \mathbb{I}_{i=j}\mathbb{I}_{k=l},$$

where $\mathcal{E}$ is a symmetric function of (i, j, k, l). Prove that

- If n is odd, $\mathrm{E}\big(\det(\Delta)\big) = 0$.
- If $n = 2m$,

$$\mathrm{E}\big(\det(\Delta)\big) = \frac{(-1)^m(2m)!}{m!2^m}.$$

Hint: Develop the determinant using permutations and a signature, and use the formula above to see that the part corresponding to $\mathcal{E}$ vanishes.

(e) Let $D = \Delta - xI_n$, where Δ is as in part (d). Prove that

$$\mathrm{E}(\det(D)) = (-1)^n \sum_{j=0}^{[n/2]} (-1)^j \binom{n}{2j} \frac{(2j)!}{j!2^j} x^{n-2j} = (-1)^n \overline{H}_n(x).$$

(f) Conclude.

This result, due to Delmas (2001), extends to the nonstationary case Lemma 11.7.1 of Adler and Taylor (2007).

CHAPTER 9

THE RECORD METHOD

In this chapter we present a very efficient method for the numerical computation of the distribution of the maximum of a stochastic process or a random field with a two-dimensional parameter. It is based primarily on a paper by Mercadier (2006) and the Matlab toolbox MAGP (Mercadier, 2005), which uses the Rind routine of the Matlab package WAFO (WAFO Group, 2000).

9.1. SMOOTH PROCESSES WITH ONE-DIMENSIONAL PARAMETERS

9.1.1. Main Result

The basic idea is the following: Let $\{X(t), t \in \mathbb{R}\}$ be a real-valued stochastic process with a.s. absolutely continuous sample paths, and suppose that we are looking for an expression of

$$1 - F_M(u) = \mathrm{P}\{M > u\}.$$

We denote $M_T = \sup\{X(t) : 0 \le t \le T\}$ and $M = M_1$. Instead of looking at all crossings of the level u, we will look only at those crossings that are record times. The set $\mathcal{R}$ of *record times* is defined by

$$\mathcal{R} := \{t \in [0, 1] : X(s) < X(t), \forall s \in [0, t)\},$$

with the convention that 0 is always in $\mathcal{R}$. We obtain the trivial identity

$$P\{M \geq u\} = P\{X(0) > u\} + P\{\exists t \in \mathcal{R} : X(t) = u\}. \tag{9.1}$$

The number of record times t such that $X(t) = u$ is equal to 0 or 1. The second term on the right-hand side of (9.1) is equal to the expectation of

$$\mathcal{R}(u) := N_u(X, \mathcal{R}) = \#\{t \in \mathcal{R} : X(t) = u\}.$$

This idea is the basis of the following result:

Theorem 9.1 (Rychlik's Formula). *Let $\mathcal{X} = X(t), t \in [0, 1]$ be a real-valued stochastic process with a.s. absolutely continuous sample paths such that for almost all $t \in [0, 1]$, $X(t)$ admits a density $p_{X(t)}$ and $\mathrm{E}(|X'(t)|) < \infty$. Then for every $u \in \mathbb{R}$,*

$$P\{M > u\} = P\{X(0) \geq u\} + \lim_{\delta \to 0} \frac{1}{\delta} \int_u^{u+\delta} dx \times \int_0^1 \mathrm{E}\big(X'(t)^+ \mathbf{I}_{t \in \mathcal{R}} \big| X(t) = x\big) p_{X(t)}(x)\, dt. \tag{9.2}$$

Remarks
An early version of this formula under stronger conditions is due to Rychlik (1990). The present version is due to Mercadier (2006).

1. The limit in (9.2) is, in fact, a manner of choosing a convenient version of the conditional expectation. For Gaussian processes under certain conditions, the usual conditional distributions defined by the regression formulas are convenient (see Corollary 9.3).

2. Expression (9.2) seems at first sight worthless since its right-hand side does not seem to be simpler than the left-hand side. But in the next section we deduce from formula (9.2) some upper bounds that are sharp.

Corollary 9.2. *Suppose that in addition to the conditions of Theorem 9.1, the process $X(t)$ is Gaussian with C^1-paths and satisfies:*

- *For $s, t \in [0, 1]$, $s < t$, the distribution of $(X(s), X(t))$ does not degenerate.*
- *For $t \in [0, 1]$, the distribution of $(X(t), X'(t))$ does not degenerate.*

Then

$$P\{M > u\} = P\{X(0) > u\} + \int_0^1 \mathrm{E}\big(X'(t)^+ \mathbf{I}_{t \in \mathcal{R}} \big| X(t) = u\big) p_{X(t)}(u)\, dt. \tag{9.3}$$

Proof of Theorem 9.1. The main idea is that the number $\mathcal{R}(u)$ of record points taking a particular value u is equal to 0 or 1 and that

$$\mathrm{P}\{M \geq u\} = \mathrm{P}\{X(0) \geq u\} + \mathrm{E}(\mathcal{R}(u)).$$

To compute the expectation $\mathrm{E}(\mathcal{R}(u))$, we use the Banach formula (see Exercise 3.8). Let $g(u)$ be a continuous bounded function; we have

$$\int_{\mathbb{R}} g(u)\mathcal{R}(u)\,du = \int_0^1 |X'(t)| g(X(t))\,\mathbb{I}_{t\in\mathcal{R}}\,dt.$$

Taking expectations in both sides gives

$$\int_{\mathbb{R}} g(u)\mathrm{E}[\mathcal{R}(u)]\,du = \int_{\mathbb{R}} du\, g(u) \int_0^1 dt\, \mathrm{E}[|X'(t)|\,\mathbb{I}_{t\in\mathcal{R}} \big| X(t) = u) p_{X(t)}(u),$$

showing that the two functions of u,

$$\mathrm{E}[\mathcal{R}(u)] \quad \text{and} \quad \int_0^1 \mathrm{E}[|X'(t)|\,\mathbb{I}_{t\in\mathcal{R}} \big| X(t) = u) p_{X(t)}(u)\,dt$$

are u-almost surely equal. From this we deduce that the two functions $\mathrm{P}\{M > u\}$ and $\mathrm{P}\{X(0) > u\} + \mathrm{E}[\mathcal{R}(u)]$ are u-almost surely equal. The result follows because $\mathrm{P}\{M > u\}$ is càd-làg. □

The proof of the corollary is left to the reader.

The next proposition is an easy consequence of Rychlik's formula, which is a version under weaker hypotheses of Theorem 8.12 for random fields.

Proposition 9.3. *Let $\{X(t) : 0 \leq t \leq T\}$ be a Gaussian process that satisfies the following:*

- *It is twice differentiable in quadratic mean.*
- *For all $t \in [0, T]$, $\mathrm{E}(X(t)) = 0$ and $\mathrm{Var}(X(t)) = 1$.*
- *$\mathrm{Var}(X'(t))$ is bounded away from zero. Without loss of generality we can use the unit speed transformation and suppose that $\mathrm{Var}(X'(t)) = 1$.*
- *For all $s \neq t$, $r(s,t) < 1$.*

Then for every $\delta > 0$ there exists some constant C_δ such that

$$0 \leq 1 - \Phi(u) + T\sqrt{\frac{2}{\pi}}\varphi(u) - \mathrm{P}\{M_T > u\} \leq C_\delta \exp\left[-\left(1 + \frac{1}{Z}\right)\frac{u^2(1-\delta)}{2}\right], \tag{9.4}$$

where

$$Z := \sup_{0 \le s < t \le T} \left[\frac{\mathrm{Var}(X(s) \| X(t), X'(t))}{(1 - r(s,t))^2} + \frac{(r_{0,1}^{+}(s,t))^2}{(1 - r(s,t))^2} \right] < +\infty.$$

Proof. We use a method that has been employed in the context of random fields in the proof of Theorem 8.12. Clearly, the expression in (9.4) is bounded by

$$\int_0^1 \mathrm{E}\big(X'(t)^{+} \mathbb{I}_{t \notin \mathcal{R}} \big| X(t) = u\big) p_{X(t)}(u)\, dt.$$

An application of the Hölder inequality shows that it sufficient to give bounds to

$$\mathrm{P}\{\exists s : s < t,\ X(s) \ge u \big| X(t) = u,\ X'(t) > 0\}.$$

For that purpose we write the regression of $X(s)$ on $(X(t), X'(t))$:

$$X(s) = r(s,t)X(t) + r_{0,1}(s,t)X'(t) + R^t(s).$$

The three terms on the right-hand side above are independent. Under the condition $\{X(t) = u,\ X'(t) > 0\}$, the event $X(s) \ge u$ can be written

$$\frac{R^t(s)}{1 - r(s,t)} + \frac{r_{0,1}(s,t)}{1 - r(s,t)} X'(t) \ge u.$$

It is obvious that the left-hand side in the inequality above is smaller than

$$Y^t(s) := \frac{R^t(s)}{1 - r(s,t)} + \frac{r_{0,1}^{+}(s,t)}{1 - r(s,t)} X'(t).$$

Suppose for the moment that $Y^t(s)$ is bounded and that Z is finite. Then using the Landau–Shepp–Fernique inequality (2.33), we know that for every $\delta > 0$ there exists some constant C'_δ such that

$$\mathrm{P}\{\exists s : s < t,\ X(s) \ge u \| X(t) = u,\ X'(t) > 0\} \le C'_\delta \exp\left[-\frac{u^2(1-\delta)}{2Z} \right].$$

The rest of the proof is obvious.

It remains to prove that $Y^t(s)$ is bounded and that its maximal variance is finite. The variance is the sum of the variance of the two terms. The variance can become infinite, or $Y^t(s)$ can become infinite, only for s tending to t. Using Taylor's formula at t yields

$$X(s) = X(t) + (s - t)X'(t) + \frac{(s-t)^2}{2} Q(s),$$

where $Q(s)$ is an integral remainder. It is easy to see that

$$1 - r(s,t) \approx \frac{(s-t)^2}{2}, \qquad r_{0,1}(s,t) \approx s - t$$

and that $\lim_{s \to t} R^t(s)$ is just the projection in $L^2(\Omega)$ of $Q(s)$ onto the orthogonal complement of the linear subspace generated by $X(t)$ and $X'(t)$, so that we can conclude that it is a.s. finite and has finite variance. □

9.1.2. Numerical Application

The exact implicit formula (9.3) can be turned into an explicit upper bound by means of a discretization of the condition $\{X(s) < X(t), \forall s \in [0,t)\}$. One convenient way is to use the points $\{kt/n, k = 0, \dots, n-1\}$ to get

$$\begin{aligned} \mathrm{P}\{M > u\} &\le \mathrm{P}\{X(0) > u\} \\ &\quad + \int_0^1 \mathrm{E}\big(X'(t)^+ \mathbf{1}_{X(0),\dots,X(t(n-1)/n)<u} \big| X(t) = u\big) p_{X(t)}(u)\,dt. \end{aligned} \tag{9.5}$$

On the other hand, the time discretization provides the trivial lower bound

$$\mathrm{P}\{M > u\} \ge 1 - \mathrm{P}\{X(0), \dots, X((n-1)/n) \le u\}. \tag{9.6}$$

The main point is that when the process is Gaussian, the integrals that appear in (9.5) and (9.6) can be computed using the MAGP toolbox. All details are given on the Web page of Mercadier (2005). The program is able to perform such calculations for n up to 100.

The precision of the computations of MAGP has been evaluated by Mercadier in two ways:

1. Comparing the lower bounds (9.6) with the exact theoretical value for the sine–cosine process [i.e., the centered Gaussian process with covariance $\Gamma(t) = \cos t$] given by Berman (1971b) and Delmas (2003b) (see Exercise 4.1).
2. Comparing lower and upper bounds with results in other chapters of the book. For example, Figure 9.1 compares the lower and upper bounds of this chapter with the lower and upper bounds given by two or three terms in the Rice series for the centered stationary process with covariance $\Gamma_1(t) = e^{-t^2/2}$ (see Chapter 5).

From these comparisons it appears that one can trust the result from MAGP up to 10^{-3}.

Another question is the precision of the estimation, which can be measured by the difference between (9.5) and (9.6). We will consider that the estimation given by (9.5) and (9.6) is numerically significant if it corresponds to an absolute

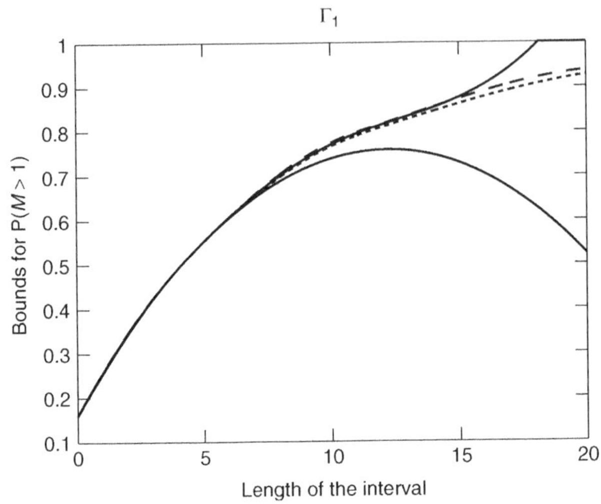

Figure 9.1. Comparison of the present bounds with those of Chapter 5. From top to bottom: upper bound from Chapter 5, upper bound (9.5), lower bound (9.6), lower bound from Chapter 5 (M stands for M_T). (From Mercadier, 2006, with permission.)

error smaller than 10^{-2} and to a relative error smaller than 10^{-1}. We concentrate ourselves on the case of stationary centered Gaussian processes with variance 1 and unit speed [$\text{Var}(X'(t)) = 1$]. The result depends, of course, on T and u. The larger u (or the smaller T), the better the results. It happens that for levels $u \geq 1$, the result is numerical significant for time intervals of sizes 20 to 25 in unit-speed measure. See Mercadier (2006) for details. One can check in Figure 9.1 that the record method implemented by MAGP is, at this stage, the most efficient for numerical computation of the distribution of the maximum.

9.2. NONSMOOTH GAUSSIAN PROCESSES

When the process has nondifferentiable paths, one way is to use smoothing, as in Chapter 5. Another way is by using Durbin's formula (Durbin, 1985), based on the pseudoderivative defined as the normalized increment $(X(t) - X(s))/(t - s)$. Mercadier has found that this method is very unstable. A better way is to use the time discretization and the lowerbound (9.6).

It remains to give bounds on the discretization error. This will be done for a process defined on [0.1], discretized at the points $k/n, k = 0, 1, \ldots, n$, which has the same irregularity as the Wiener process (Brownian motion); that is, it satisfies a.s. the law of iterated logarithm (LIL), for fixed t:

$$-1 = \liminf_{s \to 0} \frac{X(t+s) - X(t)}{\sqrt{2s \log(\log(1/s))}} \quad \text{and} \quad \limsup_{s \to 0} \frac{X(t+s) - X(t)}{\sqrt{2s \log(\log(1/s))}} = 1. \quad (9.7)$$

Generalizations to other local behaviors can be performed using similar tools, *mutatis mutandis*, that is, changing this oscillation by one of the processes being considered.

Our method is based on the following heuristic approximations (which may not actually be verified by the paths):

1. The instant t^* where the maximum is attained satisfies the lim inf part of the LIL.
2. The maximum of the discretized process is attained within point k/n at point t_n^* that is nearest to t^*.
3. $|t_n^* - t^*|$ has a uniform distribution among the possible values in $[0, 1/(2n)]$.

With all these approximations, we get

$$M - M_n \simeq \sqrt{2Z \log(\log(1/Z))},$$

where Z is uniformly distributed over $[0, 1/2n]$, which amounts to saying that

$$P\{M > u\} \simeq \int_0^{1/2n} P\{M_n > u - \sqrt{2z \log(\log(1/z))}\} 2n \, dz. \qquad (9.8)$$

Notice that the right-hand side of (9.8) is easy to compute numerically using MAGP.

Results are shown in Figure 9.2 for the maximum of processes defined on an interval $[0, T]$ and parametrized by the length T. They refer to the

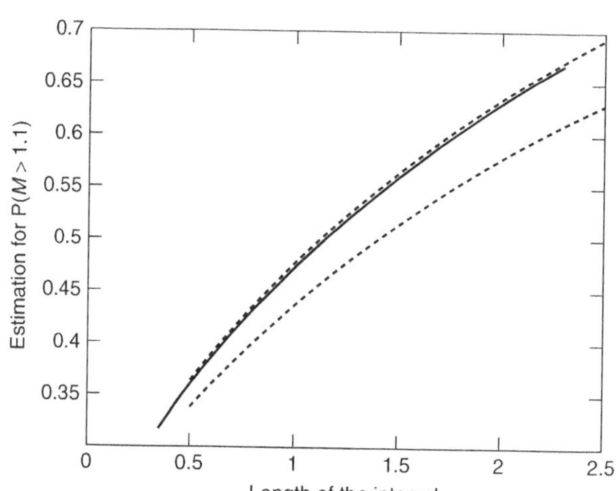

Figure 9.2. Ornstein–Uhlenbeck process: approximation (9.8) (top), lower bound (9.6) (bottom), and exact value (solid line) for $P\{M_T > 1.1\}$ as a function of T.

Ornstein–Uhlenbeck process [i.e., the centered stationary Gaussian process with covariance $\Gamma(t) := \exp(-t)$, $t \geq 0$]. These results are compared with the exact value from DeLong's (1981) paper. The figure suggests that the approximation (9.8) is very good.

9.3. TWO-PARAMETER GAUSSIAN PROCESSES

9.3.1. Main Result

This section is based on ideas similar to those of Section 9.1, adapted to two-parameter processes. We consider a continuous random field $\mathcal{X} = \{X(t) : t \in S\}$ with real values and defined on a compact subset S of $\mathbb{R}^2$ and $M_S := \max_{t \in S} X(t)$. Let us consider next a "rather high" level u and a realization such that $\{M_S > u\}$. Let us suppose that the probability that the process remains above the level u for all $t \in S$ can be neglected. Then the event $\{M_S > u\}$ is almost equivalent to

"the level curve $\mathcal{C}_u := \{t \in S : X(t) = u\}$ is not empty."

More precisely, let us choose a particular direction (say, south) and to consider the point at the southern extremity of $\mathcal{C}_u$ (which is, in general, unique). To do so, denote by $\lessdot$ the lexicographic order on $\mathbb{R}^2$; that is,

$$s = (s_1, s_2) \lessdot t = (t_1, t_2) \Leftrightarrow \{s_2 < t_2\} \text{ or } \{s_2 = t_2; s_1 < t_1\}.$$

We define the *lexicographic past* $\mathcal{L}(t)$ of a point $t \in S$ as

$$\mathcal{L}(t) := \{s \in S : s \lessdot t\}.$$

A point $t \in S$ will be called a record point if for all points, $s \in \mathcal{L}(t)$: $X(s) < X(t)$. We denote by $\mathcal{R}$ the set of record points. Obviously, there is at most one record point where the process $X(t)$ takes a particular value u, and this point is (in general) at the southern extremity of the level curve.

Eventually, the event $\{M > u\}$ is almost equivalent to

{the number of record points on $\mathcal{C}_u$ is 1 and not 0}.

We assume the following hypotheses:

(A0) The set S is compact, convex, and the parameterization $\rho : [0, L] \to \partial S$ of the boundary ∂S by its arc length is of class $\mathcal{C}^1$, except perhaps at a finite number of points where ρ is only continuous. Moreover, we assume that $\rho(0)$ is the point of ∂S that is minimal with respect to $\lessdot$.

(A1) The sample paths of the random field $Z := (X, X_{10})$ are a.s. continuously differentiable.

(A2) For $t \in S$, the distribution of $Z(t)$ does not degenerate.

(A3) For every $w \in \mathbb{R}^2$ there is almost surely no point $t \in S$ such that $Z(t) = w$ and $\det(Z'(t)) = 0$.

Recall the notation $X_{ij}(t_1, t_2) := (\partial^{i+j}/\partial t_1^i \partial t_2^j) X(t_1, t_2)$ $(i, j = 0, 1, \ldots)$.

Theorem 9.4. *Let S be a subset of $\mathbb{R}^2$ satisfying assumption (A0). Let $X(t)$ be a real-valued Gaussian process defined on some neighborhood of S and satisfying assumptions (A1) to (A3). Then for every real u:*

$$
\begin{aligned}
\mathrm{P}\{M > u\} &= \mathrm{P}\{Y(0) > u\} + \int_0^L \mathrm{E}(|Y'(\ell)| \mathbf{1}_{X(s)<u, \forall s \in \mathcal{L}(\rho(\ell))} \| Y(\ell) = u) p_{Y(\ell)}(u)\, d\ell \\
&\quad + \int_S \mathrm{E}(|X_{20}(t)^- X_{01}(t)^+| \mathbf{1}_{X(s)<u, \forall s \in \mathcal{L}(t)} \| X(t) = u, X_{01}(t) = 0) \\
&\quad \times p_{X(t), X_{01}(t)}(u, 0)\, dt, \qquad (9.9)
\end{aligned}
$$

where $Y(\ell) := X(\rho(\ell))$.

Sufficient conditions for (A3) are given in Proposition 6.5.

Proof. The proof is very close to that of Theorem 7.2 and will be sketched. Assume that the event $\{M > u\} \bigcap \{Y(0) < u\}$ occurs. Then $\mathcal{C}_u$ is nonempty and compact. The point τ that is minimal for $\lessdot$ on $\mathcal{C}_u$ is uniquely determined. We want to prove that τ is a record point.

- If τ and $\rho(0)$ have the same second coordinate, $\mathcal{L}(\tau)$ is reduced to the segment $I := [\rho(0); \tau)$. The value of the process $X(t)$ in $\rho(0)$ is less than u, and by definition, the process $X(t)$ cannot take the value u on I. $X(t)$ cannot take a value larger than u on I because of the intermediate value theorem. As a consequence, τ is a record point.
- If τ and $\rho(0)$ have distinct second coordinates $\rho(0)_2 < \tau_2$. On $\mathcal{L}(\tau)$, $X(t)$ cannot take the value u. Suppose that there exists $\widetilde{\tau}$ in $\mathcal{L}(\tau)$ such that $X(\widetilde{\tau}) > u$. The entire segment $[\rho(0), \widetilde{\tau}]$ is in S and thus in $\mathcal{L}(\tau)$ and by the intermediate value theorem, there is a point on this segment where $X(t)$ takes the value u, which is not possible.

Adding the trivial case $\{Y(0) > u\}$, we have proved that in the event $Y(0) \neq u$, which has probability 1, almost surely $\{M > u\}$ is the disjoint union of the two events

"$Y(0) > u$" and "there exists a record point with value u."

The event "there exists a record point with value u" can be split into two disjoint events, depending on whether τ belongs to ∂S or $\overset{\circ}{S}$. Since there is at most one record point, these two cases are disjoint and their probabilities are equal to the expectation of the number of record points in ∂S and $\overset{\circ}{S}$.

Let us consider a nonincreasing function $\mathcal{F}, \mathbb{R} \to \mathbb{R}$ satisfying

$$\mathcal{F}(x) = 1 \text{ if } x < -\tfrac{1}{2}, \qquad \mathcal{F}(x) = 0 \text{ if } x \geq 0.$$

Then

$$\mathcal{F}_n(x) := \mathcal{F}(nx) \uparrow \mathbf{1}_{x<0} \quad \text{as } n \uparrow +\infty.$$

We first compute the expectation of the number of zeros of $Z(t)$ on S with the weights $\mathcal{F}_n(\sup_{s\in\mathcal{L}(t)} X(s) - X(t))$ using the Rice formula (Theorem 6.4), and we then pass to the monotone limit as $n \to \infty$.

For the boundary of S,

$$\mathrm{E}(\#\{\ell \in [0, L) : \rho(\ell) \in \mathcal{R}, Y(\ell) = u\}),$$

we have to use the same kind of proof after splitting $(0, L)$ into a finite number of subintervals in which $\ell \rightsquigarrow Y(\ell)$ is C^1. We remark that $Y(\cdot)$ a.s. does not takes the value u at the extremities of these intervals. Summing up, we get the result. □

Theorem 9.5 (Bounds). *Let S be a subset of $\mathbb{R}^2$ satisfying:*

(A′0) S is compact, S and its complement are connected, and the parameterization $\rho : [0, L] \to \partial S$ of the boundary ∂S by its arc length is of class C^1 except perhaps at a finite number of points where ρ is only continuous.

Let $\mathcal{X}$ be a real-valued Gaussian process defined on some neighborhood of S, satisfying assumptions (A1) to (A3). Then, for every real u, using the notation of Theorem 9.9 gives us

$$\begin{aligned} \mathrm{P}\{M > u\} \leq \mathrm{P}\{Y(0) > u\} &+ \int_0^L \mathrm{E}(|Y'(\ell)|)\|Y(\ell) = u) p_{Y(\ell)}(u)\, d\ell \\ &+ \int_S \mathrm{E}(|X_{20}(t)^- X_{01}(t)^+|\|X(t) = u, X_{01}(t) = 0) \\ &\times p_{X(t), X_{01}(t)}(u, 0)\, dt. \end{aligned} \tag{9.10}$$

Proof. Let M_∂ be the maximum of $X(t)$ on ∂S. One has

$$\begin{aligned} \mathrm{P}\{M > u\} &= \mathrm{P}\{M_\partial > u\} + \mathrm{P}(\{M > u\} \cap \{M_\partial < u\}) \\ &\leq \mathrm{P}\{Y(0) > u\} + \mathrm{P}\{U_u^Y([0, L] > 0\} \\ &+ \mathrm{P}\{\exists t \in \overset{\circ}{S} : X(t) = u, X_{10}(t) = 0, X_{20} < 0, X_{01} > 0\}. \end{aligned} \tag{9.11}$$

The last inequality is due to the fact that if $M > u$ and $M_\partial < u$, the level curve $\mathcal{C}_u$ is contained in the interior of S. There exists at least one point on this curve with minimal second coordinate. It follows that

$$\mathrm{P}\{M > u\} \leq \mathrm{P}\{Y(0) > u\} + \mathrm{E}\{U_u^Y([0, L]\} + \mathrm{E}\big(\#\{t \in \overset{\circ}{S} : X(t) = u, X_{10}(t) = 0, X_{20} < 0, X_{01} > 0\}\big).$$

It suffices to apply the Rice formula (Theorem 6.2), remarking that under condition $X(t) = u$, $X_{10}(t) = 0$, we have $\det(Z'(t)) = X_{20}(t)X_{01}(t)$. □

9.3.2. Numerical Application

As in Section 9.1.2, the exact formula (9.9) can be transformed into an explicit upper bound by discretizing the condition $\mathbb{1}_{X(s)<u, \forall s} \in \mathcal{L}(t)$. For simplicity we now limit our attention to the case where S is the square $[0, T]^2$ and the process X is "standardized" (i.e., centered, stationary, isotropic), with variance 1 and identity speed. In that case:

- The two terms in (9.9) corresponding to the edges $0 \times [0, T]$ and $1 \times [0, T]$ are equal and equal to

$$\int_0^T \mathrm{E}((X_{10}(v, 0))^+ \mathbb{1}_{X(s)<u, \forall s \in \mathcal{L}(v,0)} \| X(v, 0) = u)\phi(u)\, dv.$$

- The term corresponding to the edge $[0, T] \times 1$ vanishes: Indeed, if there is a record point τ on this edge, the derivative X_{01} must vanish. Because of the Rice formula (or using Bulinskaya's Lemma 1.20), the expectation of the number of such points is zero.

If in formula (9.9), we replace the entire lexicographic past $\mathcal{L}(t)$ of a point t by its intersection $\mathcal{L}_n(t)$ with a grid

$$\mathcal{L}_n(t) := \mathcal{L}(t) \cap \left\{\left(\frac{kt}{n}, \frac{lt}{n}\right), k = 0, \ldots, n, l = 0, \ldots, n\right\}$$

we get the upper bound:

$$\begin{aligned}\mathrm{P}\{M > u\} \leq \overline{\Phi}(u) + \varphi(u)\Bigg[2\int_0^T \mathrm{E}(X_{01}^+(v, 0)\,\mathbb{1}_{X(s)<u, \forall s \in \mathcal{L}_n(v,0)} \big| X(v, 0) = u)\, dv \\ + \int_0^T \mathrm{E}(X_{10}^+(v, 0)\,\mathbb{1}_{X(s)<u, \forall s \in \mathcal{L}_n(v,0)} \big| X(v, 0) = u)\, dv \\ + \frac{1}{\sqrt{2\pi}} \iint_{[0,T]^2} \\ \mathrm{E}(X_{20}(t)^- X_{01}(t)^+ \mathbb{1}_{X(s)<u, \forall s \in \mathcal{L}_n(t)} | X(t) = u, X_{01}(t) = 0)\, dt\Bigg]. \quad (9.12)\end{aligned}$$

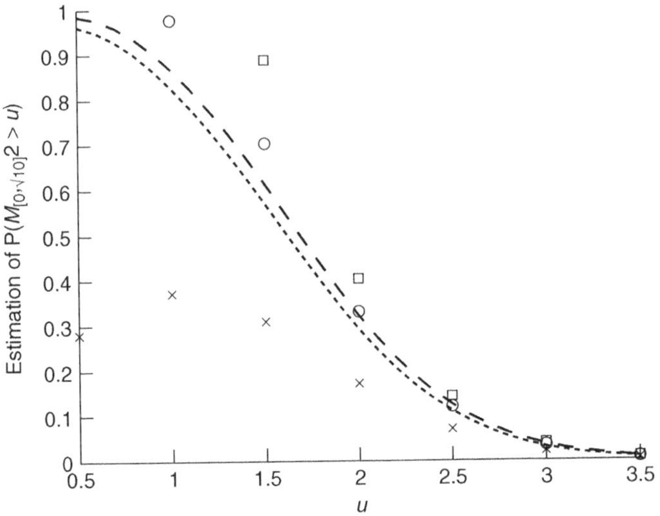

Figure 9.3. $P\{M \rangle u\}$ for the random field with Gaussian covariance on $[0, \sqrt{10}]^2$ from top to bottom: upper bound (9.10), Euler characteristic approximation, upper bound (9.12), lower bound (9.12), and equivalent by Adler(1981). (From Mercadier, 2006, with permission.)

The main point is that this upper bound can be computed by MAGP.

In the other direction, we get a lower bound using discretization:

$$P\{M > u\} \geq P\left\{\max\left(X\left(\frac{kt}{n}, \frac{lt}{n}\right), k = 0, \ldots, n, l = 0, \ldots, n\right) > u\right\}. \quad (9.13)$$

Figure 9.3 shows that these bounds compared with the equivalent in Adler (1981) and the equivalent given by the Euler–Poincaré characteristic method.

EXERCISES

9.1. Prove Corollary 9.3 using Ylvisaker's theorem and the regression method of Proposition 9.4.

9.2. Suppose that S is the square $[0, T]^2$. Suppose that the process $X(t)$ is "standardized." Show that the upper bound (9.10) takes the form

$$P\{M > u\} \leq \overline{\Phi}(u) + \sqrt{\frac{2}{\pi}} T\phi(u) + \frac{T^2 u}{2\pi}\left[c\phi(u/c) + u\Phi(u/c)\right]\phi(u)$$

with $c := \sqrt{\mathrm{Var}(X_{20}) - 1}$. Show that the difference between this bound and the equivalent given by the EPC method, $\overline{\Phi}(u) + \sqrt{(2/\pi)}\phi(u) + (T^2u/2\pi)$ $\phi(u)$, is bounded by

$$\frac{T^2 u}{(2\pi)^{3/2}} c^3 u^{-2} \phi\left(u\sqrt{\frac{1+c^2}{c^2}}\right)\left[c\phi(u/c) + u\Phi(u/c)\right]\phi(u).$$

CHAPTER 10

ASYMPTOTIC METHODS FOR AN INFINITE TIME HORIZON

In this chapter we consider asymptotic results for one-parameter processes on intervals of time with size tending to infinity. In Section 10.1 the level u tends to infinity jointly with the size of the interval so that the expectation of the number of crossings remains constant. In that case it is proven that under weak hypotheses, the asymptotic distribution of the number of crossings is Poisson. This implies that the maximum of the process converges, after renormalization, to a Gumbel distribution.

In Section 10.2 the level u is fixed and the size of the interval tends to infinity. Under certain conditions, this number of crossings satisfies a central limit theorem. The main tool is an elementary presentation of Wiener chaos decomposition.

10.1. POISSON CHARACTER OF HIGH UP-CROSSINGS

In this section we give a proof of the following theorem, originally due to Volkonskii and Rozanov (1959, 1961).

Theorem 10.1. *Let $\mathcal{X} = \{X(t) : t \in \mathbb{R}\}$ be a zero-mean stationary Gaussian process with covariance $\Gamma(\tau) = \mathrm{E}(X(t)X(t+\tau))$ satisfying the following conditions:*

- $\Gamma(0) = 1$.
- $\lambda_2 < \infty$.

- $\Gamma(\tau)\log(\tau) \to 0$ *as* $\tau \to \infty$ *(Berman's condition).*
- *If one writes* $\Gamma(\tau) = 1 - \lambda_2\tau^2/2 + \theta(\tau)$, *then, for some* $\delta > 0$, *the integral*

$$\int_0^{\delta} \frac{\theta'(\tau)}{\tau^2}\, d\tau$$

is convergent (this is Geman's condition, which we mentioned in Proposition 4.2).

Set

$$C_u := \mathrm{E}\big(U_u(X; [0, 1])\big) = \sqrt{\lambda_2}\frac{\exp(-u^2/2)}{2\pi}$$

and define, for $t \in \mathbb{R}^+$,

$$R_u(t) := U_u(X; [0, C_u^{-1}t]). \tag{10.1}$$

Then as $u \to +\infty$ *the family of point processes* $\{R_u(t) : t \geq 0\}$ *converges weakly in the Skorohod space to a standard Poisson process.*

Remark. Notice that Berman's condition implies that $|\Gamma(\tau)| < 1$ for all $\tau \neq 0$ (see Feller, 1966, Chap. XV).

Theorem 10.1 has a direct interest for modeling phenomena depending on time, such as pollution levels, floods, or other situations in which up-crossings of a threshold by a certain stochastic process imply the occurrence of a relevant event. This theorem is the mathematical explanation of a standard procedure which consists in using the Poisson process as a model to represent the sequence of these random time points whenever the threshold and the size of the time window are large enough. Of course, beyond the statement of the theorem, the fitting of such a model to empirical data is a problem of statistical nature that should be considered appropriately in each case.

In the formulation above, the process is Gaussian and stationary, has some local regularity given by Geman's condition, and some mixing (asymptotic independence for distant values of the parameter) given by Berman's condition. In fact, the mild Geman condition is not needed in its full generality. [See Leadbetter et al. (1983) for a longer proof without this condition and also for various extensions, including to non-Gaussian processes.]

We are using here only some elementary well-known properties of point processes defined on the half-line $[0, +\infty)$ (see, e.g., Neveu, 1977) . We indistinctly call *point process* the random set of points $\Psi = \{t_k\}$, the random measure μ_Ψ having a unit atom at each one of these points, which we assume to be almost surely locally finite and the càd-làg version of its cumulative distribution function, that is, for $t \geq 0$, $F_\Psi(t) = \#\{k : t_k \leq t\}$. Weak convergence of these point processes is to be understood as weak convergence of the stochastic process

$F_\Psi(\cdot)$ in the Skorohod space (see Section 4.4.1). The two properties that we will use without proving them are the following:

Proposition 10.2 (Rényi, 1967)
(a) *Let $T > 0$. The family of point processes $\{F_{\Psi_n}(\cdot)\}_{n=1,2,\ldots}$ is tight in the space $D[0, T]$ if and only if the sequence of distributions of the (integer-valued) random variables $\{F_{\Psi_n}(T)\}_{n=1,2,\ldots}$ is tight on the line.*
(b) *Assume that a point process as above verifies the following conditions for any subset B of $[0, +\infty)$, which is a union of intervals:*

$$\mathrm{E}\big(\mu_\Psi(B)\big) \le \lambda(B) \tag{10.2}$$

$$\mathrm{P}\big\{\mu_\Psi(B) = 0\big\} = \exp(-\lambda(B)). \tag{10.3}$$

Then Ψ is a standard Poisson process.

The next useful corollary of Theorem 10.1 states that after renormalization, the maximum M_T of the process converges weakly to a Gumbel distribution:

Corollary 10.3. *Under the conditions of Theorem 10.1, setting*

$$M_T := \sup_{t\in[0,T]} X_t, \qquad a_T := (2\log T)^{1/2}$$

$$b_T := (2\log T)^{1/2} + \log\left(\frac{\sqrt{\lambda_2}}{2\pi}\right)(2\log T)^{-1/2},$$

then as $T \to +\infty$,

$$\mathrm{P}\{a_T(M_T - b_T) \le x\} \to \exp(-e^{-x}).$$

Both the theorem and its corollary can be extended, under mild conditions, to constant variance nonstationary processes (see Azaïs and Mercadier, 2003).

Proof of Theorem 10.1. Without loss of generality we may assume that $\lambda_2 = 1$. Set $\rho(t) := \sup_{s>t} |\Gamma(s)|$. If not specified otherwise, all the limits in this proof are for $u \to +\infty$. Let T_0 be such that $\rho(T_0) < \frac{1}{3}$.

We break the proof into several steps.

STEP 1. Let $T(u)$ be increasing as a function of u, and tend to infinity in a controlled manner, meaning that $T(u) = O\big((C_u)^{-1/2}\big)$. Let us prove that

$$\mathrm{P}\{M_{T(u)} > u\} = \sqrt{\frac{\lambda_2}{2\pi}}\,T(u)\phi(u)\big(1 + o(1)\big). \tag{10.4}$$

Let u be large enough so that $T(u) > T_0$. We use the computation in the proof of Proposition 4.2.

For short, denote as ν_2 the second factorial moment of the up-crossings on an interval of length $T(u)$. It suffices to prove that

$$\nu_2 = T(u)\phi(u)o(1).$$

For given $\varepsilon > 0$, choose $\delta > 0$ so that

$$\int_0^\delta \frac{\theta'(\tau)}{\tau^2}\, d\tau < \varepsilon.$$

Then

$$\nu_2 \leq T(u)\phi(u)\big[\varepsilon + T_0\phi(\delta_1 u) + (T(u) - T_0)\phi^{1/2+\delta_2}(u)\big],$$

where $\delta_1, \delta_2 > 0$, which implies (10.4).

STEP 2. It is sufficient to prove the result for $t \in [0, T]$ and each $T \in \mathbb{R}^+$. We use Proposition 10.2(a) and observe that by the construction $\mathrm{E}\big(R_u(T)\big) = T$, which obviously implies the tightness of the probability distributions of the random variables $\{R_u(T)\}_{u>0}$. So the family of processes $R_u(\cdot)$ is tight, and if we denote by $R(\cdot)$ a limit process of the family, what remains is to show that $R(\cdot)$ is a standard Poisson process. For this goal we use the characterization of Poisson processes given by Proposition 10.2(b).

Clearly, Fatou's lemma implies that the process $R(\cdot)$ satisfies (10.2). The remainder of the proof is based on a discretization argument to verify condition (10.3). We do this first when B is a unique interval. Due to the stationarity of the process, we may assume that $B = (0, L]$.

Set $J_u = (0, C_u^{-1}L]$. Consider the partition of J_u into $n = n(u)$ intervals $I_1, \ldots, I_n$ of equal length $C_u^{-1}L/n$, where $n - 1$ is the integer part of $C_u^{-1/2}L$. Then each interval I_i $(i = 1, \ldots, n)$ satisfies the result of step 1.

We now divide each interval I_i into intervals of length $q = q(u)$ such that $qu \to 0$, but sufficiently slowly in a manner described in detail later. In step 3 we prove the following intermediate results:

- $\mathrm{P}\{U_u(J_u) = 0\} = \mathrm{P}\{M_{J_u} < u\} + o(1).$ (10.5)
- For all intervals $A = A_u \subset J_u$ having the form $A_u = [aq, bq]$, a and b integers,

$$0 \leq \mathrm{P}\{X(kq) \leq u : \forall k;\, kq \in A_u\} - \mathrm{P}\{M_{A_u} < u\} = \lambda(A_u)C_u o(1) \quad (10.6)$$

  with $M_{A_u} := \sup_{t\in A_u} X(t)$, where the $o(1)$ is uniform over all sequences of intervals with length bounded below by a positive number.
- $$\mathrm{P}\{X(kq) \leq u : \forall k;\, kq \in D_u\} = \prod_{i=1}^{n} \mathrm{P}\{X(kq) \leq u : \forall k;\, kq \in I_i\} + o(1).$$

(10.7)

A direct consequence of these equivalences is that

$$\begin{aligned}
\mathrm{P}\{U_u(J_u)=0\} &= \prod_{i=1}^{n}\Big(1 - C_u^{1/2}\big(1+o(1)\big)\Big) \\
&= \Big(1 - C_u^{1/2}\big(1+o(1)\big)\Big)^{C_u^{-1/2}L\left(1+o(1)\right)} v + o(1) \\
&= \exp(-L)\big(1+o(1)\big). \qquad (10.8)
\end{aligned}$$

Now use the fact that $R([0, L))$ is the weak limit of $U_u(J_u)$ through a subsequence of u's tending to $+\infty$ to conclude that (10.3) holds true when B is a single interval.

STEP 3. We prove (10.5). Clearly, $\mathrm{P}\{M_{J_u} < u\} \le \mathrm{P}\{U_u(J_u) = 0\}$. On the other hand,

$$\begin{aligned}
\mathrm{P}\{U_u(J_u)=0\} &= \mathrm{P}\{U_u(J_u)=0;\ X(0)<u\} + \mathrm{P}\{U_u(J_u)=0;\ X(0)\ge u\} \\
&\le \mathrm{P}\{M_{J_u} < u\} + 1 - \Phi(u).
\end{aligned}$$

Now consider (10.6). We define $U_u^q(A_u)$ as the number of up-crossings of the size-q discretization of X on the interval A_u. More precisely,

$$U_u^q(A_u) := \#\{k \in \mathbf{Z} :\ kq \in A_u,\ X(kq) > u,\ X\big((k-1)q\big) < u\}.$$

Then

$$\begin{aligned}
0 &\le \mathrm{P}\{X(kq) \le u : \forall k;\ kq \in A_u\} - \mathrm{P}\{M_{A_u} < u\} \\
&\le \mathrm{P}\{U_u(A_u) - U_u^q(A_u) > 0\} \le \mathrm{E}\big(U_u(A_u) - U_u^q(A_u)\big).
\end{aligned}$$

An upper bound for the right-hand side is given in the following auxiliary lemma, of which we give a proof after finishing the proof of the theorem.

Lemma 10.4. *Let A_u, q, and $U_u^q(A_u)$ be defined as above. Then*

$$\mathrm{E}[U_u(A_u) - U_u^q(A_u)] = o\big(\lambda(A_u)C_u\big),$$

with $o\big(\lambda(A_u)C_u\big)$ uniform over the class of intervals considered.

Proof of Theorem 10.1 (cont.). We now prove (10.7). We use Li and Shao's normal comparison Lemma 2.1. Let Σ be the variance matrix of $X(kq)$ for kq in J_u and Σ' the variance matrix obtained by setting to zero the extra-diagonal blocks of Σ with respect to the partition $I_1, \ldots, I_n$. Using the Li–Shao inequality in both senses, we obtain

$$\left| \mathrm{P}\{X(kq) \le u : \forall k; kq \in J_u\} - \prod_{i=1}^{n} \mathrm{P}\{X(kq) \le u : \forall k; kq \in I_i\} \right|$$
$$\le \frac{1}{4} \sum_{i,j=1,\ldots,N,i<j} |\Sigma_{ij}| \exp\left(-\frac{u^2}{1+|\Sigma_{ij}|}\right), \tag{10.9}$$

where N is the number of discretization points in J_u. Using stationarity and remarking that the function $r \rightsquigarrow r\exp\left(-u^2/(1+r)\right)$ is increasing on $(0, 1)$, we obtain the fact that the right-hand side in formula (10.9) is bounded above by

$$\frac{1}{4} \sum_{l=1,N} k(l)\rho(lq) \exp\left(-\frac{u^2}{1+\rho(lq)}\right), \tag{10.10}$$

where $k(l)$ is the difference in the number of occurrences of the quantity $\Gamma(lq)$ between Σ and Σ'. It is easy to see that

$$k(l) = l(n-1) \qquad \text{for } lq < C_u^{-1}L/n \tag{10.11}$$
$$k(l) \le C_u^{-1}L/q \qquad \text{for every } q. \tag{10.12}$$

So, using T_0, introduced in step 1, and the monotonicity of $r \rightsquigarrow r\exp\left(-u^2/(1+r)\right)$, we can bound the terms in the sum (10.10) in the following way:

- For $0 < l < T_0/q$, we use (10.11) and $\rho(lq) \le 1$.
- For $T_0/q \le l < C_u^{-1}L/nq$, we use (10.11) and $\rho(lq) \le \rho(T_0)$.
- For $C_u^{-1}L/nq \le l$, we use (10.12) and the fact that $t \rightsquigarrow C_u^{-1}L/nq$ is non-decreasing.

We conclude that the expression in formula (10.10) is bounded by

$$(\text{const}) \sum_{0<l<T_0/q} ln \exp\left(-\frac{u^2}{2}\right)$$
$$+ (\text{const}) \sum_{T_0/q \le l < C_u^{-1}L/nq} ln \exp\left(-\frac{u^2}{1+\rho(T_0)}\right)$$
$$+ (\text{const}) \frac{C_u^{-1}}{q^2} \int_{C_u^{-1}L/n}^{C_u^{-1}L} \rho(t) \exp\left(-\frac{u^2}{1+\rho(t)}\right) dt = I_1 + I_2 + I_3. \tag{10.13}$$

It is easy to see that

$$I_1 \le (\text{const}) C_u^{-1/2} q^{-2} \exp(-u^2/2)$$

$$I_2 \le (\text{const})\Big(\frac{C_u^{-1}L}{nq}\Big)^2 n \exp\left(-\frac{u^2}{1+\rho(T_0)}\right)$$
$$\le q^{-2}(\text{const}) \exp\left(\frac{3u^2}{4}\right) \exp\left(-\frac{u^2}{1+\rho(T_0)}\right)$$

and that these quantities tend to zero as soon as q does not go to zero faster than some power of u.

As for I_3,

$$I_3 \le (\text{const})C_u q^{-2} \int_{C_u^{-1}L/n}^{C_u^{-1}L} \rho(t) \exp\big(u^2\rho(t)\big)\, dt$$
$$\le (\text{const})(qu)^{-2} \int_0^L u^2\rho(tC_u^{-1}) \exp\big(u^2\rho(tC_u^{-1})\big)\, dt,$$

after a change of variables. Since $\rho(z)\log(z) \to 0$, $\rho(z) \le (\text{const})\big(\log(z)\big)^{-1}$, and $u^2\rho(tC_u^{-1})$ is bounded and converges pointwise to zero, the dominated convergence theorem implies that

$$\int_0^L u^2\rho(tC_u^{-1}) \exp\big(u^2\rho(tC_u^{-1})\big)\, dt \to 0. \tag{10.14}$$

Now it suffices to assume that $(qu)^{-2}$ grows to infinity more slowly than (10.14) and than some power of u to prove that I_1, I_2, and I_3 tend to zero.

STEP 4. Let $D = \cup_{i=1,p} D_i = \cup_{i=1,p}(a_i, b_i]$ be a union of disjoint intervals. We can apply the same arguments as above, discretization and normal comparison lemma, to show that

$$\mathrm{P}\{M_D < u\} = \prod_{i=1,p} \mathrm{P}\{M_{D_i} < u\},$$

which gives the result. The proof is simpler in the sense that terms such as I_1 and I_2 in formula (10.13) are not present. The only modification is to use a monotone convergence argument in order to replace the extremes of the intervals $(C_u^{-1}a_i, C_u^{-1}b_i]$ by multiples of q. □

Proof of Lemma 10.4. The stationarity of the process implies that

$$\mathrm{E}\big(U_u(A)\big) = \lambda(A)C_u,$$

and since A is supposed to be of the form $(kq, hq]$, $k, h \in \mathbb{N}$,

$$\mathrm{E}\big(U_u^q(A)\big) = \frac{\lambda(A)}{q}\mathrm{P}\{X(0) < u < X(q)\}.$$

Notice that

$$J_q(u) := \mathrm{P}\left\{X(0) < u < X(q)\right\} = \mathrm{P}\left\{|Y_1 - u| < \frac{q}{2} Y_2\right\},$$

where $Y_1 := [X(0) + X(q)]/2$ and $Y_2 := [X(q) - X(0)]/q$ are two independent Gaussian variables with respective variances: $\sigma_1^2 := [1 + r(q)]/2$ and $\sigma_2^2 := 2[1 - r(q)]/q^2$.

We want to prove that $J_q(u) \approx qC_u$. Toward this goal we compute

$$\begin{aligned}(C_u q)^{-1} J_q(u) &= (C_u q)^{-1} \int_0^{+\infty} dy_2 \frac{\varphi(y_2/\sigma_2)}{\sigma_2} \int_{u-(qy_2)/2}^{u+(qy_2)/2} dy_1 \frac{\varphi(y_1/\sigma_1)}{\sigma_1} \\ &= \int_0^{+\infty} \frac{y_2}{\sigma_2^2} \exp\left(-\frac{y_2^2}{2\sigma_2^2}\right) \left[\frac{\sigma_2}{2\sigma_1\sqrt{\lambda_2}}\right. \\ &\quad \left.\int_{-1}^{1} \exp\left(-\frac{u^2}{2\sigma_1^2}(1-\sigma_1^2) - \frac{uqsy_2}{2\sigma_1^2} - \frac{q^2s^2y_2^2}{8\sigma_1^2}\right) ds\right] dy_2.\end{aligned} \tag{10.15}$$

Since $1 - \sigma_1^2 \simeq \lambda_2 q^2/4$, $\sigma_1 \simeq 1$ $\sigma_2^2 \simeq \lambda_2$, we see that pointwise in s and y_2,

$$-\frac{u^2}{2\sigma_1^2}(1-\sigma_1^2) - \frac{uqsy_2}{2\sigma_1^2} - \frac{q^2 - y_2^2}{8\sigma_1^2} \to 0.$$

On the other hand, the integrand in the last term of (10.15) is bounded by

$$(\text{const}) y_2 \exp\left(-\frac{y_2^2}{2\sigma_2^2}\right) \exp\left(-\frac{u^2}{2\sigma_1^2}(1-\sigma_1^2) + \frac{uqy_2}{2\sigma_1^2}\right). \tag{10.16}$$

For $y_2 > 2uq\sigma_2^2/\sigma_1^2$ the exponent in formula (10.16) is bounded by $-y_2^2/4\sigma_2^2$, so that the integral for y_2 in $[2uq\sigma_2^2/\sigma_1^2, +\infty]$ tends, by the dominated convergence theorem, to

$$\int_0^{+\infty} \frac{y_2}{\lambda_2} \exp\left(-\frac{y_2^2}{2\lambda_2}\right) dy_2.$$

The remaining integral can be bounded by

$$\begin{aligned}&(\text{const}) \int_0^{2uq\sigma_2^2/\sigma_1^2} y_2 \exp\left(\frac{uqy_2}{2\sigma_1^2}\right) dy_2 \\ &\leq (\text{const}) \int_0^{2uq\sigma_2^2/\sigma_1^2} y_2 \exp\left(\frac{u^2q^2\sigma_2^2}{\sigma_1^4}\right) dy_2.\end{aligned}$$

Since $uq \to 0$, we see that this integral tends to zero. □

Proof of Corollary 10.3. Set

$$\tau = \exp(-x), \qquad u^2 = 2\left[\log T + x + \log\left(\frac{\sqrt{\lambda_2}}{2\pi}\right)\right].$$

We have $TC_u = \tau$. By Theorem 10.1,

$$\mathrm{P}\{M_T < u\} \approx \mathrm{P}\{R_u(\tau) = 0\} \approx \exp(-\tau).$$

Remarking that

$$u = \frac{x}{a_T} + b_T + o(a_T^{-1}),$$

we get the result. □

10.1.1. Extensions to Random Fields

There exist a series of extensions of the Volkonskii–Rozanov theorem (Theorem 10.1). We consider only two of them here and do not give proofs. Both refer to real-valued d-parameter random fields, with $d > 1$.

The first extension consists of studying, instead of the number of up-crossings of a high level, as was done above, the geometric measure of the inverse image of a high level, that is, to replace the zero-dimensional measure by the $(d-1)$-geometric measure, under an adequate normalization. This has been done by Wschebor (1986) using Rice formulas for random fields and is the subject of Exercise 10.2.

The second is based on the remark that whenever some mixing-like condition is present, one can expect that the point process of local maxima above a high level has a Poisson behavior under a renormalization similar to that of the Volkonskii–Rozanov theorem, adapted to the multiparameter case.

We restrict ourselves here to quoting Theorem 14.1 of Piterbarg's book (1996a), which implies a consequence that is close to Corollary 10.3.

Theorem 10.5 (Piterbarg). *Suppose that the real-valued Gaussian centered stationary random field $\{X(t) : t \in \mathbb{R}^d\}$ satisfies the following conditions:*

- *The covariance $\Gamma(t) = \mathrm{E}(X(s)X(s+t))$ verifies $\Gamma(0) = 1$ and $\Gamma(t) \to 0$ as $\|t\| \to \infty$.*
- *The process is three times differentiable, in the mean square sense.*
- *There exist $C > 0$, $\alpha > 1$, and $\delta > 0$ such that*

$$\lambda_d\big(t \in [0,T]^d;\ \Gamma(t)\log^{\alpha}(\|t\|) > C\big) = O(T^{d(1-\delta)}) \qquad T \to +\infty.$$

Then if $M_T = \max_{t \in [0,T]^d} X(t)$,

$$\mathrm{P}\{((M_T - l_T)l_T) < x\} \to \exp(-\exp(-x)) \qquad T \to \infty,$$

where l_T *is the largest solution in* l *of the equation*

$$T^d \sqrt{\det(\Lambda)} l^{d-1} \exp(-l^2/2) = (2\pi)^{(d+1)/2}$$

and $\Lambda = \mathrm{Var}(X'(0))$.

10.2. CENTRAL LIMIT THEOREM FOR NONLINEAR FUNCTIONALS

10.2.1. Ergodic Processes

Let $\mathcal{Y} = \{Y(t) : t \in \mathbb{R}\}$, be a real-valued stochastic process defined on a probability space $(\Omega, \mathcal{A}, \mathrm{P})$. The process is said to be *strictly stationary* if for any choice of the positive integer k and $t_1, ..., t_k, t \in \mathbb{R}$, the joint distribution (in $\mathbb{R}^k$) of $Y(t_1 + t), ..., Y(t_k + t)$ does not depend on t. Clearly, if the process is Gaussian, it is strictly stationary if and only if for any choice of $\tau \in \mathbb{R}$, the expectation and covariance $\mathrm{E}(Y(t))$ and $\mathrm{Cov}(Y(t), Y(t + \tau))$ do not depend on t.

We assume some mild regularity of the paths of the process $\mathcal{Y}$ such as that almost surely, they are Riemann integrable on every bounded interval. For example, if the paths are a.s. càd-làg, this follows easily. In fact, this condition can be replaced by some more general measurability condition without affecting what follows. We denote by $\sigma(\mathcal{Y})$ the smallest σ-algebra with respect to which all the functions $Y(t) : \Omega \to \mathbb{R},\ t \in \mathbb{R}$ are measurable. Clearly, $\sigma(\mathcal{Y}) \subset \mathcal{A}$.

Let $\mathcal{Y}$ be a real-valued strictly stationary process and η a random variable defined on $(\Omega, \mathcal{A}, \mathrm{P})$, which is also $\sigma(\mathcal{Y})$-measurable. For $t \in \mathbb{R}$ one can define the random variable $\theta_t(\eta)$, which is the image of η under the translation of size t, in the following natural way: If η has the form

$$\eta = g(Y(t_1), ..., (t_k)), \tag{10.17}$$

where $g : \mathbb{R}^k \to \mathbb{R}$ is Borel-measurable, we define

$$\theta_t(\eta) = g(Y(t_1 + t), ..., (t_k + t)).$$

For general η, we approximate it in probability by means of cylindrical functions having the form (10.17) and commute the limit with the translation, using the strict stationarity of the process. A $\sigma(\mathcal{Y})$-measurable random variable η is "invariant" if for every $t \in \mathbb{R}$, almost surely $\theta_t(\eta) = \eta$. The stochastic process $\mathcal{Y}$ is called *ergodic* when η is invariant if and only if it is almost surely constant. A famous theorem due to Maruyama (1949) states that if $\mathcal{Y}$ is a stationary Gaussian process, it is ergodic if and only if its spectral measure has no atoms.

Assume now that $\mathcal{Y}$ is a strictly stationary stochastic process and $\mathrm{E}(|Y(t)|) < +\infty$ (notice that this expectation does not depend on t). Then the classical Birkhoff–Khintchine ergodic theorem says that, almost surely, as $T \to +\infty$, the time average $(1/T)\int_0^T Y(t)\,dt$ converges to an invariant random variable with finite expectation, I_∞ with $\mathrm{E}(I_\infty) = \mathrm{E}(Y(0))$. If the process is also ergodic, this random variable is almost surely constant and equal to $\mathrm{E}(Y(0))$. This corresponds to the usual statement that for strictly stationary ergodic processes, "one can replace time averages by space averages."

A similar result holds true for the time average $(1/T)\int_{-T}^0 Y(t)\,dt$. For proofs, the reader can consult, for example, books by Cramér and Leadbetter (1967) or Brown (1976).

10.2.2. Nonlinear Functionals

Let us now turn to the main subject of this section. Let $\mathcal{X} = \{X(t) : t \in \mathbb{R}\}$ be a centered real-valued stationary Gaussian process. Without loss of generality, we assume that $\mathrm{Var}(X(t)) = 1\ \forall t \in \mathbb{R}$. We want to consider functionals having the form

$$T_t := 1/t \int_0^t F(X(s))\,ds, \tag{10.18}$$

where F is some function in $L^1(\phi(x)\,dx)$.

Set $\mu := \mathrm{E}(F(Z))$, Z being a standard normal variable. The ergodic theorem implies that a.s. the expression in (10.18) has an invariant limit as $t \to +\infty$, which is also $\sigma(\mathcal{X})$-measurable. So if the spectral measure of the process has no atoms, because of Maruyama's theorem, this limit is a.s. constant and equal to μ. Our aim is to compute the speed of convergence and establish for it a central limit theorem.

We will assume furthermore that the function F is in $L^2(\phi(x)\,dx)$. For the statement of the next result, which is not hard to prove, we need the following additional definition. The Gaussian process $\{X(t) : t \in \mathbb{R}\}$ is called *m-dependent* if $\mathrm{Cov}(X(s), X(t)) = 0$ whenever $|t - s| > m$.

Theorem 10.6 (Hoeffding and Robins, 1948). *With the notations and hypotheses above, if the process $X(t)$ is m-dependent, then*

$$\sqrt{t}\left(1/t \int_0^t F(X(s))\,ds - \mu\right) \to N(0, \sigma^2) \text{ in distribution as } t \to +\infty,$$

where σ^2 is the variance of $F(Z)$. Z stands for a standard normal variable.

Our aim is to extend this result to processes that are not m-dependent. The proof we present follows Berman (1992b) with a generalization, due to Kratz and León (2001) (see also León, 2006), to functions F in (10.18) having a Hermite rank not necessarily equal to 1. For $\varepsilon > 0$, we approximate the given process $\mathcal{X}$ by a new one, $\mathcal{X}_\varepsilon$, which is $1/\varepsilon$-dependent and estimate the error.

We need to recall some facts and prove some auxiliary ones before stating and proving the main results. $\overline{H}_n(x)$ denotes the modified Hermite's polynomials of degree n, orthogonal with respect to the standard Gaussian measure defined in Chapter 8. Recall that $\overline{H}_n$ can be defined by means of the identity

$$\exp(tx - t^2/2) = \sum_{n=0}^{\infty} \overline{H}_n(x)\frac{t^n}{n!}.$$

Since F is in $L^2(\phi(x)\,dx)$, it can be written as

$$F(x) = \sum_{n=0}^{\infty} a_n \overline{H}_n(x),$$

with

$$a_n = \frac{1}{n!}\int_{-\infty}^{\infty} F(x)\overline{H}_n(x)\phi(x)\,dx,$$

and the norm of F in $L^2(\phi(x)\,dx)$ satisfies

$$||F||_2^2 = \sum_{n=0}^{\infty} a_n^2 n!.$$

The Hermite rank of F is defined as the smallest n such that $a_n \neq 0$. For our purposes we can assume that this rank is greater than or equal to 1.

A useful standard tool to perform computations with Hermite polynomials is Mehler's formula, which we state and prove with an extension (see León and Ortega, 1989).

Lemma 10.7 (Mehler's Formula). (a) *Let (X, Y) be a centered Gaussian vector* $\mathrm{E}(X^2) = \mathrm{E}(Y^2) = 1$ *and* $\rho = \mathrm{E}(XY)$. *Then*

$$\mathrm{E}(\overline{H}_j(X)\overline{H}_k(Y)) = \delta_{j,k}\rho^j.$$

(b) *Let (X_1, X_2, X_3, X_4) be a centered Gaussian vector with variance matrix*

$$\Sigma = \begin{pmatrix} 1 & 0 & \rho_{13} & \rho_{14} \\ 0 & 1 & \rho_{23} & \rho_{24} \\ \rho_{13} & \rho_{23} & 1 & 0 \\ \rho_{14} & \rho_{24} & 0 & 1 \end{pmatrix}$$

Then, if $r_1 + r_2 = r_3 + r_4$,

$$\mathrm{E}\big(\overline{H}_{r_1}(X_1)\overline{H}_{r_2}(X_2)\overline{H}_{r_3}(X_3)\overline{H}_{r_4}(X_4)\big) = \sum_{(d_1,d_2,d_3,d_4)\in Z} \frac{r_1!\, r_2!\, r_3!\, r_4!}{d_1!\, d_2!\, d_3!\, d_4!} \rho_{13}^{d_1}\rho_{14}^{d_2}\rho_{23}^{d_3}\rho_{24}^{d_4},$$

where Z is the set of d_i's satisfying : $d_i \geq 0$:

$$d_1 + d_2 = r_1, \qquad d_3 + d_4 = r_2, \qquad d_1 + d_3 = r_3, \qquad d_2 + d_4 = r_4. \quad (10.19)$$

If $r_1 + r_2 \neq r_3 + r_4$ the expectation is equal to zero. Notice that the four equations in (10.19) are not independent and that the set Z is finite and contains, in general, more than one 4-tuple.

Proof. We give a proof of part (b), part (a) being a particular case. We have

$$\mathrm{E}\left(\prod_{i=1}^{4} \exp(t_i X_i - \frac{1}{2}t_i^2)\right) = \exp(\rho_{13}t_1t_3 + \rho_{14}t_1t_4 + \rho_{23}t_2t_3 + \rho_{24}t_2t_4). \quad (10.20)$$

First, we have, by definition,

$$\exp\left(tx - \frac{1}{2}t^2\right) = \sum_{q=0}^{\infty} \frac{t^q \overline{H}_q(x)}{q!}.$$

So the left-hand side of (10.20) is equal to

$$\sum_{r_i=0}^{\infty} \frac{t_1^{r_1} t_2^{r_2} t_3^{r_3} t_4^{r_4}}{r_1!\, r_2!\, r_3!\, r_4!} \mathrm{E}\big(\overline{H}_{r_1}(X_1)\overline{H}_{r_2}(X_2)\overline{H}_{r_3}(X_3)\overline{H}_{r_4}(X_4)\big).$$

Second, the right-hand side of (10.20) is equal to

$$\sum_{r=0}^{\infty} \frac{1}{r!}(\rho_{13}t_1t_3 + \rho_{14}t_1t_4 + \rho_{23}t_2t_3 + \rho_{24}t_2t_4)^r$$

$$= \sum_{r=0}^{\infty} \sum_{d_1+d_2+d_3+d_4=r} \frac{\rho_{13}^{d_1}\rho_{14}^{d_2}\rho_{23}^{d_3}\rho_{24}^{d_4}}{d_1!\, d_2!\, d_3!\, d_4!} t_1^{d_1+d_2} t_2^{d_3+d_4} t_3^{d_1+d_3} t_4^{d_2+d_4}. \quad (10.21)$$

Identifying both sides, it follows that the expectation $\mathrm{E}\big(\overline{H}_{r_1}(X_1)\overline{H}_{r_2}(X_2)\overline{H}_{r_3}(X_3)\overline{H}_{r_4}(X_4)\big)$ is zero if $r_1 + r_2 \neq r_3 + r_4$. In the other cases, the monomial of degree (r_1, r_2, r_3, r_4) on the right hand side of (10.20) corresponds to $r = (r_1 + r_2 + r_3 + r_4)/2$, and it can be found in a unique term in the sum $\sum_{r=0}^{\infty}$. The result follows. □

As an additional hypothesis, we will assume that the process $\mathcal{X}$ has a spectral density $f(\lambda)$. $\mathcal{X}$ has the following spectral representation:

$$X(t) = \sqrt{2}\int_0^{\infty} \left[\cos(t\lambda)\sqrt{f(\lambda)}\,dW_1(\lambda) + \sin(t\lambda)\sqrt{f(\lambda)}\,dW_2(\lambda)\right], \qquad (10.22)$$

where W_1 and W_2 are two independent Wiener processes (Brownian motions) . Indeed, using isometry properties of the stochastic integral, it is easy to see that the process given by (10.22) is centered Gaussian with covariance

$$\begin{aligned}\Gamma(t) = \mathrm{E}(X(s)X(s+t)) &= 2\int_0^{\infty} \cos(\lambda s)\cos(\lambda(t+s))f(\lambda)\,d\lambda \\ &\quad + 2\int_0^{\infty} \sin(\lambda s)\sin(\lambda(t+s))f(\lambda)\,d\lambda \\ &= 2\int_0^{\infty} \cos(\lambda t)f(\lambda)\,d\lambda\end{aligned}$$

Now define the function $\psi(\cdot)$ as the convolution $\mathbf{I}_{[-1/2,1/2]} * \mathbf{I}_{[-1/2,1/2]}$. This function is even, nonnegative, $\psi(0) = 1$, has support included in $[-1, 1]$, and has a nonnegative Fourier transform. Set $\psi_\varepsilon(t) = \frac{1}{2\pi}\psi(\varepsilon t)$ and let $\widehat{\psi}_\varepsilon$ be its Fourier transform. Define

$$X^{\varepsilon}(t) := \sqrt{2}\int_0^{\infty} \left[\cos(t\lambda)\sqrt{f * \widehat{\psi}_\varepsilon(\lambda)}\,dW_1(\lambda) + \sin(t\lambda)\sqrt{f * \widehat{\psi}_\varepsilon(\lambda)}\,dW_2(\lambda)\right], \qquad (10.23)$$

where the convolution must be understood after prolonging f as an even function on $\mathbb{R}$. The covariance function Γ_ε of $X^\varepsilon(t)$ satisfies $\Gamma_\varepsilon(t) = \Gamma(t)\psi(\varepsilon t)$. This implies that the process $X^\varepsilon(t)$ is $(1/\varepsilon)$-dependent. We have the following proposition:

Proposition 10.8. *Let $\mathcal{X}$ be a centered stationary Gaussian process with spectral density $f(\lambda)$ and covariance function Γ, with $\Gamma^\ell \in L^1(\mathbb{R})$, ℓ positive integer. Let $X_\varepsilon(t)$ be defined by (10.23). Then*

$$\lim_{\varepsilon\to 0}\lim_{t\to\infty} \mathrm{E}\left[\frac{1}{\sqrt{t}}\int_0^t \left(\overline{H}_\ell(X(s)) - \overline{H}_\ell(X^\varepsilon(s))\right) ds\right]^2 = 0. \qquad (10.24)$$

Proof. Using Mehler's formula and the change of variables $\tau = s_1 - s_2$ yields

$$\mathrm{E}\left[\frac{1}{\sqrt{t}}\int_0^t \left(\overline{H}_\ell(X(s)) - \overline{H}_\ell(X^\varepsilon(s))\right) ds\right]^2$$

$$= 2\ell!\left(\int_0^t (1-\tau/t)(\Gamma^\ell(\tau) + \Gamma_\varepsilon^\ell(\tau) - 2\rho_\varepsilon^\ell(\tau))\,d\tau\right)$$

$$= 2\ell!\left(\int_0^t (1-\tau/t)(\Gamma_\varepsilon^\ell(\tau) - \Gamma^\ell(\tau))\,d\tau\right.$$

$$\left.+ 2\int_0^t (1-\tau/t)(\Gamma^\ell(\tau) - \rho_\varepsilon^\ell(\tau))\,d\tau\right),$$

where $\rho_\varepsilon(\tau) := \mathrm{E}[X(0)X^\varepsilon(\tau)]$.

Since $|\Gamma_\varepsilon(\tau)|^\ell \leq |\Gamma(\tau)|^\ell$, we see that the first term tends to zero as t tends to infinity and then ε tends to zero, on applying the dominated convergence theorem.

As for the second, we have

$$\int_0^t (1-\tau/t)\big[\Gamma^\ell(\tau) - \rho_\varepsilon^\ell(\tau)\big]d\tau = \int_0^t (1-\tau/t)\,d\tau$$

$$\times \int_{-\infty}^{+\infty} \cos(\lambda\tau)\big[f^{*(\ell)}(\lambda) - g_\varepsilon^{*(\ell)}(\lambda)\big]\,d\lambda,$$

where g_ε is the spectral density $\lambda \rightsquigarrow \sqrt{f(\lambda)}\sqrt{(f * \widehat{\psi_\varepsilon})(\lambda)}$ and $g_\varepsilon^{*(\ell)}$ denotes the convolution of g_ε ℓ times with itself.

Using Fubini's theorem gives us

$$\int_0^t (1-\tau/t)\,d\tau \int_{-\infty}^{+\infty} \cos(\lambda\tau)\big(f^{*(\ell)}(\lambda) - g_\varepsilon^{*(\ell)}(\lambda)\big)\,d\lambda$$

$$= \int_{-\infty}^{+\infty} \frac{1-\cos\lambda t}{t\lambda^2}\big(f^{*(\ell)}(\lambda) - g_\varepsilon^{*(\ell)}(\lambda)\big)\,d\lambda$$

$$= \int_{-\infty}^{+\infty} \frac{1-\cos\lambda}{\lambda^2}\left(f^{*(\ell)}\left(\frac{\lambda}{t}\right) - g_\varepsilon^{*(\ell)}\left(\frac{\lambda}{t}\right)\right)d\lambda. \tag{10.25}$$

When ℓ is equal to 1, the function f and thus g_ε are bounded and continuous and the dominated convergence theorem implies that the limit as $t \to \infty$ of (10.25) is equal to

$$\big[f(0) - g_\varepsilon(0)\big]\int_{-\infty}^{+\infty} \frac{1-\cos\lambda}{\lambda^2}\,d\lambda.$$

This quantity tends to zero as $\varepsilon \to 0$.

When $\ell > 1$, we first prove that $g_\varepsilon^{*\ell}$ is bounded by $\|f * \psi_\varepsilon\|_\infty$. We have

$$g_\varepsilon^{*2}(\lambda) = \int_{-\infty}^{+\infty} g_\varepsilon(\lambda - \lambda_1)g_\varepsilon(\lambda_1)\,d\lambda_1$$

$$\leq \|f * \psi_\varepsilon\|_\infty \int_{-\infty}^{+\infty} f(\lambda - \lambda_1) f(\lambda_1)\, d\lambda_1 \leq \|f * \psi_\varepsilon\|_\infty, \tag{10.26}$$

because of the Cauchy–Schwarz inequality.

For $k > 2$, we use induction. Clearly,

$$\int_{-\infty}^{+\infty} (f * \widehat{\psi}_\varepsilon)(\lambda)\, d\lambda = \Gamma_\varepsilon(0) = \Gamma(0)\psi(0) = 1,$$

so that

$$\begin{aligned} g_\varepsilon^{*(k)}(\lambda) &\leq \|g_\varepsilon^{*(k-1)}\|_\infty \int_{-\infty}^{+\infty} \left((f * \psi_\varepsilon)(\lambda) f(\lambda)\right)^{1/2} d\lambda \\ &\leq \|g_\varepsilon^{*(k-1)}\|_\infty \leq \|f * \psi_\varepsilon\|_\infty. \end{aligned} \tag{10.27}$$

Now $g_\varepsilon(\cdot - \lambda/t)$ converges to $g_\varepsilon(\cdot)$ in $L^1(\mathbb{R})$, as $t \to +\infty$. This is nothing more than the continuity of the translation. The duality between $L^1(\mathbb{R})$ and $L^\infty(\mathbb{R})$ implies that $g_\varepsilon^{*(k)}(\lambda/t) \to g_\varepsilon^{*(k)}(0)$. Using (10.25) and (10.27) gives us

$$\int_0^{+\infty} \left[(1 - \tau/t)\Gamma^\ell(\tau) - \rho_\varepsilon^\ell(\tau)\right] d\tau \to \left(f^{*\ell}(0) - g^{*\ell}(0)\right) \int_{-\infty}^{+\infty} \frac{1 - \cos\lambda}{\lambda^2} d\lambda,$$

as $t \to +\infty$.

Fatou's lemma and the definition of g_ε imply that

$$\liminf_{\varepsilon \to 0} g_\varepsilon^{*(\ell)}(0) \geq f^{*(\ell)}(0). \tag{10.28}$$

On the other hand,

$$\begin{aligned} g_\varepsilon^{*\ell}(0) &= \int_{-\infty}^{\infty} g_\varepsilon(\lambda_{\ell-1}) g_\varepsilon^{*(\ell-1)}(\lambda_{\ell-1})\, d\lambda_{\ell-1} \\ &= \int_{R^{\ell-1}} g_\varepsilon(\lambda_{\ell-1}) g_\varepsilon(\lambda_{\ell-1} - \lambda_{\ell-2}) \cdots g_\varepsilon(\lambda_2 - \lambda_1) g_\varepsilon(\lambda_1)\, d\lambda_1, \ldots, d\lambda_{\ell-1} \\ &\leq \Bigg[\int_{R^{\ell-1}} (f * \widehat{\psi}_\varepsilon)(\lambda_{\ell-1})(f * \widehat{\psi}_\varepsilon)(\lambda_{\ell-1} - \lambda_{\ell-2}) \cdots \\ &\quad (f * \widehat{\psi}_\varepsilon)(\lambda_2 - \lambda_1)(f * \widehat{\psi}_\varepsilon)(\lambda_1)\, d\lambda_1, \ldots, d\lambda_{\ell-1}\Bigg]^{1/2} \\ &\quad \times \Bigg[\int_{R^{\ell-1}} f(\lambda_{\ell-1}) f(\lambda_{\ell-1} - \lambda_{\ell-2}) \cdots f(\lambda_2 - \lambda_1) \\ &\quad f(\lambda_1)\, d\lambda_1, \ldots, d\lambda_{\ell-1}\Bigg]^{1/2} \\ &= [(f * \widehat{\psi}_\varepsilon)^{*(\ell)}(0)]^{1/2} [f^{*(\ell)}(0)]^{1/2} \to f^{*(\ell)}(0) \text{ as } \varepsilon \to 0, \end{aligned} \tag{10.29}$$

using the Cauchy–Schwarz inequality and the continuity of $f^{*(\ell)}$ since Γ^ℓ is in L^1. Summing up, (10.28) and (10.29) imply that $\lim_{\varepsilon\to 0} g_\varepsilon^{*(\ell)}(0) = f^{*(\ell)}(0)$, and we are done. □

Theorem 10.9. *Let $\mathcal{X}$ be a Gaussian process satisfying the hypotheses of Proposition 10.8 and F a function in $L^2(\phi(x)dx)$ with Hermite rank $\ell \geq 1$. Then, as $t \to +\infty$,*

$$\sqrt{t}\, T_t = \frac{1}{\sqrt{t}} \int_0^t F(X(s))\, ds \to N(0, \sigma^2(F)) \text{ in distribution,}$$

where

$$\sigma^2(F) := 2 \sum_{k=\ell}^{\infty} a_k^2 k! \int_0^\infty \Gamma^k(s)\, ds.$$

Proof. Define $F_M := \sum_{n=\ell}^{M} a_n \overline{H}_n(x)$ and $T_t^M := (1/t) \int_0^t F_M(X(s))\, ds$. Let $M = M(\delta) > \ell$ such that

$$2 \sum_{k=M+1}^{\infty} a_k^2 < \delta.$$

Using Mehler's formula, we get

$$\begin{aligned} t\, \mathrm{Var}(T_t - T_t^M) &= 2 \sum_{k=M}^{\infty} c_k^2 k! \int_0^t \left(1 - \frac{s}{t}\right) \Gamma^k(s)\, ds \\ &\leq 2 \sum_{k=M}^{\infty} c_k^2 k! \int_0^\infty |\Gamma|^k(s)\, ds < \delta \int_0^\infty |\Gamma|^\ell(s)\, ds. \end{aligned}$$

Since δ is arbitrary, we only need to prove the asymptotic normality for T_t^M. Let us introduce

$$T_t^{M,\varepsilon} = \frac{1}{t} \int_0^t F_M(X^\varepsilon(s))\, ds,$$

where $X^\varepsilon(t)$ has been defined in (10.23). By Proposition 10.8, recalling that for $k \geq l$, Γ^k is in $L^1(\mathbb{R})$ since Γ^ℓ is, we obtain

$$\lim_{\varepsilon\to 0} \lim_{t\to\infty} t\, \mathrm{Var}(T_t^M - T_t^{M\varepsilon}) = 0.$$

Now Theorem 10.6 for m-dependent sequences implies that $\sqrt{t}\, T_t^{M,\varepsilon}$ is asymptotically normal. Notice that

$$\sigma_{M,\varepsilon} := \lim_{t\to\infty} t\,\mathrm{Var}(T_t^{M\varepsilon}) = 2\sum_{k=0}^{M} a_k^2 k! \int_0^{1/\varepsilon} \Gamma_\varepsilon^k(s)\,ds$$

and that $\sigma_{M\varepsilon} \to \sigma^2(F)$ when $\varepsilon \to 0$ and $M \to \infty$, giving the result. □

10.2.3. Hermite Expansion for Crossings of Regular Processes

Let $\mathcal{X}$ be a centered stationary Gaussian process. With no loss of generality for our purposes, we assume that $\Gamma(0) = -\Gamma''(0) = 1$ and $\Gamma(t) \neq \pm 1$ for $t \neq 0$. We also assume Geman's condition of Proposition 4.2:

$$\Gamma(t) = 1 - t^2/2 + \theta(t) \qquad \text{with} \int \frac{\theta'(t)}{t^2}dt \text{ converges at } 0^+.$$

We define the following expansions:

$$x^+ = \sum_{k=0}^{\infty} a_k \overline{H}_k(x), \qquad x^- = \sum_{k=0}^{\infty} b_k \overline{H}_k(x), \qquad |x| = \sum_{k=0}^{\infty} c_k \overline{H}_k(x). \quad (10.30)$$

We have $a_1 = \frac{1}{2}$, $b_1 = -\frac{1}{2}$, $c_1 = 0$, and using (8.7) and integration by parts for $k > 2$ we obtain

$$a_k = \frac{1}{k!}\int_0^{+\infty} x\overline{H}_k(x)\varphi(x)\,dx = \frac{1}{k!\sqrt{2\pi}}\overline{H}_{k-2}(0).$$

The classical properties of Hermite polynomials easily imply that for positive k,

$$a_{2k+1} = b_{2k+1} = c_{2k+1} = 0$$
$$a_{2k} = b_{2k} = \frac{(-1)^{k+1}}{\sqrt{2\pi}\,2^k k!\,(2k-1)}$$
$$c_{2k} = 2a_{2k}.$$

We have the following Hermite expansion for the number of up-crossings:

Theorem 10.10. *Under the conditions above,*

$$U_u := U_u(X,[0,T]) = \sum_{j=0}^{\infty}\sum_{k=0}^{\infty} d_j(u)a_k \int_0^T H_j(X(s))H_k(X'(s))\,ds \text{ a.s.},$$

where $d_j(u) = (1/j!)\phi(u)\overline{H}_j(u)$ and a_k is defined by (10.30). We have similar results, replacing a_k by b_k or c_k, for the number $D_u(X,[0,T])$ of down-crossings and for the total number of crossings $N_u(X,[0,T])$.

Proof. Let $g(.) \in L^2(\phi(x)dx)$ and define the functional

$$T_g^+(t) = \int_0^t g(X(s))X'^+(s)\,ds.$$

The convergence of the Hermite expansion implies that a.s.

$$T_g^+(t) = \sum_{j=0}^{\infty}\sum_{k=0}^{\infty} g_j\, a_k \int_0^t H_j(X(s))H_k(X'(s))\,ds, \tag{10.31}$$

where the g_j's are the coefficients of the Hermite expansion of g. Using the fact that for each s, $X(s)$ and $X'(s)$ are independent, we get

$$\mathrm{E}\left[\int_0^t \left[g(X(s))|X'(s)| - \sum_{j,k\geq 0:k+j\leq Q} g_j a_k H_j(X(s))H_k(X'(s))\right] ds\right]^2$$
$$\leq (\text{const})t^2 \sum_{j,k\geq 0:k+j\geq Q} j!\, g_j^2 k!\, a_k^2. \tag{10.32}$$

On the other hand, using Proposition 4.2 gives us

$$\nu_2(u,T) = \mathrm{E}\big(U_u([0,T])(U_u([0,T])-1)\big) = \int_0^T 2(T-\tau)A_{0,\tau}^+(u,u)\,d\tau,$$

with

$$A_{0,\tau}^+(u,u) = \mathrm{E}\big(X'^+(0)X'^+(\tau)\big|X(0) = X(\tau) = u\big)p_{X(0),X(\tau)}(u,u) \leq \frac{\theta'(\tau)}{\tau^2}.$$

For every T, $\nu_2(u,T)$ is a bounded continuous function of u and the same holds true for $\mathrm{E}(U_u^2)$. Let us now define

$$U_u^\delta := \frac{1}{2\delta}\int_0^T \mathbb{I}_{|X(t)-u|\leq\delta} X'^+(t)\,dt.$$

In our case, hypotheses $H_{1,u}$ of Lemma 3.1 are a.s. satisfied. This lemma can easily be extended to up-crossings, showing that

$$U_u^\delta \to U_u \text{ a.s.} \quad \text{a } \delta \to 0.$$

By Fatou's lemma,

$$\mathrm{E}\big((U_u)^2\big) \leq \liminf_{\delta\to 0} \mathrm{E}\big((U_u^\delta)^2\big).$$

To obtain an inequality in the opposite sense, we use the Banach formula (3.31) (see Exercise 3.8). To do that, notice that this formula remains valid if on the left-hand side one replaces the total number of crossings by the up-crossings and on the right-hand side $|f'(t)|$ by $f'^{+}(t)$. So, on applying it to the random path $X(\cdot)$, we see that

$$U_u^\delta = \frac{1}{2\delta}\int_{u-\delta}^{u+\delta} U_x\,dx.$$

and using Jensen's inequality, we have

$$\limsup_{\delta\to 0} \mathrm{E}\big((U_u^\delta)^2\big) \le \limsup_{\delta\to 0}\frac{1}{2\delta}\int_{u-\delta}^{u+\delta} \mathrm{E}\big((U_x)^2\big)\,dx = \mathrm{E}\big((U_u)^2\big).$$

So $\mathrm{E}\big((U_u^\delta)^2\big) \to \mathrm{E}\big((U_u)^2\big)$, and since the random variables involved are nonnegative, a standard argument of passage to the limit based on Fatou's lemma shows that $U_u^\delta \to U_u$ in L^2. We now apply (10.31) to U_u^δ:

$$U_u^\delta = \sum_{j,k=0}^{\infty} d_j^\delta(u) a_k \zeta_{jk}, \tag{10.33}$$

where $d_j^\delta(u)$ are the Hermite coefficients of the function $x \rightsquigarrow (1/\delta)\mathbb{I}_{\|x-u\|\le\delta}$ and

$$\zeta_{jk} = \int_0^T H_j(X(s))H_k(X'(s))\,ds.$$

Notice that

$$d_j^\delta(u) \to \frac{1}{j!}\phi(u)\overline{H}_j(u) = d_j(u). \tag{10.34}$$

On the other hand, let us denote by $\mathcal{S}_q$ the closed linear subspace of the L^2 of the probability space, generated by the random variables $\{\zeta_{jk} : j,k \ge 0,\ j+k=q\}$.

Direct application of Mehler's formula's, Lemma 10.7, part (b), plus Fubini's theorem shows that the subspaces $\{\mathcal{S}_q\}_{q=0,1,\dots}$ are pairwise orthogonal. So we may rewrite (10.33) in the form

$$U_u^\delta = \sum_{q=0}^{\infty} \gamma_q^\delta, \tag{10.35}$$

where

$$\gamma_q^\delta = \sum_{j+k=q} d_j^\delta(u) a_k \zeta_{jk} \to \gamma_q := \sum_{j+k=q} d_j(u) a_k \zeta_{jk}.$$

For every integer $Q > 0$, $\sum_{q=0}^{Q} \gamma_q^\delta$ is equal to $\Pi_Q(U_u^\delta)$, where Π_Q is the orthogonal projector on the space generated by the first Q spaces $\mathcal{S}_q$. Using the convergence of U_u^δ and the continuity of the projection,

$$\Pi_Q(U_u^\delta) \to \Pi_Q(U_u).$$

On the other hand,

$$\Pi_Q(U_u^\delta) \to \sum_{q=0}^{Q} \gamma_q \quad \text{as } \delta \to 0.$$

This implies that

$$U = \sum_{q=0}^{\infty} \sum_{j+k=q} d_j(u) a_k \zeta_{jk}. \tag{10.36}$$

□

Theorem 10.11. *Let $\mathcal{X} = \{X(t) : t \in \mathbb{R}\}$ be a centered stationary Gaussian process verifying the conditions at the beginning of this subsection. Furthermore, let us assume that*

$$\int_0^{+\infty} |\Gamma(t)|\, dt, \int_0^{+\infty} |\Gamma'(t)|\, dt, \int_0^{+\infty} |\Gamma''(t)|\, dt < \infty.$$

Let $\{g_k\}_{k=0,1,2,\ldots}$ be a sequence of coefficients that satisfy $\sum_0^{+\infty} g_k^2 k! < \infty$. Set

$$F_t := \frac{1}{\sqrt{t}} \sum_{k,j \geq 0} g_j a_k \int_0^t \overline{H}_j(X(s)) \overline{H}_k(X'(s))\, ds,$$

where a_k has been defined in (10.30). Then

$$F_t - \mathrm{E}(F_t) \to N(0, \sigma^2) \text{ in distribution as } t \to +\infty,$$

where

$$0 < \sigma^2 = \sum_{q=1}^{\infty} \sigma^2(q) < \infty$$

and

$$\sigma^2(q) := \sum_{k=0}^{q} \sum_{k'=0}^{q} a_k a_{k'} g_{q-k} g_{q-k'}$$
$$\int_0^{+\infty} \mathrm{E}\left[\overline{H}_{q-k}(X(0)) \overline{H}_k(X'(0)) \overline{H}_{q-k'}(X(s)) \overline{H}_{k'}(X'(s))\right] ds.$$

The integrand on the right-hand side of this formula can be computed using Lemma 10.7. Similar results exist, mutatis mutandis, for the sequences $\{b_k\}$ and $\{c_k\}$.

A consequence is

Corollary 10.12. *If the process $\mathcal{X}$ satisfies the conditions of Theorem 10.11, then as $T \to +\infty$,*

$$\frac{1}{\sqrt{t}}\left(U_u\big([0,T]\big) - T\frac{e^{-u^2/2}}{2\pi}\right) \to N(0,\sigma_1^2) \text{ in distribution}$$

$$\frac{1}{\sqrt{t}}\left(N_u\big([0,T]\big) - T\frac{e^{-u^2/2}}{\pi}\right) \to N(0,\sigma_2^2) \text{ in distribution,}$$

where σ_1^2 and σ_2^2 are finite and positive.

Remark. The result of Theorem 10.11 is, in fact, true under weaker hypotheses: namely, $\int_0^{+\infty} |\Gamma(t)|\,dt < \infty$, $\int_0^{+\infty} |\Gamma''^2(t)|\,dt < \infty$ [see Theorem 1 of Kratz and León (2001) or Kratz (2006)]. Our stronger hypotheses make it possible to make a proof self-contained and rather short.

Proof of Theorem 10.11. Since Γ is integrable, the process $\mathcal{X}$ admits a spectral density. The hypotheses and the Riemann–Lebesgue lemma imply that

$$\Gamma^{(i)}(t) \to 0 \qquad i = 0,1,2 \quad \text{as } t \to +\infty.$$

Hence, we can choose T_0 so that for $t \ge T_0$,

$$\overline{\Gamma}(t) := \sup\{|\Gamma(t)|, |\Gamma'(t)|, |\Gamma''(t)|\} \le \tfrac{1}{4}. \tag{10.37}$$

Step 1. In this step we prove that one can choose Q large enough so that F_t can be replaced with an arbitrarily small error (in the L^2 sense) by

$$F_t^Q := \frac{1}{\sqrt{t}}\sum_{q=0}^{Q} G_t^q \quad \text{with } G_t^q := \sum_{k=0}^{q} g_{q-k}a_k \int_0^t \overline{H}_{q-k}(X(s))\overline{H}_k(X'(s))\,ds.$$

Let us consider

$$\frac{1}{t}\mathrm{E}\big((G_t^q)^2\big)$$

$$= 1/t \sum_{k,k'=0}^{q} g_{q-k} a_k g_{q-k'} a_{k'} \int_0^t dt_1$$

$$\int_0^t \mathrm{E}\big(\overline{H}_{q-k}(X(t_1))\overline{H}_k(X'(t_1))\overline{H}_{q-k'}(X(t_2))\overline{H}_{k'}(X'(t_2))\big)\, dt_2. \qquad (10.38)$$

To give an upper bound for this quantity, we split it into two parts.

The part corresponding to $|t_1 - t_2| \geq T_0$ is bounded, using Lemma 10.7, by

$$(\text{const}) \sum_{k,k'=0}^{q} |g_{q-k}||a_k||g_{q-k'}||a_{k'}| \int_{T_0}^{t} \sum_{(d_1,d_2,d_3,d_4)\in Z}$$
$$\times \frac{k!\,(q-k)!\,k'!\,(q-k')!}{d_1!\,d_2!\,d_3!\,d_4!} |\Gamma(s)|^{d_1} |\Gamma'(s)|^{d_2+d_3} |\Gamma''(s)|^{d_4}$$
$$\leq (\text{const}) \sum_{k,k'=0}^{q} |g_{q-k}||a_k||g_{q-k'}||a_{k'}| \int_{T_0}^{t} \sum_{(d_1,d_2,d_3,d_4)\in Z}$$
$$\times \frac{k!\,(q-k)!\,k'!\,(q-k')!}{d_1!\,d_2!\,d_3!\,d_4!} \left(\frac{1}{4}\right)^{(q-1)} \overline{\Gamma}(t), \qquad (10.39)$$

where Z is as in Lemma 10.8, setting $r_1 = q - k, r_2 = k, r_3 = q - k'$, and $r_4 = k'$.

Remarking that $\sup_d 1/d!\,(k-d)! \leq 2^k/k!$, it follows that $\frac{k!(q-k)!\,k'!\,(q-k')!}{d_1!\,d_2!\,d_3!\,d_4!}$ in (10.39) is bounded above by $2^q(k')!\,(q-k')!$ or $2^q(k)!\,(q-k)!$, depending on the way that we group terms. As a consequence, it is also bounded above by $2^q\sqrt{(k')!\,(q-k')!\,(k)!\,(q-k)!}$ and the right-hand side of (10.39) is bounded above by

$$(\text{const}) \sum_{k,k'=0}^{q} |g_{q-k}||a_k||g_{q-k'}||a_{k'}| q 2^{-q} \sqrt{(k')!\,(q-k')!\,(k)!\,(q-k)!} \int_0^{+\infty} \overline{\Gamma}(t)\, dt$$
$$\leq (\text{const}) \sum_{k,k'=0}^{q} |g_{q-k}||a_k||g_{q-k'}||a_{k'}| \sqrt{(k')!\,(q-k')!\,(k)!\,(q-k)!}, \qquad (10.40)$$

where we have used that the number of terms in Z is bounded by q.

On the other hand, the integration region in (10.38) corresponding to $|t_1 - t_2| \leq T_0$ can be covered by at most $[t/T_0]$ squares of size $2T_0$. Using Jensen's inequality as we did for the proof of (10.32), we obtain

$$\mathrm{E}\Big(\big(G_{2T_0}^q\big)^2\Big) \leq (\text{const})\, T_0^2 \sum_{k=0}^{q} (q-k)!\,k!\, g_{q-k}^2 a_k^2. \qquad (10.41)$$

Finally,

$$\frac{1}{t}\mathrm{E}\Big(\big(G_t^q\big)^2\Big) \leq (\mathrm{const}) \sum_{k=0}^{q} (q-k)!\, k!\, g_{q-k}^2 a_k^2,$$

which is the general term of a convergent series. This also proves that σ^2 is finite.

STEP 2. Let us prove that $\sigma^2 > 0$. It is sufficient to prove that $\sigma^2(2) > 0$. Recall that $a_1 = 0$, so that

$$\begin{aligned}\sigma^2(2) &= a_0^2 g_2^2 \int_0^{+\infty} \mathrm{E}\big(\overline{H}_2(X(0))\overline{H}_2(X(s))\, ds \\ &\quad + a_2^2 g_0^2 \int_0^{+\infty} \mathrm{E}\big(\overline{H}_2(X'(0))\overline{H}_2(X'(s))\, ds \\ &\quad + 2a_0 g_2 a_2 g_0 \int_0^{+\infty} \mathrm{E}\big(\overline{H}_2(X(0))\overline{H}_2(X'(s))\, ds. \end{aligned} \tag{10.42}$$

Using the Mehler formula yields

$$\begin{aligned}\sigma^2(2) &= 2a_0^2 g_2^2 \int_0^{+\infty} \Gamma^2(s)\, ds \\ &\quad + 2a_2^2 g_0^2 \int_0^{+\infty} (\Gamma''(s))^2\, ds + 4a_0 g_2 a_2 g_0 \int_0^{+\infty} (\Gamma'(s))^2\, ds \\ &= \int_{-\infty}^{+\infty} \big(\lambda^4 a_0^2 g_2^2 + \lambda^2 2a_0 g_2 a_2 g_0 + a_0^2 g^2\big) f^2(\lambda)\, d\lambda \\ &= \int_{-\infty}^{+\infty} \big(\lambda^2 a_2 g_0 + a_0 g_2\big)^2 f^2(\lambda)\, d\lambda > 0. \end{aligned} \tag{10.43}$$

STEP 3. Set

$$F_t^{Q,\varepsilon} := \frac{1}{\sqrt{t}} \sum_{q=0}^{Q} G_t^{q,\varepsilon},$$

with

$$G_t^{q,\varepsilon} = \sum_{k=0}^{q} g_{q-k} a_k \int_0^t \overline{H}_{q-k}(X^\varepsilon(s))\overline{H}_k\big((X^\varepsilon)'(s)\big)\, ds.$$

In this step we prove that F_t^Q can be replaced, with an arbitrarily small error if ε is small enough, by $F_t^{Q,\varepsilon}$. Since the expression of F_t^Q involves only a finite number of terms having the form

$$K_{q-k,k}^0 := \frac{1}{\sqrt{t}} \int_0^t \overline{H}_{q-k}(X(s))\overline{H}_k\big(X'(s)\big)\, ds$$

if ε is small enough, one can substitute, with an arbitrarily small error,

$$K^{\varepsilon}_{q-k,k} := \frac{1}{\sqrt{t}} \int_0^t \overline{H}_{q-k}(X^{\varepsilon}(s))\overline{H}_k\big((X^{\varepsilon})'(s)\big)\, ds.$$

For that purpose we study

$$\begin{aligned} \mathrm{E}(K^0_{q-k,k} - K^{\varepsilon}_{q-k,k})^2 = 2 \int_0^t \frac{t-s}{t} \\ \times \mathrm{E}\Big[\overline{H}_{q-k}(X(0))\overline{H}_k\big(X'(0)\big)\overline{H}_{q-k}(X(s))\overline{H}_k\big(X'(s)\big)\Big] \\ + \mathrm{E}\Big[\overline{H}_{q-k}(X^{\varepsilon}(0))\overline{H}_k\big((X^{\varepsilon})'(0)\big)\overline{H}_{q-k}(X^{\varepsilon}(s))\overline{H}_k\big((X^{\varepsilon})'(s)\big)\Big] \\ - 2\mathrm{E}\Big[\overline{H}_{q-k}(X(0))\overline{H}_k\big(X'(0)\big)\overline{H}_{q-k}(X^{\varepsilon}(s))\overline{H}_k\big((X^{\varepsilon})'(s)\big)\Big]\, ds. \end{aligned} \tag{10.44}$$

Consider the computation of terms of the type

$$\int_0^t \frac{t-s}{t} \mathrm{E}\Big[\overline{H}_{q-k}(Y_1(0))\overline{H}_k\big(Y_1'(0)\big)\overline{H}_{q-k}(Y_2(s))\overline{H}_k\big(Y_2'(s)\big)\Big] ds, \tag{10.45}$$

where the processes $Y_1(t)$ and $Y_2(t)$ are chosen among $\{X(t), X^{\varepsilon}(t)\}$. It suffices to prove that all these terms have the same limit as $t \to +\infty$, and then $\varepsilon \to 0$ whatever the choice is.

Applying Lemma 10.8, the expression in (10.45) is equal to

$$\int_0^t \frac{t-s}{t} \sum_{d_1,\ldots,d_4 \in Z} \frac{(q-k)!^2 k!^2}{d_1!\, d_2!\, d_3!\, d_4!} (\rho(s))^{d_1} (\rho'(s))^{d_2} (-\rho'(s))^{d_3} (-\rho''(s))^{d_4}\, ds,$$

where $\rho(\cdot)$ is the covariance function between the processes Y_1 and Y_2, and Z is defined as in Lemma 10.8. Again, since the number of terms in Z is finite, it suffices to prove that

$$\lim_{\varepsilon \to 0} \lim_{t \to \infty} \int_0^t \frac{t-s}{t} (\rho(s))^{d_1} (\rho'(s))^{d_2+d_3} (\rho''(s))^{d_4}\, ds,$$

where $(d_1, \ldots, d_4)$ is chosen in Z, does not depend on the way to choose Y_1 and Y_2. ρ is the Fourier transform of (say) $g(\lambda)$, which is taken among $f(\lambda)$; $f * \widehat{\psi}_{\varepsilon}(\lambda)$ or $\sqrt{f(\lambda)}\sqrt{f * \widehat{\psi}_{\varepsilon}(\lambda)}$. Define $\overline{g}(\lambda) = i\lambda g(\lambda)$ and $\overline{\overline{g}}(\lambda) = -\lambda^2 g(\lambda)$. Then $(\rho(s))^{d_1} (\rho'(s))^{d_2+d_3} (\rho''(s))^{d_4}$ is the Fourier transform of the function

$$h(\lambda) = g^{*(d_1)}(\lambda) * \overline{g}^{*(d_2+d_3)} * \overline{\overline{g}}^{*(d_4)}(\lambda).$$

The continuity and boundedness of f imply that all the functions above are bounded and continuous. The same reasoning that led to (10.25) shows that

$$\int_0^t \frac{t-s}{t}\rho(s)^{d_1}\rho'(s)^{d_2+d_3}(\rho''(s))^{d_4}\,ds = \int_{-\infty}^{+\infty}\frac{1-\cos\lambda}{\lambda^2}h\left(\frac{\lambda}{t}\right).$$

As $t \to +\infty$, the right-hand side converges, using dominated convergence, to

$$\int_{-\infty}^{+\infty}\frac{1-\cos\lambda}{\lambda^2}h(0)\,d\lambda.$$

The continuity of f now gives the result, as in Proposition 10.8. □

Proof of Corollary 10.12. Some attention must be paid to the fact that the coefficients

$$d_j(u) = \frac{1}{j!}\phi(u)\overline{H}_j(u)$$

do not satisfy $\sum_{j=0}^{\infty} j!\,d_j^2(u) < \infty$. They only satisfy the relation

$$j!\,d_j^2(u) \text{ is bounded.} \tag{10.46}$$

First, we can improve the bound given by the right hand side of (10.40) by reintroducing the factor $q2^{-q}$ that had been bounded by 1. We get that in its new expression, this right-hand side is bounded by

$$\begin{aligned}
&(\text{const})q2^{-q}\sum_{k,k'=0}^{q}|d_{q-k}(u)||a_k||d_{q-k'}(u')||a_{k'}|\sqrt{(k')!\,(q-k')!\,k!\,(q-k)!}\\
&\le (\text{const})q2^{-q}\sum_{k=0}^{q}\left(d_{q-k}(u)\right)^2 a_k^2(k)!(q-k)!\\
&\le (\text{const})q2^{-q}\sum_{k=0}^{q}a_k^2 k! \le (\text{const})q2^{-q}.
\end{aligned}$$

Second, we have to replace the bound (10.41). Since the series in (10.36) is convergent, $\mathrm{E}\left(\left(G_{2T_0}^q\right)^2\right)$ is the term of a convergent series, and this in enough to conclude. □

10.2.4. Extensions to Random Fields

Some of these results can be extended to real-valued random fields to obtain convergence for the geometric measure of level sets corresponding to a fixed hight u, as well as some related functionals defined on them, when the observation window grows to the entire space. More precisely, an article by Iribarren (1989) contains a central limit theorem for integrals on the level set under some regularity and mixing conditions. The main tools are the formulas (6.9) and (6.10). This asymptotic result has been used when $d = 2$ by Cabaña (1987) to provide a method to test the isotropy of the law of the random field on the basis of the observation of level sets. The original idea is simple and fruitful: a deformation in the domain that breaks isotropy is reflected in the length of the level sets, and this can be used to estimate anisotropy. The same idea is used by Wschebor (1985, Chap. 3) for general $d \geq 1$. Some extensions may be found in an article by Kratz and León (2001).

EXERCISES

10.1. Prove Theorem 10.6. *Hint:* Partition the interval $[0, t]$ into $2n - 1$ intervals, (n being a function of t) $I_1, J_1, \ldots, J_{n-1}, I_n$, the J_i's being of size m.

10.2. Let $\{X(t) : t \in \mathbb{R}^d\}$, $d \geq 2$ be a real-valued, centered Gaussian, stationary random field with paths of class C^4 and covariance

$$\Gamma(t) = \mathrm{E}(X(s)X(s+t)), \qquad s, t \in \mathbb{R}^d.$$

We assume the normalization $\Gamma(0) = 1$ and that $-\Gamma''(0) = \mathrm{Var}(X'(t))$ is positive definite. For each $u \in \mathbb{R}$, denote by $\sigma(T, u)$ the $(d-1)$-dimensional geometric measure of the intersection of the inverse image of u with the window $T \subset \mathbb{R}^d$.

(a) Prove that for each Borel subset T of $\mathbb{R}^d$, one has

$$\mathrm{E}\big(\sigma(T, u)\big) = \lambda_d(T)\phi(u)\mathrm{E}\big(\|\xi\|\big),$$

where ξ is a centered random vector with values in $\mathbb{R}^d$, $\mathrm{Var}(\xi) = -\Gamma''(0)$.

(b) Set [see the notation in part a)]

$$c(u) = \phi(u)\mathrm{E}\big(\|\xi\|\big).$$

Under the additional (mixing-type) hypothesis that for $i = 0, 1, 2$,

$$(\log \|t\|)^{1-i/2}\Gamma^{(i)}(t) \to 0 \quad \text{as} \quad \|t\| \to +\infty,$$

prove that for each bounded Borel set $T \subset \mathbb{R}^d$, as $u \to +\infty$, one has

$$\sigma\big((c(u))^{-1/d}T, u\big) \to \lambda_d(T) \tag{10.47}$$

in probability. *Hint:* Apply Rice's formula for $k = 2$.

CHAPTER 11

GEOMETRIC CHARACTERISTICS OF RANDOM SEA WAVES

In this chapter we consider extremely simplified representations of a very complicated phenomenon, and our presentation will not go into the actual fluid dynamics and numerical problems. We consider only a set of limited questions that interest oceanographers, at least since the 1950s: say, since the founding papers of M. S. Longuett-Higgins and collaborators.

The random sea surface will be modeled using a special Gaussian stationary model that appears as a limit of the superposition of infinitely many elementary sea waves obeying the Euler model. For the random surface defined, we consider some geometrical characteristics such as wave length, crests, length, and speed of contours. The various Rice formulas are used to compute expectation or Palm distribution (see the definition below) of such quantities. Some numerical applications are presented and a brief description of some non-Gaussian models is given in Section 11.5.

11.1. GAUSSIAN MODEL FOR AN INFINITELY DEEP SEA

Let us consider a moving incompressible fluid (the water of the sea) in a domain of infinite depth. If one writes the Euler equations, after some approximations one can show that a class of solutions describing the sea level $W(t, x, y)$, where t is the time variable and x and y are space variables, is given by

$$W(t, x, y) = f \cos(\lambda_t t + \lambda_x x + \lambda_y y + \theta), \tag{11.1}$$

Level Sets and Extrema of Random Processes and Fields, By Jean-Marc Azaïs and Mario Wschebor

where f and θ and the amplitude and phase and the pulsations λ_t, λ_x, and λ_y are some parameters that satisfy the *Airy relation*

$$\kappa = \frac{\lambda_t^2}{g} \qquad \text{with } \kappa^2 := \lambda_x^2 + \lambda_y^2, \tag{11.2}$$

where g is the acceleration of gravity. In what follows we assume that units have been chosen so that $g = 1$.

For a suitable random choice of f and θ: namely, independent, f having Rayleigh distribution (see Exercise 3.12) and θ uniform in $[0, 1\pi]$, $W(t, x, y)$ is an elementary Gaussian field called the *sine–cosine process* because it can be written in the form

$$W(t, x, y) = \xi_1 \sin(\lambda_t t + \lambda_x x + \lambda_y y) + \xi_2 \cos(\lambda_t t + \lambda_x x + \lambda_y y), \tag{11.3}$$

where ξ_1 and ξ_2 are two independent standard normal random variables.

Since the Euler equation is linear, a finite sum of elementary waves having the form (11.3) is again a solution. The limit of such a sum as the number of elementary waves tends to infinity is, using the results in Chapter 1, a stationary random field having the particularity that its spectral measure $F(d\lambda_t, d\lambda_x, d\lambda_y)$ lies in the surface defined by the Airy relation (Figure 11.1). This surface is a

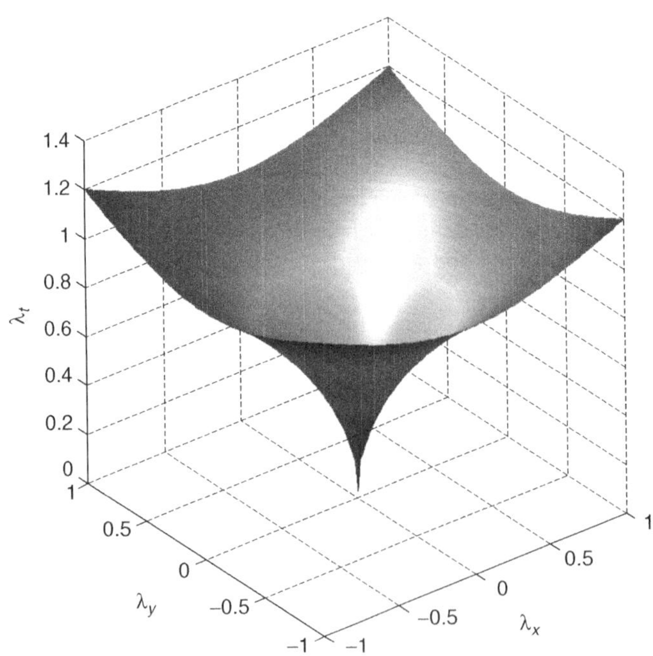

Figure 11.1. Representation of the surface on which the spectral measure lies.

paraboloid having circular sections for constant t. This will be our basic model. It is an approximation that can be valid only over short periods of time (about 1 hour) and over short geographical areas (several kilometers). It is also understood that very long-period phenomena such as tide and surge, have been removed, so that we will also assume that the process is centered.

The symmetry of the distribution implied by the Gaussian hypothesis [i.e., that the random fields $W(t, x, y)$ and $-W(t, x, y)$ have the same law] is considered by certain authors to be a drawback for adequate representation of the true behavior of the sea level. In Section 11.5 we present an extension, intending to take this problem into account.

The covariance function of the process, that is,

$$\Gamma(\Delta t, \Delta x, \Delta y) = \mathrm{E}\{W(t, x, y)W(t + \Delta t, x + \Delta x, y + \Delta y)\},$$

is the Fourier transform of the Borel measure $F(d\lambda_t, d\lambda_x, d\lambda_y)$. Since the spectral measure is symmetric with respect to 0, the lower half of the paraboloid can be removed for our calculations. If we keep only the polar variable κ and α, where α is the angle of the vector (λ_x, λ_y) with the x-axis, we can write

$$\Gamma(\Delta t, \Delta x, \Delta y) = \int_0^{+\infty} \int_0^{2\pi} \cos(\sqrt{\kappa}\, \Delta t + \kappa \cos\alpha\, \Delta x + \kappa \sin\alpha\, \Delta y)\widetilde{G}(d\kappa, d\alpha). \tag{11.4}$$

Here $\widetilde{G}$ is the measure obtained by expressing in polar coordinates the projection of the spectral measure (after removing the lower part) onto the plane (λ_x, λ_y). Notice that this measure does not need to be symmetric, in the sense that it may not be invariant under the transformation $(\kappa, \alpha) \to (\kappa, \alpha + \pi)$.

A standard form to write the spectral representation of the covariance is a slight modification of (11.4). Set $\omega = \lambda_t$ (the pulsation) and make the change of variables $\omega = \lambda_t = \sqrt{\kappa}$. Then

$$\begin{aligned}\Gamma(\Delta t, \Delta x, \Delta y) = \int_0^{+\infty} \int_0^{2\pi} & \cos\big(\omega\, \Delta t + \omega^2 \cos(\alpha\, \Delta x) \\ & + \omega^2 \sin(\alpha\, \Delta y)\big)G(d\omega, d\alpha).\end{aligned} \tag{11.5}$$

G is the spectral measure of the random wave in the sense of *wave community*. It is a nonnegative measure expressed in m^2/s, which is a unit of power. G is called the *directional power spectrum*. More details on wave modeling may be found, for example, in books by Kinsman (1965) and Ochi (1998).

11.2. SOME GEOMETRIC CHARACTERISTICS OF WAVES

For the time being, the observation of sea level is performed by indirect methods. As far as the authors know, registration of height as a function of the three variables t, x, and y is not available, and measurements are often limited to the spectrum, in the sense of (11.5), computed as solutions of certain inverse problems. So a very important problem is to deduce from these spectra some information on the geometry of the waves.

11.2.1. Time Waves

Suppose for the moment that the location (x, y) is fixed and we consider the level $W(t) = W(t, x, y)$ as a function of the time variable only. The length and height of waves can be defined in various ways. The definitions given by Lindgren and Rychlik (1995) are the following: Let t_3 be a down-crossing of zero "chosen at random" (this notion will be defined precisely later using the Palm distribution) and consider (see Figure 11.2):

- t_1, the last up-crossing of zero preceding t_3
- The point (t_2, w_2), where the maximum between t_1 and t_3 is attained
- t_5, the first up-crossing of zero following t_3
- The point (t_4, w_4), where the minimum between t_3 and t_5 is attained

Then the wave is defined as the part of the curve between t_1 and t_5, its length is $L = t_5 - t_1$, its height is $H = w_2 - w_4$, and its half length can be defined as $t_5 - t_3$ or $t_3 - t_1$.

Other definitions exist, based on local extrema [again, see Lindgren and Rychlik (1995) and references therein].

Definition 11.1 (Palm Distribution). *Let $\{T_i, M_i\}_{i=1,2,\dots}$ be a stationary marked point process. This means that $\{T_i\}_{i=1,2,\dots}$ is a point process on the real line (see Chapter 10), and to each point T_i is attached a random variable M_i, "the mark,"*

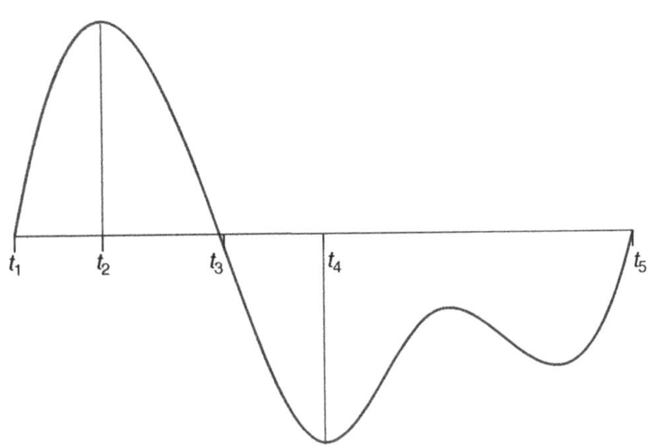

Figure 11.2. Remarkable points in the definition of wave length and wave height.

which takes its values in a measurable space E. Then for every measurable subset B of E, the Palm distribution of B is given by

$$\mathcal{P}(B) := \frac{\mathrm{E}(\#\{T_i \in [0, T] : M_i \in B\})}{\mathrm{E}(\#\{T_i \in [0, T]\})}. \tag{11.6}$$

Because of the stationarity, this quantity does not depend on the value of $T > 0$.

If the process is defined on the real line and is ergodic (see Chapter 10), then, almost surely, one has

$$\mathcal{P}(B) = \lim_{T\to\infty} \frac{(\#\{T_i \in [0, T] : M_i \in B\})}{\#\{T_i \in [0, T]\}},$$

so that the Palm measure can be estimated in a consistent way as $T \to +\infty$ by means of the quotient on the right-hand side of this formula, on the basis of the observation of the point process in the window $[0, T]$.

On the other hand, when applied to random waves, according to the definition given by (11.6), the Palm measure can be computed using one-parameter weighted Rice formulas. A basic example is the following:

Proposition 11.2 (Rychlik, 1987). *Let $\{X(t), t \in \mathbb{R}\}$ be a centered stationary Gaussian process satisfying the conditions of Theorem 6.2. The density of the Palm distribution of the half-wave period $T_5 - T_3$ is*

$$p_{T_5-T_3}(\tau) = (\mathrm{const})\, p_{X(0),X(\tau)}(0, 0) \\ \times \mathrm{E}\big(X'(0)X'(\tau)\mathbb{1}_{X(s)\le 0, \forall s\in[o,\tau]} \big| X(0) = X(\tau) = 0\big).$$

See Exercise 11.1, which contains a hint for the proof.

11.3. LEVEL CURVES, CRESTS, AND VELOCITIES FOR SPACE WAVES

Let $\mathcal{Z} = \{Z(x, y) : (x, y) \in \mathbb{R}^2\}$ be a real-valued two-parameter centered stationary Gaussian process with differentiable paths. The part of the level curve corresponding to level u contained in the Borel set S is

$$\mathcal{C}_u(Z, S) = \{(x, y) \in S : Z(x, y) = u\}.$$

Its mean length $\mathrm{E}\big(\mathcal{L}(\mathcal{C}_u(Z, S))\big)$ is given by the Rice formula for random fields (Theorem 6.8).

Theorem 11.3. *Assume that the process $\mathcal{Z}$ satisfies the conditions of Theorem 6.8. Then, with the notations above,*

$$\mathrm{E}\big(\mathcal{L}(\mathcal{C}_u(Z, S))\big) = \lambda_2(S) p_Z(u) \mathrm{E}\big(\|Z'(0, 0)\|\big) = \sqrt{\frac{2}{\pi}} \lambda_2(S) p_Z(u) \sqrt{\gamma_2}\, \mathcal{E}(k), \tag{11.7}$$

where Σ is the variance matrix of $Z'(0,0)$, $\gamma_2 > \gamma_1$ are its eigenvalues; $k^2 := (1 - \gamma_1/\gamma_2)$; $\mathcal{E}(k) := \int_0^{\pi/2} (1 - k^2 \sin^2\theta)^{1/2}\, d\theta$ is the elliptic integral of the first kind, and p_Z is the density of $Z(x, y)$.

Proof. Applying Theorem 6.8, we get

$$E\big(\mathcal{L}(\mathcal{C}_u(Z,S))\big) = \int_S E\big(\|Z'(x,y)\| \big| Z(x,y) = u\big) \times p_{Z(x,y)}(u)\, dx\, dy = \lambda_2(S) p_Z(u) E\|Z'(0,0)\|,$$

because of stationarity. This proves the first relation.

As for the second, after diagonalization of $\mathrm{Var}(Z'(0,0))$, $\|Z'(0,0)\|$ can be represented by $\|\sqrt{\gamma_1}\xi_1 + \sqrt{\gamma_2}\xi_2\|$, where ξ_1 and ξ_2 are two independent standard normal variables. Passing to polar coordinates, we have

$$\begin{aligned} E\big(\|Z'(0,0)\|\big) &= \frac{1}{2\pi}\int_0^{+\infty} d\rho \int_0^{2\pi} \sqrt{\gamma_1 \sin^2\theta + \gamma_2 \cos^2\theta}\, \rho^2 e^{-\rho^2/2}\, d\theta \\ &= \sqrt{\frac{2}{\pi}}\sqrt{\gamma_2}\int_0^{\pi/2} \sqrt{\cos^2\theta + (\gamma_1/\gamma_2)\sin^2\theta\, d\theta} \\ &= \sqrt{\frac{2}{\pi}}\sqrt{\gamma_2}\int_0^{\pi/2} \sqrt{1 + \Big(\frac{\gamma_1}{\gamma_2} - 1\Big)\sin^2\theta\, d\theta}. \end{aligned}$$

□

Remarks

1. One can find this formula in an article by Longuett-Higgins [1957, formula (2.3.13)].

2. Formula (11.7) gives a generalization to every level u of Corrsin's formula (1955). This formula was established for $u = 0$ in a different manner. It says that

$$\frac{E\,[\mathcal{L}(\mathcal{C}_u(Z,S))]}{\lambda_2(S)} = \frac{1}{4}\int_0^{2\pi} \tilde{E}(N_u^\theta)\, d\theta,$$

where $\tilde{E}(N_u^\theta)$ is the expectation per unit of space of the number of crossings in the θ direction. By Rice's formula,

$$\tilde{E}(N_u^\theta) = \sqrt{\frac{2m_{2,\theta}}{\pi}}\, p_Z(u),$$

where $m_{2,\theta}$ is the second spectral moment in the direction θ. Without loss of generality we can assume that the direction in the plane has been chosen to diagonalize the variance matrix of Z'. Then

$$\Sigma(\theta) = \begin{pmatrix} \gamma_2 & 0 \\ 0 & \gamma_1 \end{pmatrix}$$

and $m_{2,\theta} = \sqrt{\gamma_2}(1 - (1 - \gamma_1/\gamma_2)\sin^2\theta)^{1/2}$ so that the right-hand side of (11.7), for S having Lebesgue measure equal to 1, is equal to

$$p_Z(u)\sqrt{\frac{2}{\pi}}\sqrt{\gamma_2}\int_0^{\pi/2}\left(1 - \left(1 - \frac{\gamma_1}{\gamma_2}\right)\sin^2\theta\right)^{1/2} d\theta = p_Z(u)\sqrt{\frac{2}{\pi}}\sqrt{\gamma_2}\,\mathcal{E}(k).$$

11.3.1. Length of a Crest

A *crest* is defined as a local maximum in a given direction, say θ, of the sea surface modeled as in Section 11.1. First we define a static crest at a fixed time (say, $t = 0$) as

$$C^s(S, \theta) := \{(x, y) \in S;\ W'_\theta(x, y) = 0;\ W''_\theta(x, y) < 0\},$$

where W'_θ and W''_θ are, respectively, the first and second derivatives of the field $W(x, y)$ in the θ direction of the (x, y) plane at point $(x, y, 0)$. Since θ is the direction of a straight line, it can be chosen in $[0, \pi)$.

It is also possible to define a moving crest as

$$\begin{aligned} C^m(S, T, \theta) := \{&(z\cos\theta, z\sin\theta, t) \in S \times [0, T]; \\ &W'_\theta(z\cos\theta, z\sin\theta, t) = 0;\ W''_{\theta\theta}(z\cos\theta, z\sin\theta, t) < 0\}. \end{aligned}$$

See Azaïs et al. (2005) for more details.

Proposition 11.4. *To simplify the presentation, and without loss of generality, we assume that $\theta = 0$. Let us define the spectral moments*

$$m_{ijk} = \int_0^\infty \int_0^{2\pi} (\omega^2 \cos\alpha)^i (\omega^2 \sin\alpha)^j \omega^k G(d\omega, d\alpha),$$

where G is as defined in (11.5). Set $m_{ij} = m_{ij0}$, with the definitions above and if the process W'_θ satisfies the conditions of Theorem 6.8. Then

$$\mathrm{E}\big(\mathcal{L}(C^s(S, \theta))\big) = \frac{\lambda_2(S)\sqrt{\gamma_2}}{2\pi(a_{11})^{1/2}}\mathcal{E}(k),$$

where $k = \sqrt{1 - \gamma_1/\gamma_2}$; $a_{11} = \mathrm{E}[W'_x(0, 0, 0)^2] = m_{20}$ and $\gamma_2 > \gamma_1$ are the eigenvalues of Σ, the variance matrix of the gradient of W'_x:

$$\Sigma(\theta) = \begin{pmatrix} m_{40} & m_{31} \\ m_{31} & m_{22} \end{pmatrix}.$$

Proof. Let $Z(x, y) = \partial W(x, y, 0)/\partial x$. Then $C^s_{S,0}$ can be written as

$$C^s_{S,0} = \{(x, y) \in S;\ Z(x, y) = 0;\ Z'_x(x, y) < 0\},$$

where Z'_x stands for $\partial Z(x, y)/\partial x$. Thus,

$$\mathrm{E}(\mathcal{L}(C^s_{S,0})) = \mathrm{E} \int_{\mathcal{C}_0(Z,S)} \mathbb{I}_{\{Z'_x(x,y)<0\}} d\sigma,$$

where $\mathcal{C}_0(Z, S) = \{(x, y) \in S : Z(x, y) = 0\}$. Since Z' and $-Z'$ have the same distribution,

$$\mathrm{E}(\mathcal{L}(C^s_{S,0})) = \mathrm{E} \int_{\mathcal{C}_0(Z,S)} \mathbb{I}_{\{Z'_x(x,y)\geq 0\}}\, d\sigma = \frac{1}{2}\mathrm{E}(\mathcal{L}(\mathcal{C}_0(Z, S)).$$

Applying Theorem 11.3, we get

$$\mathrm{E}(\mathcal{L}(C^s_{S,0})) = \frac{1}{2\pi} \frac{\lambda_2(S)}{(a_{11})^{1/2}} \sqrt{\gamma_2}\, \mathcal{E}(k)$$

with $a_{11} = \mathrm{Var}(Z(x, y))$, and $\gamma_2 > \gamma_1$ are the eigenvalues of the variance matrix of Z'. □

Remark. When W is an elementary wave of the form

$$W(x, y) = \xi_1 \cos(\lambda_x x + \lambda_y y) + \xi_2 \sin(\lambda_x x + \lambda_y y),$$

where ξ_1 and ξ_2 are two standard normal variables, direct computations on the sine–cosine process show that

$$\mathrm{E}(\mathcal{L}(C^s_{S,0})) = \frac{\lambda_2(S)\sqrt{\lambda_x^2 + \lambda_y^2}}{2\pi}. \tag{11.8}$$

Thus, the length of the crest is a nonlinear functional of the spectrum.

11.3.2. Velocity of Contours

In this section we give a more rigorous basis to some heuristic considerations of Longuett-Higgins (1957). Other approaches to the same problem have been proposed by Podgórski et al. (2000) and Baxevani et al. (2003), where several notions of velocity are introduced, including the one used here, called *velocity in the direction of the gradient*. Our results are different in the sense that we look at the two components of the gradient while the authors cited express their results in terms of the joint distribution of the modulus and the angle.

Speed of Crossings. Let us fix y (say, $y = 0$). We want to study the speed of a crossing of a given level u chosen "at random" among all the crossings. Define S_0 as the section of S in the direction of the x-axis. Using stationarity, it is always possible to suppose that $S_0 = [0, M]$ for some value M. Also, by stationarity we can look at the speed of the sea at time 0. A crossing is a point x such that

$$W(x, 0, 0) = u. \tag{11.9}$$

The expectation of the number of crossings N_u is given by Rice's formula,

$$D := \mathrm{E}(N_u) = M\sqrt{\frac{2m_{200}}{\pi}}\, p_Z(u).$$

The speed of such crossings can be computed using the implicit function theorem. From (11.9) we get that

$$C_x(x) := \frac{dx}{dt} = -\frac{W'_t(x, 0, 0)}{W'_x(x, 0, 0)}. \tag{11.10}$$

The mean number of crossings with speed C_x in the interval $[\alpha_1, \alpha_2]$ ($\alpha_1 < \alpha_2$) can also be computed using a Rice formula. If the spectral measure S defined in (11.5) is not reduced to a unit atom, then

$$N := \mathrm{E}(N_u \mathbf{I}_{C_x \in [\alpha_1, \alpha_2]}) = \int_{\alpha_1}^{\alpha_2} dc \int_0^M dx \int_{-\infty}^{\infty} |x'| p_{W, W'_x, C_x}(u, x', c)\, dx', \tag{11.11}$$

where p_{W, W'_x, C_x} is the joint density of $(W(x, 0, 0), W'_x(x, 0, 0), C_x(x, 0, 0))$ which does not depend on x because of stationarity. As the values of the process and its derivative at a given point are independent random variables, we get

$$N = M p_W(u) \int_{\alpha_1}^{\alpha_2} dc \int_{-\infty}^{\infty} |x'| p_{W'_x, C_x}(x', c)\, dx'.$$

The probability of a crossing chosen at random to have a speed in the range $[\alpha_1, \alpha_2]$ is therefore N/D. Now divide by $\alpha_2 - \alpha_1$ and let both α_1 and α_2 tend to a common limit c, and we get that the distribution of the speed of the crossing is given by

$$\tilde{p}_{C_x}(c) = \sqrt{\frac{\pi}{2m_{200}}} \int_{-\infty}^{\infty} |x'| p_{W'_x, C_x}(x', c)\, dx',$$

where

$$p_{W'_x, W'_t}(x', t') = \frac{\Delta^{-1/2}}{2\pi} \exp\left(-\frac{1}{2\Delta}\left(m_{002}x'^2 - 2m_{101}x't' + m_{200}t'^2\right)\right)$$

with

$$\Delta = \det \begin{pmatrix} m_{200} & m_{101} \\ m_{101} & m_{002} \end{pmatrix}.$$

Making the change of variables $c = -t'/x'$, we get

$$\tilde{p}_{C_x}(c) = \tfrac{1}{2}\Delta(m_{002} + 2m_{101}c + m_{200}c^2)^{-3/2}(m_{200})^{-1/2}.$$

This is (with a slightly different notation) formula (2.5.14) of Longuett-Higgins (1957), where it is shown that it can also be written as

$$\tilde{p}_{C_x}(c) = \tfrac{1}{2}\Delta m_{200}^{-2}((c - \widehat{c}) + \Delta m_{200}^{-2})^{-3/2},$$

showing that this distribution is symmetric around its mean value $\widehat{c} = -m_{101}/m_{200}$. An important point is that this speed does not depend on the level.

11.3.3. Velocity of Level Curves

To define the normal velocity of a level curve, we fix a point $P = (0, x_0, y_0)$ such that $W(0, x_0, y_0) = u$ and consider:

- The level surface in time and space:

$$C_1 := \big\{(t, x, y) : W(t, x, y) = u; \\ \text{for } (t, x, y) \text{ in some neighborhood of } (0, x_0, y_0)\big\}.$$

- The level curve at a fixed time:

$$C_2 := \big\{(x, y) : W(0, x, y) = u; \\ \text{for } (x, y) \text{ in some neighborhood of } (x_0, y_0)\big\}.$$

In an infinitesimal interval of time the point P moves to $P' = P + dt\ \vec{v}$, where

- $\vec{v}$ is in the tangent space to C_1. So, $\vec{v}$ is orthogonal to the gradient of W, that is, (W'_t, W'_x, W'_y) (the derivatives are computed at point P).
- The t-coordinate of $\vec{v}$ is equal to 1.
- Define $\vec{V} = (V_x, V_y)$ as the orthogonal projection of $\vec{v}$ onto the x, y-plane. $\vec{V}$ is the *normal velocity to the curve* if it is orthogonal to C_2 at point P.

Then, V_x and V_y satisfy the equations

$$W'_t + V_x W'_x + V_y W'_y = 0$$
$$V_x W'_y - V_y W'_x = 0$$

and it is easy to deduce that

$$V_x = -\frac{W'_t W_x}{(W'_x)^2 + (W'_y)^2} \qquad V_y = -\frac{W'_t W_y}{(W'_x)^2 + (W'_y)^2}.$$

Following Longuett-Higgins, it is simpler to obtain first the distribution of (K_x, K_y) with $K_x = -W'_t/W'_x$ and $K_y = -W'_t/W'_y$ and then pass to the distribution of the velocity using the change-of-variables formula. As in the preceding proof, we consider two intervals $[\alpha_1, \alpha_2]$, $\alpha_1 < \alpha_2$ and $[\alpha_3, \alpha_4]$, $\alpha_3 < \alpha_4$, for $t = 0$ and define

$$D := \mathrm{E}(\mathcal{L}(\mathcal{C}_u(Z, S))) = |S| p(u) \mathrm{E}\| W_{xy}(0, 0, 0)\|,$$

where $Z(x, y) = W(0, x, y)$ and W_{xy} is the gradient limited to the variables x and y and

$$N := \mathrm{E}\left[\int_{\mathcal{C}_u(Z,S)} \mathbb{1}_{K_x \in [\alpha_1, \alpha_2]} \mathbb{1}_{K_y \in [\alpha_3, \alpha_4]} \, d\sigma\right]. \tag{11.12}$$

This expectation can be computed using Rice's formula for integrals on a level set (Theorem 6.10) as soon as the process $W(t, x, y)$ satisfies the hypotheses of Theorem 6.8.

$$\begin{aligned} N &= \lambda_2(S) p_W(u) \mathrm{E}\left[\| W_{xy}\| \mathbb{1}_{K_x \in [\alpha_1, \alpha_2]} \mathbb{1}_{K_y \in [\alpha_3, \alpha_4]}\right] \\ &= \lambda_2(S) p_W(u) \int_{\mathbb{R}^3} \sqrt{x'^2 + y'^2}\, \mathbb{1}_{\{-x'/t' \in [\alpha_1, \alpha_2]\}} \mathbb{1}_{\{-y'/t' \in [\alpha_3, \alpha_4]\}} \\ &\quad \times p_{W'_x, W'_y, W'_t}(x', y', t') \, dx' \, dy' \, dt'. \end{aligned}$$

Making the change of variables $k_x = -x'/t', k_y = -y'/t', t' = t'$ with $dx' \, dy' \, dt' = t'^2 \, dk_x \, dk_y \, dt'$, after some calculations we get

$$\begin{aligned} N &= 4\lambda_2(S) p_W(u) \pi^{-2} \Delta_2^{-1/2} \\ &\cdot \int_{\alpha_1}^{\alpha_2} \int_{\alpha_3}^{\alpha_4} dk_x \, dk_y \sqrt{k_x^2 + k_y^2} \\ &[\mu_{11} k_x^2 + 2\mu_{12} k_x k_y - 2\mu_{13} k_x + \mu_{22} k_y^2 - 2\mu_{23} k_y + \mu_{33})]^{-2}, \end{aligned}$$

where Δ_2 and μ_{ij} are, respectively, the determinant, and the entries of the inverse matrix, of

$$\begin{pmatrix} m_{200} & m_{110} & m_{101} \\ m_{110} & m_{020} & m_{011} \\ m_{101} & m_{011} & m_{002} \end{pmatrix}. \tag{11.13}$$

Letting α_1 and α_2 tend to k_x and α_3 and α_4 tend to k_y, we get the joint density of K_x and K_y:

$$\begin{aligned} \tilde{p}_{K_x,K_y}(k_x,k_y)) &= \lim_{\substack{\alpha_1,\alpha_2 \to k_x \\ \alpha_3,\alpha_4 \to k_y}} \frac{1}{(\alpha_2-\alpha_1)(\alpha_4-\alpha_3)} \frac{N}{D} \\ &= \frac{1}{\pi}(\gamma_2)^{-1/2}\Delta_2^{-1/2}(\mathcal{E}(k))^{-1}\sqrt{k_x^2+k_y^2}[\mu_{11}k_x^2 \\ &\quad + 2\mu_{12}k_xk_y - 2\mu_{13}k_x + \mu_{22}k_y^2 - 2\mu_{23}k_y + \mu_{33})]^{-2}, \end{aligned}$$

where $k = \sqrt{1-\gamma_1/\gamma_2}$, as before, but γ_1 and γ_2 are the eigenvalues of the matrix

$$\begin{pmatrix} m_{200} & m_{110} \\ m_{110} & m_{020} \end{pmatrix}.$$

Again we find the same result as in Longuett-Higgins [1957, eq. (2.6.21)].

We look now at the distribution of the velocity

$$\vec{V} = (V_x, V_y) = \left(\frac{K_x}{K_x^2+K_y^2}, \frac{K_y}{K_x^2+K_y^2}\right)$$

so that

$$dK_x dK_y = (V_x^2+V_y^2)^{-2} dV_x\, dV_y.$$

As a consequence,

$$\begin{aligned} \tilde{p}_{V_x,V_y}(v_x,v_y) &= \frac{1}{\pi}(\gamma_2)^{-1/2}\Delta_2^{-1/2}(\mathcal{E}(k))^{-1}(v_x^2+v_y^2)^{7/2} \\ &\quad \cdot [\mu_{11}v_x^2 + 2\mu_{12}v_xv_y - 2\mu_{13}v_x(v_x^2+v_y^2) + \mu_{22}v_y^2 \\ &\quad - 2\mu_{23}v_y(v_x^2+v_y^2) + \mu_{33}(v_x^2+v_y^2)^2)]^{-2}. \end{aligned}$$

Velocity of Crests. Since the distributions of W and $-W$ are the same, the mean velocity of a crest is the mean speed of the zero level set for the process W'_θ. Thus the same result holds, changing the meaning of the moments in matrix (11.13).

11.4. REAL DATA

In this section we present a numerical application from Azaïs et al. (2005). We consider two directional spectra, depicted in Figure 11.3. We now compare the geometric characteristics of the random seas corresponding to these spectra.

Figure 11.4 shows the expected length of static crests along directions, showing a maximum at approximately 1.3 rad. It is interesting to observe that, in

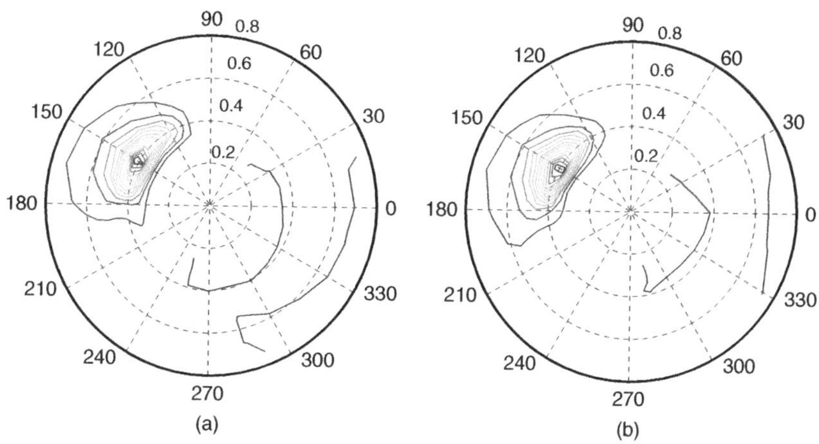

Figure 11.3. Representation of the two spectra: (a) spectrum 1; (b) spectrum 2. (Courtesy of M. Prevosto, Ifremer, Paris.)

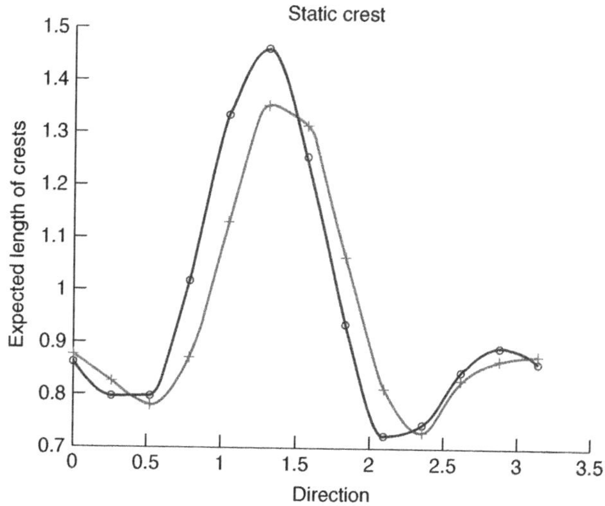

Figure 11.4. Expected length of static crests per unit area. Circles; spectrum 1; crosses, spectrum 2.

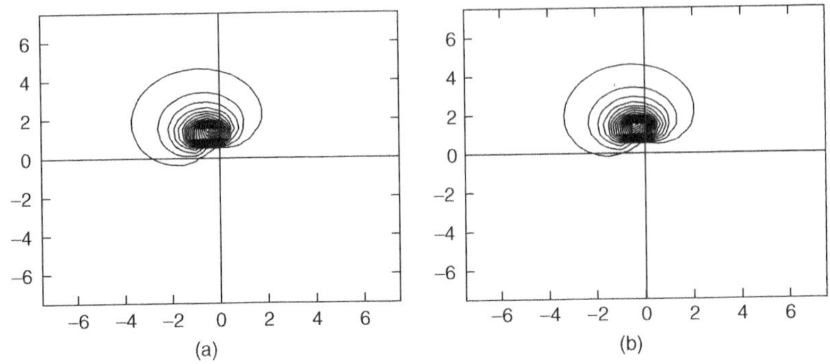

Figure 11.5. Representation of the distributions of the velocities of contours: (a) spectrum 1; (b) spectrum 2.

accordance with theoretical results, this direction is orthogonal to the direction for the maximum integral of the spectrum, which is the most probable direction for the waves.

Figure 11.5 shows the level curves for probability densities of the velocity of a level contour of $W(x, y, t)$. Both graphs show a clear asymmetry as predicted by Longuett-Higgins (1957). The distributions are clearly different, although the spectra differ only slightly.

11.5. GENERALIZATIONS OF THE GAUSSIAN MODEL

The crest–trough symmetry of the Gaussian model does not correspond exactly to reality, especially in the critical situation of very high waves. In practice, it is often observed that the crests are peaked and narrow while the troughs are wide and shallow. This can affect significantly the distribution of the slopes of the waves as well as the extremal behavior. These are two important issues used to study wave slamming on ships or offshore structures. Unfortunately, the description and understanding of non-Gaussian models appears to be very difficult, so that effective generalizations are based on "nearly Gaussian" models. We sketch two of them here: the transformed Gaussian models and the Lagrange models. We limit ourselves to one- or two-dimensional models.

11.5.1. Transformed Gaussian Models

Let us consider the elevation $W(t)$ of the sea as a function of time. We assume that it follows a model that has the form

$$W(t) = \mu + G(X(t)), \tag{11.14}$$

where $X(t)$ is a stationary Gaussian process and $G(\cdot)$ is a "nice function." Such an equation has several advantages: (1) computations are tractable because they can

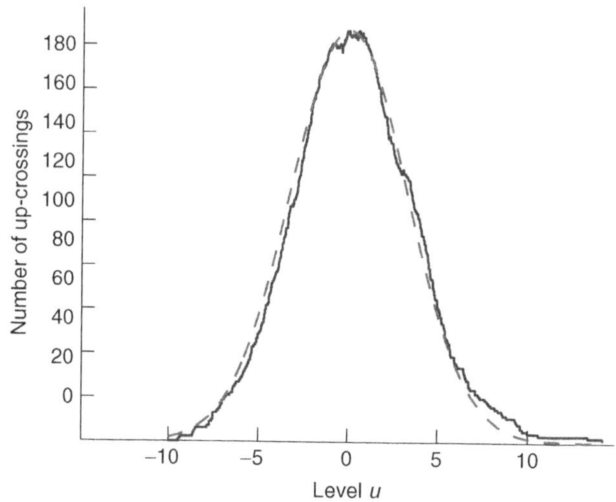

Figure 11.6. Number of up-crossings of a level for data from hurricane Camilla. The dashed line represents the expectation under the Gaussian model.

be conducted on the underlying Gaussian process $X(t)$, and (2) the transformation (11.14) modifies the extremal behavior as described by Azaïs et al. (2009). Notice that this is not the case with the Lagrange model (Section 11.5.2).

The function G can be:

- A polynomial, in which case it is convenient to use a low-degree polynomial (say, 4) and represent it in the Hermite basis (see Azaïs et al., 2009, and references therein). The estimation of $G(\cdot)$ is based on the marginal density of the process $W(t)$ and uses the method of moments.
- Nonparametric, as in Rychlik et al. (1997). In this case the function G can be estimated by the intensity of crossings. We give an example of extreme situations corresponding to a registration of hurricane Camilla in 1969. Figure 11.6 shows a small discard from normality in the high levels.

11.5.2. Lagrange Models

We return to the equations of a incompressible fluid. Under less crude assumptions than for the Euler equation, we obtain the *Lagrange model*, which has as a main characteristic that water particles have a circular movement around a mean position. Random models issued from that model are described in a paper by Lindgren (2006). The sea surface (depending on t, x) is described as a parametric surface depending on t (the time) and a dummy parameter u which is close to the location. It represents the mean position around which particles are moving.

The sea surface is written as

$$(t, u) \rightsquigarrow \big(X(t, u), W(t, u)\big),$$

where $W(t, u)$ is the height of the sea at the location $X(t, u)$ at time t. The two random fields $X(t, u)$ and $W(t, u)$ are jointly Gaussian and described by the stochastic integrals

$$W(t, u) = \int_{\mathbb{R}} \exp\big(i(\kappa(\lambda)u - \lambda t)\big)\, d\xi(\lambda)$$

$$X(t, u) = u + \int_{\mathbb{R}} i \frac{\cosh(\kappa(\lambda)h)}{\sinh(\kappa(\lambda)h)} \exp\big(i(\kappa(\lambda)u - \lambda t)\big)\, d\xi(\lambda),$$

where h is the water depth, $\kappa(\lambda)$ is defined (up to the sign, the choice of which defines different types of waves) by the relation $\lambda^2 = |\kappa| \tanh(|\kappa|h)$, and ξ is a complex spectral process with orthogonal increment satisfying $d\xi(-\lambda) = \overline{d\xi(\lambda)}$.

This model is used primarily to compute distribution of steepness of waves that differ significantly from the Euler model.

EXERCISES

11.1. Prove Proposition 11.2. Let F_n a continuous approximation of the function $\mathbb{I}_{.>0}$ as defined for Example 6.7, define $Y^t(s) = X(s)$ and

$$g(t, Y^t) = F_n[\sup_{s \in [t, t+\tau]} X(s)]$$

and use a monotone convergence argument.

11.2. Prove formula (11.8) by a direct computation. Consider now the case of a spectrum G with two atoms, which is the sum of two spectra of elementary waves. Show that the length of the crest is not a linear function of the spectrum.

11.3. Prove formula (11.11). Prove first that if the spectral measure G is not restricted to a Dirac measure, the joint distribution of the derivatives W_t and W_x does not degenerate. Second, replacing the indicator function $\mathbb{I}_{[\alpha_1, \alpha_2]}$ by a continuous approximation, prove (11.11) using Theorem 6.4. Then conclude.

11.4. Prove formula (11.12) using Theorem 6.10 and the same type of approximation as in Exercise 11.3.

11.5. Give a detailed version of the last argument of Section 11.3.3 concerning the velocity of crests.

CHAPTER 12

SYSTEMS OF RANDOM EQUATIONS

In this chapter we use Rice formula to study the number of real roots of a system of random equations. Our emphasis is on polynomial systems, even though we also give some results on nonpolynomial systems. Let us consider m polynomials in m variables with real coefficients:

$$X_i(t) = X_i(t_1, \ldots, t_m) \qquad i = 1, \ldots, m.$$

We use the notation

$$X_i(t) := \sum_{\|j\| \leq d_i} a_j^{(i)} t^j, \tag{12.1}$$

where $j := (j_1, \ldots, j_m)$ is a multi-index of nonnegative integers, $\|j\| := j_1 + \cdots + j_m$, $j! := j_1! \cdots j_m!$, $t = (t_1, \ldots, t_m) \in \mathbb{R}^m$, $t^j := t_1^{j_1} \cdots t_m^{j_m}$, $a_j^{(i)} := a_{j_1 \ldots, j_m}^{(i)}$. The degree of the ith polynomial is d_i and we assume that $d_i \geq 1\, \forall i$.

We denote by $N^X(V)$ the number of roots of the system of equations

$$X_i(t) = 0 \qquad i = 1, \ldots, m. \tag{12.2}$$

lying in the Borel subset V of $\mathbb{R}^m$. We denote $N^X = N^X(\mathbb{R}^m)$. Let us randomize the coefficients of the system. In the case of one equation in one variable, a certain number of results on the probability distribution of the number of roots

Level Sets and Extrema of Random Processes and Fields, By Jean-Marc Azaïs and Mario Wschebor

have been known for a long time, starting in the 1930s with the work of Bloch and Polya (1932), Littlewood and Offord (1938, 1939), and especially, of Kac (1943). We are not going to consider this special subject [see, e.g., the book by Bharucha-Reid and Sambandham (1986)].

Instead, when $m > 1$, little is known of the distribution of the random variables $N^X(V)$ or N^X, even for simple choices of the probability law on the coefficients. This appears to be quite different and much more difficult than one equation only, and it is this case that we consider in this chapter. In fact, we will be especially interested in large systems, in the sense that $m \gg 1$. In the last 15 years some initial progress has been made in the understanding of distributional properties of the number of roots. The first important result in this context is the Shub–Smale theorem (1993), in which the authors computed by means of a simple formula the expectation of N^X when the coefficients are Gaussian centered independent random variables with certain specified variances (see Theorem 12.1). Extensions of their work, including new results for one polynomial in one variable, can be found in the review paper by Edelman and Kostlan (1995) (see also Kostlan, 2002).

There is, of course, curiosity about the number of roots: for example, being able to answer the question as to whether the system has no real roots (i.e., $N^X = 0$), or, in the random case, what one can say about $\mathrm{P}(N^X = 0)$ or $\mathrm{P}(N^X > n)$, where n is a meaningful integer for the underlying problem. More deeply, the study of the number of roots is associated with natural questions in numerical analysis and complexity theory. Generally speaking, the complexity in solving a system of equations numerically is naturally related to the number of roots. So, understanding the mean (or probabilistic) behavior of an algorithm with respect to a family of problems of this sort is associated with the distribution of the random variable N^X. On the other hand, the condition number of a system of equations which in this case measures the difficulty for an algorithm to separate roots, is related to analogous problems and plays a central role in complexity computations. We are not going to pursue this subject here; the interested reader can consult the book by Blum et al. (1998).

It is obvious that distribution of the number of roots will depend on the probability law that we put on the coefficients of the system. So the first question is: What conditions should we require for this law? As we said above, only a restricted family of distributions has been considered until now. The Shub–Smale distribution on the coefficients is invariant under the orthogonal group of the underlying space $\mathbb{R}^m$ and is related to the H. Weyl L^2-structure in the space of polynomial systems [see also the book by Blum et al. (1998) on this subject].

In Section 12.2 we review some results that extend the computation of the expectation to some other probability laws on the coefficients, which have a centered Gaussian law that is invariant under the orthogonal group of $\mathbb{R}^m$. This allows us to extend substantially the family of examples and to show that the behavior of the expectation of the number of roots can be very different from the one in the Shub–Smale theorem.

We have also included some recent asymptotic results for variances, but only for the Shub–Smale model with equal degrees (which we call Kostlan–Shub–Smale). The main tool is Rice's formula to compute the factorial moments of the number of zeros of a random field (see Theorem 6.3), and the asymptotics is for large systems, meaning by that $m \to +\infty$. We are only giving a brief sketch of the proofs, which turn out to require lengthy calculations, at least when using the methods available. At present, a major open problem is to show weak convergence of some renormalization of N^X, under the same asymptotics.

In Section 12.3 we consider *smooth analysis*; that is, we start with a nonrandom system and perturb it with some noise. The question is: What can we say about the number of roots of the perturbed system under some reasonable hypotheses on the relationship between "signal" and "noise"? Here again, we are only able to give results having some interest when the number m of equations and unknowns becomes large.

Finally, in Section 12.4 we consider random systems having a probability law that is invariant under translations as well as orthogonal transformations of the underlying Euclidean space. This implies that the system is nonpolynomial and the expectation of N^X is infinite in nontrivial cases. So one has to localize and consider $N^X(V)$ for subsets V of $\mathbb{R}^m$ having finite Lebesgue measure. These systems are interesting by themselves, and under general conditions, one can use similar methods to compute the expected number of roots per unit volume as well as to understand the behavior of the variance as the number of unknowns m tends to infinity, which turns out to be strikingly opposite the one in the Kostlan–Shub–Smale model for polynomial systems.

All the above concerns "square" systems. We have not included results on random systems having fewer equations than unknowns. If the system has n equations and m unknowns with $n < m$, generically the set of solutions will be $(m - n)$-dimensional, and the description of the geometry becomes more complicated (and more interesting) than for $m = n$. A recent contribution to the calculation of the expected value of certain parameters describing the geometry of the (random) set of solutions is given by Bürgisser (2007).

12.1. THE SHUB–SMALE MODEL

We say that (12.2) is a Shub–Smale system if the coefficients

$$\{a_j^{(i)} : i = 1, \dots, m; \ \|j\| \le d_i\}$$

are centered independent Gaussian random variables such that

$$\operatorname{Var}\left(a_j^{(i)}\right) = \binom{d_i}{j} = \frac{d_i!}{j!\,(d_i - \|j\|)!}. \tag{12.3}$$

12.1.1. Expectation of N^X

Theorem 12.1 (Shub-Smale, 1993). *Let the system (12.2) be a Shub–Smale system. Then*

$$\mathrm{E}(N^X) = \sqrt{D}, \tag{12.4}$$

where $D = d_1 \cdots d_m$ is the Bézout number of the polynomial system.

Proof. For $i = 1, \ldots, m$, let $\tilde{X}_i$ denote the homogeneous polynomial of degree d_i in $m+1$ variables associated with X_i; that is,

$$\tilde{X}_i(t_0, t_1, \ldots, t_m) = \sum_{\sum_{h=0}^{m} j_h = d_i} a^{(i)}_{j_1,\ldots,j_m} t_0^{j_0} t_1^{j_1} \cdots t_m^{j_m}$$

and Y_i denote the restriction of $\tilde{X}_i$ to the unit sphere S^m in $\mathbb{R}^{m+1}$. It is clear that

$$N^X = \tfrac{1}{2} N^Y(S^m). \tag{12.5}$$

A simple computation using (12.3) and the independence of the coefficients shows that $\tilde{X}_1, \ldots, \tilde{X}_m$ are independent Gaussian centered random fields, with covariances given by

$$r^{\tilde{X}_i}(t, t') = \mathrm{E}(\tilde{X}_i(t)\tilde{X}_i(t')) = \langle t, t' \rangle^{d_i}, \quad t, t' \in \mathbb{R}^{m+1} \qquad i = 1, \ldots, m. \tag{12.6}$$

Here $\langle \cdot, \cdot \rangle$ denotes the usual scalar product in $\mathbb{R}^{m+1}$.

For $\mathrm{E}(N^Y(S^m))$ we apply Rice's formula to the random field Y defined on the parameter set S^m:

$$\mathrm{E}(N^Y(S^m)) = \int_{S^m} \mathrm{E}(|\det(Y'(t))| \,|\, Y(t) = 0) \frac{1}{(2\pi)^{m/2}} \sigma_m(dt), \tag{12.7}$$

where $\sigma_m(dt)$ stands for the m-dimensional geometric measure on S^m. Equation (12.7) follows easily from the fact that for each $t \in S^m$, the random variables $Y_1(t), \ldots, Y_m(t)$ are i.i.d. standard normal.

Since $\mathrm{E}(Y_i^2(t)) = 1$ for all $t \in S^m$, on differentiating under the expectation sign, we see that for each $t \in S^m$, $Y(t)$ and $Y'(t)$ are independent, and the condition can be erased in the conditional expectation on the right-hand side of (12.7).

Since the law of $Y'(t)$ is invariant under the orthogonal group of $\mathbb{R}^{m+1}$, it suffices to compute the integrand at one point of the sphere. Denote the canonical basis of $\mathbb{R}^{m+1}$ by $\{e_0, e_1, \ldots, e_m\}$. Then

$$\mathrm{E}(N^Y(S^m)) = \sigma_m(S^m) \frac{1}{(2\pi)^{m/2}} \mathrm{E}(|\det(Y'(e_0))|)$$

$$= \frac{2\pi^{(m+1)/2}}{\Gamma((m+1)/2)} \frac{1}{(2\pi)^{m/2}} \mathrm{E}\big(|\det(Y'(e_0))|\big). \qquad (12.8)$$

To compute the probability law of $Y'(e_0)$, let us write it as an $m \times m$ matrix with respect to the orthonormal basis $e_1, \ldots, e_m$ of the tangent space to S^m at e_0. This matrix is

$$\left(\left(\frac{\partial \tilde{X}_i}{\partial t_j}(e_0)\right)\right)_{i,j=1,\ldots,m}$$

and

$$\mathrm{E}\left(\frac{\partial \tilde{X}_i}{\partial t_j}(e_0)\frac{\partial \tilde{X}_{i'}}{\partial t_{j'}}(e_0)\right) = \delta_{ii'}\frac{\partial^2 r^{\tilde{X}_i}}{\partial t_j \partial t'_{j'}}\bigg|_{t=t'=e_0} = d_i \delta_{ii'} \delta_{jj'}.$$

The last equality follows computing derivatives of the function $r^{\tilde{X}_i}$ given by (12.6). So

$$\det(Y'(e_0)) = \sqrt{D}\det(G), \qquad (12.9)$$

where G is an $m \times m$ matrix with i.i.d. standard normal entries.

To finish, we need only compute $\mathrm{E}(|\det(G)|)$. One way to do it is to observe that $|\det(G)|$ is the volume (in $\mathbb{R}^m$) of the set

$$\left\{ v \in \mathbb{R}^m : v = \sum_{k=1}^{m} \lambda_k g_k, \quad 0 \le \lambda_k \le 1, k = 1, \ldots, m \right\},$$

where $\{g_1, \ldots, g_m\}$ are the columns of G. Then using the invariance of the standard normal law in $\mathbb{R}^m$ with respect to isometries, we get

$$\mathrm{E}(|\det(G)|) = \prod_{k=1}^{m} \mathrm{E}(\|\eta_k\|_{(k)}),$$

where η_k is standard normal in $\mathbb{R}^k$, and in this chapter we use the notation $\|x\|_{(k)}$ for Euclidean norm of $x \in \mathbb{R}^k$. Also, $\|x\| = \|x\|_{(m)}$. An elementary computation gives

$$\mathrm{E}(\|\eta_k\|_{(k)}) = \sqrt{2}\frac{\Gamma((k+1)/2)}{\Gamma(k/2)},$$

which implies that

$$\mathrm{E}(|\det(G)|) = \frac{1}{\sqrt{2\pi}} 2^{(m+1)/2}\Gamma((m+1)/2).$$

Using (12.9), (12.8), and (12.5), we get the result. □

Remark When the hypotheses of Theorem 12.1 are verified, and moreover, all the degrees d_i $(i = 1, \ldots, m)$ are equal, formula (12.4) was first proved by Kostlan. In what follows we call such a model the *KSS* (Kostlan–Shub–Smale) *model*.

12.1.2. Variance of the Number of Roots

We restrict this section to the KSS model. In this case, a few asymptotic results have been proved on variances, when the number m of unknowns tends to ∞. More precisely, consider the normalized random variable

$$n^X = \frac{N^X}{\sqrt{D}}.$$

It is an obvious consequence of Theorem 12.1 that $\mathrm{E}(n^X) = 1$. Let us denote $\sigma_{m,d}^2 = \mathrm{Var}(n^X)$.

We have:

Theorem 12.2. *Assume that the random polynomial system (12.1) is a KSS system with common degree equal to d, $d \geq 2$, and assume that $d \leq d_0 < \infty$, where d_0 is some constant independent of m. Then, as $m \to +\infty$:*

- *If $d = 2$, $\mathrm{Var}(n^X) \approx \frac{1}{2}\frac{\log m}{m}$.*
- *If $d = 3$, $\mathrm{Var}(n^X) \approx \frac{3}{2}\frac{\log m}{m^2}$.*
- *If $d \geq 4$, $\mathrm{Var}(n^X) \approx \frac{K_d}{m^{3\wedge(d-2)}}$ where $K_4 = \frac{15}{2}$, $K_d = \frac{3465}{64}$ if $d \geq 5$.*

Remark. A simple but interesting corollary of the fact that $\mathrm{Var}(n^X)$ tends to zero as $m \to +\infty$ is that $n^X = N^X/d^{m/2}$ tends to 1 in probability. Using a similar method, it is also possible to obtain the same type of result if we allow d to tend to infinity, slowly enough. For $d \geq 3$ one can find a proof of this weaker result in a paper by Wschebor (2005). Notice that the theorem above is more precise; it gives the equivalent of the normalized variance as $m \to +\infty$.

Proof of Theorem 12.2. We use the same notation as at the beginning of this section. We have

$$\mathrm{Var}(n^X) = \frac{\mathrm{E}((N^X)^2)}{d^m} - 1 = \frac{1}{4}\frac{\mathrm{E}(N^Y(N^Y - 1))}{d^m} + \frac{1}{2d^{m/2}} - 1. \tag{12.10}$$

We will not perform detailed computations of the proof, which turn out to be somewhat heavy, but only sketch the main steps and give some more details in the cases $d = 2$ and $d > 5$. The remaining ones are similar and the detailed computations are given by Wschebor (2007). The general scheme is the following: We show that the first term on the right-hand side of (12.10) has the form $1 + \alpha_m$, where α_m has the speed in the statement. This will be sufficient.

We use Rice's formula:

$$\mathrm{E}\big(N^Y(N^Y - 1)\big) = \int_{S^m \times S^m} \mathrm{E}\big(|\det(Y'(s)||\det(Y'(t)|\big|Y(s) = Y(t) = 0\big)$$
$$\cdot\, p_{Y(s),Y(t)}(0,0)\sigma_m(ds)\sigma_m(dt), \tag{12.11}$$

where

$$p_{Y(s),Y(t)}(0,0) = \frac{1}{(2\pi)^m\big(1 - \langle s,t\rangle)^{2d}\big)^{m/2}}.$$

Conditional Expectation

- Let $s, t \in S^m$ be linearly independent.
- $v_2, \ldots, v_m$ pairwise orthogonal, $v_k \perp s, t$ for $k = 2, \ldots, m$.
- $B_s = \{v_1', v_2, \ldots, v_m\}$ orthonormal basis of the tangent space $T_s(S^m) = s^\perp$ (in $\mathbb{R}^{m+1}$).
- $B_t = \{v_1'', v_2, \ldots, v_m\}$ orthonormal basis of $T_t(S^m) = t^\perp$ (in $\mathbb{R}^{m+1}$).

We express the derivatives $Y'(s)$ and $Y'(t)$ in the basis B_s and B_t, respectively, and compute the covariances of the pairs of coordinates. This is standard calculation. Once this has been done, we can perform the Gaussian regression of the matrices $Y'(s)$ and $Y'(t)$ on the condition $Y(s) = Y(t) = 0$ and replace the conditional expectation in (12.11) by

$$d^m\mathrm{E}\big(|\det(M^s)||\det(M^t)|\big),$$

where the matrices M^s and M^t have the following joint law:

- (M^s_{ik}, M^t_{ik}) $(i, k = 1, \ldots, m)$ are independent bivariate Gaussian centered random vectors.
- For $i = 1, \ldots, m; k = 2, \ldots, m$,

$$\mathrm{E}\big((M^s_{ik})^2\big) = \mathrm{E}\big((M^t_{ik})^2\big) = 1$$
$$\mathrm{E}\big(M^s_{ik}M^t_{ik}\big) = \langle s,t\rangle^{d-1}.$$

- $$\sigma^2 = \mathrm{E}\big((M^s_{i1})^2\big) = \mathrm{E}\big((M^t_{i1})^2\big) = 1 - \frac{d\langle s,t\rangle^{2d-2}}{1 + \langle s,t\rangle^2 + \cdots + \langle s,t\rangle^{2d-2}}$$
$$\tau = \mathrm{E}\big((M^s_{i1}M^t_{i1})\big) = \langle s,t\rangle^{d-2}\Big[1 - \frac{d}{1 + \langle s,t\rangle^2 + \cdots + \langle s,t\rangle^{2d-2}}\Big].$$

Gaussian Regression of M^t_{ik} on M^s_{ik}. For $i = 1, \ldots, m; k = 2, \ldots, m$:

$$M^t_{ik} = M^t_{ik} - \langle s, t\rangle^{d-1} M^s_{ik} + \langle s, t\rangle^{d-1} M^s_{ik} = \zeta_{ik} + \langle s, t\rangle^{d-1} M^s_{ik},$$

where $E(\zeta^2_{ik}) = 1 - \langle s, t\rangle^{2d-2}$ and ζ_{ik} is independent of all the rest.

For $i = 1, \ldots, m; k = 1$, the regression has the form

$$M^t_{i1} = M^t_{i1} - \frac{\tau}{\sigma^2} M^s_{i1} + \frac{\tau}{\sigma^2} M^s_{i1} = \zeta_{i1} + \frac{\tau}{\sigma^2} M^s_{i1},$$

with $E(\zeta^2_{i1}) = \sigma^2 - \tau^2/\sigma^2$, and again, ζ_{i1} is independent of all the rest.

Notice that if $d = 2$, one has $\sigma^2 = (1 - \langle s, t\rangle^2)/(1 + \langle s, t\rangle^2) = -\tau$, which obviously implies that $M^t_{i1} = -M^s_{i1}$ a.s. for $i = 1, \ldots, m$. So we distinguish in the computation between $d = 2$ and $d > 2$. In the last case, $|\tau| < \sigma^2$.

Case $d = 2$. Replace the results above on the right-hand side of (12.11). We get

$$E\big(N^Y(N^Y - 1)\big) = \frac{2^m}{(2\pi)^m} \iint_{S^m \times S^m} \frac{1}{(1 - \langle s, t\rangle^4)^{m/2}} \Delta\sigma_m(ds)\sigma_m(dt), \quad (12.12)$$

where:

- $\Delta = E\big(|\det(M^s)\det(M^t)|\big)$.
- $(M^s_{ik}, M^t_{ik})_{i,k=1,\ldots,m}$ are centered Gaussian independent pairs.
- $E((M^s_{ik})^2) = E((M^t_{ik})^2) = 1$ and $E(M^s_{ik}M^t_{ik}) = \langle s, t\rangle$ for $i = 1, \ldots, m; k = 2, \ldots, m$.
- $\sigma^2 = E((M^s_{i1})^2) = (1 - \langle s, t\rangle^2)/(1 + \langle s, t\rangle^2)$ and $M^t_{i1} = -M^s_{i1}$ for $i = 1, \ldots, m$.

We divide the integral in (12.12) into two parts: I_1 is the integral over the pairs $(s, t) \in S^m \times S^m$ such that $|\langle s, t\rangle| \geq \delta_m = 1/m^\delta$, where $\delta > 0$ will be chosen afterward in such a way that this part is negligible, and I_2 over the pairs such that $|\langle s, t\rangle| \leq \delta_m$. The second part is the relevant part.

Bound for I_1. We take common factor $\sqrt{1 - \langle s, t\rangle^2}/1 + \langle s, t\rangle^2$ in the first column of both matrices A^s and A^t. Since each of the resulting matrices is standard normal, we apply the Cauchy–Schwarz inequality to bound $E\big(|\det(M^s)||\det(M^t)|\big)$ and use the fact that if G is a Gaussian standard $m \times m$ matrix, then

$$E\big((\det(G))^2\big) = m!.$$

So

$$I_1 \leq \frac{2^m}{(2\pi)^m} \iint_{|\langle s,t\rangle| \geq \delta_m} \frac{1}{(1 - \langle s, t\rangle^4)^{m/2}} m! \frac{1 - \langle s, t\rangle^2}{1 + \langle s, t\rangle^2} \sigma_m(ds)\sigma_m(dt).$$

Now we use the invariance of the integrand and the measure under the orthogonal group and the form of the volume element on S^m. So, rewriting the right-hand side, we obtain

$$
\begin{aligned}
I_1 &\le \frac{2^m}{(2\pi)^m}\sigma_m(S^m)\sigma_{m-1}(S^{m-1})m! \int_{|t_0|\ge\delta_m} \frac{1}{(1+t_0^2)^{(m/2)+1}}\, dt_0 \\
&\le (\text{const})2^m m^{1-\delta} \exp\left(-\frac{1}{2}m^{1-2\delta}\right),
\end{aligned}
$$

using the usual Stirling's formula and the choice $\delta_m = 1/m^\delta$. If $0 < \delta < \frac{1}{2}$, this shows that $I_1/2^m$ goes to zero faster that any power of m.

Equivalent for I_2. This is finer than the bound for I_1. For the conditional expectation, we use the computation of the absolute value of the determinant as the volume of the parallelotope generated by the columns. Based on the invariance of the standard normal distribution under the isometries of the underlying Euclidean space and using an expression similar to the one used to obtain the bound for I_1, we see that Δ in (12.12) can be written as

$$
\begin{aligned}
\Delta &= (1-\langle s,t\rangle^2)^{(m-1)/2}\mathrm{E}(\|\xi_m\|^2)\frac{1-\langle s,t\rangle^2}{1+\langle s,t\rangle^2} \\
&\quad\cdot\left[\prod_{k=1}^{m-1}\mathrm{E}\left(\|\xi_k\|_{(k)}\left\|\eta_k + \frac{\langle s,t\rangle}{(1-\langle s,t\rangle^2)^{1/2}}\xi_k\right\|_{(k)}\right)\right]
\end{aligned}
$$

and

$$
\begin{aligned}
I_2 &= \frac{2^m}{(2\pi)^m}\sigma_m(S^m)\sigma_{m-1}(S^{m-1})m\prod_{k=1}^{m-1}\mathrm{E}(\|\xi_k\|_{(k)})^2\int_{-\delta_m}^{\delta_m}\frac{(1-t_0^2)^{(m-1)/2}}{(1+t_0^2)^{(m+2)/2}} \\
&\quad\cdot\prod_{k=1}^{m-1}\frac{\mathrm{E}\big(\|\xi_k\|_{(k)}\big\|\eta_k + (t_0/(1-t_0^2)^{1/2})\xi_k\big\|_{(k)}\big)}{[\mathrm{E}(\|\xi_k\|_{(k)})]^2}\, dt_0. \qquad (12.13)
\end{aligned}
$$

On replacing the various terms here and making the change of variables $t_0 = \theta/\sqrt{m}$, we obtain

$$
\begin{aligned}
\frac{I_2}{4.2^m} &= \frac{1}{\sqrt{\pi}}\frac{\Gamma((m+1)/2)}{\Gamma(m/2)}\, m\int_{-\delta_m}^{\delta_m}\frac{(1-t_0^2)^{(m-1)/2}}{(1+t_0^2)^{(m+2)/2}} \\
&\quad\cdot\prod_{k=1}^{m-1}\frac{\mathrm{E}\big(\|\xi_k\|_{(k)}\big\|\eta_k + (t_0/(1-t_0^2)^{1/2})\xi_k\big\|_{(k)}\big)}{[\mathrm{E}(\|\xi_k\|_{(k)})]^2}\, dt_0
\end{aligned}
$$

$$= C_m \int_{-m^{(1/2)-\delta}}^{m^{(1/2)-\delta}} \frac{(1-\theta^2/m)^{(m-1)/2}}{(1+\theta^2/m)^{(m+2)/2}}$$

$$\cdot \prod_{k=1}^{m-1} \frac{\mathrm{E}\big(\|\xi_k\|_{(k)} \big\| \eta_k + \big[(\theta/\sqrt{m})/(1-\theta^2/m)^{1/2}\big]\xi_k \big\|_{(k)}\big)}{[\mathrm{E}(\|\xi_k\|_{(k)})]^2}\, d\theta, \quad (12.14)$$

where

$$C_m = \frac{1}{\sqrt{\pi m}} \frac{\Gamma((m+1)/2)}{\Gamma(m/2)}.$$

We want to write this expression for $I_2/4.2^m$ in the form $1+\alpha_m$. This is based on the following lemmas, some of which are well known, and the remaining ones are proved by standard computations.

Lemma 12.3. *We have the expansion, valid for real z, $z \to +\infty$ [see Erdelyi et al. (1953), p. 57]:*

$$\Gamma(z) = e^{-z} z^{z-1/2} (2\pi)^{1/2} \Big(1 + \frac{1}{12z} + \frac{1}{288z^2} - \frac{139}{51840z^3} - \cdots\Big).$$

Lemma 12.4. *For $c \in \mathbb{R}$ and $k = 1, 2, \ldots$, let*

$$G_k(c) = \mathrm{E}\left(\left[(\eta_1 + c)^2 + \eta_2^2 + \cdots + \eta_k^2\right]^{1/2}\right),$$

where the random variables $\eta_1, \ldots, \eta_k$ are i.i.d. standard normal. Then

1. $G_k(0) = \sqrt{2}[\Gamma((k+1)/2)/\Gamma(k/2)]$.
2. $G_k'(0) = 0$.
3. $0 \le G''(c) \le G''(0) = (1/k)G_k(0)$ *for all c.*
4. $|G_k'''(c)| \le 3(\sqrt{2/\pi} + |c|)\mathrm{E}\big(\|\eta\|_{(k-1)}^{-3}\big)$, *where η is standard normal in $\mathbb{R}^{k-1}$. This inequality has some interest if $k \ge 5$ since otherwise the right-hand side is infinite.*

Proof of parts (1), (2), and (3) is given in Lemma 12.13.

Lemma 12.5. $C_m = \dfrac{1}{\sqrt{2\pi}}\left[1 - \dfrac{1}{4m} + \dfrac{1}{32m^2} + \dfrac{5}{128m^3} + O\left(\dfrac{1}{m^4}\right)\right].$

Lemma 12.6. *For $k = 1, 2, \ldots$; j an integer, set $m_{kj} = \mathrm{E}(\|\xi\|_k^j)$. Then*

$$m_{kj} = 2^{j/2} \frac{\Gamma((j+k)/2)}{\Gamma(k/2)}.$$

Lemma 12.7. *For fixed integer j, we have*

$$m_{kj} = k^{\frac{j}{2}} \left[1 + \frac{1}{k}\left(\frac{j^2}{4} - \frac{j}{2}\right) + \frac{1}{k^2}\left(-\frac{j^3}{12} + \frac{j^2}{4} - \frac{j}{6}\right) \right.$$
$$\left. + \frac{1}{k^3}\left(\frac{j^4}{24} - \frac{5}{24}j^3 - \frac{j^2}{12} + \frac{j}{6}\right) + O(1/k^4) \right],$$

where the bound in O depends on j.

Proof of Theorem 12.2 (cont.). With these ingredients and some additional effort, one can prove that

$$\frac{I_2}{4.2^m} = \left(1 - \frac{1}{4m} + O\left(\frac{1}{m^2}\right)\right)\left[\int_{-m^{(1/2)-\delta}}^{m^{(1/2)-\delta}} \frac{1}{\sqrt{2\pi}} e^{-\theta^2/2}\left(1 + \frac{\theta^2}{2}\frac{\log m}{m}\right) d\theta + O(1/m)\right]$$
$$= \left(1 - \frac{1}{4m} + O\left(\frac{1}{m^2}\right)\right)\left[\int_{-\infty}^{+\infty} \frac{1}{\sqrt{2\pi}} e^{-\theta^2/2}\left(1 + \frac{\theta^2}{2}\frac{\log m}{m}\right) d\theta + O(1/m)\right]$$
$$= 1 + \frac{1}{2}\frac{\log m}{m} + O(1/m). \tag{12.15}$$

This shows that

$$\operatorname{Var}(n^X) \approx \frac{1}{2}\frac{\log m}{m}$$

and finishes the computation when $d = 2$.

Case $d \geq 3$. Instead of the general formula in the case $d = 2$, we have

$$\mathrm{E}\big(N^Y(N^Y - 1)\big) = \frac{d^m}{(2\pi)^m} \iint_{S^m \times S^m} \frac{1}{(1 - \langle s, t\rangle^{2d})^{m/2}} \Delta \sigma_m(ds)\sigma_m(dt), \tag{12.16}$$

where

$$\Delta = (1 - \langle s, t\rangle^{2d-2})^{(m-1)/2}\, (\sigma^4 - \tau^2)^{1/2}$$
$$\cdot \left[\prod_{k=1}^{m-1} \mathrm{E}\Big(\|\xi_k\|_{(k)} \|\eta_k + \frac{\langle s, t\rangle^{d-1}}{(1 - \langle s, t\rangle^{2d-2})^{1/2}}\xi_k\|_{(k)}\Big)\right.$$
$$\left. \cdot\, \mathrm{E}\Big(\|\xi_m\|_{(m)} \|\eta_m + \frac{\tau}{(\sigma^4 - \tau^2)^{1/2}}\xi_m\|_m\Big)\right].$$

In each factor, ξ and η are independent standard normal vectors in $\mathbb{R}^k$. The proof that the part of the integral corresponding to the pairs $(s, t) \in S^m \times S^m$ such that $|\langle s, t\rangle| \geq \delta_m$, with $\delta_m = 1/m^\delta$, $0 < \delta < \frac{1}{2}$ is negligible is similar to the case $d = 2$ [take into account that $\sigma^2 \leq (\text{const})(1 - \langle s, t\rangle^2)$] and the question is again the equivalent of the integral over the set $|\langle s, t\rangle| < \delta_m$. The overall computation is similar to the case $d = 2$, with some minor differences.

Using as above the invariance under isometries, we have to consider the new integral:

$$\frac{I_2}{4d^m} = \frac{\Gamma((m+1)/2)}{\sqrt{\pi}\,\Gamma(m/2)} \int_{|t_0|<\delta_m} \frac{(1-t_0^{2d-2})^{(m-1)/2}}{(1-t_0^{2d})^{m/2}} (\sigma^4 - \tau^2)^{1/2} (1-t_0^2)^{(m/2)-1} H_m \, dt_0,$$

where

$$H_m = \prod_{k=1}^{m} \left[\frac{1}{(G_k(0))^2} \mathrm{E}\big(\|\xi_k\|_{(k)} \|\eta_k + \alpha_k \xi_k\|_{(k)}\big) \right]$$

$$\alpha_k = \frac{t_0^{d-1}}{(1-t_0^{2d-2})^{1/2}} \text{ if } k = 1, \ldots, m-1 \text{ and } \alpha_m = \frac{\tau}{(\sigma^4 - \tau^2)^{1/2}}.$$

Again we perform the change of variables $t_0 = \theta/\sqrt{m}$:

$$\frac{I_2}{4d^m} = C_m \int_{-m^{1/2-\delta}}^{m^{1/2-\delta}} \frac{(1-(\theta^2/m)^{d-1})^{(m-1)/2}}{(1-(\theta^2/m)^d)^{m/2}} (\sigma^4 - \tau^2)^{1/2} \left(1 - \frac{\theta^2}{m}\right)^{(m/2)-1} H_m \, d\theta,$$

with

$$H_m = \left(1 + \frac{1}{2} \frac{\tau^2}{\sigma^4 - \tau^2} c_m\right) \prod_{k=1}^{m-1} \left[1 + \frac{1}{2} \frac{(\theta^2/m)^{d-1}}{1-(\theta^2/m)^{d-1}} c_k\right],$$

where $c_k = m_{k3}/k m_{k1}$.

Lemma 12.8

$$c_k = 1 + \frac{1}{k} - \frac{1}{4k^2} - \frac{55}{48k^3} + O(1/k^4).$$

Lemma 12.9 *Let*

$$A_m = \frac{(1-(\theta^2/m)^{d-1})^{(m-1)/2}}{(1-(\theta^2/m)^d)^{m/2}} \left(1 - \frac{\theta^2}{m}\right)^{(m/2)-1}.$$

Then

$$A_m = \exp\left[-\frac{\theta^2}{2} + \frac{1}{m}\left(\theta^2 - \frac{1}{4}\theta^4\right) + \frac{1}{m^2}\left(\frac{1}{2}\theta^4 - \frac{1}{6}\theta^6\right) + \frac{1}{m^3}\left(\frac{1}{3}\theta^6 - \frac{1}{8}\theta^8\right) \right.$$
$$\left. - \frac{1}{2}\frac{\theta^{2d-2}}{m^{d-2}} + \frac{1}{m^{d-1}}\left(\frac{1}{2}\theta^{2d-2} + \frac{1}{2}\theta^{2d}\right) - \frac{\theta^{4d-4}}{4m^{2d-3}} + 0(1/m^{8\delta}) \right].$$

Lemma 12.10 *Let* $x = t_0^2 \leq 1/m^{2\delta} \to 0$. *Then*

$$\tau^2 = x^{d-2}\big[(d-1)^2 + d^2(x^2 + x^{2d} - 2x^{d+1} + 2x^{d+2})$$
$$- 2d(1-d)(-x + x^d - x^{d+1} + x^{2d})\big] + O(x^8)$$

$$\sigma^4 - \tau^2 = 1 - (d-1)^2 x^{d-2} + d(d-2)x^{d-1}\left[2 - x - x^{d-1} + 2x^d - x^{d+1} - x^{2d-1}\right] + O(x^8).$$

Proof of Theorem 12.2 (cont.)

CASE $d > 5$. We have:

- $\tau^2 = O_u(1/m^{8\delta})$.
- $\sigma^4 - \tau^2 = 1 + O_u(1/m^{8\delta})$.
- $H_m = 1 + O_u(1/m^{10\delta-1})$.

We choose δ so that $10\delta - 1 > 3$.

$$\frac{I_2}{4d^m} = \frac{1}{\sqrt{2\pi}}\left[1 - \frac{1}{4m} + \frac{1}{32m^2} + \frac{5}{128m^3} + O\left(\frac{1}{m^4}\right)\right]\left[1 + O_u\left(1/m^{10\delta-1}\right)\right]$$
$$\cdot \int_{-m^{(1/2)-\delta}}^{m^{(1/2)-\delta}} \exp\left[-\frac{\theta^2}{2} + \frac{1}{m}\left(\theta^2 - \frac{1}{4}\theta^4\right) + \frac{1}{m^2}\left(\frac{1}{2}\theta^4 - \frac{1}{6}\theta^6\right) + \frac{1}{m^3}\left(\frac{1}{3}\theta^6 - \frac{1}{8}\theta^8\right) + O_u\left(\frac{1}{m^{10\delta-1}}\right)\right] d\theta.$$

Notice (key point) that excluding the first term, all other terms in the exponent are $o_u(1)$ as $m \to +\infty$, that is, are uniformly small if m is large enough. Expanding and using the moments of the standard normal distribution (up to order 12), we get, for $d > 5$,

$$\frac{I_2}{4d^m} = 1 + \frac{3465}{64}\frac{1}{m^3} + O\left(\frac{1}{m^{10\delta-1}}\right),$$

so that $\mathrm{Var}(n^X) \approx (3465/64)(1/m^3)$.

The cases $d = 3, 4, 5$ can be treated in a similar way. □

12.2. MORE GENERAL MODELS

The probability law of the Shub–Smale model defined in Section 12.1 has the simplifying property of being invariant under the orthogonal group of the underlying Euclidean space $\mathbb{R}^m$. In this section we present the extension of formula (12.4) to general systems that share the same invariance property. This section follows the work of Azaïs and Wschebor (2005b).

We require the polynomial random fields X_i $(i = 1, \ldots, m)$ to be centered, Gaussian, and independent, and their covariances

$$r^{X_i}(s,t) = \mathrm{E}(X_i(s)X_i(t))$$

to be invariant under orthogonal linear transformation of $\mathbb{R}^m$ [i.e., $r^{X_i}(Us, Ut) = r^{X_i}(s, t)$ for any orthogonal transformation U and any pair $s, t \in \mathbb{R}^m$]. This implies in particular that the coefficients $a_j^{(i)}$ remain independent for different i's but can now be correlated from one j to another for the same value of i. It is easy to check that this implies that for each $i = 1, \ldots, m$, the covariance $r^{X_i}(s, t)$ is a function of the triple $\left(\langle s, t\rangle, \|s\|^2, \|t\|^2\right)$. It is somewhat more difficult but can also be proved (see Spivak, 1979) that this function is, in fact, a polynomial with real coefficients, say $Q^{(i)}$:

$$r^{X_i}(s, t) = Q^{(i)}(\langle s, t\rangle, \|s\|^2, \|t\|^2), \tag{12.17}$$

satisfying the symmetry condition

$$Q^{(i)}(u, v, w) = Q^{(i)}(u, w, v). \tag{12.18}$$

A natural question is: Which are the polynomials $Q^{(i)}$ such that the function on the right-hand side of (12.17) is a covariance (i.e., nonnegative definite)? A simple way to construct a class of covariances of this type is to take

$$Q^{(i)}(u, v, w) = P(u, vw), \tag{12.19}$$

where P is a polynomial in two variables with nonnegative coefficients. In fact, the functions $(s, t) \rightsquigarrow \langle s, t\rangle$ and $(s, t) \rightsquigarrow \|s\|^2\|t\|^2$ are covariances and the set of covariances is closed under linear combinations with nonnegative coefficients as well as under multiplication, so that $P(\langle s, t\rangle, \|s\|^2\,\|t\|^2)$ is also the covariance of some random field.

The situation becomes simpler if one considers only functions of the scalar product, that is,

$$Q(u, v, w) = \sum_{k=0}^{d} c_k u^k.$$

The necessary and sufficient condition for $\sum_{k=0}^{d} c_k \langle s, t\rangle^k$ to be a covariance is that $c_k \geq 0\ \forall\, k = 0, 1, \ldots, d$. In that case it is the covariance of the random field $X(t) := \sum_{\|j\| \leq d} a_j t^j$, where the a_j's are centered, Gaussian, independent random variables, $\operatorname{Var}(a_j) = c_{\|j\|}(\|j\|!/j!)$. (The proof of this is left to the reader.) The Shub–Smale model is the special case corresponding to the choice $c_k = \binom{d}{k}$. A general description of the polynomial covariances which are invariant under the action of the orthogonal group is given by Kostlan (2002, p. II).

We now state the extension of the Shub–Smale formula to the general case.

Theorem 12.11. *Assume that the X_i are independent centered Gaussian polynomial random fields with covariances $r^{X_i}(s, t) = Q^{(i)}(\langle s, t\rangle, \|s\|^2, \|t\|^2)$ $(i = 1, \ldots, m)$. Let us denote by $Q_u^{(i)}, Q_w^{(i)}, Q_{uv}^{(i)}, \ldots$ the partial derivatives of $Q^{(i)}$. We assume that $Q^{(i)}(x, x, x)$ and $Q_u^{(i)}(x, x, x)$ do not vanish for $x \geq 0$. Set*

$$q_i(x) := \frac{Q_u^{(i)}}{Q^{(i)}}$$

$$r_i(x) := \frac{Q^{(i)}\left(Q_{uu}^{(i)} + 2Q_{uv}^{(i)} + 2Q_{uw}^{(i)} + 4Q_{vw}^{(i)}\right) - \left(Q_u^{(i)} + Q_v^{(i)} + Q_w^{(i)}\right)^2}{(Q^{(i)})^2},$$

where the functions on the right-hand sides are always computed at the triple (x, x, x). Set

$$h_i(x) := 1 + x\frac{r_i(x)}{q_i(x)}.$$

Then for all Borel sets V, we have

$$\mathrm{E}\left(N^X(V)\right) = \kappa_m \int_V \left(\prod_{i=1}^m q_i\left(\|t\|^2\right)\right)^{1/2} E_h(\|t\|^2)\, dt. \tag{12.20}$$

In this formula,

$$E_h(x) := \mathrm{E}\left(\left(\sum_{i=1}^m h_i(x)\xi_i^2\right)^{1/2}\right),$$

where $\xi_1, \ldots, \xi_m$ are i.i.d. standard normal in $\mathbb{R}$ and

$$\kappa_m = \frac{1}{\sqrt{2\pi}}\frac{\Gamma(m/2)}{\pi^{m/2}}.$$

Proof. Let us set $K_j = \mathrm{E}(\|\eta_j\|)$ with η_j standard normal in $\mathbb{R}^j$. An elementary computation gives

$$K_m = \sqrt{2}\frac{\Gamma((m+1)/2)}{\Gamma(m/2)}.$$

We define the integral

$$J_m := \int_0^{+\infty} \frac{\rho^{m-1}}{(1+\rho^2)^{(m+1)/2}} d\rho = \sqrt{\pi/2}\frac{1}{K_m}$$

that will appear later. Consider the normalized Gaussian fields

$$Z_i(t) := \frac{X_i(t)}{\left(Q^{(i)}(\|t\|^2, \|t\|^2, \|t\|^2)\right)^{1/2}},$$

which have variance 1. Denote $Z(t) = (Z_1(t), \ldots, Z_m(t))^{\mathrm{T}}$. Applying Rice's formula for the expectation of the number of zeros of Z gives as

$$\mathrm{E}\left(N^X(V)\right)=\mathrm{E}\left(N^Z(V)\right)=\int_V \mathrm{E}\left(\left|\det(Z'(t)\right|\,\Big|Z(t)=0\right)\frac{1}{(2\pi)^{m/2}}dt,$$

where $Z'(t) := [Z_1'(t) \vdots \cdots \vdots Z_m'(t)]$ is the matrix obtained by concatenation of the vectors $Z_1'(t), \ldots, Z_m'(t)$. Note that since $\mathrm{E}\left(Z_i^2(t)\right)$ is constant, it follows that $\mathrm{E}\big(Z_i(t)(\partial Z_i(t)/\partial t_j)\big) = 0$ for all $i, j = 1, \ldots, m$. Since the field is Gaussian, this implies that $Z_i(t)$ and $Z_i'(t)$ are independent and given that the coordinate fields $Z_1, \ldots, Z_m$ are independent, one can conclude that for each t, $Z(t)$ and $Z'(t)$ are independent. So

$$\mathrm{E}\left(N^X(V)\right)=\mathrm{E}\left(N^Z(V)\right)=\frac{1}{(2\pi)^{m/2}}\int_V \mathrm{E}\left(\left|\det(Z'(t)\right|\right)dt. \tag{12.21}$$

A straightforward computation shows that the (α, β)-entry, $\alpha, \beta = 1, \ldots, m$, in the covariance matrix of $Z_i'(t)$ is

$$\mathrm{E}\left(\frac{\partial Z_i}{\partial t_\alpha}(t)\frac{\partial Z_i}{\partial t_\beta}(t)\right)=\frac{\partial^2}{\partial s_\alpha\,\partial t_\beta}r^{Z_i}(s,t)\;|_{s=t}=r_i(\|t\|^2)t_\alpha t_\beta+q_i(\|t\|^2)\delta_{\alpha\beta},$$

where $\delta_{\alpha,\beta}$ denotes the Kronecker symbol. This can be rewritten as

$$\mathrm{Var}\big(Z_i'(t)\big)=q_iI_m+r_itt^{\mathrm{T}},$$

where the functions on the right-hand side are to be computed at the point $\|t\|^2$. Let U be the orthogonal transformation of $\mathbb{R}^m$ that gives the coordinates in a basis with first vector $t/\|t\|$; we get

$$\mathrm{Var}\big(UZ_i'(t)\big)=\mathrm{diag}\big((r_i.\|t\|^2+q_i),q_i,\ldots,q_i\big),$$

so that

$$\mathrm{Var}\left(\frac{UZ_i'(t)}{\sqrt{q_i}}\right)=\mathrm{diag}\left(h_i,1,\ldots,1\right).$$

Now set

$$T_i := \frac{UZ_i'(t)}{\sqrt{q_i}}$$

and

$$T := [T_1 \vdots \ddots \vdots T_m].$$

We have

$$\left|\det\left(Z'(t)\right)\right| = \left|\det\left(T\right)\right| \prod_{i=1}^{m} q_i^{1/2}. \tag{12.22}$$

Now we write

$$T = \begin{bmatrix} W_1 \\ \cdots \\ \cdots \\ \cdots \\ W_m \end{bmatrix},$$

where the W_i are random row vectors. Because of the properties of independence of all the entries of T, we know that:

- $W_2, \ldots, W_m$ are independent standard normal vectors in $\mathbb{R}^m$.
- W_1 is independent of the other W_i, $i \geq 2$, and has a centered Gaussian distribution with variance matrix $\mathrm{diag}(h_1, \ldots, h_m)$.

Now $\mathrm{E}\big(|\det(T)|\big)$ is calculated as the expectation of the volume of the parallelotope generated by $W_1, \ldots, W_m$ in $\mathbb{R}^m$. That is,

$$|\det(T)| = \|W_1\| \prod_{j=2}^{m} d(W_j, S_{j-1}),$$

where S_{j-1} denotes the subspace of $\mathbb{R}^m$ generated by $W_1, \ldots, W_{j-1}$ and d denotes the Euclidean distance. Using the invariance under isometries of the standard normal distribution of $\mathbb{R}^m$, we know that, conditioning on $W_1, \ldots, W_{j-1}$, the projection $P_{S_{j-1}^{\perp}}(W_j)$ of W_j on the orthogonal $S_{j-1}^{\perp}$ of S_{j-1} has a distribution that is standard normal on the space $S_{j-1}^{\perp}$, which is of dimension $m-j+1$ with probability 1. Thus, $\mathrm{E}\big(d(W_j, S_{j-1})\big|W_1, \ldots, W_{j-1}\big) = K_{m-j+1}$. By successive conditionings on W_1, W_1, W_2,..., we get

$$\mathrm{E}\big(|\det(T)|\big) = \mathrm{E}\left(\left(\sum_{i=1}^{m} h_i(x)\xi_i^2\right)^{1/2}\right) \times \prod_{j=1}^{m-1} K_j,$$

where $\xi_1, \ldots, \xi_m$ are i.i.d. standard normal in $\mathbb{R}$. Using (12.22) and (12.21), we obtain (12.20). □

12.2.1. Examples

1. Let $Q^{(i)}(u, v, w) = Q^{l_i}(u)$ for some polynomial Q. We get

$$q_i(x) = l_i q(x) = l_i \frac{Q'(x)}{Q(x)}, \qquad h_i(x) = h(x) = 1 - x\frac{Q'^2(x) - Q(x)Q''(x)}{Q(x)Q'(x)}.$$

Applying formula (12.20) with $V = \mathbb{R}^m$ and using polar coordinates yields

$$\mathrm{E}(N^X) = \frac{2}{\sqrt{\pi}} \frac{\Gamma((m+1)/2)}{\Gamma(m/2)} \sqrt{l_1 \cdots l_m} \int_0^{\infty} \rho^{m-1} q(\rho^2)^{m/2} \sqrt{h(\rho^2)}\, d\rho. \quad (12.23)$$

2. If in example 1 we put $Q(u) = 1 + u$, we get the Shub–Smale model. Replacing, an elementary computation in (12.23) reproduces (12.4).

3. A simple variant of the Shub–Smale theorem corresponds to taking $Q^{(i)}(u) = 1 + u^d$ for all $i = 1, \ldots, m$ (here all the X_i's have the same law). Even though in this case the derivative $Q^{(i)}(u)$ vanishes at zero, the reader can easily check that the conclusion of Theorem 12.11 remains valid and

$$q(x) = q_i(x) = \frac{du^{d-1}}{1+u^d}, \qquad h(x) = h_i(x) = \frac{d}{1+u^d}$$

$$\mathrm{E}(N^X) = \sqrt{\frac{2}{\pi}} K_m \int_0^{+\infty} \frac{\rho^{md-1}}{(1+\rho^{2d})^{(m+1)/2}} d\rho = d^{(m-1)/2},$$

which differs by a constant factor from the analogous Shub–Smale result for $(1+u)^d$, which is $d^{m/2}$.

4. *Linear systems with a quadratic perturbation*. Consider linear systems with a quadratic perturbation

$$X_i(s) = \xi_i + < \eta_i, s > + \zeta_i \|s\|^2,$$

where the ξ_i, ζ_i, η_i, $i = 1, \ldots, m$ are independent and standard normal in $\mathbb{R}, \mathbb{R}$, and $\mathbb{R}^m$, respectively. This corresponds to the covariance $r^{X_i}(s,t) = 1 + \langle s, t\rangle + \|s\|^2\|t\|^2$. If there is no quadratic perturbation, it is obvious that the number of roots is a.s. equal to 1.

For the perturbed system, applying Theorem 12.11 and performing the computations required in this case, we obtain

$$q(x) = \frac{1}{1+x+x^2},$$

$$r(x) = \frac{4}{1+x+x^2} - \frac{(1+2x)^2}{(1+x+x^2)^2}, \qquad h(x) = \frac{1+4x+x^2}{1+x+x^2}$$

and

$$\mathrm{E}(N^X) = \frac{H_m}{J_m} \qquad \text{with} \quad H_m = \int_0^{+\infty} \frac{\rho^{m-1}(1 + 4\rho^2 + \rho^4)^{1/2}}{(1 + \rho^2 + \rho^4)^{(m/2)+1}}\, d\rho.$$

An elementary computation shows that $\mathrm{E}(N^X) = o(1)$ as $m \to +\infty$ (see example 5 for more precise behavior). In other words, the probability that the perturbed system has no solution tends to 1 as $m \to +\infty$.

5. *More general perturbed systems*. Let us consider the covariances given by the polynomials

$$Q^i(u, v, w) = Q(u, v, w) = 1 + 2u^d + (vw)^d.$$

This corresponds to adding a perturbation depending on the product of the norms of s, t to the modified Shub–Smale systems considered in our third example. We know that for the unperturbed system, one has $\mathrm{E}(N^X) = d^{(m-1)/2}$. Notice that the factor 2 in Q has only been added for computational convenience and does not modify the random variable N^X of the unperturbed system. For the perturbed system, we get

$$q(x) = \frac{2dx^{d-1}}{(1 + x^d)^2}, \qquad r(x) = \frac{2d(d-1)x^{d-2}}{(1 + x^d)^2}, \qquad h(x) = d.$$

Therefore,

$$\begin{aligned} \mathrm{E}\left(N^X\right) &= \sqrt{\frac{2}{\pi}} K_m \int_0^{+\infty} \rho^{m-1} \left(\frac{2d\rho^{2(d-1)}}{(1 + \rho^{2d})^2}\right)^{m/2} \sqrt{d}\, d\rho \\ &= \sqrt{\frac{2}{\pi}} K_m 2^{m/2} d^{(m+1)/2} \int_0^{+\infty} \frac{\rho^{md-1}}{\left(1 + \rho^{2d}\right)^m} d\rho. \end{aligned} \tag{12.24}$$

The integral can be evaluated by an elementary computation and we obtain

$$\mathrm{E}(N^X) = 2^{-(m-2)/2} d^{(m-1)/2},$$

which shows that the mean number of zeros is reduced by the perturbation at a geometrical rate as m grows.

6. *Polynomial in the scalar product, real roots*. Consider again the case in which the polynomials $Q^{(i)}$are all equal and the covariances depend only on the scalar product [i.e., $Q^{(i)}(u, v, w) = Q(u)$]. We assume further that the roots of Q, which we denote $-\alpha_1, \ldots, -\alpha_d$, are real ($0 < \alpha_1 \le \cdots \le \alpha_d$). We get

$$q(x) = \sum_{h=1}^{d} \frac{1}{x + \alpha_h},$$

$$r(x) = \sum_{h=1}^{d} \frac{1}{(x + \alpha_h)^2}, \qquad h(x) = \frac{1}{q_i(x)} \sum_{h=1}^{d} \frac{\alpha_h}{(x + \alpha_h)^2}.$$

It is easy now to write an upper bound for the integrand in (12.20) and compute the remaining integral, thus obtaining the inequality

$$\mathrm{E}(N^X) \le \sqrt{\frac{\alpha_d}{\alpha_1}}\, d^{m/2},$$

which is sharp if $\alpha_1 = \cdots = \alpha_d$.

If we assume further that $d = 2$, with no loss of generality $Q(u)$ has the form $Q(u) = (u+1)(u+\alpha)$ with $\alpha \in [0, 1]$. Replacing q by $1/(x+1) + 1/(x+\alpha)$ in formula (12.23), we get

$$\mathrm{E}(N^X) = \sqrt{2/\pi}\, K_m \cdot \int_0^{\infty} \rho^{m-1} \left(\frac{1}{1+\rho^2} + \frac{1}{\alpha+\rho^2} \right)^{(m-1)/2} \left(\frac{1}{(1+\rho^2)^2} + \frac{\alpha}{(\alpha+\rho^2)^2} \right)^{1/2} d\rho. \tag{12.25}$$

One can compute the limit on the right-hand side as $\alpha \to 0$. For this purpose, notice that the function $\alpha \to \alpha/(\alpha + \rho^2)^2$ attains its maximum at $\alpha = \rho^2$ and is dominated by $1/4\rho^2$. We divide the integral on the right-hand side of (12.25) into two parts, setting for some $\delta > 0$,

$$I_{\delta,\alpha} := \int_0^{\delta} \rho^{m-1} \left(\frac{1}{1+\rho^2} + \frac{1}{\alpha+\rho^2} \right)^{(m-1)/2} \left(\frac{1}{(1+\rho^2)^2} + \frac{\alpha}{(\alpha+\rho^2)^2} \right)^{1/2} d\rho,$$

$$J_{\delta,\alpha} := \int_{\delta}^{+\infty} \rho^{m-1} \left(\frac{1}{1+\rho^2} + \frac{1}{\alpha+\rho^2} \right)^{(m-1)/2} \left(\frac{1}{(1+\rho^2)^2} + \frac{\alpha}{(\alpha+x\rho^2)^2} \right)^{1/2} d\rho.$$

By dominated convergence,

$$J_{\delta,\alpha} \to \int_{\delta}^{+\infty} \left(\frac{2\rho^2+1}{\rho^2+1} \right)^{(m-1)/2} \frac{d\rho}{1+\rho^2}$$

as $\alpha \to 0$. On the other hand,

$$I_{\delta,\alpha}^{-} \le I_{\delta,\alpha} \le I_{\delta,\alpha}^{+},$$

where

$$I_{\delta,\alpha}^{-} := \int_0^{\delta} \left(\frac{\rho^2}{1+\rho^2} + \frac{\rho^2}{\alpha+\rho^2} \right)^{(m-1)/2} \frac{\sqrt{\alpha}}{\rho^2+\alpha} d\rho$$

$$= \int_0^{\delta/\alpha} \left(\frac{\alpha z^2}{1+\alpha z^2} + \frac{\alpha z^2}{\alpha(z^2+1)} \right)^{(m-1)/2} \frac{dz}{z^2+1} \to J_m \tag{12.26}$$

as $\alpha \to 0$, and

$$I^{+}_{\delta,\alpha} := \int_0^{\delta} \left(\frac{\rho^2}{1+\rho^2} + \frac{\rho^2}{\alpha+\rho^2} \right)^{(m-1)/2} \left(\frac{1}{1+\rho^2} + \frac{\sqrt{\alpha}}{\rho^2+\alpha} \right) d\rho$$

$$\to \int_0^{\delta} \left(\frac{2\rho^2+1}{\rho^2+1} \right)^{(m-1)/2} \frac{d\rho}{1+\rho^2} + J_m \tag{12.27}$$

as $\alpha \to 0$. Since δ is arbitrary, the integral on the right-hand size of (12.27) can be chosen arbitrarily small. Using the identity $K_m J_m = \sqrt{\pi/2}$, we get

$$\mathrm{E}(N^X) \to \upsilon := 1 + \frac{1}{J_m} \int_0^{+\infty} \left(\frac{2\rho^2+1}{\rho^2+1} \right)^{(m-1)/2} \frac{d\rho}{1+\rho^2}$$

as $\alpha \to 0$. Since $2\rho^2/(\rho^2+1) < (2\rho^2+1)/(\rho^2+1) < 2$,

$$1 + 2^{(m-1)/2} < \upsilon < 1 + \frac{2^{(m-1)/2}}{J_m} \frac{\pi}{2}.$$

12.3. NONCENTERED SYSTEMS (SMOOTHED ANALYSIS)

The aim of this section is to remove the hypothesis that the coefficients have zero expectation. Let us start with a nonrandom system

$$P_i(t) = 0 \qquad i = 1, \ldots, m, \tag{12.28}$$

and perturb it with a polynomial noise $\{X_i(t) : i = 1, \ldots, m\}$; that is, we consider the new system

$$P_i(t) + X_i(t) = 0 \qquad i = 1, \ldots, m.$$

What can one say about the number of roots of the new system? Of course, to obtain results on $E(N^{P+X})$ we need a certain number of hypotheses on both the noise X and the polynomial "signal" P, especially the relation between the size of P and the probability distribution of X.

Some of these hypotheses are technical in nature, allowing us to perform the computations. Beyond this, roughly speaking, Theorem 12.12 says that if the relation signal over noise is neither too big nor too small, in a sense that we make precise later, there exist positive constants C, θ, $0 < \theta < 1$ such that

$$\mathrm{E}(N^{P+X}) \le C\,\theta^m \mathrm{E}(N^X). \tag{12.29}$$

Inequality (12.29) becomes of interest if the starting nonrandom system (12.28) has a large number of roots, possibly infinite, and m is large. In this situation, the effect of adding polynomial noise is a reduction at a geometric rate of the expected number of roots compared to the centered case. In formula (12.29), $E(N^X)$ can be computed or estimated using the results in Sections 12.1 and 12.2, and bounds for the constants C and θ can be deduced explicitly from the hypotheses.

Before the statement we need to introduce some additional notations and hypotheses: (H1) and (H2) concern only the noise; (H3) and (H4) include relations between noise and signal. The noise will correspond to polynomials $Q^{(i)}(u, v, w) = \sum_{k=0}^{d_i} c_k^{(i)} u^k$, $c_k^{(i)} \geq 0$, considered in Section 12.2, (i.e., the covariances are only a function of the scalar product). Also, each polynomial $Q^{(i)}$ has effective degree d_i:

$$c_{d_i}^{(i)} > 0 \qquad i = 1, \ldots, m$$

and does not vanish for $u \geq 0$, which amounts to saying that for each t the distribution of $X_i(t)$ does not degenerate. An elementary calculation then shows that for each polynomial $Q^{(i)}$, as $u \to +\infty$,

$$q_i(u) \sim \frac{d_i}{1+u} \tag{12.30}$$

$$h_i(u) \sim \frac{c_{d_i-1}^{(i)}}{d_i c_{d_i}^{(i)}} \frac{1}{1+u}. \tag{12.31}$$

Since we are interested in the large m asymptotics, the polynomials P and Q can vary with m and we will require somewhat more than relations (12.30) and (12.31), as specified in the following hypotheses:

(H_1) h_i is independent of i ($i = 1, \ldots, m$) (but may vary with m). We set $h = h_i$.

(H_2) There exist positive constants D_i, E_i ($i = 1, \ldots, m$) and $\underline{q}$ such that

$$0 \leq D_i - (1+u)q_i(u) \leq \frac{E_i}{1+u} \quad \text{and} \quad (1+u)q_i(u) \geq \underline{q} \tag{12.32}$$

for all $u \geq 0$, and moreover,

$$\max_{1 \leq i \leq m} D_i, \ \max_{1 \leq i \leq m} E_i$$

are bounded by constants $\overline{D}$ and $\overline{E}$, respectively, which are independent of m. $\underline{q}$ is also independent of m.

Also, there exist positive constants $\underline{h}$ and $\overline{h}$ such that

$$\underline{h} \leq (1+u)h(u) \leq \overline{h} \tag{12.33}$$

for $u \geq 0$.

Notice that the auxiliary functions q_i, r_i, h $(i = 1, \ldots, m)$ will also vary with m. To simplify the notation somewhat we are dropping the parameter m in P, Q, q_i, r_i, and h. However, in (H2) the constants $\underline{h}$ and $\overline{h}$ do not depend on m. One can check that these conditions imply that $h(u) \geq 0$ when $u \geq 0$.

Let us now describe the second set of hypotheses. Let P be a polynomial in m real variables with real coefficients having degree d and Q a polynomial in one variable with nonnegative coefficients, also having degree d, $Q(u) = \sum_{k=0}^{d} c_k u^k$. We assume that Q does not vanish on $u \geq 0$ and $c_d > 0$. Define

$$H(P, Q) := \sup_{t \in \mathbb{R}^m} \left\{ (1 + \|t\|) \cdot \left\| \nabla \left(\frac{P}{\sqrt{Q(\|t\|^2)}} \right)(t) \right\| \right\}$$

$$K(P, Q) := \sup_{t \in \mathbb{R}^m \setminus \{0\}} \left\{ \left(1 + \|t\|^2\right) \cdot \left| \frac{\partial}{\partial \rho} \left(\frac{P}{\sqrt{Q\left(\|t\|^2\right)}} \right)(t) \right| \right\}$$

where $\partial/\partial\rho$ denotes the derivative in the direction defined by $t/\|t\|$ at each point $t \neq 0$.

For $r > 0$, set

$$L(P, Q, r) := \inf_{\|t\| \geq r} \frac{P(t)^2}{Q(\|t\|^2)}.$$

One can check by means of elementary computations that for each pair P and Q as above, one has

$$H(P, Q) < \infty, \qquad K(P, Q) < \infty.$$

With these notations, we introduce the following hypotheses on the systems P and Q, as m grows:

$$\text{(H}_3\text{)} \qquad A_m = \frac{1}{m} \sum_{i=1}^{m} \frac{H^2(P_i, Q^{(i)})}{i} = o(1) \quad \text{as } m \to +\infty \tag{12.34}$$

$$B_m = \frac{1}{m} \sum_{i=1}^{m} \frac{K^2(P_i, Q^{(i)})}{i} = o(1) \quad \text{as } m \to +\infty. \tag{12.35}$$

(H$_4$) There exist positive constants r_0 and ℓ such that if $r \geq r_0$,

$$L(P_i, Q^{(i)}, r) \geq \ell \qquad \text{for all } i = 1, \ldots, m.$$

Theorem 12.12 (Armentano and Wschebor, 2008). *Under hypotheses* (H_1) *to* (H_4), *one has*

$$\mathrm{E}(N^{P+X}) \le C\,\theta^m \mathrm{E}(N^X), \tag{12.36}$$

where C *and* θ *are positive constants,* $0 < \theta < 1$.

Remarks on the Statement of Theorem 12.12

1. It is obvious that our problem does not depend on the order in which the equations

$$P_i(t) + X_i(t) = 0 \qquad i = 1, \ldots, m$$

appear. However, conditions (12.34) and (12.35) in hypothesis (H3) do depend on the order. One can restate them saying that there exists an order $i = 1, \ldots, m$ on the equations such that (12.34) and (12.35) hold true.

2. Condition (H3) can be interpreted as a bound on the quotient signal over noise. In fact, it concerns the gradient of this quotient. Appearing in (12.35) is the radial derivative, which happens to decrease faster as $\|t\| \to \infty$ than the other components of the gradient. Clearly, if $H(P_i, Q^{(i)})$ and $K(P_i, Q^{(i)})$ are bounded by fixed constants, (12.34) and (12.35) hold true. Also, some of them may grow as $m \to +\infty$ provided that (12.34) and (12.35) remain satisfied.

3. Hypothesis (H4) goes in the opposite direction: For large values of $\|t\|$ we need a lower bound of the relation signal over noise.

4. A result of the type of Theorem 12.12 cannot be obtained without putting some restrictions on the relation signal over noise. In fact, consider the system

$$P_i(t) + \sigma\, X_i(t) = 0 \qquad i = 1, \ldots, m, \tag{12.37}$$

where σ is a positive real parameter. As $\sigma \downarrow 0$ the expected value of the number of roots of (12.37) tends to the number of roots of $P_i(t) = 0 (i = 1, \ldots, m)$, for which no a priori bound is available. In this case, the relation signal over noise tends to infinity. On the other hand, if we let $\sigma \to +\infty$, the relation signal over noise tends to zero and the number of roots expected will tend to $E(N^X)$.

Proof of Theorem 12.12. We follow the same lines of the proof of Theorem 12.11. Let

$$Z_j(t) = \frac{P_j(t) + X_j(t)}{\sqrt{Q^{(j)}(\|t\|^2)}} \qquad j = 1, \ldots, m$$

and $Z = (Z_1, .., Z_m)^{\mathrm{T}}$. Clearly,

$$N^{P+X}(V) = N^Z(V)$$

for any subset V of $\mathbb{R}^m$.

$\{Z_j(t) : t \in \mathbb{R}^m\}$ $(j = 1, \dots, m)$ are independent centered Gaussian processes, $\mathrm{E}(Z_j^2(t)) = 1$, for all $j = 1, \dots, m$ and all $t \in \mathbb{R}^m$. This implies that $Z_j(t)$ and $\nabla Z_j(t)$ are independent for each $t \in \mathbb{R}^m$. We apply Rice's formula to compute $E(N^Z(V))$; that is,

$$\mathrm{E}(N^Z(V)) = \int_V \mathrm{E}\left(\left|\det\left(Z'(t)\right)\right| \,\middle|\, Z(t) = 0\right) p_{Z(t)}(0)\, dt.$$

Using the independence between $Z'(t)$ and $Z(t)$, one gets

$$\begin{aligned}\mathrm{E}(N^Z(V)) &\\ = \int_V \mathrm{E}\left(\left|\det\left(Z'(t)\right)\right|\right) \frac{1}{(2\pi)^{m/2}} \exp\left[-\frac{1}{2}\left(\frac{P_1(t)^2}{Q^{(1)}(\|t\|^2)} + \cdots + \frac{P_m(t)^2}{Q^{(m)}(\|t\|^2)}\right)\right] dt &\end{aligned} \tag{12.38}$$

and our main problem consists of the evaluation of $\mathrm{E}\left(\left|\det\left(Z'(t)\right)\right|\right)$.

As in the centered case, we have

$$\mathrm{Cov}\left(\frac{\partial Z_i}{\partial t_\alpha}(t), \frac{\partial Z_j}{\partial t_\beta}(t)\right) = \delta_{ij}\left[r_i\left(\|t\|^2\right) t_\alpha t_\beta + q_i\left(\|t\|^2\right)\delta_{\alpha\beta}\right]$$

for $i,\ j,\ \alpha,\ \beta = 1, \dots, m$. □

For each $t \neq 0$, let U_t be an orthogonal transformation of $\mathbb{R}^m$ that takes the first element of the canonical basis into the unit vector $t/\|t\|$. Then

$$\mathrm{Var}\left(\frac{U_t \nabla Z_j(t)}{\sqrt{q_j(\|t\|^2)}}\right) = \mathrm{Diag}\left(h(\|t\|^2), 1 \dots, 1\right), \tag{12.39}$$

where we denote the gradient $\nabla Z_j(t)$ as a column vector.

$\mathrm{Diag}(\lambda_1, \dots, \lambda_m)$ denotes the $m \times m$ diagonal matrix with elements $\lambda_1, \dots, \lambda_m$ in the diagonal. So we can write

$$\frac{U_t \nabla Z_j(t)}{\sqrt{q_j(\|t\|^2)}} = \zeta_j + \alpha_j \qquad j = 1 \dots, m,$$

where ζ_j is a Gaussian centered random vector in $\mathbb{R}^m$ having covariance given by (12.39), $\zeta_1, \dots, \zeta_m$ are independent, and α_j is the nonrandom vector

$$\alpha_j = \frac{U_t \nabla\left(P_j(t)/\sqrt{Q^{(j)}(\|t\|^2)}\right)}{\sqrt{q_j(\|t\|^2)}} = (\alpha_{1j}, \dots, \alpha_{mj}) \qquad j = 1 \dots, m. \tag{12.40}$$

We denote by T the $m \times m$ random matrix having columns $\zeta_j + \alpha_j (j=1, \ldots, m)$. We have

$$\left|\det\left(Z'(t)\right)\right| = |\det(T)| \prod_{i=1}^{m} \left(q_i\left(\|t\|^2\right)\right)^{1/2},$$

so that

$$\mathrm{E}\left(\left|\det\left(Z'(t)\right)\right|\right) = \mathrm{E}\left(|\det(T)|\right) \prod_{i=1}^{m} \left(q_i\left(\|t\|^2\right)\right)^{1/2}. \tag{12.41}$$

Denote by $\eta_1, \ldots, \eta_m$ the columns of T,

$$\eta_j = \zeta_j + \alpha_j \qquad j = 1, \ldots, m,$$

where the ζ_{ij} are Gaussian centered independent and

$$\mathrm{Var}(\zeta_{ij}) = \begin{cases} 1 & \text{for } i = 2, \ldots, m;\ j = 1, \ldots, m \\ h(\|t\|^2) & \text{for } i = 1;\ j = 1, \ldots, m. \end{cases}$$

Proceeding as in the centered case to compute the volume of the associated parallelotope, we obtain the bound

$$\mathrm{E}\left(|\det T|\right) \le \sqrt{h\left(\|t\|^2\right)} \prod_{j=1}^{m} \mathrm{E}\left(\|\xi_j + c_j(t)\|_j\right), \tag{12.42}$$

where $\|\cdot\|_j$ denotes Euclidean norm in $\mathbb{R}^j$, $(\|\cdot\| = \|\cdot\|_m)$, ξ_j is a random vector with normal standard distribution in $\mathbb{R}^j$, and $c_j(t)$ is a nonrandom vector in $\mathbb{R}^j$ having norm $\|\tilde{\alpha}_j\|$, $j = 1, \ldots, m$, where $\tilde{\alpha}_j = \left(\alpha_{1j}/\sqrt{h(\|t\|^2)}, \alpha_{2j}, \ldots, \alpha_{mj}\right)^{\mathrm{T}}$ and the α_{ij} are given in (12.40). We denote

$$\gamma_j(c) = \mathrm{E}\left(\|\xi_j + c\|_j\right),$$

where $c \in \mathbb{R}^j$ is nonrandom. We have (see the auxiliary Lemma 12.13 following this proof):

$$\gamma_j(c) \le \left(1 + \|c\|_j^2 \frac{1}{2j}\right) \gamma_j(0).$$

Substituting in (12.42) and using (12.38) and (12.41), we get

$$\mathrm{E}(N^Z) \le \frac{1}{(2\pi)^{m/2}} L_m \int_{\mathbb{R}^m} \left\{ \sqrt{h(\|t\|^2)} \right.$$

$$\cdot \left(\prod_{i=1}^{m} q_i(\|t\|^2)\right)^{1/2} \exp\left[-\frac{1}{2}\sum_{i=1}^{m}\frac{P_i(t)^2}{Q^{(i)}(\|t\|^2)} + \frac{1}{2}\sum_{j=1}^{m}\|c_j(t)\|_j^2\,\frac{1}{j}\right]\Bigg\}, \tag{12.43}$$

where

$$L_m = \prod_{i=1}^{m} E\left(\|\xi_i\|_i\right) = \frac{1}{\sqrt{2\pi}}\, 2^{(m+1)/2}\,\Gamma\left(\frac{m+1}{2}\right).$$

Our final task is to obtain an adequate bound for the integral in (12.43). For $j = 1, \dots, m$ [use (H1)],

$$|\tilde{\alpha}_{1j}| = \frac{1}{\sqrt{h(\|t\|^2)\, q_j(\|t\|^2)}} \cdot \left|\frac{\partial}{\partial \rho}\frac{P_j(\|t\|^2)}{\sqrt{Q^{(j)}(\|t\|^2)}}\right| \le \frac{1}{\sqrt{\underline{h}\,\underline{q}}}\, K(P_j, Q^{(j)})$$

and

$$\|\alpha_j\| = \frac{\left\|\nabla\left(P_j(t)/\sqrt{Q^{(j)}(\|t\|^2)}\right)\right\|}{\sqrt{q_j(\|t\|^2)}} \le \frac{1}{\sqrt{\underline{q}}}\, H(P_j, Q^{(j)}).$$

Then if we bound $\|\tilde{\alpha}_j\|^2$ by

$$\|\tilde{\alpha}_j\|^2 \le |\tilde{\alpha}_{1j}|^2 + \|\alpha_j\|^2,$$

we obtain

$$\|\tilde{\alpha}_j\|^2 \le \frac{1}{\underline{h}\,\underline{q}}\, K^2(P_j, Q^{(j)}) + \frac{1}{\underline{q}}\, H^2(P_j, Q^{(j)}),$$

which implies that

$$\sum_{j=1}^{m} \|c_j\|_j^2 \cdot \frac{1}{j} \le \frac{1}{\underline{q}}\, m\, A_m + \frac{1}{\underline{h}\,\underline{q}}\, m\, B_m.$$

Substituting in (12.43), we get the bound

$$\mathrm{E}(N^Z) \le s_m H_m,$$

where

$$s_m = \left(\frac{\overline{h}}{\underline{h}}\right)^{1/2} \exp\left(\frac{1}{2}\left(\frac{1}{\underline{q}}\, m\, A_m + \frac{1}{\underline{h}\,\underline{q}}\, m\, B_m\right)\right) = e^{o(m)} \quad \text{as } m \to +\infty$$

and

$$H_m = \frac{1}{\pi^{(m+1)/2}} \Gamma\left(\frac{m}{2}\right) \int_{\mathbb{R}^m} \left(\prod_{i=1}^{m} q_i(\|t\|^2) \right)^{1/2}$$
$$\sqrt{h(\|t\|^2)}\, \mathrm{E}(\|\xi_m\|)\, e^{-1/2 \sum_{i=1}^{m} (P_i(t)^2 / Q^{(i)}(\|t\|^2))}\, dt. \tag{12.44}$$

The integrand in (12.44) is the same as in the expectation in the centered case except for the exponential, which will help for large values of $\|t\|$.

Let us write H_m as

$$H_m = H_m^{(1)}(r) + H_m^{(2)}(r),$$

where $H_m^{(1)}(r)$ corresponds to integrating on $\|t\| \le r$ and $H_m^{(2)}(r)$ on $\|t\| > r$ instead of the entire $\mathbb{R}^m$ in formula (12.44). We first choose r large enough so that the condition in hypothesis (H4) is satisfied. Then

$$H_m^{(2)}(r) \le e^{-\ell m/2}\, \mathrm{E}(N^X). \tag{12.45}$$

We now turn to $H_m^{(1)}(r)$. We have, bounding the exponential in the integrand by 1 and using hypothesis (H2),

$$H_m^{(1)}(r) \le \frac{1}{\pi^{(m+1)/2}} \Gamma\left(\frac{m}{2}\right) \overline{h}^{1/2}\, \mathrm{E}(\|\xi_m\|) \left(\prod_{i=1}^{m} D_i^{1/2} \right) \sigma_{m-1} \int_0^r \frac{\rho^{m-1}}{(1+\rho^2)^{(m+1)/2}}\, d\rho, \tag{12.46}$$

where σ_{m-1} is the $(m-1)$-dimensional area measure of S^{m-1}. The integral on the right-hand side is bounded by

$$\frac{\pi}{2} \left(\frac{r^2}{1+r^2} \right)^{(m-1)/2}.$$

Again using (H2), we have the lower bound:

$$\mathrm{E}(N^X)$$
$$\ge \frac{1}{\pi^{(m+1)/2}} \Gamma\left(\frac{m}{2}\right) \underline{h}^{1/2}\, \mathrm{E}(\|\xi_m\|)$$
$$\int_0^{+\infty} \left[\prod_{i=1}^{m} \left(\frac{D_i}{1+\|t\|^2} - \frac{E_i}{(1+\|t\|^2)^2} \right)^{1/2} \right] \frac{1}{(1+\|t\|^2)^{1/2}}\, dt$$

$$= \frac{1}{\pi^{(m+1)/2}} \Gamma\left(\frac{m}{2}\right) \underline{h}^{1/2} \mathrm{E}(\|\xi_m\|) \left(\prod_{i=1}^{m} D_i^{1/2}\right) \sigma_{m-1}$$

$$\int_0^{+\infty} \frac{\rho^{m-1}}{(1+\rho^2)^{(m+1)/2}} \prod_{i=1}^{m} \left(1 - \frac{F_i}{1+\rho^2}\right)^{1/2} d\rho,$$

where we have denoted $F_i = E_i/D_i (i = 1, \ldots, m)$, which implies that

$$F_i \le \frac{\max_k E_k}{\underline{q}} \qquad i = 1, \ldots, m.$$

Now choose $\tau > 1$ large enough to have

$$\lambda^2 = \frac{r^2}{1+r^2} < 1 - \frac{\max_{1 \le i \le m} F_i}{1+\tau^2 r^2} = \nu^2$$

and we get for $\mathrm{E}(N^X)$ the lower bound:

$$\mathrm{E}(N^X) \ge \frac{1}{\pi^{(m+1)/2}} \Gamma\left(\frac{m}{2}\right) \underline{h}^{1/2} E(\|\xi_m\|) \left(\prod_{i=1}^{m} D_i^{1/2}\right) \sigma_{m-1}\, \nu^m$$

$$\int_{\tau r}^{+\infty} \frac{\rho^{m-1}}{(1+\rho^2)^{(m+1)/2}} d\rho. \tag{12.47}$$

Now, compare (12.46) and (12.47) and use the elementary equivalence for each $a > 0$:

$$\int_a^{+\infty} \frac{\rho^{m-1}}{(1+\rho^2)^{(m+1)/2}} d\rho \simeq \sqrt{\frac{\pi}{2m}} \quad \text{as } m \to +\infty.$$

We get

$$H_m^1(r) \le C_1 \lambda_1^m \mathrm{E}(N^X),$$

where C_1 is a positive constant and $\lambda/\nu < \lambda_1 < 1$. This implies that

$$\mathrm{E}(N^{P+X}) \le s_m \left[C_1 \lambda_1^m + e^{-\ell m/2}\right] \mathrm{E}(N^X) \le C\,\theta^m \mathrm{E}(N^X)$$

for positive constants C, θ, and $0 < \theta < 1$.

More precisely, we can obtain first θ and then m_0 and the constant C in such a way that whenever $m \ge m_0$, inequality (12.36) holds true. The reader can verify, following the proof step by step, that a possible choice is the following. Choose r_0 and ℓ from (H4),

$$\theta_1 = \max\left\{\frac{r_0}{\sqrt{r_0^2 + \frac{1}{2}}}, e^{-\ell/2}\right\}, \qquad \theta = \frac{1+\theta_1}{2}.$$

Set $F_i = E_i/D_i (i = 1, \ldots, m)$ and $\bar{F} = \max\{F_1, \ldots, F_m\}$. From the hypotheses, one has $\bar{F} \leq \bar{E}/\underline{q}$. Let $\tau > 0$ such that

$$\frac{\bar{F}}{1 + \tau^2 r_0^2} < \frac{1}{2} \frac{1}{1 + r_0^2}.$$

Choose m_0 [using (H3)] so that if $m \geq m_0$, one has

$$\exp\left(\frac{1}{2}\left(\frac{mA_m}{\underline{q}} + \frac{mB_m}{\underline{hq}}\right)\right) \theta_1^m \sqrt{m} \leq \theta^m$$

$$\pi \left(\frac{\tau^2 r_0^2}{1 + \tau^2 r_0^2}\right)^{(m-1)/2} < \frac{e^{-2}}{\sqrt{m}}.$$

Then (12.36) is satisfied for $m \geq m_0$, with

$$C = 30 \left(\frac{\bar{h}}{\underline{h}}\right) \frac{\sqrt{1 + r_0^2}}{r_0}.$$

□

12.3.1. Auxiliary Lemma

Lemma 12.13. *Let $\gamma : \mathbb{R}^k \to \mathbb{R}$, $k \geq 1$ be defined as $\gamma(c) = E(\|\xi + c\|)$, where ξ is a standard normal random vector in $\mathbb{R}^k$ and $c \in \mathbb{R}^k$. Then*

(i) $\gamma(0) = \sqrt{2}\,\dfrac{\Gamma((k+1)/2)}{\Gamma(k/2)}$.

(ii) *γ is a function of $\|c\|$ and verifies*

$$\gamma(c) \leq \gamma(0)\left(1 + \frac{1}{2k}\|c\|^2\right). \tag{12.48}$$

Proof

(i) This follows on integrating in polar coordinates.

(ii) That γ is a function of $\|c\|$ is a consequence of the invariance of the distribution of ξ under the orthogonal group of $\mathbb{R}^k$. For $k = 1$, (12.48) follows from the exact computation

$$\gamma(c) = \sqrt{2/\pi}\, e^{-(1/2)c^2} + c \int_{-c}^{c} \frac{1}{\sqrt{2\pi}} e^{-(1/2)x^2}$$

and a Taylor expansion at $c = 0$, which gives

$$\gamma(c) \leq \sqrt{2/\pi}\left(1 + \frac{1}{2}c^2\right).$$

For $k \geq 2$, we write

$$\gamma(c) = \mathrm{E}\left(\left[(\xi_1 + a)^2 + \xi_2^2 + \cdots + \xi_k^2\right]^{1/2}\right) = G(a),$$

where $a = \|c\|$ and $\xi_1, \ldots, \xi_k$ are independent standard normal variables. Differentiating under the expectation sign, we get

$$G'(a) = \mathrm{E}\left(\frac{\xi_1 + a}{[(\xi_1 + a)^2 + \xi_2^2 + \cdots + \xi_k^2]^{1/2}}\right),$$

so that $G'(0) = 0$, due to the symmetry of the distribution of ξ.

One can differentiate formally once more, obtaining

$$G''(a) = \mathrm{E}\left(\frac{\xi_2^2 + \cdots + \xi_k^2}{\left[(\xi_1 + a)^2 + \xi_2^2 + \cdots + \xi_k^2\right]^{3/2}}\right). \tag{12.49}$$

For the validity of equality (12.49) for $k \geq 3$, one can use the fact that if $d \geq 2$, $1/\|x\|$ is integrable in $\mathbb{R}^d$ with respect to the Gaussian standard measure. For $k = 2$ one must be more careful, but it holds true and is left to the reader. The other ingredient of the proof is that one can verify that G'' has a maximum at $a = 0$. Hence, on applying Taylor's formula, we get

$$G(a) \leq G(0) + \tfrac{1}{2} a^2 G''(0).$$

Check that $G''(0) = (\sqrt{2/k})[\Gamma((k+1)/2)/\Gamma(k/2)]$, which together with (i) gives

$$\frac{G''(0)}{G(0)} = \frac{1}{k},$$

which implies (ii). □

12.3.2. Examples

1. *Shub–Smale noise*. Assume that the noise follows the Shub–Smale model. If the degrees d_i are uniformly bounded, one can easily check that (H1) and (H2) are satisfied. For the signal, we give two simple examples. Let

$$P_i(t) = \|t\|^{d_i} - r^{d_i},$$

where d_i is even and $r > 0$ remains bounded as m varies. One has

$$\frac{\partial}{\partial\rho}\left(\frac{P_i}{\sqrt{Q^{(i)}}}\right)(t) = \frac{d_i\,\|t\|^{d_i - 1} + d_i\,r^{d_i}\,\|t\|}{(1 + \|t\|^2)^{(d_i/2)+1}} \leq \frac{d_i(1 + r^{d_i})}{(1 + \|t\|^2)^{3/2}}$$

$$\nabla \left(\frac{P_i}{\sqrt{Q^{(i)}}} \right)(t) = \frac{d_i \, \|t\|^{d_i - 2} + d_i \, r^{d_i}}{(1 + \|t\|^2)^{(d_i/2)+1}} t,$$

which implies that

$$\left\| \nabla \left(\frac{P_i}{\sqrt{Q^{(i)}}} \right)(t) \right\| \leq \frac{d_i(1 + r^{d_i})}{(1 + \|t\|^2)^{3/2}}.$$

So since the degrees $d_1, \ldots, d_m$ are bounded uniformly, (H3) follows. (H4) also holds under the same hypothesis.

Notice that an interest in this choice of the P_i's lies in the fact that obviously the system $P_i(t) = 0$ $(i = 1, \ldots, m)$ has infinite roots (all points in the sphere of radius r centered at the origin are solutions), but the expected number of roots of the perturbed system is geometrically smaller than the Shub–Smale expectation $\sqrt{D}$ when m is large.

Our second example of signal is as follows. Let T be a polynomial of degree d in one variable that has d distinct real roots. Define

$$P_i(t_1, \ldots, t_m) = T(t_i) \qquad i = 1, \ldots, m.$$

One can easily check that the system verifies our hypotheses, so that there exist C and θ positive constants, $0 < \theta < 1$, such that

$$\mathrm{E}(N^{P+X}) \leq C\theta^m d^{m/2},$$

where we have used the Kostlan–Shub–Smale formula. On the other hand, it is clear that $N^P = d^m$, so that the diminishing effect of the noise on the number of roots can be observed.

2. $Q^{(i)} = Q$, *only real roots*. Assume that all the $Q^{(i)}$'s are equal, $Q^{(i)} = Q$, and Q has only real roots. Since Q does not vanish on $u \geq 0$, all the roots should be strictly negative, say $-\alpha_1, \ldots, -\alpha_d$, where $0 < \alpha_1 \leq \alpha_2 \leq \cdots \leq \alpha_d$. With no loss of generality, we may assume that $\alpha_1 \geq 1$.

We will assume again that the degree d (which can vary with m) is bounded by a fixed constant $\overline{d}$ as well as the roots $\alpha_k \leq \overline{\alpha}$ $(k = 1, \ldots, d)$ for some constant $\overline{\alpha}$. One verifies (12.32), choosing $D_i = d$, $E_i = d \max_{1 \leq k \leq d} (\alpha_k - 1)$. Similary, a direct computation gives (12.33).

Again let us consider the particular example of signals:

$$P_i(t) = \|t\|^{d_i} - r^{d_i},$$

where d_i is even and r is positive and remains bounded as m varies:

$$\left| \frac{\partial}{\partial \rho} \left(\frac{P_i}{\sqrt{Q^{(i)}}} \right) \right| \leq d_i(\overline{\alpha} + r^{d_i}) \frac{1}{(1 + \|t\|^2)^{3/2}},$$

so that $K(P_i, Q^{(i)})$ is bounded uniformly. A similar computation shows that $H(P_i.Q^{(i)})$ is bounded uniformly. Finally, it is obvious that

$$L(P_i, Q^{(i)}, r) \geq \left(\frac{1}{1+\overline{\alpha}}\right)^{\overline{d}}$$

for $i = 1, \ldots, m$ and any $r \geq 1$. So the conclusion of Theorem 12.12 can be applied.

One can check that the second signal in the preceding example also works with respect to this noise.

3. *Some other examples*. Assume that the covariance of the noise has the form of example 1 in 12.2.1. Q is a polynomial in one variable having degree ν and positive coefficients, $Q(u) = \sum_{k=0}^{\nu} b_k u^k$. Q may depend on m as well as the exponents $l_1, \ldots, l_m$. Notice that $d_i = \nu l_i$ $(i = 1, \ldots, m)$. One easily verifies that (H1) is satisfied.

We assume that the coefficients $b_0, \ldots, b_\nu$ of the polynomial Q verify the conditions

$$b_k \leq \frac{\nu - k + 1}{k} b_{k-1} \qquad k = 1, 2, \ldots, \nu.$$

Moreover, $l_1, \ldots, l_m, \nu$ are bounded by a constant independent of m and there exist positive constants $\underline{b}$ and $\overline{b}$ such that

$$\underline{b} \leq b_0, b_1, \ldots, b_\nu \leq \overline{b}.$$

Under these conditions, one can check that (H2) holds true, with $D_i = d_i$ $(i = 1, \ldots, m)$. For the relation signal over noise, conditions are similar to those in the preceding example.

Notice that if $\nu = 2$ and we choose for Q the fixed polynomial

$$Q(u) = 1 + 2au + bu^2$$

with $0 < a \leq 1$, $\sqrt{b} > a \geq b > 0$, then the conditions in this example are satisfied, but the polynomial Q (hence Q^{d_i}) does not have real roots, so that it is not included in the previous example.

12.4. SYSTEMS HAVING A LAW INVARIANT UNDER ORTHOGONAL TRANSFORMATIONS AND TRANSLATIONS

In this section we assume that $X_i : \mathbb{R}^m \to \mathbb{R}$, $i = 1, \ldots, m$ are independent centered Gaussian random fields with covariances having the form

$$r^{X_i}(s,t) = \gamma_i(\|t-s\|^2) \qquad i = 1, \ldots, m. \tag{12.50}$$

We assume that γ_i is of class C^2 and, with no loss of generality, that $\gamma_i(0) = 1$.

The computation of the expectation of the number of roots belonging to a Borel set V can be done using Rice's formula (12.21), obtaining

$$\mathrm{E}\left(N^X(V)\right) = (2\pi)^{-m/2}\mathrm{E}\left(|\det(X'(0))|\right)\lambda_m(V). \tag{12.51}$$

To prove (12.51) we take into account that the law of the random field is invariant under translations and for each t, $X(t)$ and $X'(t)$ are independent. Compute for $i, \alpha, \beta = 1, \dots, m$,

$$\mathrm{E}\left(\frac{\partial X_i}{\partial t_\alpha}(0)\frac{\partial X_i}{\partial t_\beta}(0)\right) = \left.\frac{\partial^2 r^{X_i}}{\partial s_\alpha\, \partial t_\beta}\right|_{t=s} = -2\gamma_i'(0)\delta_{\alpha\beta},$$

which implies, using a method similar to the one in the proof of Theorem 12.1, that

$$\mathrm{E}\left(|\det(X'(0))|\right) = \frac{1}{\sqrt{\pi}}2^m\Gamma((m+1)/2)\prod_{i=1}^m |\gamma_i'(0)|^{1/2}$$

and substituting in (12.51) yields

$$\mathrm{E}\left(N^X(V)\right) = \frac{1}{\sqrt{\pi}}\left(\frac{2}{\pi}\right)^{m/2}\Gamma((m+1)/2)\left[\prod_{i=1}^m |\gamma_i'(0)|^{1/2}\right]\lambda_m(V). \tag{12.52}$$

Next, let us consider the variance. One can prove that under certain additional technical conditions, the variance of the normalized number of roots,

$$n^X(V) = \frac{N^X(V)}{\mathrm{E}\left(N^X(V)\right)},$$

which obviously has a mean value equal to 1, grows exponentially when the dimension m tends to infinity. This establishes a striking difference with respect to the results in Section 12.1. In other words, one should expect to have large fluctuations of $n^X(V)$ around its mean for systems having large m.

Our additional requirements are the following:

1. All the γ_i coincide, $\gamma_i = \gamma$, $i = 1, \dots, m$.
2. The function γ is such that $(s,t) \rightsquigarrow \gamma(\|t-s\|^2)$ is a covariance for all dimensions m.

It is well known (Schoenberg, 1938) that γ satisfies requirement 2 and $\gamma(0) = 1$ if and only if there exists a probability measure G on $[0, +\infty)$ such that

$$\gamma(x) = \int_0^{+\infty} e^{-xw}G(dw) \qquad \text{for all } x \geq 0. \tag{12.53}$$

Theorem 12.14 (Azaïs-Wschebor, 2005b). *Let* $r^{X_i}(s,t) = \gamma(\|t-s\|^2)$ *for* $i = 1, \ldots, m$, *where* γ *is of the form (12.53). We assume further that*

(1) *G is not concentrated at a single point and*

$$\int_0^{+\infty} x^2 G(dx) < \infty.$$

(2) $\{V_m\}_{m=1,2\ldots}$ *is a sequence of Borel sets,* $V_m \subset \mathbb{R}^m$, $\lambda_m(\partial V_m) = 0$ *and there exist two positive constants* δ *and* Δ *such that for each* m, V_m *contains a ball with radius* δ *and is contained in a ball with radius* Δ.

Then

$$Var\left(n^X(V_m)\right) \to +\infty, \tag{12.54}$$

exponentially fast as $m \to +\infty$.

To compute the variance of $N^X(V)$, we start as in the case of the KSS model:

$$\mathrm{Var}\left(N^X(V)\right) = \mathrm{E}\left(N^X(V)\left(N^X(V)-1\right)\right) + \mathrm{E}\left(N^X(V)\right) - \left[\mathrm{E}\left(N^X(V)\right)\right]^2, \tag{12.55}$$

so that to prove (12.54), it suffices to show that

$$\frac{\mathrm{E}\left(N^X(V)\left(N^X(V)-1\right)\right)}{\left[\mathrm{E}\left(N^X(V)\right)\right]^2} \to +\infty \tag{12.56}$$

exponentially fast as $m \to +\infty$. The denominator in (12.56) is given by formula (12.52). For the numerator, we can use Rice's formula for the second-order factorial moment:

$$\mathrm{E}\left(N^X(V)\left(N^X(V)-1\right)\right)$$
$$= \iint_{V\times V} \mathrm{E}\left(\left|\det(X'(s))\det(X'(t))\right| \,\middle|\, X(s) = X(t) = 0\right) p_{X(s),X(t)}(0,0)\,ds\,dt. \tag{12.57}$$

Next, we compute the ingredients of the integrand in (12.57). Because of invariance under translations, the integrand is a function of $\tau = t - s$. We denote by $\tau_1, \ldots, \tau_m$ the coordinates of τ.

The Gaussian density is immediate:

$$p_{X(s),X(t)}(0,0) = \frac{1}{(2\pi)^m}\frac{1}{\left[1-\gamma^2(\|\tau\|^2)\right]^{m/2}}. \tag{12.58}$$

Let us turn to the conditional expectation in (12.57). We set

$$\mathrm{E}\left(\left|\det(X'(s))\det(X'(t))\right| \,\middle|\, X(s)=X(t)=0\right) = \mathrm{E}\left(\left|\det(A^s)\det(A^t)\right|\right),$$

where $A^s = ((A^s_{i\alpha}))$, $A^t = ((A^t_{i\alpha}))$ are $m \times m$ random matrices having as joint (Gaussian) distribution the conditional distribution of the pair $X'(s)$ and $X'(t)$ given that $X(s) = X(t) = 0$. So, to describe this joint distribution we must compute the conditional covariances of the elements of the matrices $X'(s)$ and $X'(t)$ given the condition $\mathcal{C} : \{X(s) = X(t) = 0\}$. This is easily done using standard regression formulas:

$$\mathrm{E}\left(\frac{\partial X_i}{\partial s_\alpha}(s)\frac{\partial X_i}{\partial s_\beta}(s)\middle|\mathcal{C}\right) = \left.\frac{\partial^2 r}{\partial s_\alpha\, \partial t_\beta}\right|_{t=s} - \frac{1}{1-(r(s,t))^2}\frac{\partial r}{\partial s_\alpha}(s,t)\frac{\partial r}{\partial s_\beta}(s,t)$$

$$\mathrm{E}\left(\frac{\partial X_i}{\partial s_\alpha}(s)\frac{\partial X_i}{\partial t_\beta}(t)\middle|\mathcal{C}\right) = \frac{\partial^2 r}{\partial s_\alpha \partial t_\beta}(s,t) + \frac{1}{1-(r(s,t))^2}\frac{\partial r}{\partial s_\alpha}(s,t)\frac{\partial r}{\partial t_\beta}(s,t)r(s,t).$$

Substituting in our case, we obtain

$$\mathrm{E}\left(A^s_{i\alpha}A^s_{i\beta}\right) = \mathrm{E}\left(A^t_{i\alpha}A^t_{i\beta}\right) = -2\gamma'(0)\delta_{\alpha\beta} - 4\frac{\gamma'^2\tau_\alpha\tau_\beta}{1-\gamma^2}, \tag{12.59}$$

$$\mathrm{E}\left(A^s_{i\alpha}A^t_{i\beta}\right) = -4\gamma''\tau_\alpha\tau_\beta - 2\gamma'\delta_{\alpha\beta} - 4\frac{\gamma\gamma'^2\tau_\alpha\tau_\beta}{1-\gamma^2}, \tag{12.60}$$

and for every $i \neq j$,

$$\mathrm{E}\left(A^s_{i\alpha}A^s_{j\beta}\right) = \mathrm{E}\left(A^t_{i\alpha}A^t_{j\beta}\right) = \mathrm{E}\left(A^s_{i\alpha}A^t_{j\beta}\right) = 0,$$

where $\gamma = \gamma(\|\tau\|^2)$, $\gamma' = \gamma'(\|\tau\|^2)$, and $\gamma'' = \gamma''(\|\tau\|^2)$.

Now take an orthonormal basis of $\mathbb{R}^m$ having the unit vector $\tau/\|\tau\|$ as the first element. Then the variance $2m \times 2m$ matrix of the pair A^s_i and A^t_i (the i-th rows of A^s and A^t, respectively) takes the following form:

$$T = \left[\begin{array}{ccccc|ccccc} U_0 & \cdots & . & . & & U_1 & \cdots & . & . \\ . & V_0 & \cdots & . & & . & V_1 & \cdots & . \\ & & \ddots & & & & & \ddots & \\ . & . & & . & & . & . & & . \\ . & . & \cdots & V_0 & & . & . & \cdots & V_1 \\ \hline U_1 & \cdots & . & . & & U_0 & \cdots & . & . \\ . & V_1 & \cdots & . & & . & V_0 & \cdots & . \\ & & \ddots & & & & & \ddots & \\ . & . & & . & & . & . & & . \\ . & . & \cdots & V_1 & & . & . & \cdots & V_0 \end{array}\right],$$

where

$$U_0 = U_0(\|\tau\|^2) = -2\gamma'(0) - 4\frac{\gamma'^2 \|\tau\|^2}{1-\gamma^2}$$

$$V_0 = -2\gamma'(0)$$

$$U_1 = U_1(\|\tau\|^2) = -4\gamma'' \|\tau\|^2 - 2\gamma' - 4\frac{\gamma\gamma'^2\|\tau\|^2}{1-\gamma^2}$$

$$V_1 = V_1(\|\tau\|^2) = -2\gamma',$$

and there are zeros outside the diagonals of each of the four blocks. Let us perform a second regression of $A^t_{i\alpha}$ on $A^s_{i\alpha}$, that is, write the orthogonal decompositions

$$A^t_{i\alpha} = B^{t,s}_{i\alpha} + C_\alpha A^s_{i\alpha} \qquad i, \alpha = 1, m,$$

where $B^{t,s}_{i\alpha}$ is centered Gaussian independent of the matrix A^s, and

$$\text{for } \alpha = 1: \quad C_1 = \frac{U_1}{U_0}, \ \mathrm{Var}(B^{t,s}_{i1}) = U_0\left(1 - \frac{U_1^2}{U_0^2}\right)$$

$$\text{for } \alpha > 1: \quad C_\alpha = \frac{V_1}{V_0}, \ \mathrm{Var}(B^{t,s}_{i\alpha}) = V_0\left(1 - \frac{V_1^2}{V_0^2}\right).$$

Conditioning, we have

$$\mathrm{E}\left(|\det(A^s)||\det(A^t)|\right) = \mathrm{E}\left[|\det(A^s)|\mathrm{E}\left(|\det((B^{t,s}_{i\alpha} + C_\alpha A^s_{i\alpha})_{i,\alpha=1,\dots,m})|\,\big|A^s\right)\right]$$

with obvious notation. For the inner conditional expectation, we can proceed in the same way as we did in the proof of Theorem 12.11 to compute the determinant, obtaining a product of expectations of Euclidean norms of noncentered Gaussian vectors in $\mathbb{R}^k$ for $k = 1, \dots, m$. Now we use the well-known inequality

$$\mathrm{E}\left(\|\xi + v\|\right) \geq \mathrm{E}\left(\|\xi\|\right)$$

valid for ξ standard normal in $\mathbb{R}^k$ and v any vector in $\mathbb{R}^k$, and it follows that

$$\mathrm{E}\left(|\det(A^s)||\det(A^t)|\right) \geq \mathrm{E}\left(|\det(A^s)|\right)\mathrm{E}\left(|\det(B^{t,s})|\right).$$

Since the elements of A^s (respectively, $B^{t,s}$) are independent centered Gaussian with known variance, we obtain

$$\mathrm{E}\left|\det(A^s)\det(A^t)\right| \geq U_0 V_0^{m-1}\left(1 - \frac{U_1^2}{U_0^2}\right)^{1/2}\left(1 - \frac{V_1^2}{V_0^2}\right)^{(m-1)/2} L_m^2.$$

Going back to (12.56) and because of (12.52) and (12.57), we have

$$\frac{\mathrm{E}\left(N^X(V)\left(N^X(V)-1\right)\right)}{\mathrm{E}\left(N^X(V)\right)^2} \geq (\lambda_m(V))^{-2} \iint_{V\times V} ds\,dt \left[\frac{1-V_1^2V_0^{-2}}{1-\gamma^2}\right]^{m/2} H(\|\tau\|^2). \tag{12.61}$$

Let us put $V = V_m$ in (12.61) and study the integrand on the right-hand side. The function

$$H(x) = \left(\frac{U_0^2(x) - U_1^2(x)}{V_0^2 - V_1^2(x)}\right)^{1/2}$$

is continuous for $x > 0$. Let us show that it does not vanish if $x > 0$.

It is clear that $U_1^2 \leq U_0^2$ on applying the Cauchy–Schwarz inequality to the pair of variables A_{i1}^s and A_{i1}^t. The equality holds if and only if the variables A_{i1}^s and A_{i1}^t are linearly dependent. This would imply that the distribution in $\mathbb{R}^4$ of the random vector

$$\zeta := (X(s), X(t), \partial_1 X(s), \partial_1 X(t))$$

would degenerate for $s \neq t$ (we have denoted ∂_1 differentiation with respect to the first coordinate). We will show that this is not possible. Notice first that for each $w > 0$, the function

$$(s,t) \rightsquigarrow e^{-\|t-s\|^2 w}$$

is positive definite; hence it is the covariance of a centered Gaussian stationary field defined on $\mathbb{R}^m$, say $\{Z^w(t) : t \in \mathbb{R}^m\}$, whose spectral measure has the nonvanishing density

$$f^w(x) = (2\pi)^{-m/2}(2w)^{-m/2} \exp\left(-\frac{\|x\|^2}{4w}\right) \quad (x \in \mathbb{R}^m).$$

The field $\{Z^w(t) : t \in \mathbb{R}^m\}$ satisfies the conditions of Proposition 3.1 of Azaïs and Wschebor (2005a) so that the distribution of the 4-tuple

$$\zeta^w := \left(Z^w(s), Z^w(t), \partial_1 Z^w(s), \partial_1 Z^w(t)\right)$$

does not degenerate for $s \neq t$. Based on (12.53) we have

$$\mathrm{Var}(\zeta) = \int_0^{+\infty} \mathrm{Var}(\zeta^w) G(dw),$$

where integration of the matrix is integration term by term. This implies that the distribution of ζ does not degenerate for $s \neq t$ and that $H(x) > 0$ for $x > 0$.

We now show that for $\tau \neq 0$,

$$\frac{1 - V_1^2(\|\tau\|^2)V_0^{-2}}{1 - \gamma^2(\|\tau\|^2)} > 1,$$

which is equivalent to

$$-\gamma'(x) < -\gamma'(0)\gamma(x) \qquad \forall x > 0. \tag{12.62}$$

The left-hand side of (12.62) can be written as

$$-\gamma'(x) = \frac{1}{2}\iint_0^{+\infty} (w_1 \exp(-xw_1) + w_2 \exp(-xw_2))\, G(dw_1)G(dw_2)$$

and the right-hand side as

$$-\gamma'(0)\gamma(x) = \frac{1}{2}\iint_0^{+\infty} (w_1 \exp(-xw_2) + w_2 \exp(-xw_1))\, G(dw_1)G(dw_2),$$

so that

$$\begin{aligned} &-\gamma'(0)\gamma(x) + \gamma'(x) \\ &= \frac{1}{2}\iint_0^{+\infty} (w_2 - w_1)\,(\exp(-xw_1) - \exp(-xw_2))\, G(dw_1)G(dw_2), \end{aligned}$$

which is ≥ 0 and is equal to zero only if G is concentrated at a point, which is not the case. This proves (12.62). Now, using the hypotheses on the inner and outer diameter of V_m, the result follows by a compactness argument. □

CHAPTER 13

RANDOM FIELDS AND CONDITION NUMBERS OF RANDOM MATRICES

Let A be an invertible $n \times n$ real matrix and $b \in \mathbb{R}^n$. We are interested in understanding how the solution $x \in \mathbb{R}^n$ of the linear system of equations

$$Ax = b \tag{13.1}$$

is affected by perturbations in the input (A, b).

Early work by von Neumann and Goldstine (1947) and Turing (1948) identified that the key quantity is

$$\kappa(A) = \|A\| \cdot \|A^{-1}\|,$$

where $\|A\|$ denotes the operator norm of A defined in the usual way:

$$\|A\| = \max_{\|x\|=1} \|Ax\|.$$

Here $\|v\|$ denotes the Euclidean norm of $v \in \mathbb{R}^n$. Of course, other norms can be considered, but in this chapter we restrain ourselves to the Euclidean norm. If A is singular, we put $\kappa(A) = +\infty$. Turing called $\kappa(A)$ the *condition number* of A. The first meaning of $\kappa(A)$ is a consequence of the following property.

Let $x + \Delta x$ be the solution of system (13.1) when the input is $(A + \Delta A, b + \Delta b)$ instead of (A, b). If $\kappa(A)\|\Delta A\|/\|A\| < 1$, then

$$\frac{\|\Delta x\|}{\|x\|} \le \frac{\kappa(A)}{1 - \kappa(A)(\|\Delta A\|/\|A\|)} \left(\frac{\|\Delta A\|}{\|A\|} + \frac{\|\Delta b\|}{\|b\|} \right). \tag{13.2}$$

In fact,

$$(A + \Delta A)\, \Delta x = \Delta b - (\Delta A)x,$$

which implies, whenever $\|A^{-1}(\Delta A)\| < 1$, that

$$\Delta x = (A + \Delta A)^{-1}(\Delta b - (\Delta A)x) = (I + A^{-1}(\Delta A))^{-1} A^{-1}(\Delta b - (\Delta A)x)$$

and taking norms yields

$$\begin{aligned} \|\Delta x\| &\le \|(I + A^{-1}(\Delta A))^{-1}\| \|A^{-1}\| (\|\Delta b\| + \|\Delta A\| \|x\|) \\ &\le \|(I + A^{-1}(\Delta A))^{-1}\| \|A^{-1}\| \left(\frac{\|\Delta b\|}{\|b\|} + \frac{\|\Delta A\|}{\|A\|} \right) \|A\| \|x\|, \end{aligned}$$

from which (13.2) follows.

Notice that the factor $\kappa(A)/[1 - \kappa(A)(\|\Delta A\|/\|A\|)]$ tends to $\kappa(A)$ when $\|\Delta A\| \to 0$. Thus, $\kappa(A)$ is a bound for the amplification of the relative error between output and input in the system (13.1) when the last one is small. The reader may also easily check that, in addition, $\kappa(A)$ is sharp in the sense that no smaller number will satisfy a similar inequality for all increments ΔA and Δb. In other words, if one thinks in terms of binary floating-point arithmetic, $\log_2 \kappa(A)$ measures the loss of precision due to error in the input. So for numerical analysis purposes, it is usual that the relevant function of the matrix A appears to be $\log_2 \kappa(A)$.

Matrices A with $\kappa(A)$ small are said to be *well-conditioned*; those with $\kappa(A)$ large are said to be *ill-conditioned*. The set $\Sigma = \{A : \kappa(A) = +\infty\}$ is called the set of *ill-posed matrices*. The distance of a matrix A to the set Σ is closely related to $\kappa(A)$, as shown in the next theorem. For a proof, see Blum et al. (1998).

Theorem 13.1 (Eckart-Young, 1936). *For any $n \times n$ real matrix A, one has*

$$\kappa(A) = \frac{\|A\|}{d_F\,(A, \Sigma)}.$$

Here d_F means distance in $\mathbb{R}^{n^2}$ with respect to the Frobenius norm $\|A\|_F = \sqrt{\sum a_{ij}^2}$.

The relationship between conditioning and distance to ill-posedness is a recurrent theme in numerical analysis (see, e.g., Demmel, 1987). $\kappa(A)$ appears in more elaborate round-off analysis of algorithms, in which errors may occur in all the operations. As an example, let us mention such an analysis for Cholesky's method (see Wilkinson, 1963). If A is symmetric and positive definite, we may solve the linear system $Ax = b$ by using Cholesky's factorization. Assume that the length of the mantissa in the binary representation used in the computation is equal to ℓ. Then, if ℓ is large enough, one can prove that

$$\frac{\|\Delta x\|}{\|x\|} \leq 3n^3 \cdot 2^{-\ell} \kappa(A).$$

The interested reader can find a variety of related subjects in the books by Higham (1996) and Trefethen and Bau (1997).

Next, we introduce some notation. Given A, an $n \times n$ real matrix, we denote by $\nu_1, \ldots, \nu_n$, $0 \leq \nu_1 \leq \cdots \leq \nu_n$, the squares of the singular values of A, that is, the eigenvalues of $A^{\mathrm{T}}A$. If $X : S^{m-1} \to \mathbb{R}$ is the quadratic polynomial $X(x) = x^{\mathrm{T}}A^{\mathrm{T}}Ax$, then:

- $\nu_n = \|A\|^2 = \max_{x \in S^{n-1}} X(x)$.
- In case A is nonsingular, it follows that $\nu_1 = 1/\left\|\mathrm{A}^{-1}\right\|^2 = \min_{x \in S^{n-1}} X(x)$.

Hence,

$$\kappa(A) = \left(\frac{\nu_n}{\nu_1}\right)^{1/2}$$

when $\nu_1 > 0$ and $\kappa(A) = +\infty$ if $\nu_1 = 0$. Notice also that $\kappa(A) \geq 1$ and $\kappa(rA) = \kappa(A)$ for any real r, $r \neq 0$. That is, the computation of the condition number of an $n \times n$ matrix A is a problem about the spectrum of the nonnegative definite matrix $A^{\mathrm{T}}A$ or, more precisely, about the largest and the smallest singular values of A.

Suppose that one is interested in the analysis of a certain algorithm in which the condition number $\kappa(A)$ plays a role. Typically, the condition number will be a component appearing in some bound for the cost of the algorithm, in which, as in the example above, the size of the problem and the length of the mantissa in the floating-point computation will also be present. A natural setting consists of imagining that the algorithm is applied to a problem that is drawn at random from a certain family of problems, which in our case amounts to choosing at random the coefficients of the system of equations. The cost of the algorithm and $\kappa(A)$ become random variables. One can try to compute, or give bounds, for the expectation or the higher moments of the cost, or estimate its distribution function, and this will depend on the moments or the probability distribution of $\kappa(A)$ itself. Thus, in our case, the question of probabilistic analysis of algorithms, becomes a problem on the spectrum of random matrices.

There is a large body of knowledge on random matrices: specifically, on their singular values or eigenvalues. Our general reference is Mehta's book (2004), which we referred to in Chapter 8. See also Girko's book (1996).

Since the 1930s, motivation and methods have come from different sources, perhaps beginning with Fisher in 1939 in multivariate statistics (see Muirhead,

1982; Kendall et al., 1983). By the mid-1950s, Wigner's work (e.g., 1958, 1967) was followed by great interest in random matrices in some areas of mathematical physics, which continues until the present time (see, e.g., Soshnikov, 1998; Tracy and Widom, 1999; Davidson and Szarek, 2001). A third source of interest has been numerical analysis, motivated by the aforementioned condition number problems, which may have begun with a paper by Smale (1985). A very interesting survey of analytical methods, originally inspired by numerical analysis problems but including a diversity of applications, is that of Edelman and Rao (2005).

In this chapter we are considering only a very small part of the subject, which concerns the condition number $\kappa(A)$. One reason to include this topic here is that the methods we present are based on random fields and Rice formulas, which is the core of this book. In some cases, these methods do not give optimal results, as happens in the canonical case of matrices having i.i.d. standard normal entries. However, for noncentered Gaussian matrices, our methods still provide the best bounds for the tails of the probability distribution of $\kappa(A)$ of which the authors are aware.

We start in Section 13.1 with some elementary upper bounds for $\mathrm{E}(\log \kappa(A))$ when A is a random matrix with i.i.d. (not necessarily Gaussian) entries. The methods are ad hoc, but we are not aware of the existence of better general bounds. The remainder of the chapter concerns Gaussian matrices. Section 13.2 covers centered matrices and Section 13.3, noncentered matrices.

13.1. CONDITION NUMBERS OF NON-GAUSSIAN MATRICES

The results of this section are extracted from Cuesta-Albertos and Wschebor (2003). Throughout the section we assume that $A = ((a_{ij}))$, $i, j = 1, 2, \ldots, n$ is an $n \times n$ matrix, where the a_{ij}'s are i.i.d. real-valued random variables defined on some probability space $(\Omega, \mathcal{A}, \mathrm{P})$. We denote by μ the common distribution measure of the a_{ij}'s.

13.1.1. General Bound for E(log $\kappa(A)$) for Symmetric Entries

Theorem 13.2. *We assume that the distribution μ satisfies the following conditions:*

(1) *For any pair α, β of real numbers, $\alpha < \beta$, one has*

$$\mu\left([\alpha, \beta]\right) \leq \mu\left(\left[-\frac{\beta-\alpha}{2}, \frac{\beta-\alpha}{2}\right]\right). \tag{13.3}$$

(2) $\mathrm{E}[|a_{1,1}|^r] = 1$, *for some $r > 0$.*

(3) *There exist positive numbers C and γ such that*

$$\mu([-\alpha, \alpha]) \leq C\alpha^{\gamma} \qquad \textit{for all } \alpha > 0.$$

Then

$$\mathrm{E}[\log \kappa(A)] \leq \left(1 + \frac{2}{r}\right) \log n + \frac{1}{r} + \frac{1}{\gamma}\{[(2+\gamma)\log n + \log C]^{+} + 1\}. \quad (13.4)$$

Proof. Notice that $\|A\| \leq \left(\sum_{i,j=1}^{n} a_{i,j}^2\right)^{1/2}$. So, with the only assumption that the random entries are identically distributed, one has, for $t > 0$,

$$\mathrm{P}\left[\|A\| > t\right] \leq \mathrm{P}\left[n^2 \sup_{i,j=1,\ldots,n} a_{i,j}^2 > t^2\right] \leq \mathrm{P}\left[\bigcup_{i,j=1}^{n} \left\{|a_{i,j}| > \frac{t}{n}\right\}\right]$$

$$\leq n^2 \mathrm{P}\left[|a_{1,1}| > \frac{t}{n}\right]. \quad (13.5)$$

Hence, for $\alpha_n \geq 0$,

$$\mathrm{E}[\log \|A\|] \leq \alpha_n + \int_{\alpha_n}^{\infty} \mathrm{P}[\log \|A\| > x]\, dx$$

$$= \alpha_n + \int_{\alpha_n}^{\infty} \mathrm{P}[\|A\| > e^x]\, dx$$

$$\leq \alpha_n + \int_{\alpha_n}^{\infty} n^2 \mathrm{P}\left[|a_{1,1}| > \frac{e^x}{n}\right] dx$$

$$\leq \alpha_n + \int_{\alpha_n}^{\infty} n^2 \left(\frac{n}{e^x}\right)^r dx = \alpha_n + n^{2+r} \frac{1}{r} e^{-r\alpha_n}, \quad (13.6)$$

where the last inequality follows from Markov inequality and assumption (2).

Choose $\alpha_n \geq 0$ to minimize the right-hand side of (13.6) [i.e., $\alpha_n = (1 + 2/r) \log n$] and it follows that

$$\mathrm{E}[\log \|A\|] \leq \left(1 + \frac{2}{r}\right) \log n + \frac{1}{r}. \quad (13.7)$$

We now consider the factor $\|A^{-1}\|$. Denote $A^{-1} = ((b_{i,j}))_{i,j=1,\ldots,n}$; that is,

$$b_{i,j} = \frac{a^{i,j}}{\det(\mathrm{A})} \qquad i, j = 1, \ldots, n,$$

where $a^{i,j}$ is the cofactor of the position (i, j) in the matrix A.

Clearly, the random variables $|b_{i,j}|, i, j = 1, \ldots, n$ are identically distributed so that we may apply (13.5) to the matrix A^{-1} instead of A:

$$\mathrm{P}\left[\|A^{-1}\| > t\right] \le n^2 \mathrm{P}\left[|b_{1,1}| > \frac{t}{n}\right] = n^2 \mathrm{P}\left[\left|\frac{a^{1,1}}{\sum_{j=1}^{n} a_{1,j} a^{1,j}}\right| > \frac{t}{n}\right]$$

$$= n^2 \mathrm{P}\left[\left|a_{1,1} + \sum_{j=2}^{n} a_{1,j} \frac{a^{1,j}}{a^{1,1}}\right| < \frac{n}{t}\right].$$

The random variables

$$a_{1,1} \quad \text{and} \quad \eta = \sum_{j=2}^{n} a_{1,j} \frac{a^{1,j}}{a^{1,1}}$$

are independent, so that for each $\alpha > 0$, denoting by P_η the probability distribution of η, and using Fubini's theorem and assumption (1), we have

$$\mathrm{P}\left[|a_{1,1} + \eta| < \alpha\right] = \int_{-\infty}^{\infty} \mu[(-\alpha - y, \alpha - y)] \mathrm{P}_\eta(dy)$$

$$\le \int_{-\infty}^{\infty} \mu[(-\alpha, \alpha)] \mathrm{P}_\eta(dy) = \mu[(-\alpha, \alpha)].$$

Hence, by assumption (3),

$$\mathrm{P}\left[\|A^{-1}\| > t\right] \le n^2 \mu\left(\left[-\frac{n}{t}, \frac{n}{t}\right]\right) \le n^2 C \left(\frac{n}{t}\right)^\gamma, \tag{13.8}$$

and with $\beta_n \ge 0$,

$$\mathrm{E}[\log \|A^{-1}\|] \le \beta_n + \int_{\beta_n}^{\infty} \mathrm{P}\left[\|A^{-1}\| > e^x\right] dx$$

$$\le \beta_n + \int_{\beta_n}^{\infty} C n^{2+\gamma} e^{-\gamma x}\, dx = \beta_n + C \frac{n^{2+\gamma}}{\gamma} e^{-\gamma \beta_n}.$$

Choosing $\beta_n = (1/\gamma)[(2+\gamma)\log n + \log C]^+$, one obtains

$$\mathrm{E}[\log \|A^{-1}\|] \le \frac{1}{\gamma} \left\{[(2+\gamma)\log n + \log C]^+ + 1\right\}, \tag{13.9}$$

and putting (13.7) and (13.9) together, (13.4) follows. □

Remarks on the Statement of Theorem 13.2. It is not difficult to see that assumption (1) implies that the measure μ is symmetric around zero. In particular, this implies that in case the random variables $a_{i,j}, i, j = 1, \ldots, n$ are integrable, their common expectation must be zero. Since $\kappa(\lambda A) = \kappa(A)$ for any nonzero real number λ and any nonsingular matrix A, in case $m_r = \int_{-\infty}^{\infty} |x|^r \mu(dx) < \infty$ it is possible to replace A by $m_r^{-1/r} A$ so that assumption (2) holds true, without modifying the condition number. Of course, in this case one must, accordingly, change the constant C in assumption (3). In this sense, assumption (2) is not more restrictive than the finiteness of the rth moment of the probability measure μ.

13.1.2. Examples

1. *Density*. Assume that μ has a density function f, that f is even and nonincreasing on $[0, \infty)$, and that $m_r = \int_{-\infty}^{\infty} |x|^r f(x)\, dx < \infty$ for some $r > 0$. We replace the original density f by $m_r^{1/r} f(m_r^{1/r} x)$ so that assumption (2) is satisfied without changing $\kappa(A)$; assumption (3) is verified with $\gamma = 1$ and $C = 2m_r^{1/r} f(0)$. Inequality (13.4) becomes

$$\mathrm{E}[\log(\kappa(A)] \le \left(1 + \frac{2}{r}\right) \log n + \frac{1}{r} + \left[3 \log n + \frac{1}{r} \log m_r + \log(2f(0))\right]^+ + 1. \tag{13.10}$$

2. *Uniform distribution*. Let μ be the uniform distribution on $[-H, H]$, $H > 0$. In this case, $m_r = H^r (r+1)^{-1}$ and (13.10) holds true for any $r > 0$. Letting $r \to +\infty$, we obtain

$$\mathrm{E}[\log \kappa(A)] \le 4 \log n + 1. \tag{13.11}$$

3. *Strong concentration near the mean*. Assume that the density of μ has the form

$$\frac{1}{2} \frac{\gamma}{|x|^{1-\gamma}} \mathbf{1}_{[-1,1]}(x)$$

for some $\gamma, 0 < \gamma < 1$.

One has $m_r = \gamma/(r+\gamma)$ for each $r > 0$ and easily checks that introducing the modification suggested above, assumptions (1), (2), and (3) are satisfied with $C = m_r^{\gamma/r}$. Hence, Theorem 13.2 implies that for any $r > 0$,

$$\mathrm{E}[\log \kappa(A)] \le \left(1 + \frac{2}{r}\right) \log n + \frac{1}{r} + \frac{1}{\gamma} \left\{ \left[(2+\gamma) \log n + \frac{\gamma}{r} \log \frac{\gamma}{r+\gamma} \right]^+ + 1 \right\},$$

and letting $r \to +\infty$ it follows that

$$\mathrm{E}[\log \kappa(A)] \leq \left(2 + \frac{2}{\gamma}\right) \log n + \frac{1}{\gamma}.$$

4. *Particular distributions*. The bound in Theorem 13.2 can be improved by using the actual distribution μ instead of the Markov inequality in (13.6) or the bound in (13.8). This is, for example, the case for symmetric exponential or standard normal distributions. However, in the latter case, this method is not sharp enough: The precise behavior of $\mathrm{E}[\log \kappa(A)]$ as $n \to +\infty$ was given by Edelman (1988) using analytic methods. It is the following:

$$\mathrm{E}[\log \kappa(A)] = \log n + C_0 + \varepsilon_n,$$

where C_0 is a known constant ($C_0 \backsimeq 1537$) and $\varepsilon_n \to 0$.

13.1.3. Smoothed Analysis

We consider now the condition number when the random variables in the matrix $A = (a_{i,j})_{i,j=1,\ldots,n}$ have the form

$$a_{i,j} = m_{i,j} + \psi_{i,j} \qquad i, j = 1, \ldots, n,$$

where $M = (m_{i,j})_{i,j=1,\ldots,n}$ is nonrandom and $(\psi_{i,j})_{i,j=1,\ldots,n}$ are i.i.d. random variables with common distribution μ satisfying assumptions (1), (2), and (3) in Theorem 13.2. This has been called *smoothed analysis* and corresponds to the idea of exploring what happens to the condition number when a nonrandom matrix is perturbed with a noise having a law that verifies a certain number of requirements. We will be looking only to the effect on the moments of the loss of precision $\log \kappa(A)$ when performing this operation. [See papers by Spielman and Teng (2002) and Tao and Vu (2007), where similar questions are considered: in the last case, allowing the measure μ to be purely atomic.]

Theorem 13.3. *Under the conditions of Theorem 13.2, if we assume*

$$m_n = \sup_{i,j=1,\ldots,n} |m_{i,j}| \leq n^{2/r},$$

then

$$\mathrm{E}[\log \kappa(A)] \leq \left(1 + \frac{2}{r}\right) \log n + \log 2 + \frac{1}{r} + \frac{1}{\gamma} \left\{[(2+\gamma)\log n + \log C]^+ + 1\right\}. \tag{13.12}$$

Proof. The proof (as well as the result) is very similar to that of Theorem 13.2. For $t > 0$ one has

$$\mathrm{P}[\|A\| > t] \leq \mathrm{P}\left[\sum_{i,j=1}^{n} a_{i,j}^2 > t^2\right] \leq \sum_{i,j=1}^{n} \mathrm{P}\left[a_{i,j}^2 > \frac{t^2}{n^2}\right]$$

$$= \sum_{i,j=1}^{n} \mathrm{P}\left[|m_{i,j} + \psi_{i,j}| > \frac{t}{n}\right] \leq n^2 \mathrm{P}\left[|\psi_{1,1}| > \frac{t}{n} - m_n\right].$$

Now choose $\alpha_n = (1 + 2/r)\log n + \log 2$. If $x > \alpha_n$, then

$$\frac{e^x}{n} - m_n > \frac{1}{2n}e^x.$$

Thus,

$$\mathrm{E}[\log \|A\|] \leq \alpha_n + \int_{\alpha_n}^{\infty} \mathrm{P}[\|A\| > e^x]\,dx$$

$$\leq \alpha_n + n^2 \int_{\alpha_n}^{\infty} \mathrm{P}\left[|\psi_{1,1}| > \frac{1}{2n}e^x\right] dx$$

$$\leq \alpha_n + n^2 \int_{\alpha_n}^{\infty} \frac{1}{((1/2n)e^x)^r}\,dx$$

$$= \left(1 + \frac{2}{r}\right)\log n + \log 2 + \frac{1}{r}.$$

On the other hand, with the same notation as in the proof of Theorem 13.2, $A^{-1} = \left(b_{i,j}\right)_{i,j=1,\ldots,n}$ and

$$\mathrm{P}\left[\|A^{-1}\| > t\right] \leq \sum_{i,j=1}^{n} \mathrm{P}\left[|b_{i,j}| > \frac{t}{n}\right].$$

For each term in this sum it is possible to repeat exactly the same computations as in the proof of Theorem 13.2 to bound $\mathrm{P}\left[|b_{1,1}| > t/n\right]$ and obtain the same bound as there for $\mathrm{E}[\log \|A^{-1}\|]$. This completes the proof. □

For higher-order moments, one can obtain upper bounds for $\mathrm{E}\left[(\log \kappa(A))^k\right]$, $k = 2, 3, \ldots$ in much the same way that we did for $k = 1$. We consider here the centered case; for smoothed analysis, the situation is similar. Since $\log \kappa(A) \geq 0$, we have that

$$\mathrm{E}[(\log \kappa(A))^k] \leq 2^k [\mathrm{E}\{(\log^+ \|A\|)^k\} + \mathrm{E}\{(\log^+ \|A^{-1}\|)^k\}].$$

Using the same estimates as in the case $k = 1$ for the tails of the probability distributions of $\|A\|$ and $\|A^{-1}\|$, after an elementary computation, it is possible to obtain that if k satisfies the inequalities $2 \le k \le 1 + (2 + \gamma \wedge r) \log n$, then

$$\mathrm{E}\left[(\log \kappa(A))^k\right] \le (2 \log n)^k \left[\left(1 + \frac{2}{r}\right)^k (1 + k) + \left(1 + \frac{2}{\gamma}\right)^k (1 + Ck)\right].$$

13.2. CONDITION NUMBERS OF CENTERED GAUSSIAN MATRICES

The purpose of the present section is to prove the following:

Theorem 13.4 (Azaïs-Wschebor, 2005c). *Assume that* $A = ((a_{ij}))_{i,j=1,\ldots,n}$, $n \ge 3$ *and that the* a_{ij}*'s are i.i.d. standard normal random variables. Then there exist universal positive constants* c *and* C *such that for* $x > 1$,

$$\frac{c}{x} < \mathrm{P}(\kappa(A) > nx) < \frac{C}{x}. \tag{13.13}$$

Remarks

1. The limiting distribution of $\kappa(A)/n$ as $n \to \infty$ has been computed in Edelman's thesis (1989). The interest in this theorem lies in the uniformity of the statement and in the relationship that the proof below establishes with Rice's formulas.

2. This theorem, and related ones, can be considered as results on the Wishart matrix $A^{\mathrm{T}}A$. Introducing some minor changes, it is possible to use the same methods to study the condition number of $A^{\mathrm{T}}A$ for rectangular $n \times m$ matrices A having i.i.d. standard normal entries, $n > m$.

3. We will see below that $c = 0.13$ and $C = 5.60$ satisfy (13.13) for every $n = 3, 4, \ldots$. Using the same methods, one can obtain more precise upper and lower bounds for each n. Improved values for the constants, as well as extensions to rectangular matrices and to other canonical non-Gaussian distributions can be found in a paper by Edelman and Sutton (2005), where the proofs are based on the analytic theory of random matrices. In particular, for the constant C these authors show evidence for the value $C = 2$. See the numerical application in the next section.

Proof of Theorem 13.4. Recall the notation in the introduction of this chapter. It is easy to see that, almost surely, the eigenvalues of $A^{\mathrm{T}}A$ are pairwise different. We introduce the following additional notation:

- $\{e_1, \ldots, e_n\}$ is the canonical basis of $\mathbb{R}^n$.
- $B = A^{\mathrm{T}}A = ((b_{ij}))_{i,j=1,\ldots,n}$.
- $X(t) = t^{\mathrm{T}}Bt$.

- For $s \neq 0$ in $\mathbb{R}^n$, $\pi_s : \mathbb{R}^n \to \mathbb{R}^n$ denotes the orthogonal projection onto $\{s\}^\perp$, the orthogonal complement of s in $\mathbb{R}^n$.
- For a differentiable function F defined on a smooth manifold M embedded in some Euclidean space, $F'(s)$ and $F''(s)$ are the first and second derivatives of F, which we will represent, in each case, with respect to an appropriate orthonormal basis of the tangent space.

Instead of (13.13), we prove an equivalent statement: For $x > n$,

$$\frac{cn}{x} < \mathrm{P}(\kappa(A) > x) < \frac{Cn}{x}. \tag{13.14}$$

We break the proof into several steps. Our main task is to estimate the joint density of the pair (ν_n, ν_1); this will be done in step 4.

STEP 1. For $a, b \in \mathbb{R}$, $a > b$, one has, almost surely,

$$\{\nu_n \in (a, a+da), \nu_1 \in (b, b+db)\} \\ = \left\{ \begin{array}{c} \exists s, t \in S^{n-1}, <s,t> = 0, X(s) \in (a, a+da), X(t) \in (b, b+db), \\ \pi_s(Bs) = 0, \pi_t(Bt) = 0, X''(s) \prec 0, X''(t) \succ 0 \end{array} \right\}. \tag{13.15}$$

The random integer $N_{a,b,da,db}$ of pairs (s,t) belonging to the right-hand side of (13.15) is equal to 0 or to 4, so that

$$\mathrm{P}(\nu_n \in (a, a+da), \nu_1 \in (b, b+db)) = \tfrac{1}{4}\mathrm{E}\left(N_{a,b,da,db}\right). \tag{13.16}$$

STEP 2. In this step we give a bound for $\mathrm{E}\left(N_{a,b,da,db}\right)$ using a Rice-type formula. Let

$$V = \left\{(s,t) : s, t \in S^{n-1}, <s,t> = 0\right\}.$$

V is a C^∞-differentiable manifold without boundary, embedded in $\mathbb{R}^{2n}$, $\dim(V) = 2n - 3$. We denote by $\tau = (s,t)$ a generic point in V and by $\sigma_V(d\tau)$ the geometric measure on V. We will need the total measure $\sigma_V(V)$, which is a particular case of the following lemma. (We use the full statement in the next section.)

Lemma 13.5 *Let $a, b > 0$. We define*

$$V_{a,b} = \{(s,t) \in \mathbb{R}^n \times \mathbb{R}^n : \|s\|^2 = a, \|t\|^2 = b, \langle s,t\rangle = 0\}.$$

Denote by $\mu_{a,b}$ the geometric measure of the compact C^∞-manifold $V_{a,b}$ embedded in $\mathbb{R}^{2n}$. Then

$$\mu_{a,b} = (a+b)^{1/2}(ab)^{(n-2)/2}\sigma_{n-1}\sigma_{n-2},$$

where σ_{n-1} denotes the surface area of $S^{n-1} \subset \mathbb{R}^n$; that is, $\sigma_{n-1} = 2\pi^{n/2}/\Gamma(n/2)$.

Proof. Notice that for each point $(s, t) \in V_{a,b}$, the triplet

$$\left(\frac{s}{\|s\|}, 0\right), \left(0, \frac{t}{\|t\|}\right), \frac{1}{\sqrt{\|s\|^2 + \|t\|^2}}(t, s)$$

is an orthonormal basis of the normal space to $V_{a,b}$ at (s, t); these correspond, respectively, to the unit vectors orthogonal to each one of the $(2n-1)$-dimensional manifolds in $\mathbb{R}^{2n}$ given by

$$\begin{aligned} \|s\|^2 &= a \\ \|t\|^2 &= b \\ \langle s, t\rangle &= 0. \end{aligned} \tag{13.17}$$

So, as $\delta \downarrow 0$, the $2n$-dimensional Lebesgue measure of the set

$$E_\delta = \{(s, t) \in \mathbb{R}^{2n} : \sqrt{a} - \delta < \|s\| < \sqrt{a} + \delta, \sqrt{b} - \delta < \|t\| < \sqrt{b} + \delta, |\langle s, t\rangle| < \delta\sqrt{a+b}\}$$

is equivalent to $(2\delta)^3 \mu_{a,b}$.

On the other hand,

$$\lambda_{2n}(E_\delta) = \int_{\{\sqrt{a}-\delta<\|s\|<\sqrt{a}+\delta\}} ds \int_{\{\sqrt{b}-\delta<\|t\|<\sqrt{b}+\delta, |\langle s,t\rangle|<\delta\sqrt{a+b}\}} dt. \tag{13.18}$$

Using polar coordinates in each iterate of the double integral in (13.17), the result follows. □

Proof of Theorem 13.4 (cont.). Since $V = V_{1,1}$, Lemma 13.5 implies that $\sigma_V(V) = \sqrt{2}\,\sigma_{n-1}.\sigma_{n-2}$. On V we define the random field $Y : V \to \mathbb{R}^{2n}$ by means of

$$Y(s, t) = \begin{pmatrix} \pi_s(Bs) \\ \pi_t(Bt) \end{pmatrix}.$$

For $\tau = (s, t)$ a given point in V, we have that

$$Y(\tau) \in \{(t, -s)\}^\perp \cap \{\{s\}^\perp \times \{t\}^\perp\} = W_\tau$$

for any value of the matrix B, where $\{(t, -s)\}^\perp$ is the orthogonal complement of the point $(t, -s)$ in $\mathbb{R}^{2n}$. In fact, $(t, -s) \in \{s\}^\perp \times \{t\}^\perp$ and

$$\begin{aligned} <Y(s, t), (t, -s)>_{\mathbb{R}^{2n}} &= <\pi_s(Bs), t> - <\pi_t(Bt), s> \\ &= <Bs - <s, Bs> s, t> - <Bt - <t, Bt> t, s> = 0 \end{aligned}$$

since $<s, t> = 0$ and B is symmetric. Notice that $\dim(W_\tau) = 2n - 3$.

We also set

$$\Delta(\tau) = [\det[(Y'(\tau))^{\mathrm{T}} Y'(\tau)]]^{1/2}.$$

For $\tau = (s, t) \in V$, F_τ denotes the event

$$F_\tau = \left\{X(s) \in (a, a + da), X(t) \in (b, b + db), X''(s) \prec 0, X''(t) \succ 0\right\}$$

and $p_{Y(\tau)}(\cdot)$ is the density of the random vector $Y(\tau)$ in the $(2n-3)$-dimensional subspace W_τ of $\mathbb{R}^{2n}$. Applying Rice's formula (Theorem 6.2 with adaptation to manifolds as in Section 6.1.3) gives us

$$\mathrm{E}(N_{a,b,da,db}) = \int_a^{a+da} dx \int_b^{b+db} dy \tag{13.19}$$

$$\int_V \mathrm{E}(\Delta(s,t)\, \mathbb{1}_{\{X''(s) \prec 0, X''(t) \succ 0\}} | X(s) = x, X(t) = y, Y(s,t) = 0)$$

$$\cdot p_{X(s),X(t),Y(s,t)}(x, y, 0) \sigma_V(d(s,t)).$$

The invariance of the law of A with respect to the orthogonal group of $\mathbb{R}^n$ implies that the integrand in (13.18) does not depend on $(s, t) \in V$. Hence, we have proved that the joint law of λ_n and λ_1 has a density $g(a, b)$, $a > b$, and

$$g(a,b) = \frac{\sqrt{2}}{4} \sigma_{n-1}.\sigma_{n-2} \tag{13.20}$$

$$\mathrm{E}(\Delta(e_1, e_2)\, \mathbb{1}_{\{X''(e_1) \prec 0, X''(e_2) \succ 0\}} \big| X(e_1) = a, X(e_2) = b, Y(e_1, e_2) = 0)$$

$$\cdot p_{X(e_1),X(e_2),Y(e_1,e_2)}(a, b, 0).$$

Step 3. Next, we compute the ingredients in the right-hand side of (13.20). We take as orthonormal basis for the subspace $W_{(e_1,e_2)}$,

$$\{(e_3, 0), \ldots, (e_n, 0), (0, e_3), \ldots, (0, e_n), \frac{1}{\sqrt{2}}(e_2, e_1)\} = L_1.$$

We have

$$X(e_1) = b_{11}$$

$$X(e_2) = b_{22}$$

$$X''(e_1) = B_1 - b_{11} I_{n-1}$$

$$X''(e_2) = B_2 - b_{22} I_{n-1},$$

where B_1 (respectively, B_2) is the $(n-1) \times (n-1)$ matrix obtained by suppressing the first (respectively, the second) row and column in B.

$$Y(e_1, e_2) = (0, b_{21}, b_{31}, \ldots, b_{n1}, b_{12}, 0, b_{32}, \ldots, b_{n2}, b_{12})^{\mathrm{T}},$$

so that it has the expression in the orthonormal basis L_1:

$$Y(e_1, e_2) = \sum_{i=3}^{n} \left(b_{i1}(e_i, 0) + b_{i2}(0, e_i)\right) + \sqrt{2}b_{12}\left(\frac{1}{\sqrt{2}}(e_2, e_1)\right).$$

So the joint density of $X(e_1), X(e_2), Y(e_1, e_2)$ appearing in (13.20) in the space $\mathbb{R} \times \mathbb{R} \times W_{(e_1,e_2)}$ is the joint density of the random variables $b_{11}, b_{22}, \sqrt{2}b_{12}, b_{31}, \ldots, b_{n1}, b_{32}, \ldots, b_{n2}$ at the point $(a, b, 0, \ldots, 0)$. To compute this density, we first compute the joint density q of $b_{31}, \ldots, b_{n1}, b_{32}, \ldots, b_{n2}$, given $a_1$and a_2, where a_j denotes the jth column of A, which is standard normal in $\mathbb{R}^n$.

q is the normal density in $\mathbb{R}^{2(n-2)}$, centered with variance matrix

$$\begin{pmatrix} \|a_1\|^2 I_{n-2} & <a_1, a_2> I_{n-2} \\ <a_1, a_2> I_{n-2} & \|a_2\|^2 I_{n-2} \end{pmatrix}.$$

Set $a'_j = \frac{a_j}{\|a_j\|}$ $j = 1, 2$. Now we compute the density of the triplet

$$(b_{11}, b_{22}, b_{12}) = (\|a_1\|^2, \|a_2\|^2, \|a_1\|\|a_2\| < a'_1, a'_2>)$$

at the point $(a, b, 0)$.

Since $<a'_1, a'_2>$ and $(\|a_1\|, \|a_2\|)$ are independent, the density of the triplet at $(a, b, 0)$ is equal to

$$\chi_n^2(a)\chi_n^2(b)(ab)^{-1/2}p_{<a'_1,a'_2>}(0),$$

where $\chi_n^2(\cdot)$ denotes the χ^2 density with n degrees of freedom.

Let $\xi = (\xi_1, \ldots, \xi_n)^{\mathrm{T}}$ be standard normal in $\mathbb{R}^n$. Clearly, $<a'_1, a'_2>$ has the same distribution as $\xi_1/\|\xi\|$, because of the invariance under the orthogonal group:

$$\begin{aligned}\frac{1}{2t}\mathrm{P}\{| < a'_1, a'_2 > | \le t\} &= \frac{1}{2t}\mathrm{P}\left\{\frac{\xi_1^2}{\chi_{n-1}^2} \le \frac{t^2}{1-t^2}\right\} = \frac{1}{2t}\mathrm{P}\left\{F_{1,n-1} \le \frac{t^2(n-1)}{1-t^2}\right\} \\ &= \frac{1}{2t}\int_0^{t^2(n-1)/1-t^2} f_{1,n-1}(x)\,dx,\end{aligned}$$

where $\chi_{n-1}^2 = \xi_2^2 + \cdots + \xi_n^2$ and $F_{1,n-1}$ has the Fisher distribution with $(1, n-1)$ degrees of freedom and density $f_{1,n-1}$. Letting $t \to 0$, we obtain

$$p_{<a'_1,a'_2>}(0) = \frac{1}{\sqrt{\pi}}\frac{\Gamma(n/2)}{\Gamma\left((n-1)/2\right)}.$$

Summing up, the density in (13.20) is equal to

$$\frac{1}{\sqrt{2}}(2\pi)^{2-n}\pi^{-1/2}\frac{1}{\Gamma(n/2)\Gamma\big((n-1)/2\big)}2^{-n}\frac{1}{\sqrt{ab}}\exp\Big(-\frac{a+b}{2}\Big). \tag{13.21}$$

We now consider the conditional expectation in (13.20). The tangent space to V at the point (s,t) is parallel to the orthogonal complement in $\mathbb{R}^n \times \mathbb{R}^n$ of the triplet of vectors $(s,0)$; $(0,t)$; (t,s). This is immediate from the definition of V.

To compute the associated matrix for $Y'(e_1,e_2)$, take the set

$$\{(e_3,0),\ldots,(e_n,0),(0,e_3),\ldots,(0,e_n),\frac{1}{\sqrt{2}}(e_2,-e_1)\} = L_2$$

as the orthonormal basis in the tangent space and the canonical basis in $\mathbb{R}^{2n}$. Direct calculation gives

$$Y'(e_1,e_2) = \begin{pmatrix} -v^{\mathrm{T}} & 0_{1,n-2} & -\frac{1}{\sqrt{2}}b_{21} \\ w^{\mathrm{T}} & 0_{1,n-2} & \frac{1}{\sqrt{2}}(-b_{11}+b_{22}) \\ B_{12}-b_{11}I_{n-2} & 0_{n-2,n-2} & \frac{1}{\sqrt{2}}w \\ 0_{1,n-2} & -w^T & \frac{1}{\sqrt{2}}(-b_{11}+b_{22}) \\ 0_{1,n-2} & v^T & \frac{1}{\sqrt{2}}b_{21} \\ 0_{n-2,n-2} & B_{12}-b_{22}I_{n-2} & -\frac{1}{\sqrt{2}}v \end{pmatrix},$$

where $v^{\mathrm{T}} = (b_{31},\ldots,b_{n1})$, $w^{\mathrm{T}} = (b_{32},\ldots,b_{n2})$, $0_{i,j}$ is a null matrix with i rows and j columns, and B_{12} is obtained from B by suppressing the first and second rows and columns. The columns represent the derivatives in the directions of L_2 at the point (e_1,e_2). The first n rows correspond to the components of $\pi_s(Bs)$, the last n ones to those of $\pi_t(Bt)$. Thus, under the conditioning in (13.20),

$$Y'(e_1,e_2) = \begin{pmatrix} 0_{1,n-2} & 0_{1,n-2} & 0 \\ 0_{1,n-2} & 0_{1,n-2} & \frac{1}{\sqrt{2}}(b-a) \\ B_{12}-aI_{n-2} & 0_{n-2,n-2} & 0_{n-2,1} \\ 0_{1,n-2} & 0_{1,n-2} & \frac{1}{\sqrt{2}}(b-a) \\ 0_{1,n-2} & 0_{1,n-2} & 0 \\ 0_{n-2,n-2} & B_{12}-bI_{n-2} & 0_{n-2,1} \end{pmatrix}$$

and

$$[\det[(Y'(e_1, e_2))^{\mathrm{T}} Y'(e_1, e_2)]]^{1/2}$$
$$= |\det(B_{12} - aI_{n-2})||\det(B_{12} - bI_{n-2})|(a - b).$$

STEP 4. Notice that $B_1 - aI_{n-1} \prec 0 \Rightarrow B_{12} - aI_{n-2} \prec 0$, and similarly, $B_2 - bI_{n-1} \succ 0 \Rightarrow B_{12} - bI_{n-2} \succ 0$, and that for $a > b$, under the conditioning in (13.20), there is equivalence in these relations.

It is also clear that since $B_{12} \succ 0$, one has

$$|\det(B_{12} - aI_{n-2})| \, \mathbb{I}_{B_{12} - aI_{n-2} \prec 0} \leq a^{n-2},$$

and it follows that the conditional expectation in (13.20) is bounded by

$$a^{n-1}\mathrm{E}\big(|\det(B_{12} - bI_{n-2})| \, \mathbb{I}_{B_{12} - bI_{n-2} \succ 0} | b_{11} = a, b_{22} = b, b_{12} = 0, b_{i1} = b_{i2} = 0$$
$$\cdot (i = 3, \ldots, n)\big). \tag{13.22}$$

We further condition on a_1 and a_2. Since unconditionally $a_3, \ldots, a_n$ are i.i.d. standard normal vectors in $\mathbb{R}^n$, under the conditioning, their joint law becomes the law of i.i.d. standard normal vectors in $\mathbb{R}^{n-2}$ and independent of the condition. That is, (13.22) is equal to

$$a^{n-1}\mathrm{E}\big(|\det(M - bI_{n-2})| \, \mathbb{I}_{M - bI_{n-2} \succ 0}\big), \tag{13.23}$$

where M is an $(n-2) \times (n-2)$ random matrix with entries $M_{ij} = \langle v_i, v_j \rangle$ $(i, j = 1, \ldots, n-2)$, and the vectors $v_1, \ldots, v_{n-2}$ are i.i.d. standard normal in $\mathbb{R}^{n-2}$. The expression in (13.23) is bounded by

$$a^{n-1}\mathrm{E}\big(\det(M)\big) = a^{n-1}(n-2)!.$$

The last equality is contained in the following lemma (see, e.g., Mehta, 2004). We include a proof for completeness.

Lemma 13.6 *Let $\xi_1, \ldots, \xi_m$ be i.i.d. random vectors in $\mathbb{R}^p$, $p \geq m$, their common distribution being Gaussian centered with variance I_p. Denote by $W_{m,p}$ the matrix*

$$W_{m,p} = ((\langle \xi_i, \xi_j \rangle))_{i,j=1,\ldots,m}$$

and by

$$D(\lambda) = \det\left(W_{m,p} - \lambda I_m\right)$$

its characteristic polynomial. Then

$$\mathrm{E}\left(\det\left(W_{m,p}\right)\right) = p(p-1)\cdots(p-m+1) \tag{13.24}$$

$$\mathrm{E}\left(D(\lambda)\right) = \sum_{k=0}^{m}(-1)^k\binom{m}{k}\frac{p!}{(p-m+k)!}\lambda^k \tag{13.25}$$

Proof. Fix the values of $\xi_1, \ldots, \xi_{m-1}$, which are linearly independent with probability 1. Let $\{w_1, \ldots, w_p\}$ be an orthonormal basis of $\mathbb{R}^p$ such that $w_1, \ldots, w_{m-1} \in V_{m-1}$, where V_j denotes the subspace generated by $\xi_1, \ldots, \xi_j$. Observe that $\det\left(W_{m,p}\right)$ is the square of the volume, in $V_m \subset \mathbb{R}^p$, of the parallelotope $\left\{\sum_{i=1}^{m} c_i\xi_i, 0 \le c_i \le 1, i = 1, \ldots, m\right\}$. So

$$\det\left(W_{m,p}\right) = d^2(\xi_m, V_{m-1})\det\left(W_{m-1,p}\right),$$

where $d(\xi_m, V_{m-1})$ is the Euclidean distance from ξ_m to V_{m-1}. Because of rotational invariance of the standard normal distribution in $\mathbb{R}^p$, the conditional distribution of $d^2(\xi_m, V_{m-1})$ given $\xi_1, \ldots, \xi_{m-1}$ is independent of the condition and χ^2_{p-m+1}. Hence,

$$\begin{aligned}\mathrm{E}\left(\det\left(W_{m,p}\right)\right) &= \mathrm{E}\left(\mathrm{E}\left(d^2(\xi_m, V_{m-1})\det\left(W_{m-1,p}\right)|\xi_1, \ldots, \xi_{m-1}\right)\right)\\ &= (p-m+1)\mathrm{E}\left(\det\left(W_{m-1,p}\right)\right).\end{aligned}$$

Iterating the procedure, we get (13.24).

Let us prove (13.25). Clearly,

$$D(\lambda) = \sum_{k=0}^{m}\frac{D^{(k)}(0)}{k!}\lambda^k. \tag{13.26}$$

Standard differentiation of the determinant with respect to λ shows that for $k = 1, \ldots, m-1$ one has

$$D^{(k)}(\lambda) = (-1)^k\sum\det\left(W_{m,p}^{i_1,\ldots,i_k} - \lambda I_{m-k}\right),$$

where the sum is over all k-tuples $i_1, \ldots, i_k$ of pairwise different nonnegative integers that are smaller or equal than m, and the $(m-k)\times(m-k)$ matrix $W_{m,p}^{i_1,\ldots,i_k}$ is obtained by suppressing in $W_{m,p}$ the rows and columns numbered $i_1, \ldots, i_k$. Hence, applying (13.24) to each term in this sum, and based on the number of terms, we get

$$\mathrm{E}\left(D^{(k)}(0)\right) = (-1)^k m(m-1)\cdots(m-k+1)p(p-1)\cdots(p-(m-k)+1).$$

To finish, take expectations in (13.26), and notice that

$$D^{(m)}(\lambda) = (-1)^m m!, \qquad \mathrm{E}\left(D^{(0)}(0)\right) = p(p-1)\cdots(p-m+1). \qquad \square$$

Proof Theorem 13.4 (cont.). Summing up this part, after substitution in (13.20), we get

$$g(a,b) \leq C_n \frac{\exp\left(-(a+b)/2\right)}{\sqrt{ab}} a^{n-1}, \tag{13.27}$$

where

$$C_n = \frac{1}{4(n-2)!}.$$

STEP 5. Now we prove the upper-bound part in (13.14). One has, for $x > 1$,

$$\mathrm{P}\{\kappa(A) > x\} = \mathrm{P}\left\{\frac{\nu_n}{\nu_1} > x^2\right\} \leq \mathrm{P}\left\{\nu_1 < \frac{L^2 n}{x^2}\right\} + \mathrm{P}\left\{\frac{\nu_n}{\nu_1} > x^2, \nu_1 \geq \frac{L^2 n}{x^2}\right\}, \tag{13.28}$$

where L is a positive number to be chosen later. For the first term in (13.28), we need some auxiliary lemmas that we take from Sankar et al. (2002) in a modified form included in Cuesta-Albertos and Wschebor (2004):

Lemma 13.7 *Assume that* $A = (a_{i,j})_{i,j=1,\ldots,n}$, $a_{i,j} = m_{i,j} + g_{i,j}$ $(i, j = 1, \ldots, n)$, *where the* $m_{i,j}$*'s are nonrandom and the* $g_{i,j}$*'s are i.i.d. standard normal random variables. Let* $v \in S^{n-1}$. *Then, for* $x > 0$,

$$\mathrm{P}\left[\|A^{-1}v\| > x\right] = \mathrm{P}\left(|\xi| < \frac{1}{x}\right) < \left(\frac{2}{\pi}\right)^{1/2} \frac{1}{x},$$

where ξ *is a standard normal random variable.*

Proof. Because of the orthogonal invariance of the standard normal distribution, it suffices to prove the result for $v = e_1$. Denote by $a_1, \ldots, a_n$ the rows of A and by α_1 the first column of A^{-1}. Clearly, α_1 is orthogonal to $a_2, \ldots, a_n$ and $\langle \alpha_1, a_1 \rangle = 1$, so that $\alpha_1 = \gamma w$, where $|\gamma| = \|\alpha_1\|$ and the unit vector w is measurable with respect to $a_2, \ldots, a_n$ and orthogonal to each of these vectors. Also, $\|\alpha_1\| |\langle w, a_1 \rangle| = 1$. Observe now that the conditional distribution of the real-valued random variable $\langle w, a_1 \rangle$ given the random vectors $a_2, \ldots, a_n$ is standard normal, so that, under this condition, $\|\alpha_1\|$ can be written as $\|\alpha_1\| = 1/|\xi|$, where ξ is standard normal. The equalities

$$\mathrm{P}\left[\|A^{-1}v\| > x\right] = \mathrm{P}\left[\|\alpha_1\| > x\right] = \mathrm{P}\left(|\xi| < \frac{1}{x}\right)$$

finish the proof of the lemma. $\square$

Lemma 13.8 *Let $U = (U_1, \dots, U_n)$ be an n-dimensional vector chosen uniformly on S^{n-1} and let t_{n-1} be a real-valued random variable with a Student distribution with $n-1$ degrees of freedom. Then, if $c \in (0, n)$, we have that*

$$\mathrm{P}\left[U_1^2 > \frac{c}{n}\right] = \mathrm{P}\left[t_{n-1}^2 > \frac{n-1}{n-c}c\right].$$

Proof. U can be written as

$$U = \frac{V}{\|V\|},$$

where $V = (V_1, \dots, V_n)$ is an n-dimensional random vector with standard normal distribution. To simplify the notation, let us denote $K = V_2^2 + \cdots + V_n^2$. Then the statement

$$\frac{V_1^2}{V_1^2 + K} > \frac{c}{n}$$

is equivalent to

$$\frac{V_1^2}{K} > \frac{c}{n-c},$$

and we have

$$\mathrm{P}\left[U_1^2 > \frac{c}{n}\right] = \mathrm{P}\left[\frac{(n-1)V_1^2}{K} > \frac{n-1}{n-c}c\right] = \mathrm{P}\left[t_{n-1}^2 > \frac{n-1}{n-c}c\right],$$

where t_{n-1} is a real-valued random variable having Student's distribution with $n-1$ degrees of freedom. □

Lemma 13.9 *Let C be an $n \times n$ real nonsingular matrix. Then there exists $w \in \mathbb{R}^n$, $\|w\| = 1$ such that for every $u \in \mathbb{R}^n$, $\|u\| = 1$, one has*

$$\|Cu\| \geq \|C\| \cdot |\langle w, u\rangle|.$$

Proof. Since $\|C\|^2$ is the maximum eigenvalue of the symmetric matrix $C^{\mathrm{T}}C$, take for w an eigenvector, $\|w\| = 1$, so that $\|Cw\| = \|C\|$. Then, if $\|u\| = 1$,

$$\|C\|^2\langle w, u\rangle = \langle Cw, Cu\rangle,$$

which implies that

$$\|C\|^2\langle w, u\rangle \leq \|C\| \cdot \|Cu\|.$$

□

Lemma 13.10 *Assume that* $A = (a_{i,j})_{i,j=1,\ldots,n}$, $a_{i,j} = m_{i,j} + g_{i,j}$ $(i, j = 1, \ldots, n)$, *where the* $g_{i,j}$*'s are i.i.d. standard normal random variables, and* $M = \left(m_{i,j}\right)_{i,j=1,\ldots,n}$ *is nonrandom. Then, for each* $x > 0$,

$$\mathrm{P}[\|A^{-1}\| \geq x] \leq C_2(n)\frac{n^{1/2}}{x}, \tag{13.29}$$

where

$$\begin{aligned} C_2(n) &= \left(\frac{2}{\pi}\right)^{1/2}\left(\sup_{c\in(0,n)} \sqrt{c}\,\mathrm{P}\left[t_{n-1}^2 > \frac{n-1}{n-c}c\right]\right)^{-1} \\ &\leq C_2(\infty) = C_2 \simeq 2.34737\ldots. \end{aligned}$$

Proof. Let U be a random vector, independent of A with uniform distribution on S^{n-1}. Using Lemma 13.7, we have that

$$\mathrm{P}\left[\|A^{-1}U\| > x\right] = \mathrm{E}\left\{\mathrm{P}\left[\|A^{-1}U\| > x|U\right]\right\} \leq \left(\frac{2}{\pi}\right)^{1/2}\frac{1}{x}. \tag{13.30}$$

Now, since if w_A, $\|w_A\| = 1$ satisfies $\|A^{-1}w_A\| = \|A^{-1}\|$, and $\|u\| = 1$, then

$$\|A^{-1}u\| \geq \|A^{-1}\| \times | < w_A, u > |,$$

we have that, if $c \in (0, n)$,

$$\begin{aligned} \mathrm{P}\left[\|A^{-1}U\| \geq x\left(\frac{c}{n}\right)^{1/2}\right] &\geq \mathrm{P}\left[\left\{\|A^{-1}\| \geq x\right\} \text{ and } \left\{| < w_A, U > | \geq \left(\frac{c}{n}\right)^{1/2}\right\}\right] \\ &= \mathrm{E}\left\{\mathrm{P}\left[\{\|A^{-1}\| \geq x\} \text{ and } \right.\right. \\ &\qquad \left.\left.\{| < w_A, U > | \geq \left(\frac{c}{n}\right)^{1/2}\}\middle|A\right]\right\} \\ &= \mathrm{E}\left\{\mathbf{1}_{\{\|A^{-1}\|\geq x\}}\mathrm{P}\left[| < w_A, U > | \geq \left(\frac{c}{n}\right)^{1/2}\middle|A\right]\right\} \\ &= \mathrm{E}\left\{\mathbf{1}_{\{\|A^{-1}\|\geq x\}}\mathrm{P}\left[t_{n-1}^2 > \frac{n-1}{n-c}c\right]\right\} \\ &= \mathrm{P}\left[t_{n-1}^2 > \frac{n-1}{n-c}c\right]\mathrm{P}[\|A^{-1}\| \geq x], \end{aligned}$$

where we have used Lemma 13.8. From here and (13.29) we have that

$$\mathrm{P}[\|A^{-1}\| \geq x] \leq \frac{1}{\mathrm{P}\left[t_{n-1}^2 > [(n-1)/(n-c)]c\right]}\left(\frac{2}{\pi}\right)^{1/2}\frac{1}{x}\left(\frac{n}{c}\right)^{1/2}.$$

To end the proof, notice that if g is a standard normal random variable, then

$$
\begin{aligned}
\sup_{c\in(0,n)} c^{1/2}\mathrm{P}\left[t_{n-1}^2 > \frac{n-1}{n-c}c\right] &\geq \sup_{c\in(0,1)} c^{1/2}\mathrm{P}\left[t_{n-1}^2 > \frac{n-1}{n-c}c\right] \\
&\geq \sup_{c\in(0,1)} c^{1/2}\mathrm{P}\left[t_{n-1}^2 > c\right] \\
&\geq \sup_{c\in(0,1)} c^{1/2}\mathrm{P}\left[g^2 > c\right] \qquad (13.31) \\
&\geq 0.565^{1/2}\mathrm{P}\left[g^2 > 0.565\right] \simeq 0.3399.
\end{aligned}
$$

□

Proof of Theorem 13.4 (cont.). For the first term on the right-hand side of (13.27), using the auxiliary lemmas we obtain

$$
\mathrm{P}\left\{\nu_1 < \frac{L^2 n}{x^2}\right\} = \mathrm{P}\left\{\|A^{-1}\| > \frac{x}{L\sqrt{n}}\right\} \leq C_2(n)\frac{Ln}{x}.
$$

Here,

$$
C_2(n) = \left(\frac{2}{\pi}\right)^{1/2}\left[\sup_{0<c<n} \sqrt{c}\,\mathrm{P}\left(t_{n-1}^2 > \frac{(n-1)c}{n-c}\right)\right]^{-1} \leq C_2(+\infty) \simeq 2.3473,
$$

where t_{n-1} is a random variable having Student's distribution with $n-1$ degrees of freedom.

Let us now turn to the second term on the right-hand side of (13.28),

$$
\mathrm{P}\left\{\frac{\nu_n}{\lambda_1} > x^2, \nu_1 \geq \frac{L^2 n}{x^2}\right\} = \int_{L^2 n x^{-2}}^{+\infty} db \int_{bx^2}^{+\infty} g(a,b)\,da \leq G_n(x^2)
$$

with

$$
G_n(y) = C_n \int_{L^2 n y^{-1}}^{+\infty} db \int_{by}^{+\infty} \frac{\exp\big(-(a+b)/2\big)}{\sqrt{ab}} a^{n-1}\,da,
$$

using (13.27). We have

$$
\begin{aligned}
G_n'(y) = C_n \Bigg[&-\int_{L^2 n y^{-1}}^{+\infty} \exp(-b/2)\sqrt{b}\exp\big(-(by)/2\big)(by)^{n-3/2}\,db \\
&+ L^2 n y^{-2} \int_{L^2 n}^{+\infty} \exp\left(-\frac{1}{2}\left(a + \frac{L^2 n}{y}\right)\right) \\
&a^{n-3/2}L^{-1}n^{-1/2}y^{1/2}\,da\Bigg], \qquad (13.32)
\end{aligned}
$$

which implies that

$$-G_n'(y) \leq C_n y^{n-3/2} \int_{L^2 n y^{-1}}^{+\infty} \exp\Big(-\frac{b(1+y)}{2}\Big) b^{n-1}\, db$$

$$= \frac{y^{-3/2}}{4(n-2)!}\Big(\frac{y}{1+y}\Big)^n 2^n \int_{(L^2 n/2y)(1+y)}^{+\infty} e^{-z} z^{n-1}\, dz$$

$$\leq \frac{y^{-3/2}}{4(n-2)!} 2^n \int_{L^2 n/2}^{+\infty} e^{-z} z^{n-1}\, dz.$$

Set $I_n(a) = \int_a^{+\infty} e^{-z} z^{n-1}\, dz$. On integrating by parts, we get

$$I_n(a) = e^{-a}\big[a^{n-1} + (n-1)a^{n-2} + (n-1)(n-2)a^{n-3} + \cdots + (n-1)!\big],$$

so that for $a > 2.5m$,

$$I_n(a) \leq \frac{5}{3} e^{-a} a^{n-1}.$$

If $L^2 > 5$, we obtain the bound

$$-G_n'(y) \leq D_n y^{-3/2} \quad \text{with } D_n = \frac{5}{6}\frac{n^{n-1}}{(n-2)!} L^{2(n-1)} \exp\Big(-\frac{L^2 n}{2}\Big).$$

We now apply Stirling's formula (Abramovitz and Stegun, 1968); that is, for all $x > 0$,

$$\Gamma(x+1)\exp\Big(-\frac{1}{12x}\Big) \leq \Big(\frac{x}{e}\Big)^x \sqrt{2\pi}\, x \leq \Gamma(x+1),$$

to get

$$D_n \leq \frac{5\sqrt{2}}{12\sqrt{\pi}\, L^2} \frac{n}{\sqrt{n-2}} \exp\Big(-n\frac{L^2 - 4\log(L) - 2}{2}\Big) \leq \frac{5\sqrt{2}}{12\sqrt{\pi}\, L^2} n.$$

if we choose for L the only root larger than 1 of the equation, $L^2 - 4\log(L) - 2 = 0$ (check that $L \approx 2.3145$). To finish,

$$0 \leq G_n(y) = \int_y^{+\infty} -G_n'(t)\, dt < D_n \int_y^{+\infty} \frac{dt}{t^{3/2}} = 2D_n y^{-1/2}.$$

Substituting y for x^2 and performing the numerical evaluations, the upper bound in (13.14) follows and we get for the constant C the value 5.60.

STEP 6. We consider now the lower bound in (13.14). For $\gamma > 0$ and $x > 1$, we have

$$\mathrm{P}\{\kappa(A) > x\} = \mathrm{P}\left\{\frac{\nu_n}{\nu_1} > x^2\right\} \geq \mathrm{P}\left\{\frac{\nu_n}{\nu_1} > x^2, \nu_1 < \frac{\gamma^2 n}{x^2}\right\}$$
$$= \mathrm{P}\left\{\nu_1 < \frac{\gamma^2 n}{x^2}\right\} - \mathrm{P}\left\{\frac{\nu_n}{\nu_1} \leq x^2, \nu_1 < \frac{\gamma^2 n}{x^2}\right\}. \quad (13.33)$$

A lower bound for the first term on the right-hand side of (13.33) is obtained using the following inequality, which we state as a new auxiliary lemma. The reader can find similar statements in articles by Edelman (1988) and Szarek (1991).

Lemma 13.11 *If $0 < a < 1/n$, then $\mathrm{P}\{\nu_1 < a\} \geq \beta\sqrt{an}$, where we can choose for β the value $\beta = \left(\frac{2}{3}\right)^{\frac{3}{2}} e^{-1/3}$.*

Proof. Define the index $i_X(t)$ of a critical point $t \in S^{n-1}$ of the function X as the number of negative eigenvalues of $X''(t)$. For each $a > 0$, put

$$N_i(a) = \#\{t \in S^{n-1} : X(t) = t^{\mathrm{T}} Bt < a, X'(t) = 0, i_X(t) = i\},$$

for $i = 0, 1, \ldots, n-1$. One easily checks that if the eigenvalues of B are $\nu_1, \cdots, \nu_n$, $0 < \nu_1 < \cdots < \nu_n$, then

- If $a \leq \nu_1$, then $N_i(a) = 0$ for $i = 0, 1, \ldots, n-1$.
- If $\nu_i < a \leq \nu_{i+1}$ for some $i = 0, 1, \ldots, n-1$, then $N_k(a) = 2$ for $k = 0, \ldots, i-1$ and $N_k(a) = 0$ for $k = i, \ldots, n-1$.
- If $\nu_n < a$, then $N_i(a) = 2$ for $i = 0, 1, \ldots, n-1$.

Now consider

$$M(a) = \sum_{i=0}^{n-1} (-1)^i N_i(a).$$

$M(a)$ is the Euler–Poincaré characteristic of the set $S = \{t \in S^{n-1} : X(t) < a\}$ (see, e.g., Adler, 1981).

- If $N_0(a) = 0$, then $N_i(a) = 0$ for $i = 1, \ldots, n-1$, hence $M(a) = 0$.
- If $N_0(a) = 2$, then $M(a) = 0$ or 2.

Then, in any case, we have the inequality $M(a) \leq N_0(a)$. Hence,

$$\mathrm{P}\{\nu_1 < a\} = \mathrm{P}\{N_0(a) = 2\} = \tfrac{1}{2}\mathrm{E}\big(N_0(a)\big) \geq \tfrac{1}{2}\mathrm{E}\big(M(a)\big). \quad (13.34)$$

Given the definition of $M(a)$, its expectation can be written using the following Rice formula:

$$\begin{aligned}\mathrm{E}\big(M(a)\big) &= \int_0^a dy \int_{S^{n-1}} \mathrm{E}\big[\det\big(X''(t)\big)|X(t) = y, X'(t) = 0\big] \\ &\quad p_{X(t),X'(t)}(y, 0)\sigma_{n-1}(dt) \\ &= \int_0^a \sigma_{n-1}(S^{n-1})\mathrm{E}\big[\det\big(X''(e_1)\big)|X(e_1) = y, X'(e_1) = 0\big] \\ &\quad p_{X(e_1),X'(e_1)}(y, 0)\, dy,\end{aligned}$$

where we have again used invariance under isometries. Applying a Gaussian regression similar to the one we used in step 4, we obtain

$$\mathrm{E}\big(M(a)\big) = \int_0^a \mathrm{E}\big[\det\big(Q - yI_{n-1}\big)\big]\frac{\sqrt{2\pi}}{2^{n-1}}\Gamma^{-2}\big(n/2\big)\frac{\exp(-y/2)}{\sqrt{y}}dy, \qquad (13.35)$$

where Q is an $(n-1) \times (n-1)$ random matrix with entry i, j equal to $<v_i, v_j>$ and $v_1, \ldots, v_{n-1}$ are i.i.d. standard normal in $\mathbb{R}^{n-1}$. We now use part (ii) of Lemma 13.6:

$$\mathrm{E}\big[\det\big(Q - yI_{n-1}\big)\big] = (n-1)!\sum_{k=0}^{n-1}\binom{n-1}{k}\frac{(-y)^k}{k!}. \qquad (13.36)$$

Under condition $0 < a < n^{-1}$, since $0 < y < a$, as k increases the terms of the sum on the right-hand side of (13.36) have decreasing absolute value, so that

$$\mathrm{E}\big[\det\big(Q - yI_{n-1}\big)\big] \geq (n-1)![1 - (n-1)y].$$

Substituting into the right-hand side of (13.35), we get

$$\mathrm{E}\big[M(a)\big] \geq \frac{\sqrt{2\pi}}{2^{n-1}}\frac{(n-1)!}{\Gamma^2(n/2)}J_n(a),$$

where, again using $0 < a < n^{-1}$, we obtain

$$\begin{aligned}J_n(a) &= \int_0^a \big(1 - (n-1)y\big)\frac{\exp(-y/2)}{\sqrt{y}}dy \\ &\geq \int_0^a \frac{1 - (n-1)y}{\sqrt{y}}(1 - y/2)\, dy \geq \frac{4}{3}\sqrt{a},\end{aligned}$$

by an elementary computation. Going back to (13.35), applying Stirling's formula, and remarking that $\big(1 + 1/n\big)^{n+1} \geq e$, we get

$$\mathrm{P}\{\nu_1 < a\} \geq \big(\tfrac{2}{3}\big)^{3/2}e^{-1/3}\sqrt{an}.$$

This proves the lemma. □

Proof of Theorem 13.4 (cont.). Using Lemma 13.11, the first term on the right-hand side of (13.33) is bounded below by $\beta\gamma(n/x)$. To obtain a bound for the second term, we use again our upper bound (13.27) on the joint density $g(a, b)$, as follows:

$$\begin{aligned} P\left\{\frac{\nu_n}{\nu_1} \le x^2, \nu_1 < \frac{\gamma^2 n}{x^2}\right\} &= \int_0^{\gamma^2 n x^{-2}} db \int_b^{bx^2} g(a, b)\, da \\ &\le C_m \int_0^{\gamma^2 n x^{-2}} db \int_b^{bx^2} \frac{\exp\left(-(a+b)/2\right)}{\sqrt{ab}} a^{n-1}\, da \\ &\le C_n \int_0^{\gamma^2 n x^{-2}} b(x^2-1) b^{-1/2} (bx^2)^{n-3/2}\, db \\ &\le \frac{1}{4(n-2)!} \frac{x^2-1}{x^3} \gamma^{2n} n^{n-1} \le \frac{\sqrt{2}}{8\sqrt{\pi}} e^n \gamma^{2n} \frac{n}{x} \end{aligned} \qquad (13.37)$$

on applying Stirling's formula. Now choosing $\gamma = 1/e$, we see that the hypothesis of Lemma 13.11 is satisfied and

$$P\left\{\frac{\nu_n}{\nu_1} \le x^2, \nu_1 < \frac{\gamma^2 n}{x^2}\right\} \le \frac{\sqrt{2}}{8\sqrt{\pi}} e^{-3} \frac{n}{x}.$$

Substituting into (13.33), we obtain the lower bound in (13.13) with

$$c = \left(\frac{2}{3}\right)^{3/2} e^{-4/3} - \frac{\sqrt{2}}{8\sqrt{\pi}} e^{-3} \approx 0.138. \qquad \square$$

13.2.1. Monte Carlo Experiment

To study the tail of the distribution of the condition number of Gaussian matrices of various size, we used the following Matlab functions: `normrnd` to simulate normal variables, and `cond` to compute the condition number of matrix A. Results over 40,000 simulations using Matlab are given in Table 13.1 and Figure 13.1. Taking into account the simulation variability, the table suggests that the constants c and C should take values smaller than 0.88 and larger than 2.00, respectively.

13.3. NONCENTERED GAUSSIAN MATRICES

Let $A = ((a_{ij}))_{i,j=1,\ldots,n}$ be a random matrix. Throughout this section we assume that the α_{ij}'s are independent random variables with Gaussian distribution having expectations $m_{ij} = \mathrm{E}(a_{ij})$ and common variance σ^2. We denote $M = ((m_{ij}))$. The aim of the section is to prove Theorem 13.2, which gives a bound for the tails of probability distribution of $\kappa(A)$. One way of looking at this result is as follows: We begin with a nonrandom matrix M and add noise by putting independent centered Gaussian random variables with variance σ^2 at each location. We ask

TABLE 13.1. Values of the Estimations P{κ(A) > mx} by the Monte Carlo Method over 40,000 Simulations

	Value of x								
Probability	1	2	3	5	10	20	30	50	100
Lower bound: $0.13/x$	0.13	0.065	0.043	0.026	0.013	0.007	0.004	0.003	0.001
Upper bound: $5.6/x$	1	1	1	1	0.56	0.28	0.187	0.112	0.056
$m = 3$	0.881	0.57	0.41	0.26	0.13	0.067	0.044	0.027	0.013
$m = 5$	0.931	0.66	0.48	0.30	0.16	0.079	0.053	0.033	0.016
$m = 10$	0.959	0.71	0.52	0.34	0.17	0.088	0.059	0.035	0.017
$m = 30$	0.974	0.75	0.56	0.36	0.19	0.096	0.063	0.038	0.019
$m = 100$	0.978	0.77	0.58	0.38	0.20	0.098	0.066	0.040	0.019
$m = 300$	0.982	0.77	0.58	0.38	0.20	0.101	0.069	0.041	0.022
$m = 500$	0.980	0.77	0.59	0.38	0.20	0.100	0.066	0.039	0.020

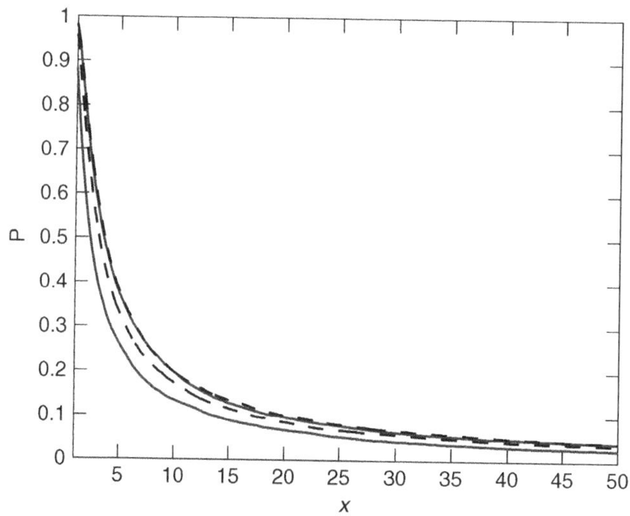

Figure 13.1. Values of P{$\kappa(A) > mx$} as a function of x for $m = 3$ (down), and 10, 100, and 500 (up). (From Azaïs and Wschebor, 2005c, with permission.)

for the condition number of the perturbed matrix. Notice that the starting matrix M can have an arbitrarily large (or infinite) condition number, but for the new one it turns out that we are able to give an upper bound for $P(\kappa(A) > x)$, which has a form similar to the one in the centered case.

Theorem 13.12. *Under the foregoing hypotheses on A, one has, for $x > 0$,*

$$P(\kappa(A) > nx) < \frac{1}{x}\left(\frac{1}{4\sqrt{2\pi n}} + C(M, \sigma, n)\right), \tag{13.38}$$

where

$$C(M,\sigma,n) = 7\left(5 + \frac{4\,\|M\|^2\,(1+\log n)}{\sigma^2 n}\right)^{1/2}.$$

Remarks

1. Theorem 13.12 implies that if $0 < \sigma \leq 1$ and $\|M\| \leq 1$, then for $x > 0$,

$$P(\kappa(A) > nx) < \frac{20}{\sigma x}. \tag{13.39}$$

   This is an immediate consequence of the statement in the theorem.
2. With calculations similar to the ones that we will perform for the proof of Theorem 13.12, one can improve somewhat the constants in (13.38) and (13.39).

Proof of Theorem 13.12. Due to the homogeneity of $\kappa(A)$, with no loss of generality we may assume that $\sigma = 1$, changing the expected matrix M by $(1/\sigma)M$ in the final result. We follow closely the proof of Theorem 13.4, with some changes to adapt it to the present conditions. In exactly the same way, we apply Rice's formula and prove that the joint density $g(a,b)$, $a > b$ of the random variables ν_n and ν_1 is given by

$$g(a,b) = \int_V \mathrm{E}\left(\Delta(s,t)\,\mathbf{1}_{\{X''(s)\prec 0,\,X''(t)\succ 0\}}|X(s)=a,\,X(t)=b,\,Y(s,t)=0\right) \cdot p_{X(s),X(t),Y(s,t)}(a,b,0)\sigma_V(d(s,t)). \tag{13.40}$$

where the notation is also borrowed from the preceding section.

Next, we compute the ingredients on the right-hand side of (13.40). This has some differences with the centered case. Put $a_{ij} = m_{ij} + g_{ij}$ with the g_{ij}'s i.i.d. standard normal and $G = ((g_{ij}))$. For each $(s,t) \in V$, we take an orthonormal basis of $\mathbb{R}^n$ so that its first two elements are, respectively, s and t, say $\{s, t, w_3, \ldots, w_m\}$. When expressing the linear transformation $x \rightsquigarrow Ax$ $(x \in \mathbb{R}^n)$ in this new basis, we denote by $A^{s,t}$ the associated matrix and by $a_{ij}^{s,t}$ its i, j entry. In a similar way we get $G^{s,t}$, $M^{s,t}$, and $B^{s,t}$. Notice that $G^{s,t}$ has the same law as G, but the nonrandom part $M^{s,t}$ can vary with the point (s,t).

We denote by $B_1^{s,t}$ (respectively, $B_2^{s,t}$) the $(n-1)\times(n-1)$ matrix obtained from $B^{s,t}$ by suppressing the first (respectively, the second) row and column. $B_{1,2}^{s,t}$ denotes the $(n-2)\times(n-2)$ matrix obtained from $B^{s,t}$ by suppressing the first and second row and column. To get an estimate for the right-hand side of (13.40), we start with the density $p_{X(s),X(t),Y(s,t)}(a,b,0)$. We denote $B^{s,t} = (b_{ij}^{s,t})$

(and similarly for the other matrices). We have

$$X(s) = b_{11}^{s,t}$$
$$X(t) = b_{22}^{s,t}$$
$$X''(s) = B_1^{s,t} - b_{11}^{s,t} I_{n-1}$$
$$X''(t) = B_2^{s,t} - b_{22}^{s,t} I_{n-1}.$$

Take the following orthonormal basis of the subspace $W_{(s,t)}$:

$$\left\{(w_3, 0), \ldots, (w_n, 0), (0, w_3), \ldots, (0, w_n), \frac{1}{\sqrt{2}}(t, s)\right\} = L_{s,t}.$$

Since the expression of $Y(s,t)$ in the canonical basis of $\mathbb{R}^{2n}$ is

$$Y(s,t) = (0, b_{21}^{s,t}, b_{31}^{s,t}, \ldots, b_{n1}^{s,t}, b_{12}^{s,t}, 0, b_{32}^{s,t}, \ldots, b_{n2}^{s,t}, b_{12}^{s,t})^{\mathrm{T}},$$

it is written in the orthonormal basis $L_{s,t}$ as the linear combination

$$Y(s,t) = \sum_{i=3}^{n} \left[b_{i1}^{s,t}(w_i, 0) + b_{i2}^{s,t}(0, w_i)\right] + \sqrt{2}\, b_{12}^{s,t}\left[\frac{1}{\sqrt{2}}(t, s)\right].$$

It follows that the joint density of $X(s)$, $X(t)$, $Y(s,t)$ appearing in (13.40) in the space $\mathbb{R} \times \mathbb{R} \times W_{(s,t)}$ is the joint density of the random variables

$$b_{11}^{s,t}, b_{22}^{s,t}, \sqrt{2}b_{12}^{s,t}, b_{31}^{s,t}, \ldots, b_{n1}^{s,t}, b_{32}^{s,t}, \ldots, b_{n2}^{s,t}$$

at the point $(a, b, 0)$. To compute this density, first compute the joint density q of

$$b_{31}^{s,t}, \ldots, b_{n1}^{s,t}, b_{32}^{s,t}, \ldots, b_{n2}^{s,t},$$

given $a_1^{s,t}$ and $a_2^{s,t}$, where $a_j^{s,t}$ denotes the jth column of $A^{s,t}$, with the additional conditions that

$$\|a_1^{s,t}\| = b_{11}^{s,t} = a, \qquad \|a_2^{s,t}\| = b_{22}^{s,t} = b, \qquad \langle a_1^{s,t}, a_2^{s,t}\rangle = b_{12}^{s,t} = 0.$$

q is the normal density in $\mathbb{R}^{2(n-2)}$, with the same variance matrix as in the centered case; that is,

$$\begin{pmatrix} aI_{n-2} & 0 \\ 0 & bI_{n-2} \end{pmatrix},$$

but not necessarily centered.

So the conditional density q is bounded above by

$$\frac{1}{(2\pi)^{n-2}} \frac{1}{(ab)^{(n-2)/2}}. \tag{13.41}$$

Our next task is to obtain an upper bound useful for our purposes for the density of the triplet

$$(b_{11}^{s,t}, b_{22}^{s,t}, b_{12}^{s,t}) = (\|a_1^{s,t}\|^2, \|a_2^{s,t}\|^2, <a_1^{s,t}, a_2^{s,t}>)$$

at the point $(a, b, 0)$, which together with (13.41) will provide an upper bound for $p_{X(s),X(t),Y(s,t)}(a, b, 0)$. We do this in the next lemma, which we will apply afterward with $\xi = a_1^{s,t}$ and $\eta = a_2^{s,t}$.

Lemma 13.13. *Let ξ and η be two independent Gaussian vectors in $\mathbb{R}^n$ $(n \geq 2)$, $\mathrm{E}(\xi) = \mu$, $\mathrm{E}(\eta) = \nu$, $\mathrm{Var}(\xi) = \mathrm{Var}(\eta) = I_n$. Then the density p of the random triplet $\left(\|\xi\|^2, \|\eta\|^2, \langle \xi, \eta \rangle\right)$ satisfies the following inequality, for $a \geq 4\|\mu\|^2$:*

$$p(a, b, 0) \leq \frac{1}{4(2\pi)^n} \sigma_{n-1}\sigma_{n-2}(ab)^{(n-3)/2} \exp\left(-\frac{a}{8}\right) \qquad a, b > 0. \tag{13.42}$$

Proof. Let $F : \mathbb{R}^n \times \mathbb{R}^n \to \mathbb{R}^3$ be the function

$$F(x, y) = \left(\|x\|^2, \|y\|^2, \langle x, y \rangle\right)^{\mathrm{T}}.$$

According to the co-area formula (6.30), density p at the point $(a, b, 0)$ can be written as

$$p(a, b, 0) = \int_{F^{-1}(a,b,0)} \left(\det\left[F'(x, y)\left(F'(x, y)\right)^{\mathrm{T}}\right]\right)^{-1/2} \frac{1}{(2\pi)^n} e^{-1/2\left[\|x-\mu\|^2 + \|y-\nu\|^2\right]} \, d\gamma(x, y),$$

where γ denotes the geometric measure on $F^{-1}(a, b, 0)$. Recall the manifold $V_{a,b}$ given by the set of equations

$$\|x\|^2 = a, \qquad \|y\|^2 = b, \qquad \langle x, y \rangle = 0.$$

Using Lemma 13.5 yields

$$\gamma(V_{a,b}) = (a + b)^{\frac{1}{2}} \sigma_{n-1}\sigma_{n-2}(ab)^{(n-2)/2}.$$

On the other hand,

$$F'(x, y) = \begin{pmatrix} 2x^{\mathrm{T}} & 0 \\ 0 & 2y^{\mathrm{T}} \\ y^{\mathrm{T}} & x^{\mathrm{T}} \end{pmatrix},$$

so that if $(x, y) \in F^{-1}(a, b, 0)$, one gets

$$\det\left[F'(x, y)\left(F'(x, y)\right)^{\mathrm{T}}\right] = 16ab(a + b).$$

Substituting in (13.43) and taking into account condition $a \geq 4\,\|\mu\|^2$, the result follows. □

Proof of Theorem 13.12 (cont.). Summing up this part, (13.41) plus (13.42) imply that

$$p_{X(s),X(t),Y(s,t)}(a, b, 0) \leq \frac{1}{2^{2n-3/2}\pi^{n-2}} \frac{1}{\Gamma(n/2)\,\Gamma((n-1)/2)} \frac{\exp(-a/8)}{\sqrt{ab}}. \tag{13.43}$$

We now consider the conditional expectation in (13.40). First, observe that the $(2n-3)$-dimensional tangent space to V at point (s, t) is parallel to the orthogonal complement in $\mathbb{R}^n \times \mathbb{R}^n$ of the triplet of vectors $(s, 0)$; $(0, t)$; (t, s). To compute the associated matrix for $Y'(s, t)$, take the set

$$\{(w_3, 0), \ldots, (w_n, 0), (0, w_3), \ldots, (0, w_n), \frac{1}{\sqrt{2}}(t, -s)\} = K_{s,t}$$

as an orthonormal basis in the tangent space. As for the image of Y, we take the canonical basis in $\mathbb{R}^{2n}$. Direct calculation gives

$$Y'(s, t) = \begin{pmatrix} -v^{\mathrm{T}} & 0_{1,n-2} & -1/\sqrt{2}b_{21}^{s,t} \\ w^{\mathrm{T}} & 0_{1,n-2} & 1/\sqrt{2}(-b_{11}^{s,t} + b_{22}^{s,t}) \\ B_{12}^{s,t} - b_{11}^{s,t} I_{n-2} & 0_{n-2,n-2} & 1/\sqrt{2}w \\ 0_{1,n-2} & -w^{\mathrm{T}} & 1/\sqrt{2}(-b_{11}^{s,t} + b_{22}^{s,t}) \\ 0_{1,n-2} & v^{\mathrm{T}} & 1/\sqrt{2}b_{21}^{s,t} \\ 0_{n-2,n-2} & B_{12}^{s,t} - b_{22}^{s,t} I_{n-2} & -1/\sqrt{2}v \end{pmatrix},$$

where $v^{\mathrm{T}} = (b_{31}^{s,t}, \ldots, b_{n1}^{s,t})$, $w^{\mathrm{T}} = (b_{32}^{s,t}, \ldots, b_{n2}^{s,t})$ and $0_{i,j}$ is a null matrix with i rows and j columns. The columns represent the derivatives in the directions of

$K_{s,t}$ at point (s, t). The first n rows correspond to the components of $\pi_s(Bs)$, the last n to those of $\pi_t(Bt)$. Thus, under the conditioning in (13.39),

$$Y'(s,t) = \begin{pmatrix} 0_{1,n-2} & 0_{1,n-2} & 0 \\ 0_{1,n-2} & 0_{1,n-2} & 1/\sqrt{2}(b-a) \\ B_{12}^{s,t} - aI_{n-2} & 0_{n-2,n-2} & 0_{m-2,1} \\ 0_{1,n-2} & 0_{1,n-2} & 1/\sqrt{2}(b-a) \\ 0_{1,n-2} & 0_{1,n-2} & 0 \\ 0_{n-2,n-2} & B_{12}^{s,t} - bI_{n-2} & 0_{n-2,1} \end{pmatrix}$$

and

$$\left[\det\left[\left(Y'(s,t)\right)^{\mathrm{T}} Y'(s,t)\right]\right]^{1/2} = |\det(B_{12}^{s,t} - aI_{n-2})||\det(B_{12}^{s,t} - bI_{n-2})|(a-b).$$

Since $B_{12}^{s,t} \succ 0$ one has

$$|\det(B_{12}^{s,t} - aI_{n-2})|\mathbf{1}_{B_{12}^{s,t} - aI_{n-2} \prec 0} \leq a^{n-2},$$

and the conditional expectation in (13.40) is bounded by

$$a^{n-1}\mathrm{E}\left[\left|\det(B_{12}^{s,t} - bI_{n-2})\right| \mathbf{1}_{B_{12}^{s,t} - bI_{n-2} \succ 0} \right.$$
$$\left. \left| b_{11}^{s,t} = a, b_{22}^{s,t} = b, b_{12}^{s,t} = 0, b_{i1}^{s,t} = b_{i2}^{s,t} = 0 \ (i = 3, \ldots, n)\right]. \right. \quad (13.44)$$

We further condition on $a_1^{s,t}$ and $a_2^{s,t}$, with the additional requirement that $\|a_1^{s,t}\|^2 = a$, $\|a_2^{s,t}\|^2 = b$, $\langle a_1^{s,t}, a_2^{s,t}\rangle = 0$. Since unconditionally, $a_3, \ldots, a_n$ are independent Gaussian vectors in $\mathbb{R}^n$, each having variance equal to 1 and mean smaller or equal to $\|M\|$, under the conditioning their joint law becomes the law of $(n-2)$ Gaussian vectors in $\mathbb{R}^{n-2}$, independent of the condition and also having variance equal to 1 and mean with Euclidean norm smaller than or equal to $\|M\|$.

As a consequence, the conditional expectation in (13.44) is bounded by $\mathrm{E}\left(\det(C^{s,t})\right)$, where $C^{s,t}$ is an $(n-2) \times (n-2)$ random matrix, $C^{s,t} = ((c_{ij}^{s,t}))$, $c_{ij}^{s,t} = <u_i^{s,t}, u_j^{s,t}> (i, j = 3, \ldots, n)$,

$$u_i^{s,t} = \zeta_i + \mu_i^{s,t} \qquad i = 3, \ldots, n,$$

$\zeta_3, \ldots, \zeta_n$ are i.i.d. standard normal in $\mathbb{R}^{n-2}$, and $\left\|\mu_i^{s,t}\right\| \leq \|M\|$ for $i = 3, \ldots, n$.

The usual argument to compute $\det(C^{s,t})$ as the square of the volume in $\mathbb{R}^{n-2}$ of the set of linear combinations of the form $\sum_{i=3}^{i=n} \nu_i u_i^{s,t}$ with $0 \leq \nu_i \leq 1$ $(i = 3, \ldots, n)$ shows that

$$\begin{aligned} \mathrm{E}\left(\det(C^{s,t})\right) &\leq \left(1 + \|M\|^2\right)\left(2 + \|M\|^2\right)\cdots\left(n - 2 + \|M\|^2\right) \\ &= (n-2)! \prod_{i=1}^{i=n-2} \left(1 + \frac{\|M\|^2}{i}\right) \\ &\leq (n-2)! \left[\left(1 + \|M\|^2 \frac{1 + \log n}{n}\right)\right]^n, \end{aligned}$$

where we have bounded the geometric mean by the arithmetic mean.

Substituting in (13.44) and based on the bound (13.43), we get from (13.40) the following bound for the joint density, valid for $a \geq 4\|M\|^2$:

$$g(a, b) \leq \overline{C}_n \frac{e^{-a/8}}{\sqrt{ab}} a^{n-1}, \tag{13.45}$$

where

$$\overline{C}_n = \frac{1}{4(n-2)!} \left[1 + \|M\|^2 \frac{1 + \log n}{n}\right]^n.$$

We now turn to the proof of (13.38), One has, for $x > 1$,

$$\mathrm{P}(\kappa(A) > x) = P\left(\frac{\nu_n}{\nu_1} > x^2\right) \leq \mathrm{P}\left(\nu_1 < \frac{L^2 n}{x^2}\right) + \mathrm{P}\left(\frac{\nu_n}{\nu_1} > x^2, \nu_1 \geq \frac{L^2 n}{x^2}\right), \tag{13.46}$$

where L is a positive number to be chosen later.

For the first term in (13.46), we use Lemma 13.10:

$$\mathrm{P}\left(\nu_1 < \frac{L^2 n}{x^2}\right) = \mathrm{P}\left(\|A^{-1}\| > \frac{x}{L\sqrt{n}}\right) \leq C_2 \frac{Ln}{x}. \tag{13.47}$$

First impose on L the condition $L^2 n \geq 4\|M\|^2$, so that for the second term in (13.46) we can make use of the bound (13.45) on the joint density $g(a, b)$:

$$\mathrm{P}\left(\frac{\nu_n}{\nu_1} > x^2, \nu_1 \geq \frac{L^2 n}{x^2}\right) = \int_{L^2 n x^{-2}}^{+\infty} db \int_{bx^2}^{+\infty} g(a, b)\, da \leq H_n(x^2) \tag{13.48}$$

with

$$K_n(y) = \overline{C}_n \int_{L^2 n y^{-1}}^{+\infty} db \int_{by}^{+\infty} \frac{\exp(-a/8)}{\sqrt{ab}} a^{n-1}\, da.$$

We have

$$K'_n(y) = \overline{C}_n \left[-\int_{L^2ny^{-1}}^{+\infty} \exp\left(-\frac{by}{8}\right)(by)^{n-1}\frac{db}{\sqrt{y}} \right.$$
$$\left. + \frac{Ln^{1/2}}{y^{3/2}} \int_{L^2n}^{+\infty} \exp\left(-\frac{a}{4}\right) a^{n-3/2} da, \right]$$

which implies that

$$-K'_n(y) \le \overline{C}_n y^{n-3/2} \int_{L^2ny^{-1}}^{+\infty} \exp\left(-\frac{by}{8}\right) b^{n-1}\, db$$
$$\le \frac{\overline{C}_n}{y^{3/2}} 8^n \int_{\frac{L^2n}{8}}^{+\infty} e^{-z} z^{n-1} dz \le \frac{\overline{C}_n}{y^{3/2}} 8^n \frac{5}{3} e^{-L^2n/8} \left(\frac{L^2n}{8}\right)^{n-1} = \overline{D}_n \frac{1}{y^{3/2}}$$

if we choose $L^2 > 20$. So

$$K_n(y) = -\int_y^{+\infty} K'_n(s)\, ds \le \overline{D}_n \int_y^{+\infty} \frac{ds}{s^{3/2}} \le 2\overline{D}_n \frac{1}{y^{1/2}}, \tag{13.49}$$

where

$$\overline{D}_n \le \frac{10}{3\sqrt{2\pi}\, L^2} \frac{n}{\sqrt{n-2}} \exp\left[\left(1 - \frac{L^2}{8} + \log L^2 + \log\theta\right) n\right],$$

where $\theta = 1 + \|M\|^2 (1 + \log n)/n$. Choosing $L = 2\sqrt{2}(1 + 4\theta)^{1/2}$, conditions $L^2 > 20$ and $L^2 n \ge 4\|M\|^2$ are verified and $1 - L^2/8 + \log L^2 + \log\theta < 0$. Hence,

$$2\overline{D}_n \le \frac{1}{4}\sqrt{\frac{n}{2\pi}}.$$

Based on (13.47, (13.48), and (13.49), substituting in the right-hand side of (13.46), inequality (13.38) in the statement of the theorem follows. □

REFERENCES AND SUGGESTED READING

Abrahams, J. (1982). The zero-crossing problem for some non-stationary Gaussian processes. *IEEE Trans. Inform. Theory*, 28, 677–678.

——— (1986). A survey of recent progress on level-crossing problems for random processes, in *Communications and Networks: A Survey of Recent Advances*, I. F. Blake and H. V. Poor, eds. Springer-Verlag, New York, pp. 6–25.

Abramovitz, M., and Stegun, I. (1968). *Handbook of Mathematical Functions with Formulas, Graphs and Mathematical Tables*. National Bureau of Standards, Washington, DC.

Adler, R. J. (1976a). On generalizing the notion of up-crossings to random fields. *Adv. Appl. Probab.*, 8, 789–805.

——— (1976b). Excursions above a fixed level by n-dimensional random fields. *J. Appl. Probab.*, 13, 276–289.

——— (1977). Excursions above high levels by Gaussian random fields. *Stochastic Processes Appl.*, 5, 21–25.

——— (1981). *The Geometry of Random Fields*. Wiley, New York.

——— (1984). The supremum of a particular Gaussian field. *Ann. Probab.*, 12, 436–444.

——— (1990). *An Introduction to Continuity, Extrema and Related Topics for General Gaussian Processes*. IMS, Hayward, CA.

——— Adler, R. A. (2000). On excursion sets, tube formulae and maxima of Gaussian fields. *Ann. Appl. Probab.*, 10, 1–74.

——— (2004). Gaussian random fields on manifolds. *Stochastic Anal. Random Fields Appl.*, 4, 3–20.

Adler, R. J., and Firman, D. (1981). A non-Gaussian model for random surfaces. *Trans. Roy. Soc. London A*, 303, 433–462.

Adler, R., and Hasofer, A. M. (1976). Level crossings of random fields. *Ann. Probab.*, 1, 1–12.

Adler, R. J., and Samorodnisky, G. (1987). Tail behaviour for the suprema of Gaussian processes with applications to empirical processes. *Ann. Probab.*, 15, 1339–1351.

Adler, R. A., and Samorodnisky, G. (1997). Level crossings of absolutely continuous stationary symmetric α-stable processes. *Ann. Appl. Probab.*, 6, 460–493.

Adler, R. A., Samorodnisky, G., and Gadrich T. (1993). The expected number of level crossings for stationary harmonisable symmetric stable processes. *Ann. Appl. Probab.*, 2, 553–575.

Adler, R. J., and Taylor, J. (2007). *Random Fields and Geometry*. Springer, New York.

Adler, R. J., and Wilson, R. J. (1982). The structure of Gaussian fields near a level crossing. *Adv. Appl. Probab.*, 14, 543–565.

Andrews, D. W. K., and Ploberger, W. (1994). Optimal tests when a nuisance parameter is present only under the alternative. *Econometrica*, 62, 1383–1414.

Armentano, D., and Wschebor, M. (2008). Random systems of polynomial equations: the expected number of roots under smooth analysis. *Bernoulli*. To appear.

Aronowich, M., and Adler, R. J. (1985). Behaviour of χ^2-processes at extrema. *Adv. Appl. Probab.*, 17, 280–297.

Azaïs, J.-M. (1989). Approximation des trajectoires et temps local des diffusions. *Ann. Inst. H. Poincaré Probab. Statist.*, 25(2), 175–194.

Azaïs, J.-M., Bardet, J.-M., and Wschebor, M. (2002). On the tails of the distribution of the maximum of a smooth stationary Gaussian process. *ESAIM: Probab. Statist.*, 6, 177–184.

Azaïs, J.-M., and Cierco-Ayrolles, C. (2002). An asymptotic test for quantitative gene detection. *Ann. Inst. H. Poincaré Probab. Statist.*, 38, 1087–1092.

Azaïs, J.-M., Cierco-Ayrolles, C., and Croquette, A. (1999). Bounds and asymptotic expansions for the distribution of the maximum of a smooth stationary Gaussian process. *ESAIM: Probab. Statist.*, 3, 107–129.

Azaïs, J.-M., Déjean, S., León, J., and Zwolska, F. (2009). Transformed Gaussian stationary models for ocean waves. Submitted.

Azaïs, J.-M., and Delmas, C. (2002). Asymptotic expansions for the distribution of the maximum of Gaussian random fields. *Extremes*, 5(2), 181–212.

Azaïs, J.-M., Gassiat, E., and Mercadier, C. (2006). Asymptotic distribution and power of the likelihood ratio test for mixtures: bounded and unbounded case. *Bernoulli*, 12(5), 775–799.

——— (2008). The likelihood ratio test for general mixture models with or without structural parameter. *ESAIM: Probab. Statist.* To appear.

Azaïs, J.-M., León, J., and Ortega, J. (2005). Geometrical characteristic of Gaussian sea waves. *J. Appl. Probab.*, 42, 1–19.

Azaïs, J.-M., and Mercadier, C. (2003). Asymptotic Poisson character of extremes in non-stationary Gaussian models. *Extremes*, 6(4), 301–318.

Azaïs, J.-M., and Wschebor, M. (1996). Almost sure oscillation of certain random processes. *Bernoulli*, 2(3), 257–270.

——— (1997a). Oscillation presque sûre de martingales continues, in *Seminars in Probability XXXI*. Lecture Notes in Mathematics 1655. Springer-Verlag, New York, pp. 69–76.

——— (1997b). Une formule pour calculer la distribution du maximum d'un processus stochastique. *C.R. Acad. Sci. Paris*, 324, Ser. I, 225–230.

——— (1999). Régularité de la loi du maximum de processus gaussiens réguliers. *C.R. Acad. Sci. Paris*, 328, Ser. I, 333–336.

——— (2001). On the regularity of the distribution of the maximum of one-parameter Gaussian processes. *Probab. Theory Related Fields*, 119, 70–98.

——— (2002). The distribution of the maximum of a Gaussian process: Rice method revisited, in *In and Out of Equilibrium: Probability with a Physical Flavour*. Progress in Probability. Birkhaüser, Boston, pp. 321–348.

——— (2005a). On the distribution of the maximum of a Gaussian field with *d* parameters. *Ann. Appl. Probab.*, 15(1A), 254–278.

——— (2005b). On the roots of a random system of equations: the theorem of Shub and Smale and some extensions. *Found. Comput. Math.*, 125–144.

——— (2005c). Upper and lower bounds for the tails of the distribution of the condition number of a Gaussian matrix. *SIAM J. Matrix Anal. Appl.*, 26(2), 426–440.

——— (2006). A self contained proof of the Rice formula for random fields. http://www.lsp.ups-tlse.fr/Azais/publi/completeproof.pdf.

——— (2008). A general formula for the distribution of the maximum of a Gaussian field and the approximation of the tail. *Stochastic Processes Appl.*, 118(7), pp. 1190–1218.

Bar David, I., and Nemirovsky, A. (1972). Level crossings of non-differentiable shot processes. *IEEE Trans. Inform. Theory*, 18, 27–34.

Baxevani, A., Podgórski, K., and Rychlik, I. (2003). Velocities for random surfaces. *Probab. Eng. Mech.*, 18, 251–271.

Belayev, Y. (1961). Continuity and Hölder conditions for sample functions of stationary Gaussian processes. *Proceedings of the Fourth Berkeley Symposium on Mathematical Statistics and Probability*, 2, 23–33.

——— (1966). On the number of intersections of a level by a Gaussian stochastic process. *Theory Probab. Appl.*, 11, 106–113.

——— (1967). Bursts and shines of random fields. *Soviet Math. Dokl.*, 8(5), 1107–1109.

——— (1968). On the number of exits across the boundary of a region by a vector stochastic process. *Theory Probab. Appl.*, 13, 320–324.

——— (1970). Distribution of the maximum of a random field and its application to reliability problems. *Eng. Cybernetics*, N 2, 269–276.

——— (1972a). Point processes and first passage problems. *Proceedings of the Sixth Berkeley Symposium on Mathematical Statistics and Probability*, 3, 1–17.

——— (1972b). On the mean number of crossings of a level by a stationary Gaussian stochastic process. *Proceedings of the Sixth Berkeley Symposium Mathematical Statistics and Probability*, 3, 1–17.

Belayev, Y., and Malevich, T. L. (1968). Asymptotic normality of the number of intersections of a fixed line by a stationary Gaussian process. *Theory Probab. Appl.*, 13, 543–544.

Belayev, Y., and Piterbarg, V. I. (1972). Asymptotics of the average number of overshot of a Gaussian field beyond a high level. *Soviet Math. Dokl.*, 13(2), 309–313.

Benzaquen, S., and Cabaña, E. M. (1982). The expected measure of the level set of a regular stationary Gaussian process. *Pacific J. Math.*, 103(1), 9–16.

Berman, S. M. (1964). Limit theorems for the maximum term in stationary sequences. *Ann. Math. Statist.*, 35, 502–516.

Berman, S. M. (1971a). Asymptotic independence of the numbers of high and low level crossings of stationary Gaussian processes. *Ann. Math. Statist.*, 42(3), 927–945.

_____ (1971b). Excursions above high levels for stationary Gaussian processes. *Pacific J. Math.*, 3, 663–679.

_____ (1974). Sojourns and extremes of Gaussian processes. *Ann. Probab.*, 2, 999–1026.

_____ (1979). High level sojourns for strongly dependent Gaussian processes. *Z. Wahrsch.*, 50, 223–236.

_____ (1985a). An asymptotic formula for the distribution of the maximum of a Gaussian process with stationary increments. *J. Appl. Probab.*, 22, 454–460.

_____ (1985b). The maximum of a Gaussian process with nonconstant variance. *Ann. Inst. H. Poincaré Probab. Statist.*, 21, 383–391.

_____ (1985c). Sojourns above a high level for a Gaussian process with a point of maximum variance. *Commun. Pure Appl. Math.*, 38(5), 519–528.

_____ (1986). The supremum of a process with stationary independent and symmetric increments. *Stochastic Processes Appl.*, 23(2), 281–290.

_____ (1987). Extreme sojourns of a Gaussian process with a point of maximum variance. *Probab. Theory Related Fields*, 74(1), 113–124.

_____ (1992a). *Sojourns and Extremes of Stochastic Processes*. Probability Series. Wadsworth/Brooks, Belmont, CA.

_____ (1992b). A central limit theorem for the renormalized self-intersection local time of a stationary vector Gaussian process. *Ann. Probab.*, 20, 61–81.

Berman, S. M., and Kôno, N. (1989). The maximum of a Gaussian process with non-constant variance: a sharp bound for the distribution of the tail. *Ann. Probab.*, 17, 632–650.

Berzin, C., and León, J. R. (2005). Convergence in fractional models and applications. *Electron. J. Probab.*, 10, 326–370.

Besson, J.-L., and Wschebor, M. (1983). Sur la finitude des moments du nombre des passages par un niveau d'une fonction aléatoire. *C.R. Acad. Sci. Paris,* 297, Ser. I, 361–363.

Bharucha-Reid, A. T., and Sambandham, M. (1986). *Random Polynomials*. Probability and Mathematical Statistics. Academic Press, Orlando, FL.

Billingsley, P. (1999). *Convergence of Probability Measures*, 2nd ed., Wiley Series in Probability and Statistics. Wiley, New York.

Bloch, A., and Polya, G. (1932). On the number of real roots of a random algebraic equation. *Proc. Cambridge Philos. Soc.*, 33, 102–114.

Blum, L., Cucker, F., Shub, M., and Smale, S. (1998). *Complexity and Real Computation*. Springer-Verlag, New York.

Borell, C. (1975). The Brunn-Minkowski inequality in Gauss space. *Invent. Math.*, 30, 207–216.

_____ (2003). The Ehrhard inequality. *C.R. Acad. Sci.Paris*, 337, Ser. I, 663–666.

Breuer, J., and Major, P. (1983). Central limit theorems for non-linear functionals of Gaussian fields. *J. Multivariate Ann.*, 13, 425–444.

Brillinger D. R. (1972). On the number of solutions of systems of random equations. *Ann. Math. Statist.*, 43, 534–540.

Brown, J. R. (1976). *Ergodic Theory and Topological Dynamics*. Academic Press, New York.

Bulinskaya, E. V. (1961). On the mean number of crossings of a level by a stationary Gaussian stochastic process. *Theory Probab. Appl.*, 6, 435–438.

Bürgisser, P. (2007). Average Euler characteristic of random real algebraic varieties. *C. R. Acad. Sci., Paris Math.*, 345(9), 507–512.

Bürgisser, P., Cucker, F., and Lotz, M. (2006). Smoothed analysis of complex conic condition numbers. *J. Math. Pures Appl.*, 86, 293–309.

Cabaña, E. M. (1985). Esperanzas de integrales sobre conjuntos de de nivel aleatorios. *Actas del segundo Congreso Latinoamericano de Probabilidad y Estadística Matemática*, Caracas, Venezuela, pp. 65–81.

——— (1987). Affine process a test of isotropy based in level sets. *SIAM J. Appl. Math.*, 47, 886–891.

——— (1991). A Gaussian process with parabolic covariance. *J. Appl. Probab.*, 28, 898–902.

Cabaña, E. M., and Wschebor, M. (1981). An estimate for the tails of the distribution of the supremum for a class of stationary multiparameter Gaussian processes. *J. Appl. Probab.*, 18, 536–541.

Cartwright, D. E., and Longuett-Higgins, M. S. (1956). The statistical distribution of the maxima of a random function. *Proc. Roy. Soc. Ser. A*, 237(1209), 212–232.

Caussinus, H., and Ruiz, A. (1995). Metrics for finding typical structures by means of principal component analysis. *Data Sci. Appl.*, 117–192.

Cierco, C. (1996). Problèmes statistiques liés à la détection et à la localisation d'un gène à effet quantitatif. Ph.D. dissertation. University of Toulouse, France.

Cierco-Ayrolle, C., Croquette, A., and Delmas, C. (2003). Computing the distribution of the maximum of Gaussian random processes. *Methodol. Comput. Appl. Probab.*, 5, 4, 427–438.

Corrsin, S. (1955). A measure of the area of a homogeneous random surface in space. *Quart. Appl. Math.*, 12, 404–408.

Cramér, H. (1965). A limit theorem for the maximum values of certain stochastic processes. *Theory Probab. Appl.*, 10, 126–128.

Cramér, H., and Leadbetter, M. R. (1965). The moments of the number of crossings of a level by a stationary normal process. *Ann. Math. Statist.*, 36, 1656–1663.

——— (1967). *Stationary and Related Stochastic Processes*. Wiley, New York.

Cressie, N (1980). The asymptotic distribution of the scan statistic under uniformity. *Ann. Probab.*, 8, 828–840.

Cressie, N., and Davies, R. W. (1981). The supremum distribution of another Gaussian process. *J. Appl. Probab.*, 18, 131–138.

Cucker, F., and Wschebor, M. (2003). On the expected condition number of linear programming problems. *Numer. Math.*, 94(3), 419–478.

Cucker, F., Diao, H., and Wei, Y. (2005). Smoothed analysis of some condition numbers. *Numer. Linear Alg. Appl.*, 13, 71–84.

Cucker, F., Krick, T., Malajovich, G., and Wschebor, M. (2008). Real polynomial systems: II. Randomization and condition number. Preprint.

Cuesta-Albertos, J., and Wschebor, M. (2003). Some remarks on the condition number of a real random square matrix. *J. Complexity*, 19(4), 548–554.

_____ (2004). Condition numbers and extrema of random fields. *IV Ascona Seminar on Stochastic Analysis, Random Fields and Applications*. Progress in Probability. Birkhaüser, Boston, pp. 69–82.

Cuzick, J. (1975). Conditions for finite moments of the number of zero crossings for Gaussian processes. *Ann. Probab.*, 3, 849–858.

_____ (1981). Boundary crossing probabilities for stationary Gaussian processes and Brownian motion. *Trans. Amer. Math. Soc.*, 263(2), 469–492.

Dacunha-Castelle, D., and Gassiat, E. (1997). Testing in locally conic models, and application to mixture models. *ESAIM: Probab. Statist.*, 1, 285–317.

_____ (1999). Testing the order of a model using locally conic parameterization: population mixtures and stationary ARMA processes. *Ann. Statist.*, 27(4), 1178–1209.

Darling, D. A. (1983). On the supremum of certain Gaussian processes. *Ann. Probab.*, 11, 803–806.

Davidson, K. R., and Szarek, S. J. (2001). Local operator theory, random matrices and Banach spaces. In *Handbook of the Geometry of Banach Spaces*, Vol. I. North-Holland, Amsterdam, pp. 317–366.

Davies, R. B. (1977). Hypothesis testing when a nuisance parameter is present only under the alternative. *Biometrika*, 64, 247–254.

_____ (1987). Hypothesis testing when a nuisance parameter is present only under the alternative. *Biometrika*, 74, 33–43.

_____ (2002). Hypothesis testing when a nuisance parameter is present only under the alternative: linear model case. *Biometrika*, 89, 484–489.

Davis, J. D. (1975). *Interpolation and Approximation*. Dover, New York.

Delmas, C. (2001). Distribution du maximum d'un champ aléatoire et applications statistiques. Ph.D. dissertation. Université P. Sabatier, Toulouse, France.

_____ (2003a). On likelihood ratio test in Gaussian mixture models. *Sankya*, 65(3), 513–531.

_____ (2003b). Projection on spherical cones, maximum of Gaussian fields and Rice's method. *Statist. Probab. Lett.*, 64(3), 263–270.

DeLong, D. M. (1981). Crossing probabilities for a square root boundary by a Bessel process. *Commun. Statist. Theory Methods*, A10, 2197–2213.

de Maré, J. (1977). Reconstruction of a stationary Gaussian process from its sign changes *J. Appl. Probab.*, 14, 38–57.

Dembo, A., Poonen, B., Shao, Q.-M., and Zeitouni, O. (2002). Random polynomials having few or no real zeros. *J. Amer. Math. Soc.*, 15(4), 857–892.

Demmel, J. (1987). On condition numbers and the distance to the nearest ill-posed problem. *Numer. Math.*, 51, 251–289.

_____ (1988). The probability that a numerical analysis problem is difficult. *Math. Comput.*, 50, 449–480.

Diebolt, J. (1979). Sur la loi du maximum de certains processus stochastiques. *Publ. Inst. Statist. Univ. Paris*, 24(3–4), 31–67.

——— (1981). Sur la loi du maximum de certain processus stochastiques sur le tore. *Ann. Inst. Henri Poincaré B*, 17(2), 165–179.

Diebolt, J., and Posse, C. (1996). On the density of the maximum of smooth Gaussian processes. *Ann. Probab.*, 24, 1104–1129.

Dietrich, C. R., and Newsam G. N. (1997). Fast and exact simulation of stationary Gaussian processes through circulant embedding of the covariance matrix. *SIAM J. Sci. Comput.*, 18(4), 1088–1107.

Dobric, V., Marcus, M. B., and Weber, M. (1988). The distribution of large values of the supremum of a Gaussian process. *Astérisque*, 157(8), 95–128.

Dudley, R. M. (1967). The sizes of compact subsets of Hilbert space and continuity of Gaussian processes. *J. Funct. Anal.*, 1, 290–330.

——— (1973). Sample functions of the Gaussian process. *Ann. Probab.*, 1, 66–103.

——— (1989). *Real Analysis and Probability*. Chapman & Hall, London.

Durbin, J. (1985). The first-passage density of a continuous Gaussian process to a general boundary. *J. Appl. Probab.*, 22, 99–122.

Eckart, C., and Young, G. (1936). The approximation of one matrix by another of lower rank. *Psycometrika*, 1, 211–218.

Edelman, A. (1988). Eigenvalues and condition numbers of random matrices. *SIAM J. Matrix Anal. Appl.*, 9, 543–556.

——— (1989). Eigenvalues and condition numbers of random matrices. Ph.D. dissertation. MIT, Cambridge, MA.

Edelman, A., and Kostlan, E. (1995). How many zeros of a random polynomial are real? *Bull. Amer. Math. Soc. (N.S.)*, 32(1), 1–37.

Edelman, A., and Rao, N. R. (2005). Random matrix theory. *Acta Numer.*, 14, 233–297.

Edelman, A., and Sutton, B. D. (2005). Tails of condition number distributions. *SIAM J. Matrix Anal. Appl.*, 27(2), 547–560.

Ehm, W. (1981). Sample properties of multiparameter stable processes. *Z. Wahrsch. Verw. Gebiete.*, 56, 195–228.

Ehrhard, A. (1981). Une démonstration de l'inégalité de Borell. *Ann. Sci. Univ. Clermont-Ferrand*, 69, 165–184.

——— (1983). Symétrisation dans l'espace de Gauss. *Math. Scand.*, 53, 281–301.

——— (1986). Éléments extrémaux pour les inégalités de Brunn-Minkowski Gaussiennes. *Ann. Inst. H. Poincaré*, 22, 149–168.

Ein-Gal, M., and Bar-Davis, I. (1975). Passages and maxima for a particular Gaussian process. *Ann. Probab.*, 3, 549–556.

Erdelyi, A., Magnus, W., Oberhettinger, F., and Tricomi F. G. (1953). Higher trascendental functions, in *The Bateman Project*, Vol. 1. McGraw-Hill, New York.

Federer, H. (1969). *Geometric Measure Theory*. Springer-Verlag, New York.

Feller, W. (1966). *An Introduction to Probability Theory and Its Applications*, Vol. II. Wiley, New York.

Fernique, X. (1964). Continuité des processus Gaussiens. *C. R. Acad. Sci. Paris*, 6058–6060.

——— (1971). Régularité des processus gaussiens. *Invent. Math.*, 12, 304–321.

_____ (1974). Régularité des trajectoires des fonctions aléatoires gaussiennes, in *Ecole d'Eté de Probabilités de Saint Flour*. Lecture Notes in Mathematics 480. Springer-Verlag, New York.

_____ (1983). *Régularité de fonctions aléatoires non-gaussiennes*. Lecture Notes in Mathematics 976, Springer-Verlag, New York, pp. 1–74.

Fyodorov, Y. (2006). Complexity of random energy landscapes, glass transition and absolute value of spectral determinant of random matrices. *Phys. Rev. Lett.*, 92, 240. Erratum, ibid., 93, 149.

Gassiat, E. (2002). Likelihood ratio inequalities with application to various mixtures. *Ann. Inst. H. Poincaré Probab. Statist.*, 38, 897–906.

Geman, D. (1972). On the variance of the number of zeros of stationary Gaussian process. *Ann. Math. Statist.*, 43(3), 977–982.

Ghosh, J., and Sen, P. (1985). On the asymptotic performance of the log-likelihood ratio statistic for the mixture model and related results, in *Proceedings of the Berkeley Conference in Honor of Jerzy Neyman and Jack Kiefer*, L. M. Le Cam and R. A. Olshen, eds.

Girko, V. L. (1996). Random matrices, in *Handbook of Algebra*, vol. 1. North-Holland, Amsterdam, pp. 27–78.

Goldman, M. (1971). On the first passage of the integrated Wiener process. *Ann. Math. Statist.*, 42(6), 2150–2155.

Gordon, Y. (1985). Some inequalities for Gaussian processes and applications. *Israel J. Math.*, 50, 265–289.

_____ (1988). Gaussian processes and almost spherical sections of convex bodies. *Ann. Probab.*, 16, 180–188.

Haldane, J. B. S. (1919). The combination of linkage values, and the calculation of distances between the loci of linked factors. *J. Genet.*, 8, 299–309.

Hansen, B. E. (1996). Inference when a nuisance parameter is not identified under the null hypothesis. *Econometrica*, 64(2), 413–430.

Hasofer, A. M. (1974). The up-crossing rate of a class of stochastic processes, in *Studies in Probability and Statistics*, E. J. Williams, ed. North-Holland, Amsterdam, pp. 153–159.

_____ (1976). The mean number of maxima above a high level in Gaussian random fields. *J. Appl. Probab.*, 13, 377–379.

_____ (1978). Up-crossings of random fields. *Adv. Appl. Probab.*, 10, 14–21.

Higham, N. (1996). *Accuracy and Stability of Numerical Algorithms*. SIAM, Philadelphia, PA.

Hoeffding, W., and Robbins, H. (1948). The central limit theorem for dependent random variables. *Duke Math. J.*, 15, 773–780.

Ibragimov, A., Sudakov, V. N., and Tsirelson, B. S. (1976). Norms of Gaussian sample functions. *Proceedings of the 3rd Japan–USSR Symposium on Probability Theory*. Lecture Notes in Mathematics 550. Springer-Verlag, New York, pp. 20–41.

Ikeda, N., and Watanabe, S. (1981). *Stochastic Differential Equations and Diffusion Processes*. North-Holland, New York.

Iribarren, I. (1989). Asymptotic behaviour of the integral of a function on the level set of a mixing random field. *Probab. Math. Statist.*, 10(1), 45–56.

Ito, K. (1964). The expected number of zeros of continuous stationary Gaussian processes. *J. Math. Kyoto Univ*, 3(2), 207–221.

Jacod, J. (1998). Rates of convergence to the local time of a diffusion. *Ann. Inst. H. Poincaré Ser. B*, 34, 505–544.

——— (2000). Non-parametric kernel estimation of the diffusion coefficient of a diffusion. *Scand. J. Statist.*, 27(1), 83–96.

Kac, M. (1943). On the average number of real roots of a random algebraic equation. *Bull. Amer. Math. Soc.*, 49, 314–320.

Kac, M., and Slepian, D. (1959). Large excursions of Gaussian processes. *Ann. Math. Statist.*, 30, 1215–1228.

Kahane, J.-P. (1985). *Some Random Series of Functions*, 2nd. ed. Cambridge University Press, New York.

——— (1986). Une inégalité du type de Slepian et Gordon sur les processus gaussiens. *Israel J. Math.*, 55, 109–110.

Karatzas, I., and Shreve, S. E. (1998). *Brownian Motion and Stochastic Calculus*. Springer-Verlag, New York, 1998.

Kendall, M., Stuart, A., and Ord, J. K. (1983). *The Advanced Theory of Statistics*, Vol. 3, *Design and Analysis and Time Series*, 4th ed. Macmillan, New York.

Kinsman, B. (1965). *Wind Waves: Their Generation and Propagation on the Ocean Surface*, Prentice-Hall, Englewood Cliffs, NJ. Reprinted by Dover Phoenix Editions, New York, 2002.

Kobelkov, G. S. (2005). The ruin problem for the stationary Gaussian process. *Theory Probab. Appl.*, 155–163.

Kôno, N. (1980). Sample path properties of stochastic processes. *J. Math. Kyoto Univ.*, 20, 295–313.

Kostlan, E. (2002). On the expected number of real roots of a system of random polynomial equations, in *Foundations of Computational Mathematics (Hong Kong, 2000)*. World Scientific, River Edge, NJ, pp. 149–188.

Kratz, M. F. (2006). Level crossings and other level functionals of stationary Gaussian processes. *Probab. Surveys*, 3, 230–288.

Kratz, M. F., and León, J. R. (1997). Hermite polynomial expansion for non-smooth functionals of stationary Gaussian processes: crossings and extremes. *Stochastic Processes Appl.*, 66, 237–252.

——— (2001). Central limit theorems for level functionals of stationary Gaussian processes and fields. *J. Theor. Probab.*, 14, 639–671.

——— (2006). On the second moment of the number of crossings by a stationary Gaussian process. *Ann. Probab.*, 34(4).

Krée, P., and Soize, C. (1983). *Mécanique aléatoire*. Dunod, Paris.

Lachal, A. (1991). Sur le premier instant de passage de l'intégrale du mouvement Brownien. *Ann. Inst. H. Poincaré Probab. Statist.*, 27, 385–405.

Landau, H. J., and Shepp, L. A. (1970). On the supremum of a Gaussian process. *Sankya Ser. A*, 32, 369–378.

Lander, E., and Bolstein, D. (1989). *Mapping Mendelian Factors Underlying Quantitative Traits Using rflp Linkage Maps*. Genetics Society of America, Bethesda, MD.

Latala, R. (1996). A note on the Ehrhard inequality. *Stud. Math.*, 118, 169–174.

——— (2002). On some inequalities for Gaussian measures. *ICM* 2002, vol. II, pp. 813–822.

Leadbetter, M. R. (1966). On crossings of levels and curves by a wide class of stochastic processes. *Ann. Math. Statist.*, 37, 260–267.

_____ (1967). A note on the moments of the number of axis crossings by a stochastic process. *Bull. Amer. Math. soc.*, 73, 129–132.

_____ (1968). On three basic results in the theory of stationary point processes. *Proc. Amer. Math. Soc.*, 115–117.

_____ (1972). Point processes generated by level crossings, in *Stochastic Point Processes*, P. A. W. Lewis, ed. Wiley, New York, pp. 436–467.

Leadbetter, M. R., and Rootzén, H. (1988). Extremal theory for stochastic processes. *Ann. Probab.*, 16, 431–478.

Leadbetter, M. R., Lindgren, G., and Rootzén, H. (1983). *Extremes and Related Properties of Random Sequences and Processes*. Springer-Verlag, New York.

Ledoux, M. (1996). Isoperimetry and Gaussian analysis, in *Ecole d'Eté de Probabilités de Saint-Flour 1994*. Lecture Notes in Mathematics 1648. Springer-Verlag. New York, pp. 165–264.

_____ (2001). *The Concentration of Measure Phenomenon*. American Mathematical Society, Providence, RI.

Ledoux, M., and Talagrand, M. (1991). *Probability in Banach Spaces*. Springer-Verlag, New York.

León, J. (2006). A note on Breuer-Major CLT for non-linear functionals of continuous time stationary Gaussian processes. Preprint.

León, J., and Ortega, J. (1989). Weak convergence of different types of variation for biparametric Gaussian processes. *Colloq. Math. Soc. J. Bolyai*, 57.

Li, W. V., and Shao, Q.-M. (2001). Gaussian processes: inequalities, small ball probabilities and applications, in *Stochastic Processes: Theory and Methods*, C. R. Rao and D. Shanbhag, eds., North-Holland, Amsterdam, pp. 533–598.

Li, W. V., and Shao, Q.-M. (2002). A normal comparison inequality and its applications. *Probab. Theory Related Fields*, 122(4), 494–508.

Li, W. V., and Shao, Q.-M. (2004). Lower tail probabilities for Gaussian processes. *Ann. Probab.*, 32(1A), 216–242.

Lifshits, M. A. (1987). On the distribution of the maximum of a Gaussian process. *Theory Probab. Appl.*, 31, 125–132.

_____ (1995). *Gaussian Random Functions*. Kluwer, Dordrecht, The Netherlands.

Lindgren, G. (1970). Some properties of a normal process near a local maximum. *Ann. Math. Statist.*, 41, 1870–1883.

_____ (1972). Wave-length and amplitude in Gaussian noise. *Adv. Appl. Probab.*, 4, 81–108.

_____ (1980). Extreme values and crossings for the c2-process and other functions of multidimensional Gaussian processes with reliability applications. *Adv. Appl. Probab.*, 12, 746–774.

_____ (1983). Use and structure of Slepian model processes for prediction and detection in crossing and extreme value theory. *Report 1983:4*. Department of Mathematical Statistics University of Lund and Lund Institute of Technology, Sweden.

_____ (2006). Slepian models for the stochastic shape of individual Lagrange sea waves. *Adv. Appl. Probab.*, 38, 430–450.

Lindgren, G., and Rychlik, I. (1991). Slepian models and regression approximations in crossings and extreme value theory. *Internat. Statist. Rev.*, 59(2), 195–225.

——— (1995). Wave analysis toolbox: a tutorial. Preprint. University of Lund.

Littlewood, J. E., and Offord, A. C. (1938). On the number of real roots of a random algebraic equation. *J. London Math. Soc.*, 13, 288–295.

——— (1939). On the roots of certain algebraic equation. *Proc. London Math. Soc.*, 35, 133–148.

Longuett-Higgins, M. S. (1957). The statistical analysis of a random moving surface, *Philos. Trans. Roy. Soc. London Ser. A*, 249, 321–387.

——— (1962a). The statistical geometry of random surfaces. *Hydraul. Instabil. Symp. Appl. Math.*, 13, 105–143.

——— (1962b). The distribution of intervals between zeros of a stationary random function, *Philos. Trans. Roy. Soc. London Ser. A*, 254, 557–599.

Malevich, T. L. (1979). On conditions for the moments of the number of zeros of Gaussian stationary processes to be finite. *Theory Probab. Appl.*, 24(4), 741–754.

——— (1984). On conditions for finiteness of the factorial moments of the number of zeros of Gaussian stationary processes. *Theory Probab. Appl.*, 29(3), 534–545.

Malevich, T. L., and Volodina, L. N. (1980). Upper and lower bounds for factorial moments of the number of zeros of a Gaussian process(in Russian). *Izv. Akad. Nauk UzUSSR*, 4, 26–33.

Malmquist, S. (1954). On certain confidence contours for distribution functions. *Ann. Math. Statist.*, 25, 523–533.

Marcus, M. (1970). A bound for the distribution of the maximum of continuous Gaussian processes. *Ann. Math. Statist.*, 41(1), 305–309.

——— (1972). Upper bounds for the asymptotic maxima of continuous Gaussian processes. *Ann. Math. Statist.*, 43(2), 522–533.

——— (1973). Continuity of Gaussian processes and random Fourier series. *Ann. Probab.*, 1, 968–981.

——— (1977). Level crossings of a stochastic process with absolutely continuous sample paths. *Ann. Probab.*, 5, 52–71.

——— (1989). Some bounds for the expected number of level crossings of symmetric harmonizable p-stable processes. *Stochastic Processes Appl.*, 33, 217–231.

Marcus, M. B., and Shepp, L. A. (1972). Sample behaviour of Gaussian processes. *Proceedings of the 6th Berkeley Symposium on Mathematical Statistics and Probability*, 2, 423–442.

Maruyama, G. (1949). The harmonic analysis of stationary stochastic processes. *Mem. Fac. Sci. Kyusyu Univ.*, 4, 45–106.

McKean, H. P. (1963). A winding problem for a resonant driven by a white noise. *J. Math. Kyoto Univ.*, 2, 227–235.

——— (1969). *Stochastic Integrals*. Academic Press, New York.

Mehta, M. L. (2004). *Random Matrices*, 3rd ed. Academic Press, San Diego, CA.

Mercadier, C. (2002). Caractère asymptotiquement Gaussien des extrèmes en modèles non-stationnaires. *Rapport de DEA*. Université Paul Sabatier, Toulouse, France.

——— (2005). MAGP tooolbox. http://math.univ-lyon1.fr/ mercadier/.

_____ (2006). Numerical bounds for the distribution of the maximum of one- and two-dimensional processes. *Adv. Appl. Probab.* 38(1), 149–170.

Milnor, J. W. (1965). *Topology from the Differentiable Viewpoint*. The Univerity Press of Virginia, Charlottesville, VA.

Miroshin, R. N. (1974). Rice series in the theory of random functions. *Vestn. Leningrad Univ. Math.*, 1, 143–155.

_____ (1977). Condition for finiteness of moments of the number of zeros of stationary Gaussian processes. *Theory Probab. Appl.*, 22, 615–624.

_____ (1983). The use of Rice series. *Theory Probab. Appl.*, 28, 714–726.

_____ (1985). A simple criterion for finiteness of the moments of the number of zeros of a Gaussian stationary process. *Theory Probab. Appl.*, 29, 566–569.

Muirhead, R. J. (1982). *Aspects of Multivariate Statistical Theory*. Wiley Series in Probability and Mathematical Statistics. Wiley, New York.

Neveu, J. (1977). *Processus ponctuels*. Lecture Notes in Mathematics, 598. Springer-Verlag, New York.

Nisio, M. (1967). On the extreme values of Gaussian processes. *Osaka J. Math.*, 4, 313–326.

Nosko, V. P. (1969). Local structure of Gaussian random fields in the vicinity of high level light sources. *Soviet Math. Dokl.*, 10(6), 1481–1484.

_____ (1973). On the possibility of using the Morse inequalities for the estimation of the number of excursions of random field in a domain. *Theory Probab. Appl.*, 18, 821–822.

_____ (1979a). On the definition of the number of excursions of a random field above a fixed level. *Theory Probab. Appl.*, 24(3), 598–602.

_____ (1979b). On the average number of distinct level curve points of a random field on the plane. *Theory Probab. Appl.*, 24, 181–183.

_____ (1986). The local structure of a homogeneous Gaussian random field in a neighbourhood of high level points. *Theory Probab. Appl.*, 30, 767–782.

_____ (1987). On the distribution of the arc length of a high-level excursion of a stationary Gaussian process. *Theory Probab. Appl.*, 31, 521–523.

Nualart, D., and Vives, J. (1988). Continuité absolue de la loi du maximum d'un processus continu. *C.R. Acad. Sci. Paris*, 307, Ser. I, 349–354.

Nualart, D., and Wschebor, M. (1991). Intégration par parties dans l'espace de Wiener et approximation du temps local. *Probab. Theory Related Fields*, 90, 83–109.

Ochi, M. (1998). *Ocean Waves*. Cambrige University Press, New York.

Orsingher, E. (1979). Level crossings of transformations of stationary Gaussian processes. *Metron*, 37, 81–100.

_____ (1980). Planar Gaussian motions generated by crossings events. *Publ. Inst. Statist. Univ. Paris*, 25, 61–86.

_____ (1981). Upcrossings and conditional behaviour of the lognormal process. *Boll. Un. Mat. Ital.*, 18B, 1017–1034.

Palm C. (1943). Intensitâtsschwankungen in Fernsprechverkehr. *Ericson Technics*, 1–189.

Paranjape, S. R., and Park, C. (1973). Distribution of the supremum of the two-parameter Yeh–Wiener process on the boundary. *J. Appl. Probab.*, 10(4), 875–880.

Perera, G., and Wschebor, M. (1998). Crossings and occupation measures for a class of semi-martingales. *Ann. Probab.*, 26(1), 253–266.

——— (2002). Inference on the variance and smoothing of the paths of diffusions. *Ann. Inst. H. Poincaré*, 38(6), 1009–1022.

Pickands, J., III (1969a). Upcrossing probabilities for stationary Gaussian processes. *Trans. Amer. Math. Soc.*, 145, 51–73.

——— (1969b). Asymptotic properties of the maximum in a stationary Gaussian process. *Trans. Amer. Math. Soc.*, 145, 75–86.

Piterbarg, V. I. (1978). The central limit theorem for the number of level crossings for Gaussian processes. *Theory Probab. Appl.*, 23(1), 178–181.

——— (1981). Comparison of distribution functions of maxima of Gaussian processes. *Theory Probab. Appl.*, 26, 687–705.

——— (1996a). *Asymptotic Methods in the Theory of Gaussian Processes and Fields*. American Mathematical Society, Providence, RI.

——— (1996b). Rice's method for large excursions of Gaussian random fields. *Technical Report 478.* University of North Carolina. Translation of Rice's method for Gaussian random fields.

Plackett, R. I. (1954). A reduction formula for normal multivariate integrals. *Biometrika*, 41, 351–360.

Podgórski, K., Rychlik, I., and Sjö, E. (2000). Statistics for velocities of Gaussian waves. *Internet. J. Offshore Polar Eng.*, 10, 91–98.

Qualls, C. (1968). On a limit distribution of high level crossings of a stationary Gaussian process. *Ann. Math. Statist.*, 39, 2108–2113.

——— (1969). On the joint distribution of crossings of high multiple levels by a stationary Gaussian process. *Ark. Mat.*, 8(15), 129–137.

Qualls, C., and Watanabe, H. (1973a). Asymptotic properties of Gaussian random fields. *Trans. Amer. Math. Soc.*, 177, 155–171.

——— (1973b). Asymptotic properties of Gaussian process. *Ann. Math. Statist.*, 43, 580–596.

Radcenko, A. N. (1982). On geometric measures of the level set of a random field. *Theory Probab. Math. Statist.*, 25, 125–135.

Rényi, A. (1967). Remarks on the Poisson process. *Stud. Sci. Math. Hung.*, 2, 119–123.

Revuz, D., and Yor, M. (1998). *Continuous Martingales and Brownian Motion*, 3rd ed. Springer-Verlag, New York.

Rice, S. O. (1939). The distribution of the maxima of a random curve. *Amer. J. Math.*, 61, 409–416

——— (1944, 1945). Mathematical analysis of random noise. *Bell Syst. Tech. J.*, 23, 282–332; 24, 45–156.

——— (1958). Distribution of the duration of fades in radio transmission. *Bell Syst. Tech. J.*, 37, 620–623, 630–631.

Rickard, J. T. (1977). The zero crossing interval statistic of the smoothed random telegraph signal. *Inform Sci.*, 13, 253–268.

Rickard, J. T., and Pawula, R. F. (1975). The zero crossing variance of the smoothed random telegraph signal. *Int. J. Control*, 21, 743–752.

Rychlik, I. (1987). A note on Durbin's formula for the first-passage density. *Statist. Probab. Lett.*, 5, 425–428.

_____ (1990). New bounds for the first passage, wavelength and amplitude density. *Stochastic Processes Appl.*, 34, 313–339.

Rychlik, I., Johannesson, P., and Leadbetter, M. R. (1997). Modelling and statistical analysis of ocean-wave data using transformed Gaussian processes. *Marine Struct.*, 10, 13–47.

Sankar, A., Spielman, D. A., and Teng, S. H. (2002). Smoothed analysis of the condition numbers and growth factors and matrices. Preprint.

Sard, A. (1942). The measure of the critical points of differentiable maps. *Bull. Amer. Math. Soc.*, 48, 883–890.

Schoenberg, I. J. (1938). Metric spaces and completely monotone functions. *Ann. Math. (2)*, 39(4), 811–841.

Shepp, L. A. (1971). First passage time for a particular Gaussian process. *Ann. Math. Statist.*, 42, 946–951.

_____ (1979). The joint density of the maximum and its location for a Wiener process with drift. *J. Appl. Probab.* 16, 423–427.

Shepp, L. A., and Slepian, D. (1976). First-passage time for a particular stationary periodic Gaussian process. *J. Appl. Probab.*, 13, 27–38.

Shub, M., and Smale, S. (1993). Complexity of Bezout's theorem: II. Volumes and probabilities, in *Computational Algebraic Geometry (Nice, 1992)*. Progress in Mathematics 109. Birkhäuser, Boston, pp. 267–285.

Slepian, D. (1961). First passage time for a particular Gaussian process. *Ann. Math. Statist.*, 32, 610–612.

_____ (1962). The one-sided barrier problem for Gaussian noise. *Bell Syst. Tech. J.*, 42, 463–501.

_____ (1963). *On the Zeros of Gaussian Noise Time Series Analysis*, Wiley, New York.

_____ (1983). Estimation of the Gauss–Markov process from observation of its signs. *Stochastic Processes Appl.*, 14, 249–265.

Slud, E. (1991). Multiple Wiener–Itô integral expansions for level-crossing-count functionals. *Probab. Theory Related. Fields*, 87, 349–364.

Smale, S. (1985). On the efficiency of algorithms of analysis. *Bull. Amer. Math. Soc. (N.S.)*, 13(2), 87–121.

_____ (1997). Complexity theory and numerical analysis, in *Acta Numerica*, A. Iserles, ed. Cambridge University Press, New York, pp. 523–551.

Soshnikov, A. (1998). Level spacings distribution for large random matrices: Gaussian fluctuations. *Ann. Math.* (2), 148(2), 573–617.

Spielman, D. A., and Teng, S. H. (2002). Smoothed analysis of algorithms. *Proc. ICM, I*, 597–606.

Spivak, M. (1979). *A Comprehensive Introduction to Differential Geometry*, vol. V, 2nd ed. Publish or Perish, Wilmington, DE.

Sudakov, V. N., and Tsirelson, B. S. (1974). Extremal properties of half spaces for spherically invariant measures (in Russian). *Zap. Nauchn. Sem. LOMI*, 45, 75–82.

Sun, J. (1993). Tail probabilities of the maxima of Gaussian random fields. *Ann. Probab.*, 21, 34–71.

Switzer, P. (1976). Geometrical measures of the smoothness of random functions. *J. Appl. Probab.*, 13, 86–95.

Szarek, J. (1991). Condition numbers of random matrices. *J. Complexity*, 7(2), 131–149.

Talagrand, M. (1987). Regularity of Gaussian processes. *Acta Math.*, 159, 99–149.

——— (1988). Small tails for the supremum of a Gaussian process. *Ann. Inst. H. Poincaré Ser. B*, 24(2), 307–315.

——— (1990). Sample boundedness of stochastic processes under increment conditions. *Ann. Probab.*, 18, 1–49.

——— (1994). The supremum of some canonical processes. *Amer. J. Math.*, 116, 284–325.

——— (1995). Concentration of measure and isoperimetric inequalities in product spaces. *IHES*, 81, 73–205.

——— (1996). Majorising measures: the general chaining. *Ann. Probab.*, 24, 1049–1103.

——— (2001). Majorizing measures without measures. *Ann. Probab.*, 29(1), 411–417.

——— (2005). The generic chaining, in *Upper and Lower Bounds of Stochastic Processes*. Springer-Verlag, New York.

Tao, T., and Vu, V. (2007). The condition number of a randomly perturbed matrix, *J. ACM*, 248–255.

Tao, T., and Van, Wu (2007). The condition number of a randomly perturbed matrix, in *Proc. Annual Symposium on Theory of Computing (STOC 2007)*. ACM Press, New York.

Taylor, J. E., and Adler, R. J. (2003). Euler characteristics for Gaussian fields on manifolds. *Ann. Probab.*, 31, 533–563.

Taylor, J. E., Takemura, A., and Adler, R. J. (2005). Validity of the expected Euler characteristic heuristic. *Ann. Probab.*, 33(4), 1362–1396.

Tracy, C. A., and Widom, H. (1999). Universality of the distribution functions of random matrix theory, in *Integrable Systems from Classical to Quantum: Proceedings of the 38th Session of the Séminaire de mathématiques supérieures*, Université de Montréal.

Trefethen, L., and Bau, D., III (1997). *Numerical Linear Algebra*. SIAM, Philadelphia, PA.

Tsirelson, V. S. (1975). The density of the maximum of a Gaussian process. *Theory Probab. Appl.*, 20, 817–856.

Turing, A. (1948). Rounding-off errors in matrix processes. *Quart. J. Mech. Appl. Math.*, 1, 287–308.

Volkonskii, V. A., and Rozanov, Y. A. (1959, 1961). Some limit theorems for random functions I and II. *Teor. Veroyatnost. Primenen.*, 4, 186–207; 6, 202–215.

Von Neumann, J., and Goldstine, H. (1947). Numerical inverting matrices of high order. *Bull. Amer. Math. Soc.*, 53, 1021–1099.

WAFO Group (2000). WAFO—A Matlab Toolbox for Analysis of Random Waves and *Loads: A Tutorial*. Mathematical Statistics. Center for Mathematical Science, Lund University, Lund, Sweden. http://www.maths.lth.se/matstat/wafo.

Weber, M. (1985). Sur la densité du maximum d'un processus gaussien. *J. Math. Kyoto Univ.*, 25, 515–521.

Wilkinson, J. H. (1963). *Rounding Errors in Algebraic Processes*. Prentice-Hall, Englewood Cliffs, NJ.

Wigner, E. (1958). On the distribution of the roots of certain symmetric matrices, *Ann. Math.*, 67, 325–327.

_____ (1967). Random matrices in physics. *SIAM Rev.*, 9, 1–23.

Wong, E. (1966). Some results concerning the zero-crossings of Gaussian noise. *SIAM J. Appl. Math.*, 14, 1246–1254.

Wschebor, M. (1982). Formule de Rice en dimension *d*. *Z. Wahrsch. Verw. Gebiete*, 60, 393–401.

_____ (1983). On crossings of Gaussian fields. *Stochastic Processes Appl.*, 14(2), 147–155.

_____ (1985). *Surfaces aléatoires: mesure géometrique des ensembles de niveau*. Lecture Notes in Mathematics 1147. Springer-Verlag, New York.

_____ (1986). Random lengths for high levels. *Acta Cient. Venezolana*, 37(6), 633–635.

_____ (1992). Sur les accroissements du Processus de Wiener. *C.R. Acad. Sci. Paris*, 315, Ser. I 1293–1296.

_____ (2000). Sur la loi du sup de certains processus Gaussiens non bornés. *C.R. Acad. Sci. Paris*, 331, Ser. I, 823–826.

_____ (2004). Smoothed analysis of K(A). *J. Complexity*, 20(1), 97–107.

_____ (2005). On the Kostlan–Shub–Smale model for random polynomial sytems: variance of the number of roots. *J. Complexity*, 21, 773–789.

_____ (2006). Smoothing and the approximation of the occupation measure of random processes. *Ann. Fac. Sci. Toulouse Math.* (*6*), 15(1), 125–156.

Wschebor, M. (2007). Large random polynomial systems. Preprint.

_____ (2008). Random systems of equations, a review of some recent results, in *In and Out of Equilibrium 2*. Progress in Probability 60. Birkhaüser, Boston, pp. 559–574.

Ylvisaker, D. (1965). The expected number of zeros of a stationary Gaussian process. *Ann. Math. Statist.*, 1043–1046.

_____ (1966). On a theorem of Cramér and Leadbetter. *Ann. Math. Statist.*, 37, 682–685.

_____ (1968). A note on the absence of tangencies in Gaussian sample paths. *Ann. Math. Statist.*, 39, 261–262.

NOTATION

A^C	complement of the set A
ϕ	density of the standard normal distribution in $\mathbb{R}$
Φ	cumulative distribution function of the standard normal distribution in $\mathbb{R}$
χ_d^2	chi-square distribution with d degrees of freedom
λ, λ_d	depending on the context: Lebesgue measure, in $\mathbb{R}^d$; spectral moment of order d
(const)	positive constant; its value can change from one occurrence to another
$p_\xi(x)$	density of the random variable or vector ξ at point x
$B(u; r)$	open ball with center u and radius r
$\overline{B}(u; r)$	closed ball with center u and radius r
$N_u(f, T)$ or N_u	number of roots of $f(t) = u$ that belong to the set T
$U_u(f, T)$ or U_u	number of up-crossings of the level u by the function f on the set T
$D_u(f, T)$ or D_u	number of down-crossings of the level u by the function f on the set T
M_T^X or M or M_T	$\sup_{t \in T} X_t$
$F_M(u)$	$P\{M \leq u\}$
$\mu(X)$	median of the distribution of the real-valued random variable X
$r(s, t)$	covariance of a (often Gaussian) stochastic process

$r_{ij}(s,t)$	$\partial^{i+j}r/\partial s^i\partial t^j$
$\Gamma(t)$	covariance function of a stationary process: $\mathrm{Cov}(X(z), X(z+t))$
càd-làg	French acronym for *continue à droite et limiteé à gauche*: right continuous with left limits
z^+	$\sup(0, z)$
z^-	$-\inf(0, z)$
$z^{[k]}$	$z(z-1)\cdots(z-k+1)$, z and k positive integers, $k \leq z$
$\nu_k(u, T)$	$\mathrm{E}(U_u^{[k]})$
$\widetilde{\nu}_k(u, T)$	$\mathrm{E}(U_u^{[k]}\mathbb{1}_{X(0)<u})$
C_u	level set: $\{t \in S : X(t) = u\}$
$\mathcal{C}_u$	$\sqrt{\lambda_2}\,\frac{\exp(-u^2/2)}{2\pi}$ (*see* Chapter 10)
$\Vert\cdot\Vert_p$	L^p-norm, $p > 0$
$\Vert\cdot\Vert_\infty$	sup norm
$\Vert\cdot\Vert_k$	Euclidean norm in $\mathbb{R}^k$
$\approx$	equivalence of two functions
$\simeq$	numerical approximative equality
$\stackrel{D}{=}$	equality in distribution
$\{W(t) : t \geq 0\}$	Wiener process or Brownian motion
$\{B_H(t) : t \geq 0\}$	fractional Brownian motion
$M \succ 0$	symmetric square matrix M is positive definite
$M \prec 0$	symmetric square matrix M is negative definite
M^{T}	transpose of the matrix M
$M_{u,1}(X, S)$,	number of local maxima of the random field X, having value bigger than u and belonging to the set S
$M_{u,2}(X, S)$,	number of critical points of the random field X, having value bigger than u and belonging to the set S
$H_n(x)$,	Hermite polynomials (*see* p. 210)
$\bar{H}_n(x)$,	modified Hermite polynomials (*see* p. 211)
GOE,	Gaussian orthogonal ensemble
I_n,	identity $n \times n$ real matrix
σ_d	geometric measure of size d
$\{e_1, \ldots, e_n\}$	canonical basis
$\sigma_u(Z, B)$	for a random field $\mathbb{R}^d \to \mathbb{R}^{d'}$, $(d-d')$—dimensional measure of the level set

INDEX